Chemistry of Protein Conjugation and Cross-Linking

Author

Shan S. Wong, Ph.D.
Department of Pathology and Laboratory Medicine
University of Texas Health Science Center
Houston, Texas

CRC Press, Inc.
Boca Raton   Ann Arbor   Boston   London

Library of Congress Cataloging-in-Publication Data

Wong, Shan S.
   Chemistry of protein conjugation and cross-linking / Shan S. Wong.
   p.m.
   Includes bibliographical references and index.
   ISBN 0-8493-5886-8
   1. Proteins—Cross-linking. I. Title.
   [DNLM: 1. Cross-linking Reagents. 2. Proteins—chemistry.
   3. Proteins—metabolism.   QU W872c]
   QP551.W596   1991
   574.19′245—dc20
   DNLM/DLC
   for Library of Congress   90-15166
                                                                                   CIP

   Direct all inquiries to CRC Press, Inc., 2000 Corporate Blvd., N.W., Boca Raton, Florida, 33431.

© 1993 by CRC Press, Inc.

International Standard Book Number 0-8493-5886-8

Library of Congress Card Number 90-15166

Printed in the United States   3 4 5 6 7 8 9 0

Printed on acid-free paper

To all those who appreciate my sole existence
especially my wife Lee-Jun
and my children
Inyork and Hansie

and
Dr. L. Maximilian Buja
who has given me inspiration

PREFACE

"It is indeed a pleasure to acquire knowledge and, as you go on acquiring, to put into practice what you have acquired. A greater pleasure still is when friends of congenial minds come from afar to seek you because of your attainments. But he is truly a wise and good man who feels no discomposure even when he is not noticed by men."

Confucius

In science, as in society, achievements, advances, and discoveries are the main themes of reports and publications. Details on noteworthy processes and the basis for their procedures are generally condensed into a few words or sentences discernible only by the experts. This is particularly so in the broad field of chemical modification and its application in conjugation and cross-linking of biological components. Any novice will have difficulty in perceiving the mechanism of reaction apart from utilization of this technique in his/her applications. In light of the fact that chemical cross-linking reagents have attained great practical use in industry as well as in basic research, and that an understanding of their fundamental principles of reaction is paramount to their applications, I have undertaken the challenge to put together a text that hopefully bridges the gap between scientific literature and practice. Although many reviews have been written, none has completely achieved that goal.

This book offers an explanation of the underlying mechanism of chemical modification, surveys all the bifunctional reagents used in bioconjugation and cross-linking, and provides a review of practical applications of these reagents in various areas of biochemistry, molecular biology, biotechnology, nucleic acid chemistry, immunochemistry, and diagnostic and biomedical disciplines. It is hoped that the coverage will be broad enough to stimulate intellectual curiosity in the development and application of chemical cross-linking reagents.

During the preparation of this book, a major journal, *Bioconjugate Chemistry*, published by the American Chemical Society, has appeared. This journal specifically addresses "the joining of two different molecular functions by chemical or biological means". Coverage emphasizes "the conjugation of antibodies and their fragments, nucleic acids and their analogs, liposomal components, and other biologically active molecules (receptor-binding proteins, hormones, peptides) with each other or with any molecular groups that add useful properties, including drugs, radionuclides, toxins, fluorophores, photoprobes, inhibitors, enzymes, haptens and ligands." The publication of this multidisciplinary journal further demands the understanding of the chemical basis of cross-linking. I sincerely hope that this book will also provide useful information for the readers of this periodical.

It is a miracle that this book is finished in time for publication. I wish to thank all those who, with their constant encouragement and stimulation, have helped me make this possible. My special thanks goes to Drs. Hann P. Wang and Dennis Johnson, who have taken their time to read my draft and given their invaluable suggestions, to Ms. Jody Grady for her help in obtaining many of the references, and to my wife who has not only reviewed the text but also drawn all the figures. I am also grateful to all the scientists and researchers who have contributed so much to the development and application of cross-linking reagents. Without their efforts, I would not have had anything to write about. I therefore apologize to those whose publications I failed to quote. Any mistake that may be found in the book is a result of my oversight. I welcome comments and suggestions for improvement and for correction of any error.

Shan S. Wong
Houston, Texas

THE AUTHOR

Shan S. Wong, Ph.D., is Associate Director of Clinical Chemistry at Hermann Hospital and Lyndon B. Johnson General Hospital in Houston, Texas, and is Associate Professor of Pathology at the University of Texas Health Science Center at Houston, Texas.

Dr. Wong graduated in 1970 from the Oregon State University, Corvallis, Oregon, with a B.S. in Chemistry and obtained his Ph.D. in 1974 from the Department of Chemistry of The Ohio State University, Columbus, Ohio. After doing postdoctoral work at Temple University, Philadelphia, he served as an instructor of Chemistry at The Ohio State University and a Visiting Assistant Professor at Denison University, Granville, Ohio. In 1978, he joined the University of Lowell, Lowell, Massachusetts, as an Assistant Professor of Biochemistry. He was promoted to Associate Professor in 1982 and again to Full Professor in 1984. While he was there, he instituted a Biochemistry Ph.D. Program and initiated a program in Biotechnology. It was in 1990 that he assumed his present position.

Dr. Wong is certified in Clinical Chemistry by the American Board of Clinical Chemistry. He is a member of the American Society of Biochemistry and Molecular Biology, the American Association for Clinical Chemistry, the American Chemical Society, the National Academy of Clinical Biochemistry, the Sigma Xi Scientific Research Society, the Gamma Alpha Graduate Scientific Society and the Honor Society of Phi Kappa Phi. He was the recipient of the Albert L. Henne Award for the best first-year graduate student in 1971, the McPherson Fellowship (The Ohio State University, 1972—1973), the Samuel A. Talbot Memorial Travel Award of the Biophysical Society in 1974, The American Association for Clinical Chemistry Travel Award in 1986 and the Young Investigator Award of the Academy of Clinical Laboratory Physicians and Scientists in 1986.

Dr. Wong has published extensively in various scientific journals and has presented numerous lectures. He has received research grants from the National Institutes of Health, the National Science Foundation, the American Chemical Society, the Research Corporation, the Department of Energy and the Department of the Army. His current major research interests include the mechanism and regulation of coagulation and fibrinolysis in the area of cardiovascular disease.

TABLE OF CONTENTS

Chapter 1

INTRODUCTION

Chemical cross-linking of biological components is a recent outgrowth of chemical modification of proteins.[1] The history of chemical modifications dates back to the 1920s when enzymes were proved to be proteins.[2] At this early stage, the technique was used mainly to identify the particular amino acids responsible for their catalytic activity. Due to the lack of instrumentation and methods of analysis, the progress in the development of new procedures and reagents was initially slow. It was only during and immediately after World War II that significant advances started to take place. The first reviews on the subject were published in 1947.[3,4] Since then, the application of chemical modification has grown exponentially hand-in-hand with the development of biochemical analysis techniques. For example, an effective procedure for the determination of amino acid sequences was introduced by Edman and Begg in 1956.[5] In 1960, automated amino acid analyzers became available.[6] Various new ion-exchange and gel filtration chromatography media were developed and different forms of gel electrophoresis also became commonplace during the same time period. Numerous review articles and specialized monographs began to appear during the last 20 years.[7-15] New directions such as photoaffinity-labeling[10,16,17] and active-site-directed affinity reagents were developed.[10,18,19] More recently, great advances were made in mechanism-based irreversible modifying agents.[18,20] The very powerful method of site mutagenesis is also providing results beyond those attainable by chemical modification, especially for the determination of the function of individual residues in the structure and reactivity relationship of a molecule.[21]

Cross-linking as a special form of chemical modification has particular capabilities of its own, unparalleled by *in vitro* mutagenesis.[14,22,23] The process involves joining of two molecular components by a covalent bond achieved through the use of cross-linking reagents. The components may be proteins, peptides, drugs, nucleic acids, or solid particles. The chemical cross-linkers are bifunctional reagents containing two reactive functional groups derived from classical chemical modification agents. The reagents are capable of reacting with the side chains of the amino acids of proteins. They may be classified into homobifunctional, heterobifunctional, and zero-length cross-linkers. The zero-length cross-linkers are essentially group-activating reagents which cause the formation of a covalent bond between the components without incorporation of any extrinsic atoms. Thus, dicyclohexyl carbodiimide, which has been used extensively to bring about the formation of amide bonds between carboxyl and amino groups in peptide synthesis, is an example of a zero-length cross-linker. The homobifunctional reagents consist of two identical functional groups and the heterobifunctional reagents contain two different types of reactive functional moieties. They therefore form bridges between the reactive amino acid side chains in proteins. Homobifunctional reagents, such as dialkyl halides and *bis*-imidoesters, were among the early cross-linkers developed,[24-27] although formaldehyde and other reagents had been used in the tanning industry many years prior without known chemical reactions. Since the first application of a bifunctional reagent by Zahn in the 1950's,[24-26] research in this area has flourished, particularly during the last two decades. The introduction of photoactivatable aryl azides marked the beginning of heterobifunctional reagents.[17,28-30] Further advancement in the application of these reagents has led to the design and synthesis of cleavable bifunctional compounds.[31,32] Over three hundred cross-linkers have now been synthesized, and more are forthcoming. The diversity of these molecules is as complicated as organic chemistry itself, limited only by the creative imaginations of the researchers. Many of the excellent reviews that have appeared will attest to the complexity of these reagents and their usefulness in various applications.[22,33-40]

The application of chemical cross-linking is multidisciplinary, ranging from basic protein biochemistry to applied biotechnology and engineering, and from immunology to medicine. These reagents have been used to stabilize tertiary structures of proteins,[41,42] to study protein-protein interactions of subunits in oligomeric proteins,[43] and in complex structures such as ribosomes,[44] to determine distances between reactive groups within or between protein subunits,[45,46] to attach ligands to solid supports,[47] and to identify membrane receptors.[38,48] Applications in the pharmaceutical area have led to the coupling between target-specific proteins and metal-chelating agents for diagnostics in *in vivo* imaging of human patients,[49,50] as well as toxins and enzymes for therapeutics.[51-55] With the advent of enzyme-linked immunoassay and genetic probe technology, chemical cross-linking has provided the great demand for enzyme-immunoglobulin conjugates and DNA probes.[56-58] These applications will be reviewed and summarized in the second half of this book.

Although cross-linking and conjugation are often used interchangeably, there is a fine distinction in connotation between these two terms. Cross-linking usually refers to the joining of two molecular species that have some sort of affinity between them, that is, they either exist as an aggregate or can associate under certain conditions. Thus, the chemical bonding between a ligand and its receptor is usually referred to as cross-linking. Similarly, cross-linking is used for the covalent bonding between subunits of enzymes. Conjugation, on the other hand, denotes the coupling of two unrelated species. For example, the linking between an enzyme and an immunoglobulin is conjugation. The product is referred to as a conjugate, and in this case, an immunoconjugate.

No matter whether it is conjugation or cross-linking, two types of products usually result. One is derived from intramolecular cross-linking, the other as a consequence of intermolecular joining of two or more species. The possible chemical reactions of a cross-linking reagent with a molecule are illustrated in Figure 1. As a general case, the molecule will exist as a monomer in dilute solutions, but might associate or interact with each other at high concentrations. When the monomeric form reacts with the reagent, intramolecular cross-linking will take place as the molecules are far apart. At high concentrations, the molecules will be in close proximity or will associate to form dimeric or polymeric aggregates. Under these conditions, the reagent will provide intermolecular cross-linking. Thus, at very low concentrations intramolecular bonding prevails, whereas intermolecular coupling is important at high concentrations. Intramolecular cross-linkages have been used to determine distances between two reactive groups in a protein that is close in space, particularly those at the active sites of enzymes.[59-62] Intermolecular cross-linking may conjugate molecules of the same kind or of different kinds to form homopolymers or heteropolymers, respectively, thus providing a means for the preparation of high-molecular weight complexes. Intermolecular coupling of different kinds of proteins also provides a tool for the study of antibody-antigen interactions, multi-enzyme complexes, membrane protein structures, hormone and receptor recognitions and other protein-protein interactions at the quaternary structural level.

To help understand the basic principles of reactions of these cross-linking reagents, this book begins with a review of the chemical reactivity of amino acid side chains and their reactions with specific chemical reagents. With the background of chemical modification, bifunctional reagents are introduced. All the existing cross-linking reagents are surveyed and classified into homo-, hetero- and zero-length cross-linkers. Various specific applications of these reagents are mentioned. Examples of conjugation are provided as well as the conditions for the reaction. The reader should find this book useful not only as a reference for the basic information about cross-linking reagents but also as a handbook for experimental application of these reagents.

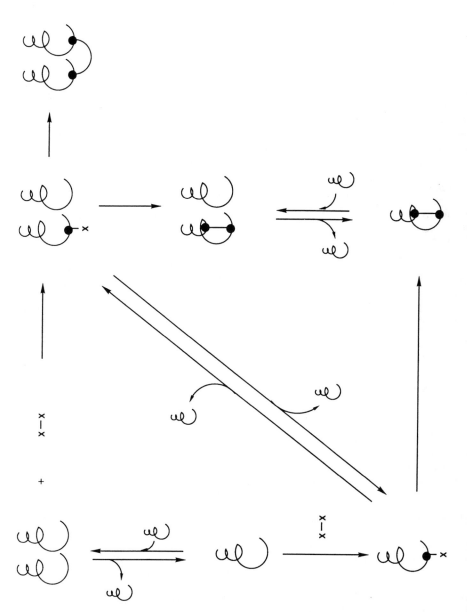

FIGURE 1. Cross-linking reactions of a molecule. Reaction of the cross-linker, X-X, with the monomeric form usually yields intramolecular cross-linking such as in the case of dilute solutions. At high concentrations or at conditions where the molecules associate, intermolecular cross-linking will result.

REFERENCES

1. **Means, G. E. and Feeney, R. E.**, Chemical modification of proteins: history and applications, *Bioconj. Chem.*, 1, 2, 1990.
2. **Sumner, J. B. and Graham, V. A.**, The nature of insoluble urease, *Proc. Soc. Exp. Biol. Med.*, 22, 504, 1925.
3. **Olcott, H. S. and Fraenkel-Conrat, H.**, Specific group reagents for proteins, *Chem. Rev.*, 41, 151, 1947.
4. **Herriott, R. M.**, Reactions of native proteins with chemical reagents, *Adv. Protein Chem.*, 3, 161, 1947.
5. **Edman, P. and Begg, G.**, A protein sequenator, *Eur. J. Biochem.*, 1, 80, 1967.
6. **Moore, S. and Stein, W. H.**, Chromatographic determination of amino acids by use of automatic recording equipment, *Methods Enzymol.*, 6, 819, 1963.
7. **Means, G. E. and Feeney, R. E.**, *Chemical Modification of Proteins*, Holden-Day, San Francisco, 1971.
8. **Hirs, C. H. W. and Timasheff, S. N., Eds.**, Enzyme structure (Part B), *Methods Enzymol.*, 25, 1972.
9. **Glazer, A. N., DeLange, R. J., and Sigman, D. S.**, *Chemical Modification of Proteins: Selected Methods and Analytical Procedures*, North-Holland, Amsterdam, 1975.
10. **Jakoby, W. B. and Wilcheck, M., Eds.**, Affinity labeling, *Methods Enzymol.*, 46, 1977.
11. **Hirs, C. H. W. and Timasheff, S. N., Eds.**, Enzyme structure (Part E), *Methods Enzymol.*, 47, 1977.
12. **Hirs, C. H. W. and Timasheff, S. N., Eds.**, Enzyme structure (Part I), *Methods Enzymol.*, 91, 1983.
13. **Lundblad, R. L. and Noyes, C. M.**, *Chemical Reagents for Protein Modification*, Vols. 1 and 2, CRC Press, Boca Raton, FL, 1984.
14. **Eyzaguirre, J., Ed.**, *Chemical Modification of Enzymes: Active Site Studies*, Ellis Horwood, Chichester, England, 1987.
15. **Feeney, R. E.**, Chemical modification of proteins: comments and perspectives, *Int. J. Peptide Protein Res.*, 29, 145, 1987.
16. **Dhowdhry, V. and Westheimer, F. H.**, Photoaffinity labeling of biological systems, *Annu. Rev. Biochem.*, 48, 293, 1979.
17. **Baley, H.**, *Photogenerated Reagents in Biochemistry and Molecular Biology*, Elsevier, New York, 1983.
18. **Silverman, R. B.**, Mechanism-based enzyme inactivators for medical uses, in *Protein Tailoring for Food and Medical Uses*, Feeney, R. E. and Whitaker, J. R., Eds., Marcel Dekker, New York, 1986, 215.
19. **Baker, B. R.**, *Design of Active-Site-Directed Irreversible Enzyme Inhibitor: The Organic Chemistry of the Enzymic Active Site*, John Wiley & Sons, New York, 1967.
20. **Walsh, C. T.**, Suicide substrates, mechanism-based enzyme inactivators: recent developments, *Annu. Rev. Biochem.*, 53, 493, 1984.
21. **Wetzel, R.**, Medical applications of protein engineering, in *Protein Tailoring for Food and Medical Uses*, Feeney, R. E. and Whitaker, J. R., Eds., Marcel Dekker, New York, 1986, 181.
22. **Friedman, M., Ed.**, *Protein Crosslinking, Biochemical and Molecular Aspects*, Plenum Press, New York, 1977.
23. **Scouten, W. H., Eds.**, *Solid Phase Biochemistry, Analytical and Synthetic Aspects*, Wiley & Sons, New York, 1983.
24. **Zahn, H.**, Bridge reactions in amino acids and fibrous proteins, *Angew. Chem.*, 67, 561, 1955.
25. **Zahn, H.**, Cross-linking reactions with amino acid and fibrous proteins, *Makromol. Chem.*, 18, 201, 1955.
26. **Zahn, H. and Meienhofer, J.**, Reactions of 1,5-difluoro-2,4-dinitrobenzene with insulin. I. Synthesis of medial compounds, *Makromol. Chem.*, 26, 126, 1958.
27. **Hartmann, F. C. and Wold, F.**, Cross-linking of bovine pancreatic RNase-A with dimethyl-adipimidate, *Biochemistry*, 6, 2439, 1967.
28. **Fleet, G. W. J., Porter, R. R., and Knowles, J. R.**, Affinity labeling of antibodies with aryl nitrene as reactive group, *Nature*, 224, 511, 1969.
29. **Fleet, G. W. J., Knowles, J. R., and Porter, R. R.**, The antibody binding site: labeling of a specific antibody against the photo-precursor of an aryl nitrene, *Biochem. J.*, 128, 499, 1972.
30. **Knowles, J. R.**, Photogenerated reagents for biological receptor-site labels, *Acc. Chem. Res.*, 5, 155, 1972.
31. **Wang, K. and Richards, F. M.**, Reaction of dimethyl-3,3'-dithio-bis-propionimidate with intact human erythrocytes, *J. Biol. Chem.*, 250, 6622, 1975.
32. **Coggins, J. R., Hooper, E. A., and Perham, R. N.**, Use of DMS and novel periodate-cleavable *bis*-imidioesters to study the quaternary structure of the pyruvate dehydrogenase multienzyme complex of *E. coli*, *Biochemistry*, 15, 2527, 1976.
33. **Fasold, H., Klappenberger, J., Meyer, C., and Remold, H.**, Bifunctional reagents for the crosslinking of proteins, *Angew. Chem. Int.*, 10, 795, 1971.
34. **Wold, F.**, Bifunctional Reagents, *Methods Enzymol.*, 25, 623, 1972.
35. **Peters, K. and Richards, F. M.**, Chemical cross-linking: reagents and problems in studies of membrane structure, *Annu. Rev. Biochem.*, 46, 523, 1977.

36. **Freedman, R. B.,** Cross-linking reagents and membrane organization, *TIBS,* 4, 193, 1979.
37. **Das, M. and Fox, C. F.,** Chemical cross-linking in biology, *Annu. Rev. Biophys. Bioeng.,* 8, 165, 1979.
38. **Ji, T. H.,** Bifunctional reagents, *Methods Enzymol.,* 91, 580, 1983.
39. **Han, K. K., Richard, C., and Delacourte, A.,** Chemical cross-links of proteins by using bifunctional reagents, *Int. J. Biochem.,* 16, 129, 1984.
40. **Lundblad, R. L. and Noyes, C. M.,** Chemical reagents for proteins modification, Vol. 2, CRC Press, Boca Raton, FL, 1984, chap. 5.
41. **Alber, T.,** Mutational effects on protein stability, *Annu. Rev. Biochem.,* 58, 765, 1989.
42. **Wold, F.,** The reaction of bovine serum albumin with the bifunctional reagent *p,p'*-difluoro-*m,m'*-dinitro-diphenylsulfone, *J. Biol. Chem.,* 236, 106, 1961.
43. **Davies, G. E. and Stark, G. R.,** Use of dimethyl suberimidate, a cross-linking reagent, in studying the subunit structure of oligomeric proteins, *Proc. Natl. Acad. Sci. U.S.A.,* 66, 651, 1970.
44. **Sun, T. T., Traut, R. R., and Kahan, L.,** Protein-protein proximity in the association of ribosomal subunits of Escherichia coli. Crosslinking of 30-S protein S16 to 50-S proteins by glutaraldehyde or formaldehyde, *J. Mol. Biol.,* 87, 509, 1974.
45. **Fasold, H.,** Decomposition of azoglobin, separation and identification of single peptide bridges, *Biochem. Z.,* 342, 295, 1965.
46. **Hartman, F. C. and Wold, F.,** Cross-linking of bovine pancreatic ribonuclease A with dimethyl adipimidate, *Biochemistry,* 6, 2439, 1967.
47. **Falb, R. D.,** Covalent linkage. I. Enzyme immobilization by covalent linkage on insolubilized support, in *Biomedical Applications of Immobilized Enzymes and Proteins,* Vol. 1, Chang, T. M. S., Ed., Plenum Publishing, New York, 1977, 7.
48. **Ji, T. H.,** Crosslinking of lectins and receptors in membranes with heterobifunctional cross-linking reagents, in *Membrane and Neoplasia: New Approaches and Strategies,* Marchisi, V. T., Ed., Alan R. Liss, New York, 1976, 171.
49. **Meares, C. J.,** Attaching metal ions to antibodies, in *Protein Tailoring for Food and Medical Uses,* Feeney, R. E. and Whitaker, J. R., Eds., Marcel Dekker, New York, 1986, 339.
50. **Meares, C. F., McCall, M. J., Deshpande, S. V., DeNardo, S. J., and Goodwin, D. A.,** Chelate radiochemistry: cleavable linkers lead to altered levels of radioactivity in the liver, *Int. J. Cancer,* 2, 99, 1988.
51. **Neville, D. M., Jr.,** Immunotoxins: current use and future prospects in bone marrow transplantation and cancer treatment, *CRC Crit. Rev. Therapeutic Drug Carrier Systems,* 2, 329, 1986.
52. **Poznansky, M. J. and Juliano, R. L.,** Biological approaches to the controlled delivery of drugs: a critical review, *Pharm. Rev.,* 36, 277, 1984.
53. **Maeda, H., Matsumura, Y., Oda, T., and Sasamoto, K.,** Cancer selective macromolecular therapeusis: tailoring of an antitumor protein drug, in *Protein Tailoring for Food and Medical Uses,* Feeney, R. E. and Whitaker, J. R., Eds., Marcel Dekker, New York, 1986, 353.
54. **Marsh, J. W. and Neville, D. M., Jr.,** Immunotoxins: chemical variables affecting cell killing efficiencies, in *Proteins Tailoring for Food and Medical Uses,* Feeney, R. E. and Whitaker, J. R., Eds., Marcel Dekker, New York, 1986, 291.
55. **Frankel, A. E., Ed.,** Immunotoxins, Kluwer Academic Publishers, Boston, MA, 1988.
56. **Ngo, T. T. and Lenhoff, H. M., Eds.,** *Enzyme-Mediated Immunoassay,* Plenum Press, New York, 1985.
57. **Ngo, T. T., Ed.,** *Nonisotopic Immunoassay,* Plenum Press, New York, 1988.
58. **Keller, G. H. and Manak, M. M.,** *DNA Probes,* Stockton Press, New York, 1989.
59. **Husain, S. S. and Lowe, G.,** Evidence for histidine in the active site of papain, *Biochem. J.,* 108, 855, 1968.
60. **Moore, J., Jr. and Fenselau, A.,** Reaction of glyceraldehyde-3-phosphate dehydrogenase with dibromoacetone, *Biochemistry,* 11, 3753, 1972.
61. **Husain, S. S., Ferguson, J. B., and Fruton, J. S.,** Bifunctional inhibitors of pepsin, *Proc. Natl. Acad. Sci. U.S.A.,* 68, 2765, 1971.
62. **Lundblad, R. L. and Stein, W. H.,** On the reaction of diazoacetyl compounds with pepsin, *J. Biol. Chem.,* 244, 154, 1969.

Chapter 2

REACTIVE GROUPS OF PROTEINS AND THEIR MODIFYING AGENTS

I. INTRODUCTION

Chemical cross-linking and conjugation of proteins depend on the reactivities of the constituents of proteins as well as the specificities of cross-linkers used. In most cases, the biological activities of the individual proteins in the conjugated products have to be preserved. These circumstances dictate that only those amino acids not involved in the biological functions may be modified. In addition, the three-dimensional structure of a protein must remain intact during the process of chemical modification. Disturbances of protein structures and properties may occur with reagents that change the charge, size, and other characteristics of the modified amino acid residues. For example, rat liver glycine methyltransferase is completely inactivated on introduction of a large and anionic 2-nitro-5-thiobenzoate, while a smaller and neutral cyano group has no effect.[1] Similar results have been observed following the addition of large or charged groups to the cysteine residues of many proteins. Thus, only those amino acid residues that are not situated at the active centers or settings critical to the integrity of the tertiary structures of proteins may be targets for chemical modification. Such amino acids are ideally located on the surface of the molecule. It follows, therefore, that the identity of the reactive functional groups on the exterior of a protein is the most important factor that controls its reactivity towards cross-linking reagents. By knowing which functional groups are located at the protein-solvent interface, the protein may be modified without sacrificing its biological activity. However, this is not as straightforward as one would like it to be. Proteins vary in their three-dimensional structures as well as their surface compositions. An amino acid may occur both buried in and on the surface of a protein. This may or may not be true in another protein. In addition, the chemical properties of an amino acid side chain may be influenced by its nearby residues with which it interacts. In fact, such differences in reactivity may be used to evaluate the microenvironment of the residue.[2,3] In order to understand the principles that govern the reactivity of a protein toward chemical reagents, it is necessary to consider the general properties of the amino acid side chains.

Since cross-linking reagents are chemical modifying agents, their efficiency in linking two proteins depends on their specificity toward the particular amino acid side chains. The choice of the reagent is critical to the success of cross-linking and conjugation. In cases where a specific amino acid is known to react, a logical choice of cross-linker may be made. Such choice requires an understanding of the specificity of the reagents. In this connection, this chapter will also review the basic factors of protein modification.

II. COMPOSITIONS OF PROTEINS

A. AMINO ACIDS

All proteins are composed of amino acids. There are twenty common amino acids with side chains of different sizes, shapes, charges, and chemical reactivity. Their degree of hydrophobicity and hydrophilicity is one of the major determinants of the three-dimensional structure of proteins. Glycine, alanine, valine, leucine, isoleucine, methionine, and proline have nonpolar aliphatic side chains while phenylalanine and tryptophan have nonpolar aromatic side groups. These hydrophobic amino acids are generally found in the interior of proteins forming the so called hydrophobic core of many molecules. Other amino acids, arginine, aspartic acid, glutamic acid, cysteine, histidine, lysine, and tyrosine have ionizable side chains. Together with asparagine, glutamine, serine, and threonine which contain non-ionic polar groups, they are usually located on the protein surface where they can intereact strongly with the aqueous environment. While it can be assumed in general that hydrophobic side chains are buried within the protein and that hydrophilic amino acids are exposed, nonpolar groups may be found on the surface and polar groups may be protected.[4] This is particularly true for amino acids such as methionine, tryptophan, and tyrosine that have both hydrophilic and hydrophobic moieties. Thus, the reactivity of a given protein, in terms of its ability to be chemically modified, will be determined largely by its amino acid composition and the sequence location of the individual amino acids in the three-dimensional structure of the molecule. The diversity of amino acid composition and its conformation imparts many of the different chemical reactivities of a protein. Invariably, however, since lysine residues are usually the most abundant amino acid found in proteins, nucleophilic amino groups will be found on the surface of a protein. The question is how many.

B. PROSTHETIC GROUPS

In addition to the various amino acids, some proteins also contain tightly bound prosthetic groups. These include metal ions, porphyrin groups, coenzymes such as biotin, and other nonpeptidyl moieties. Although many of these structures may contribute to the chemical reactivity of the protein, the most important prosthetic group in the consideration of the protein conjugation is that of carbohydrates. Glycoproteins may contain up to 50% or more of carbohydrates by weight. These polysaccharides are generally short, frequently branched sugar chains of 15 residues or less. They are covalently attached to proteins through *O*-glycosidic linkages to the hydroxyl groups of serine, threonine or hydroxylysine or through *N*-glycosidic linkages to the amide nitrogen of asparagine. The asparagine-linked oligosac-charides are better understood and seem to be more common in glycoproteins.

All asparagine-linked oligosaccharides have in common a mannose-*N*-acetylglucosamine (Man-GlcNac) core of three mannose (Man) and two *N*-acetylglucosamine (GlcNAc) residues through which the oligosaccharide is linked to asparagine (Figure 1).[5] The anomeric carbon of GlcNAc forms the α-*N*-glycosidic linkage with the amide nitrogen of asparagine. Ad-ditional mannose residues may be attached to this common core to form the high mannose type oligosaccharide. In the complex type, sialic acid (Sia), galactose (Gal), GlcNAc and L-fucose (Fuc) residues are built on the core. A hybrid type is formed when these sugars and mannose residues are added to the Man-GlcNAc core (Figure 1).

Although the function of the carbohydrate moiety of most glycoproteins is unknown, oligosaccharide units provide a useful site for chemical modification and cross-linking of proteins. The principle and method of these reactions will be considered later in this chapter.

III. FUNCTIONAL GROUPS OF PROTEINS

A. REACTIVE AMINO ACID SIDE CHAINS

In the final analysis, the chemical reactivities of proteins depend on the side chains of their amino acid compositions as well as the free amino and carboxyl groups of the N- and C-terminal residues, respectively. Studies of chemical modification have revealed that only a few of the amino acid side chains are really reactive.[6] Of the twenty amino acids, the alkyl side chains of the hydrophobic residues are for all interests and purposes chemically inert. The aliphatic hydroxyl groups of serine and threonine can be considered as water derivatives and therefore have a low reactivity. Only eight of the hydrophilic side chains are chemically active. These are the guanidinyl group of arginine, the γ- and β-carboxyl groups of glutamic and aspartic acids, respectively, the sulfhydryl group of cysteine, the imidazolyl group of histidine, the ε-amino group of lysine, the thioether moiety of methi-onine, the indolyl group of tryptophan and the phenolic hydroxyl group of tyrosine (Figure 2). Table 1 summarizes the various chemical modification reactions of these active side

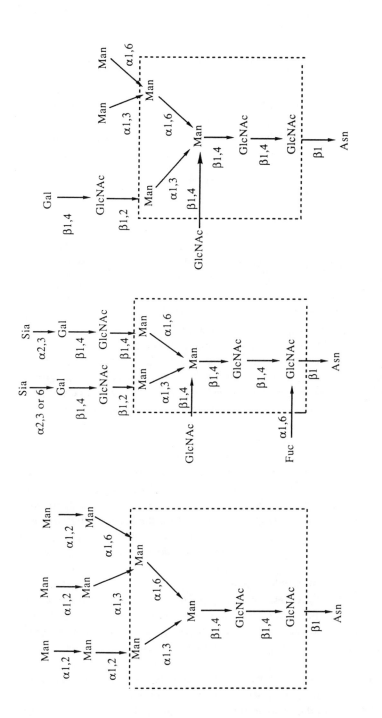

FIGURE 1. Types of asparagine-linked oligosaccharides in glycoproteins. The common core of three mannose (Man) and two *N*-acetylglucosamine (GlcNAc) residues linked to Asn are boxed. When additional sugars such as Man, Sia (sialic acid), Gal (galatose), GlcNAc and Fuc (fucose), are attached to the core, many different patterns are formed, referred to as complex, high mannose and hybrid types. The sugar linkages are also indicated in the figure. (Adapted from Kornfeld, R. and Kornfeld, S., *Annu. Rev. Biochem.*, 54, 631, 1985.).

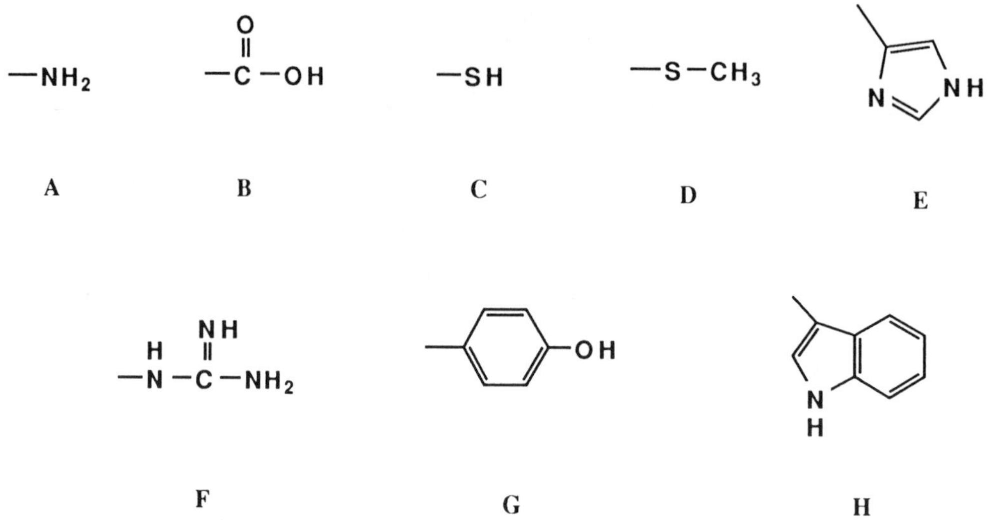

FIGURE 2.   Reactive groups of amino acid side chains. Functional groups A to F are the six most reactive entities. G and H are less reactive. (A) Amino groups of N-terminal amino acids and ε-amino groups of lysines; (B) carboxyl groups of aspartic, glutamic acids and C-terminal amino acids; (C) sulfhydryl of cysteine; (D) thioether of methionine; (E) imidazolyl group of histidine; (F) guanidinyl group of arginine; (G) phenolic group of tyrosine; and (H) indolyl of tryptophan.

chains. The most important reactions are alkylation and acylation. In alkyiation, an alkyl group is transferred to the nucleophilic atom, whereas in acylation, an acyl group is bonded.

Since methionine and tryptophan are generally buried in the interior of proteins and are thereby protected from reagents dissolved in the solvent, they show only some selected reactivity in intact proteins. The other ionizable groups are normally exposed on the surface of proteins. They are therefore the target of protein cross-linking and conjugation.

1. Relationship between Nucleophilicity and Reactivity

Most of the protein modification reactions are nucleophilic reactions, involving a direct displacement of a leaving group by the attacking nucleophile, which in this case is the amino acid side chain (Figure 3). The rate of such a bimolecular nucleophilic substitution reaction, the S_N2 mechanism, depends on at least two factors: the ability of the leaving-group to leave and the nucleophilicity of the attacking group. The easier it is for the leaving group to come off, the faster will be the reaction.[7,8] Similarly, the greater the nucleophilicity, the more expeditiously will the product be formed.[9,10] In terms of protein modification, the relative chemical reactivity is basically a function of the nucleophilicity of the amino acid side chains.

A nucleophile is any species which has an unshared pair of electrons, whether it is neutral or negatively charged, i.e., any Lewis base. The relative availability of the electrons in attacking the positive centers of the substrate determines its relative reactivity.[11] Although nucleophilicity is influenced by various factors such as solvation, size, and bond strength,[9] there are three basic rules of thumb which govern the nucleophilicity of a species:[9,10]

1.   A negatively charged nucleophile is always more powerful than its conjugate acid. Thus ArO⁻ is more powerful than ArOH, and OH⁻ more powerful than H_2O.
2.   Going across the same row of the periodic table, nucleophilicity is roughly proportional to the basicity.[12-14] An approximate order of nucleophilicity is: $NH_2^- > RO^- > OH^- > ArO^- > RNH_2 > NH_3 > H_2O$.
3.   Going down the same column of the periodic table, nucleophilicity increases. Thus sulfur is a more powerful nucleophile than its oxygen analogs.

Table 1

CHEMICAL MODIFICATION OF AMINO ACID SIDE CHAINS

Amino Acid: Side Chain	Alkylation or Arylation	Acylation	Oxidation	Others*
cysteine: —CH₂SH	+	+	+	a,d,f,h
lysine: —NH₂	+	+	−	c,e,g
methionine: —S—CH₃	+	−	+	i
histidine: (imidazole)	+	+	+	a,c
tyrosine: (—C₆H₄—OH)	+	+	+	a,b,c,d
tryptophan: (indole)	+	−	+	h
aspartic and glutamic acids: —COOH	−	+	−	d,e
arginine: —NH—C(=NH)—NH₂	−	−	−	g

* Other reactions include (a) iodination, (b) nitration, (c) diazotization, (d) esterification, (e) amidation, reaction with (f) mercurials, (g) dicarbonyls, (h) sulfenyl halides and (i) cyanogen bromide

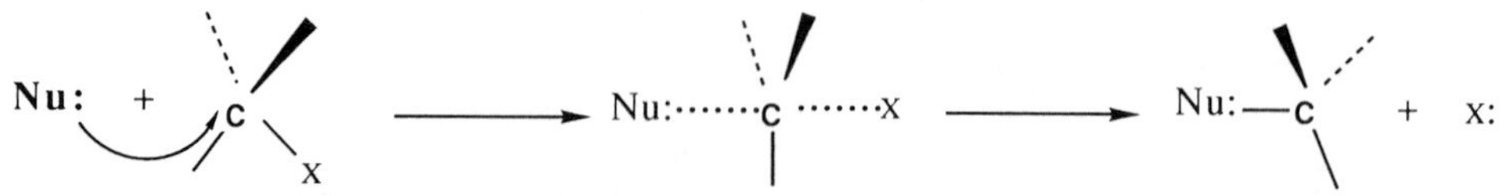

FIGURE 3. Nucleophilic substitution reaction, the S_N2 mechanism. The nucleophile (Nu:) attacks an electron deficient center displacing the good leaving group (X).

Although these rules do not always hold, Edwards and Pearson[10] have formulated an overall nucleophilicity order: $RS^- > ArS^- > I^- > CN^- > OH^- > N_3^- > Br^- > ArO^- > Cl^- >$ pyridine $> AcO^- > H_2O$. With this formulation, it may be deduced that the sulfhydryl group of cysteine is the most potent nucleophile in the protein, particularly in its thiolate form. The nitrogen, as in the amino group, is considerably less potent followed by oxygen and carbon. It should be pointed out that the aliphatic hydroxyl groups of serine and threonine, having about the same nucleophilicity as water, are generally unreactive in aqueous solutions.[15] This is particularly so considering the high concentration of water molecules (55 M) against which the aliphatic alcohols will have to compete. A very reactive reagent will favorably undergo hydrolysis before it reacts with the hydroxyl group containing amino acids. In nonaqueous solvents, however, the hydroxyl group may react effectively.

2. Effects of pH

Because protonation decreases the nucleophilicity of a species, the pH of the medium affects the rate of many nucleophilic reactions. The relationship between protonation and the pH depends on the pK_a of the nucleophile. Table 2 lists the pK_as of the reactive groups in free amino acids and in model peptides. Because the pK_a is a function of temperature, ionic strength, and microenvironment of the ionizable group,[16,17] the table reflects only the approximate values of these groups in proteins. Using these values, the ratio of protonated to deprotonated species at a certain pH can be calculated by the Henderson-Hesselbalch equation:

$$pH = pK_a + \log\{[A^-]/[AH]\}$$

It can be shown from such calculations that the following general rules hold:

1. At one pH unit below its pK_a, the species is 91% protonated.
2. At two pH units below its pK_a, the species is 99% protonated.
3. At one pH unit above its pK_a, the species is 91% deprotonated.
4. At two pH units above its pK_a, the species is 99% deprotonated.
5. When the pH is the same as pK_a, 50% protonation occurs.

Thus, at a fixed pH, the most reactive group is usually the one with the lowest pK_a. Because of their differences in the pK_a values, the degree of protonation of these amino acid side-chain groups at a certain pH provides a basis for differential modification. For example, at neutrality, the amino groups are protonated rendering them unreactive. On the other hand, carboxyl and imidazolyl groups are unprotonated and thus would be most reactive. At pH 5, imidazolyl group will be over 90% protonated, leaving only the carboxyl group in the ionic form. For a selective reaction with the carboxyl group, such as with diazoacetate, the condition of an acidic pH should be selected. At higher pHs, other nucleophiles, particularly the sulfhydryl, will react. As a consequence, it should be obvious that changing the pH also provides a means to control the course of a chemical reaction. For example, since it is known that phenolate anion is the reactive species in the iodination reaction,[18,19]

Table 2
pK$_a$ OF REACTIVE GROUPS IN PROTEINS

Functional Group	Amino Acid Residue	pK$_a$ in Free Amino Acid	pKa in Model Peptides
α—COOH	C-terminal	1.8 - 2.6	3.1 - 3.7
β—COOH	Aspartic acid	3.9	4.4 - 4.6
γ—COOH	Glutamic acid	4.3	4.4 - 4.6
imidazole (Histidine ring)	Histidine	6.0	6.5 - 7.0
α——NH$_3^+$	N-terminal	8.8 - 10.8	7.6 - 8.0
—SH	Cysteine	8.3	8.5 - 8.8
ε——NH$_3^+$	Lysine	10.8	10.0 - 10.2
phenol —OH	Tyrosine	10.9	9.6 - 10.0
—NH——C——NH$_2$ ‖ NH$_2^+$	Arginine	12.5	> 12
——OH	Serine & Threonine	> 13	

the rate of iodination of tyrosine increases with increasing pH as the tyrosine anion concentration rises. The unionized residue reacts very slowly or not at all. Consequently, the relative reactivity of the individual side chains to modifying reagents might be considered as the overriding control of the route and extent of modification.

Since the thioether group of methionine is usually not protonated, pH has little effect on its reactivity. Also, the high pK_a of hydroxyl groups of serine and threonine residues is the chemical basis why these residues are normally inert at neutral pH, unless they are activated by neighboring groups as in serine proteases discussed below.

3. Effects of Microenvironment

The reactivities of amino acid side chains in proteins vary considerably depending on their sequence locations which constitute a different microenvironment. Few, if any, of the amino acid side chains are totally free from interacting with their neighboring groups. This is particularly true for those situated in the interior of the protein where they interact through hydrogen bonding, electrostatic attraction, van der Waals, and hydrophobic forces. Those on the surface may also be involved in such interactions with their neighbors or with solvent molecules. In addition, surface polarities around a functional group may affect its chemical and physical properties. Steric hindrance by neighboring groups will obviously reduce accessibility and therefore the reactivity of a particular group. Of particular importance is the effect on the pK_a of the dissociable side chains. For example, the pK_a of acetic acid is affected by the presence of ethanol. In water alone, the pK_a of acetic acid is 4.7, whereas in absolute ethanol its pK_a is 10.3. In 80% aqueous ethanol, it is 6.9.[20] Similar environmental effects are operative in proteins. Differences in local microenvironment will render different pK_a values for identical groups at various sequence positions.[2,3] As a consequence, the pK_a of a group in one protein may not be the same in another protein and, in fact, in the same protein but at a different sequence location. This is revealed by the ranges of pK_a values of the ionizable groups in peptide model compounds as shown in Table 2. Thus the pK_a values of the free amino acids are not definitive but only indicative of their reactivity in a protein. In general, however, most of the pK_as of a protein fall within those expected values shown in Table 2, but there are dramatic exceptions. Table 3 depicts some of the examples.[21,22]

One of the most remarkable cases of deviation from the expected pK_as is the activation of hydroxyl side chain in serine proteases.[23] As a free entity, serine is relatively inert because of its high pK_a. However, at the active site of serine proteases, chymotrypsin for example, the hydroxyl proton of serine is hydrogen bonded to the nitrogen of an imidazolyl moiety of a histine residue which is polarized by a buried aspartic acid in a charge-relay system as shown in Figure 4.[24] The abstraction of the proton converts the hydroxyl group to an excellent nucleophilic alkoxide.

B. CHEMICALLY INTRODUCED REACTIVE GROUPS

In addition to the side chains of the twenty amino acids that compose proteins, special reactive groups can be introduced into proteins. These extrinsic moieties are obtained by chemical modification of the existing functional groups. The addition of such new functionalities serves many purposes. In some cases, inactive units such as carbohydrates are activated to active functional groups for further chemical reactions. In other cases spacer arms are incorporated to remove the reactive groups from the protein surface. This will not only reduce the steric hindrance caused by other amino acid side chains, but will also decrease the influence of the local environment. Still in other cases functional groups are converted into one another to either change their specificity or increase their reactivity. The importance of these manipulations is clearly demonstrated by the many examples found in the literature.

Table 3

pK$_a$ s OF IONIZABLE GROUPS IN SOME PROTEINS

Group	pK$_a$ (expected)	Protein	pKa (found)
—COOH	3.1 - 4.6	Ribonuclease	4.7
		β–lactoglobulin	4.8 (49 out of 51 groups)
			7.3 (2 out of 51 groups)
		Insulin	4.7
		Serum albumin	4.0
(imidazole)	6.5 - 7.0	Ribonuclease	6.5
		β– lactoglobulin	7.4
		Insulin	6.4
		Myoglobin	6.6 (6 out of 12 groups)
(phenol OH)	9.6 - 10.0	Ribonuclease	9.95 (3 out of 6 groups)
			>12 (3 out of 6 groups)
		Insulin	9.6
		Serum albumin	10.4
		Chymotrypsinogen	9.7 (1 out of 4 groups)
			10.4 (1 out of 4 groups)
ε—NH$_3^+$	10.0 - 10.2	Serum albumin	9.8
		Lysozyme	10.4
		Ribonuclease	10.2

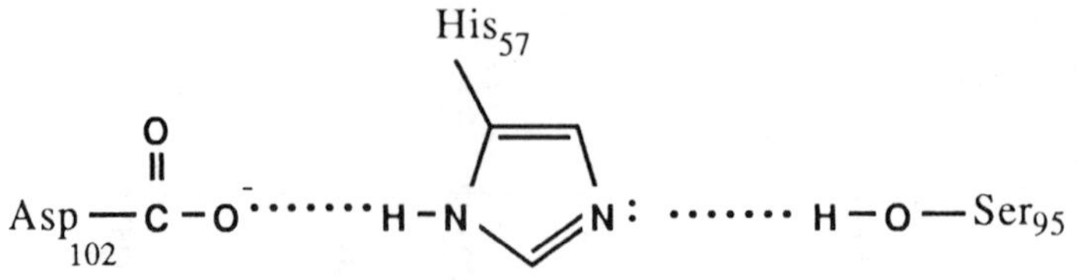

FIGURE 4.   Activation of the serine hydroxyl group in serine proteases. The charge-relay system of chymotrypsin converts the hydroxyl group into an excellent nucleo-philic alkoxide.

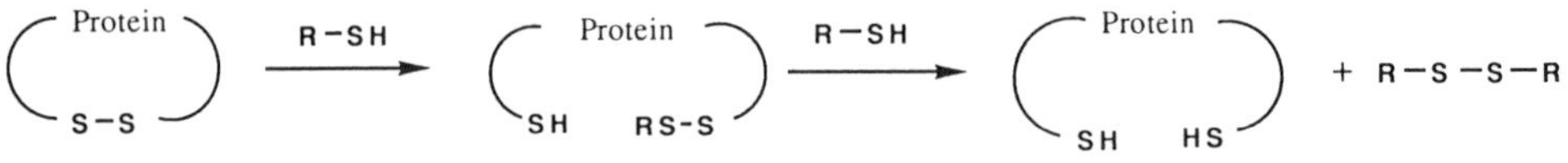

FIGURE 5.   Reduction of disulfide bonds. The reaction involves protein disulfide interchange reaction with a free thiol.

1. Reduction of Disulfide Bonds

Disulfide bonds formed from two cysteine residues cross-link two portions of a protein where these amino acids are located. This oxidized form of sulfur is relatively unreactive but can be easily activated by reduction to the sulfhydryl group. Any thiol-containing reagents such as dithiothreitol, dithioerythritol, 2-mercaptoethanol and 2-mercaptoethylamine can serve as reducing agents.[25,26] The reaction is specific for disulfide bonds and involves disulfide interchange as shown in Figure 5. Complete conversion of disulfide to thiol can be achieved with excess reducing agents. With dithiothreitol, low level is enough to drive the reaction to completion because of the thermodynamically favored formation of a 6-membered ring product.[27] It should be mentioned that these mild reagents will generally reduce only the exposed disulfide bonds but not those that are buried inside the protein. In this case, the integrity of the protein will be preserved. Other stronger reducing agents such as sodium borohydride and lithium aluminum hydride can also reduce disulfide bonds, but these reagents are usually used for complete reduction of proteins after denaturation.

The reduction of disulfide bonds has been used to prepare immunoglobulin fragments for various conjugation reactions.[28,29] These reactions will be discussed in Chapters 10 and 11.

2. Interconversion of Functional Groups

Introduction of new functional groups through the modification of existing amino acid side chains provides another diversity in the application of cross-linking reagents. Under certain circumstances, for example, in the preparation of immunotoxins (Chapter 11), it is desirable to convert one functional group into another. This conversion will either increase its nucleophilicity or change its specificity towards a reagent. For immunotoxins, converting an amino group into a sulfhydryl group enables the preparation of cleavable conjugates which are required for *in vivo* toxicity. The art of introduction of new functional groups is multifarious, limited only by the imagination of the researcher. The following sections will illustrate the voluminous methods used in the transformation of various functionalities.

a. Conversion of Amines to Carboxylic Acids

Amino groups of lysines and N-terminal amino acids are generally the most abundant functional groups on the surface of proteins that are susceptible for conversion to other functionalities. Reaction with dicarboxylic acid anhydrides will convert these amines to

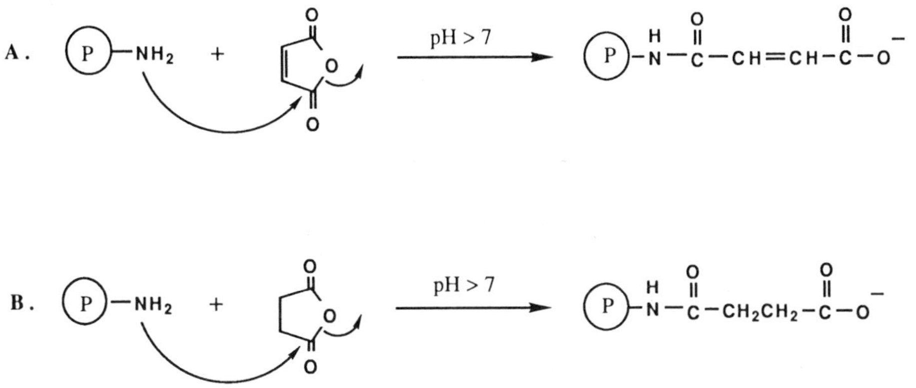

FIGURE 6. Conversion of an amine to a carboxylic acid. (A) Reaction with maleic anhydride; (B) reaction with succinic anhydride.

carboxylic acids. Succinic and maleic anhydrides are the two most commonly used dicarboxylic acid anhydrides. Although the mechanism of reaction is similar, the products are different (Figure 6). The product of maleylation is stable at neutral pH but rapidly hydrolyzes at acidic pHs.[30] Thus, succinic anhydride is the reagent of choice for introduction of carboxyl groups for the purpose of chemical cross-linking. Although succinic anhydride also reacts with tyrosyl, histidyl, cysteinyl, seryl and threonyl side-chains, the tyrosyl and histidyl derivatives are reversible, either hydrolyze spontaneously at alkaline pH or are rapidly decomposed by hydroxylamine. Ester and thioester derivatives are also susceptible to hydroxylamine cleavage at pH 10. Thus specific succinylation of amino groups is possible.[31]

Succinylation and maleylation of a protein result in a net change of a cationic amino group to an anionic acid. Such electrostatic alternation frequently brings about conformational changes and has provided a means to dissociate protein complexes.[32,33] Succinylation has been used to hinder tryptic attack at lysine residues, thus giving rise to peptides with only arginine residue at the carboxy-terminus. Its use in chemical cross-linking of proteins is limited. However, such conversion has provided a means of diversifying the surface chemistry of solid supports for the immobilization of proteins (Chapter 12).

b. Conversion of Amino to Sulfhydryl Groups

There are several approaches by which a free thiol can be linked to the amino group. Amino-specific reagents containing a free sulfhydryl group have been developed. However, since the thiol is a potent nucleophile, most of the effective reagents generally contain protected sulfhydryl groups which can be activated to generate free thiols after the amino groups have been modified. These reactions are shown in Figure 7 and will be described below.

i. Thiolation with N-acetylhomocysteine Thiolactone[34]

N-acetylhomocysteine thiolactone was introduced by Benesch and Benesch[35,36] as a thiolating agent. The thiol group is masked in the thiolactone ring. Nucleophilic attack on the carbonyl group of thiolactone opens the ring, liberating a free thiol. Direct reaction of the compound with an amino group of a protein proceeds very slowly except in a rather alkaline pH of 10 to 11. The reaction can be catalyzed by adding Ag^+ so that it can be carried out near pH 7. The resulting thiolate moiety contains an acetamido group which may interfere with the reactivity of the thiol (Figure 7A).

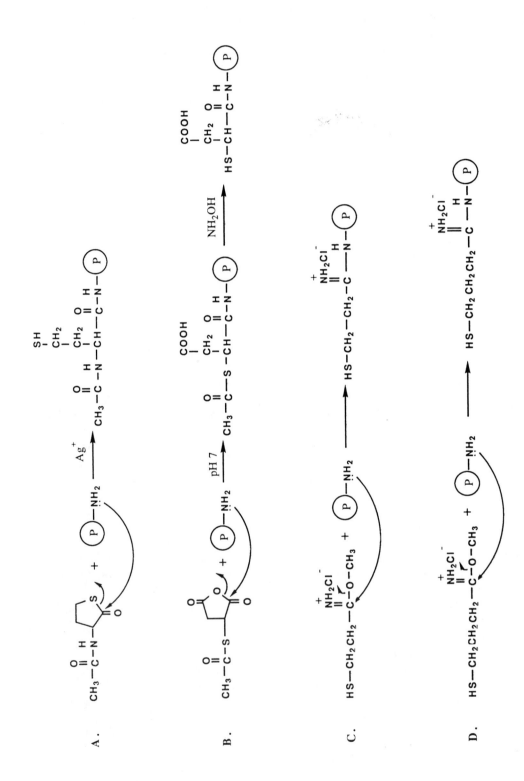

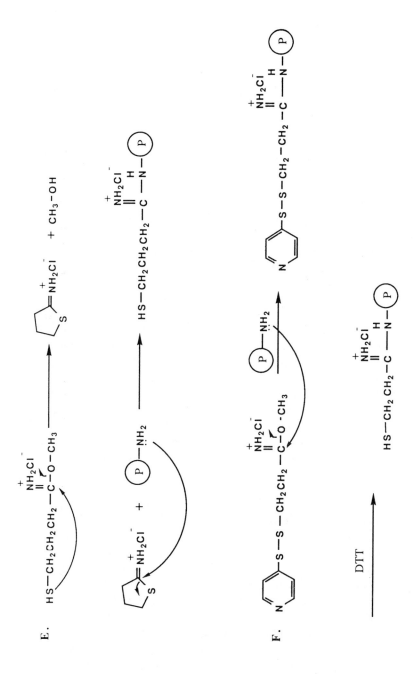

FIGURE 7. Methods of protein thiolation. (A) Thiolation with *N*-acetylhomocysteine thiolactone; (B) reaction with *S*-acetylmercaptosuccinic anhydride; (C) thiolation with methyl 3-mercaptopropionimidate; (D) reaction with methyl 4-mercaptobutyrimide; (E) cyclization of methyl 4-mercaptobutyrimide to form iminothiolane (Traut's reagent) and its thiolation reaction; (F) thiolation with methyl 3-(4-pyridyldithio)propionimidate hydrochloride.

ii. Thiolation with S-Acetylmercaptosuccinic Anhydride

Proteins can be thiolated with *S*-acetylmercaptosuccinic anhydride in a reaction involving a two-step process where the amino group first reacts with the thiol-blocked reagent.[37] The reaction occurs adequately at pH 7. At the end of the reaction, the free sulfhydryl group is generated by treatment with 0.01 to 0.05 *M* hydroxylamine at pH 7 to 8 (Figure 7B). Similar to the succinylation reaction discussed above, the resulting thiolated protein contains a side chain bearing a negative charge. Such a negative charge may affect the reactivity of the free thiol.

iii. Thiolation with Thiol-Containing Imidoesters

Several imidoesters have been used to thiolate proteins. Perham and Thomas[38] prepared methyl 3-mercaptopropionimidate hydrochloride (Figure 7C) and Traut et al.[39,40] synthesized methyl 4-mercaptobutyrimidate (Figure 7D). The latter compound gradually cyclizes on storage with elimination of methanol to form 2-iminothiolane (Traut's reagent) (Figure 7E).[41,42] The cyclic imidothioester has been used for thiolation of proteins (Figure 7E),[41,43] although it is less reactive than the corresponding open-chain methyl 4-thiobutyrimidate. It is, however, more stable and can be stored in the cold for many months. Other imidoesters slowly decompose and can only be stored for a limited amount of time. However, *S*-pyridylthio-protected methyl 3-mercaptopropionimidate is stable and can be used to thiolate proteins as other disulfide compounds (Figure 7F). These imidoesters react readily with amino groups at pH 7 to 10 and are excellent thiolating agents since they are amine-specific and do not change the net charge of the reacted protein.

iv. Thiolation with Thiol-Containing Succinimidyl Derivatives

In addition to methyl 3-(4-pyridyldithio)propionimidate hydrochloride mentioned above, amine specific disulfide compounds such as succinimidyl 3-(2-pyridyldithio)propionate (SPDP)[44] (Figure 8A) and dithiobis(succinimidyl propionate) (DSP)[45] (Figure 8B) have been used in the same reaction scheme as thiolating agents, providing additional alternatives for the introduction of spacers. Reduction of the introduced disulfide bond with dithiothreitol (DTT) produces the free sulfhydryl group. Other succinimidyl esters useful for thiolation of proteins are succinimidyl acetylthioacetate (Figure 9A)[46] and succinimidyl acetylthiopropionate (Figure 9B).[47] After reaction with amino groups, these protected thioesters are treated with hydroxylamine to liberate the free thiol group as shown in Figure 9. The *N*-hydroxysuccinimide esters are stable, crystalline compounds that react cleanly with amines.

v. Other Reactions

In a similar reaction, dithiodiglycolic acid and a water soluble carbodiimide have been used by Jou and Bankert[48] to thiolate erythrocytes. An amide bond is formed between the carboxyl group of the reagent and an amino group of the protein. The coupled component is then reduced with excess dithiothreitol to generate the free thiol group (Figure 10).

Other reagents that have been used to introduce thiol groups are 3-(3-acetylthiopropionyl)thiazolidine-2-thione and 3-(3-*p*-methoxybenzylthiopropionyl)thiazolidine-2-thione.[47,49] Amino groups readily attack the carbonyl carbon displacing the good leaving group, thiazolidine-2-thione as shown in Figure 11. Aminolysis can be easily monitored by the disappearance of the yellow color. The *S*-protecting groups can be quantitatively and quickly removed: the acetyl group by incubating with hydroxylamine,[46] and the methoxybenzyl by 1 *M* trifluoromethanesulfonic acid (TFMS)-thioanisole in trifluoroacetic acid (TFA).[50]

The introduction of a sulfhydryl group has many applications. Not only has it been used extensively in the preparation of immunotoxins (see Chapter 11), it has also been used to prepare immunoconjugates for immunoassays (see Chapter 10). For example, insulin, after thiolation, was conjugated to β-galactosidase with a thiol-specific reagent.[51,52]

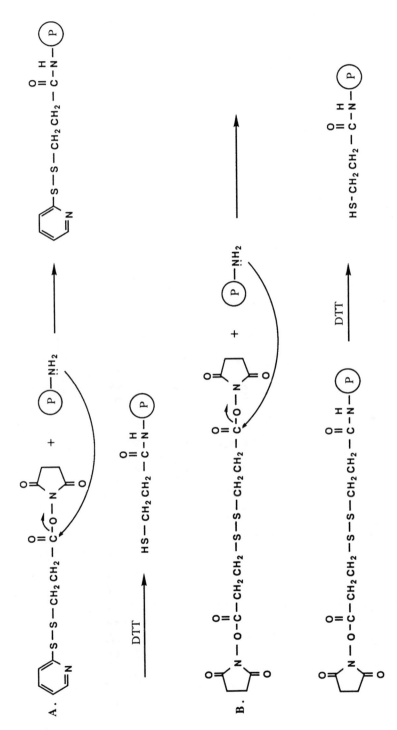

FIGURE 8. Thiolation using succinimidyl ester containing disulfide compounds. (A) Reaction with succinimidyl 3-(2-pyridyldithio)propionate (SPDP); (B) reaction with dithiobis(succinimidylpropionate) (DSP). Dithiothreitol (DTT) reduces the disulfide bond.

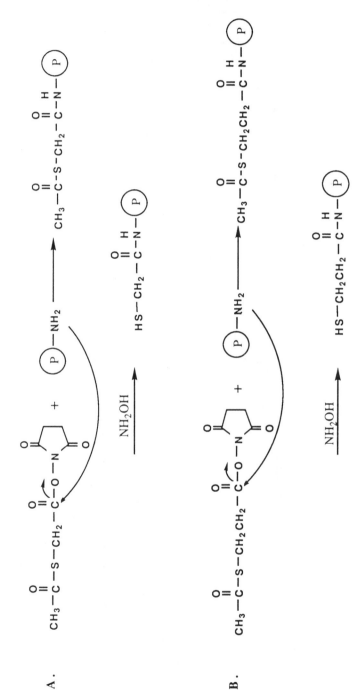

FIGURE 9.   Thiolation with thiol-protected succinimidyl esters. (A) Succinimidyl acetylthioacetate; and (B) succinimidyl acetylthiopropionate.

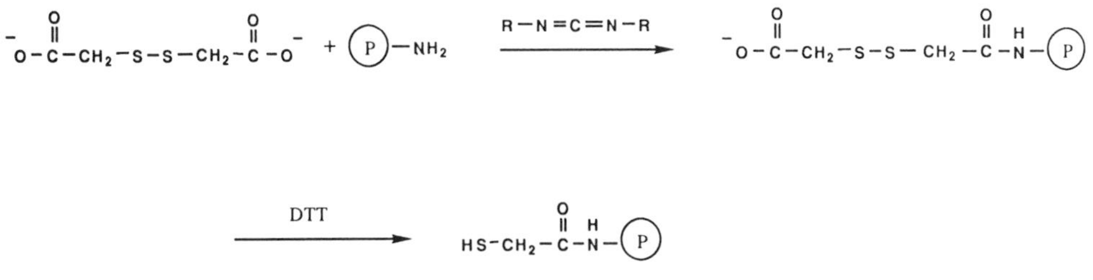

$$- \underset{O}{\overset{O}{\underset{\|}{O}}}-C-CH_2-S-S-CH_2-\overset{O}{\overset{\|}{C}}-O^- \quad + \quad (P)-NH_2 \quad \xrightarrow{\quad R-N=C=N-R \quad} \quad -\underset{O}{\overset{O}{\underset{\|}{O}}}-C-CH_2-S-S-CH_2-\overset{O}{\overset{\|}{C}}-\overset{H}{N}-(P)$$

$$\xrightarrow{\quad DTT \quad} \quad HS-CH_2-\overset{O}{\overset{\|}{C}}-\overset{H}{N}-(P)$$

FIGURE 10. Thiolation by coupling dithioglycolic acid to proteins with water soluble carbodiimides.

c. Conversion of Thiols to Carboxylic Acids

Although succinic anhydride also reacts with the sulfhydryl groups to yield carboxylic acid derivatives, the thioester bond is susceptible to hydrolysis. Carboxyalkylation is generally preferred for the modification of sulfhydryl groups (Figure 12). α-Haloacetates are commonly used as alkylating agents with the highest reactivity for sulfhydryl groups followed in a sequential order by imidazolyl, thioether and amino groups.[53]

S-Carboxymethylation has been used for amino acid quantitation of cysteine and cystine residues after reduction. The reaction occurs at neutral pH but the reactivity increases with alkalinity since the sulfur anion is the reactive species. The reactivity of the haloacids is a function of the halogen in the order of I>Br>Cl≫F. Iodoacetate reacts about twice as fast as bromoacetate and 20- to 100-fold faster than chloroacetate. Fluoroacetate is quite unreactive. Thus, by careful choice of pH and haloacetate, the sulfhydryl group can be specifically modified.

d. Conversion of Thiols to Amines

Conversion of sulfhydryl group to an amine has been achieved with ethylenimine (Figure 13A)[54] and 2-bromoethylamine (Figure 13B).[55] The reaction is quite specific for the thiol group at slightly alkaline conditions.

Aminoethylation of cysteine residues has been used to introduce new points of trypsin cleavage into polypeptide chains, although the rate of hydrolysis is much slower.[56] It may also be used for modification of surface chemistry of solid supports.

e. Conversion of Carboxylic Acids to Amines

Water-soluble carbodiimides, as will be discussed in detail later in the Chapter, have been used to promote amide formation. Using carbodiimides such as 1-ethyl-3-(3-dimethylaminopropyl)-carbodiimide (EDC) and a diamine such as ethylenediamine, the carboxyl group of proteins can be modified to a cationic amine[57] (Figure 14). The reaction is relatively mild and is best carried out at pH's between 4.5 and 5.0.

Amination of the carboxyl group has the advantage of converting it into a more potent nucleophile and at the same time provides an extended arm from the protein to avoid steric hindrance. This method has also been used to change the surface chemistry of solid particles for protein immobilization (see Chapter 12).

f. Conversion of Hydroxyl to Sulfhydryl Groups

The relatively unreactive hydroxyl group of serine and threonine can be converted to the highly reactive thiol by the reactions shown in Figure 15. Activation of the hydroxyl group is first achieved by tosyl chloride (toluenesulfonyl chloride). This tosylation step may be carried out in an aqueous phosphate buffer at pH 8 to 9 containing dioxane or pyridine. Subsequent transesterification is achieved in 0.5 M thioacetate solution at pH 5.5. Hydrolysis of the thioester to generate a free thiol is done with 0.5 N sodium methanoate. Using this

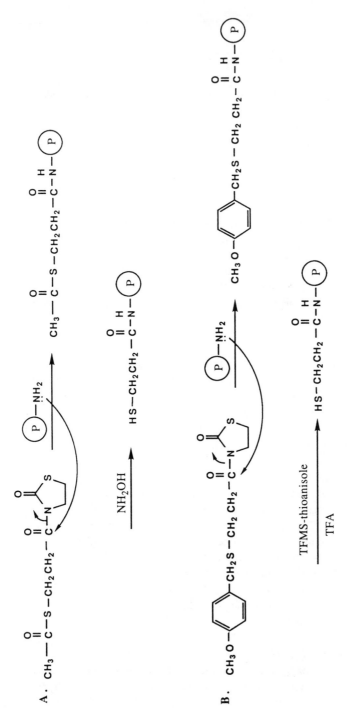

FIGURE 11.   Thiolation of amino groups with (A) 3-(3-acetylthiopropionyl)thiazolidine-2-thione and (B) 3-(3-*p*-methoxybenzylthiopropionyl)thiazolidine-2-thione.

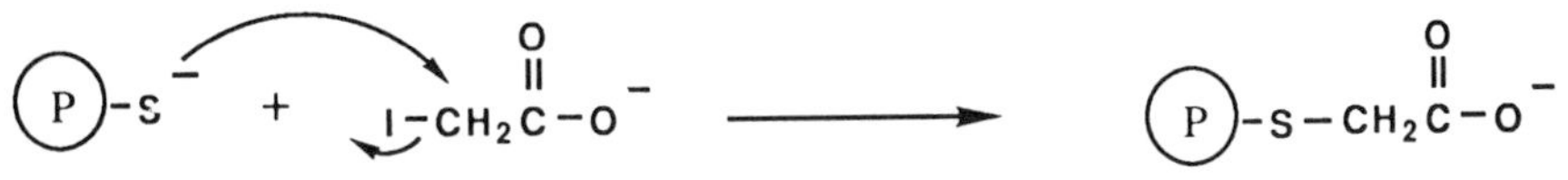

FIGURE 12. *S*-Carboxymethylation of thiol group with iodoacetate.

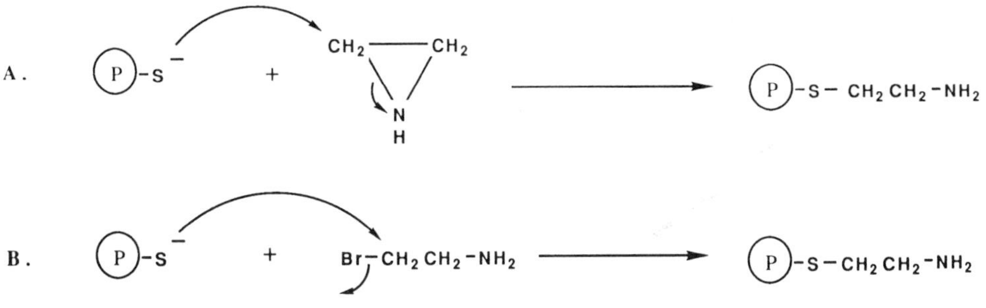

FIGURE 13. Conversion of thiols to amines. Reaction of the sulfhydryl group with (A) ethylenimine and (B) 2-bromoethylamine.

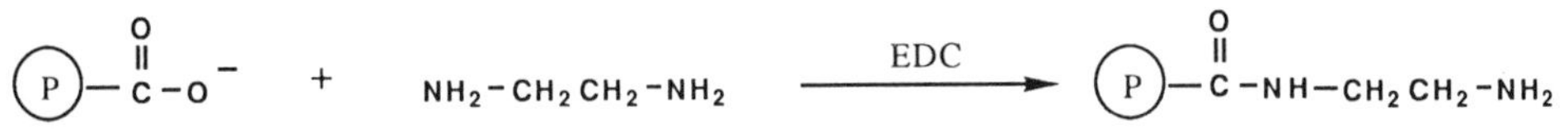

FIGURE 14. Amination of carboxylic acids. The reaction involves coupling of diamines to carboxyl groups of proteins with water soluble carbodiimide, EDC.

procedure, Ebert et al.[58] were able to form disulfide cross-links in various proteins. This procedure should also be useful for altering the surface chemistry of many carbohydrate based solid supports.

g. *Conversion of Tyrosine to Aminotyrosine and Other Derivatives*

Modification of tyrosine in proteins with tetranitromethane results in the nitration at the 3 position of the benzene ring.[6] The phenolic moiety of 3-nitrotyrosine has a pK_a of about 7 which is much lower than that of tyrosine itself, resulting in an increase in the ionized form at neutral pH.[59] Nitrotyrosine can be further reduced to aminotyrosine by sodium hydrosulfite (Figure 16).[60] The amino group of 3-aminotyrosine has a pK_a near 4.8 and is susceptible to a number of nucleophilic reactions at low pHs where other protein groups are inactive.

As will be discussed later in the chapter, diazotization of proteins provides another useful means of modification of tyrosine residues.[6] Various functional groups may be introduced through the reaction with diazonium salts. The most common precursor of diazonium salts are the derivatives of phenylamine. On reaction with nitrous acid, phenylamines will be converted to aryl diazonium salts which reacts with tyrosine (Figure 17). Such process enables the introduction of many functional groups into tyrosine residues of proteins. For example, various carbohydrate moieties, such as *p*-aminophenyl-β-galactoside or β-lactoside have been introduced.[61,62] Nucleotides such as 3' and 5'-thymidine-*p*-aminophenylphosphate have also been coupled.[63]

3. Introduction of Carbohydrate Prosthetic Groups

Simple proteins containing no nonproteous groups can be covalently linked to prosthetic

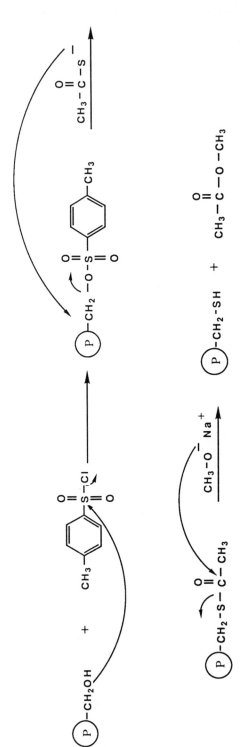

FIGURE 15. Conversion of hydroxyl to sulfhydryl groups: hydroxyl groups are first activated by tosyl chloride and then reacted with thioacetate. Subsequent hydrolysis yields free thiol.

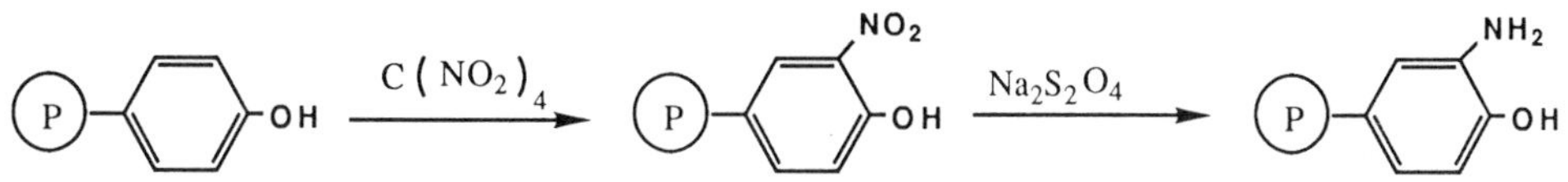

FIGURE 16.   Conversion of tyrosine to aminotyrosine.

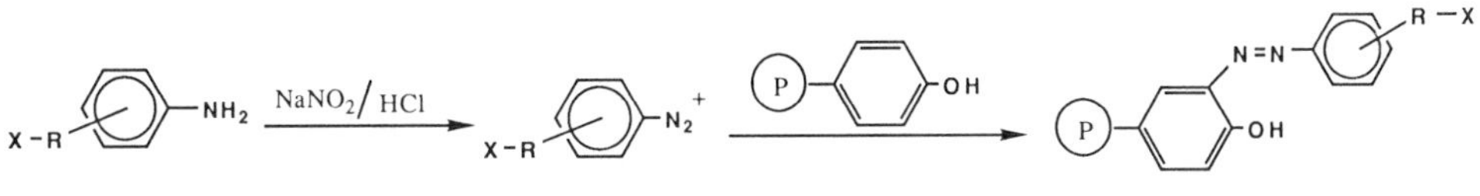

FIGURE 17.   Introduction of functional groups (X) through diazotization of tyrosine.

moieties. Carbohydrate, in particular, can be introduced through covalent bonds to proteins. Plotz and Rifai[64] have modified the soluble carbohydrate polymer Ficoll with chloroacetate, ethylenediamine, glutaric anhydride, and 2,4-dinitrophenol as shown in Figure 18A. The modified Ficoll can be covalently attached to antibodies through displacement of the dinitrophenol by an amino group of the protein.

Carbohydrates have also been coupled to proteins by diazonium, phenylisothiocyanate and mixed anhydride reactions.[61,65] Aminophenyl glycosides have been synthesized and diazotized with nitrous acid or converted to isothiocyanate with thiophosgen. These reactive carbohydrate intermediates will react with proteins on mixing. Alternatively, sugars may be activated with isobutylchloroformate. The mixed anhydride formed will react with proteins to form oligosaccharide-protein conjugates (Figure 18B). Most recently, Lee et al.[66] have used a masked heterobifunctional reagent to link glycopeptides to proteins. The amino-containing carbohydrate is first reacted with an acylazide to introduce an aldehyde. The second step involves reductive alkylation of the aldehyde with protein. Further modification of the attached carbohydrate is possible to provide additional functional groups for protein cross-linking.

Another method of attaching carbohydrates to proteins is by reductive alkylation. As will be discussed below, aldehydes derived from oxidized carbohydrates form Schiff bases with amino groups of proteins which can be reduced to form stable covalent bonds.

4. Activation of Carbohydrates by Periodate

In glycoproteins, the carbohydrate moiety contains sialic acid, galactose and relatively high content of mannose as alluded to above. These sugar consituents contain vicinal hydroxyl groups which are susceptible to oxidation with periodic acid, sodium or potassium periodate. Periodate oxidation cleaves C-C bonds bearing adjacent -OH groups, converting them to dialdehydes.[67] Cis vicinal glycols react more rapidly than trans isomers because of the formation of a cyclic intermediate during oxidation (Figure 19). Diaxial diols rigidly held at 180° are not oxidized because the cyclic intermediate is sterically impossible to form.

After periodate treatment of glycoproteins, the dialdehyde formed can react with a variety of reagents, notably with amine to form imines or Schiff bases (Figure 19). The Schiff bases may be further stabilized on reduction with sodium borohydride, sodium cyanoborohydride, or pyridine borane. The choice of the latter is preferred since it reduces only Schiff bases and is nontoxic, whereas borohydride also reduces aldehydes and disulfides.[68,69]

IV. GROUP DIRECTED REAGENTS

From the considerations presented above, it is not surprising to find that only a few

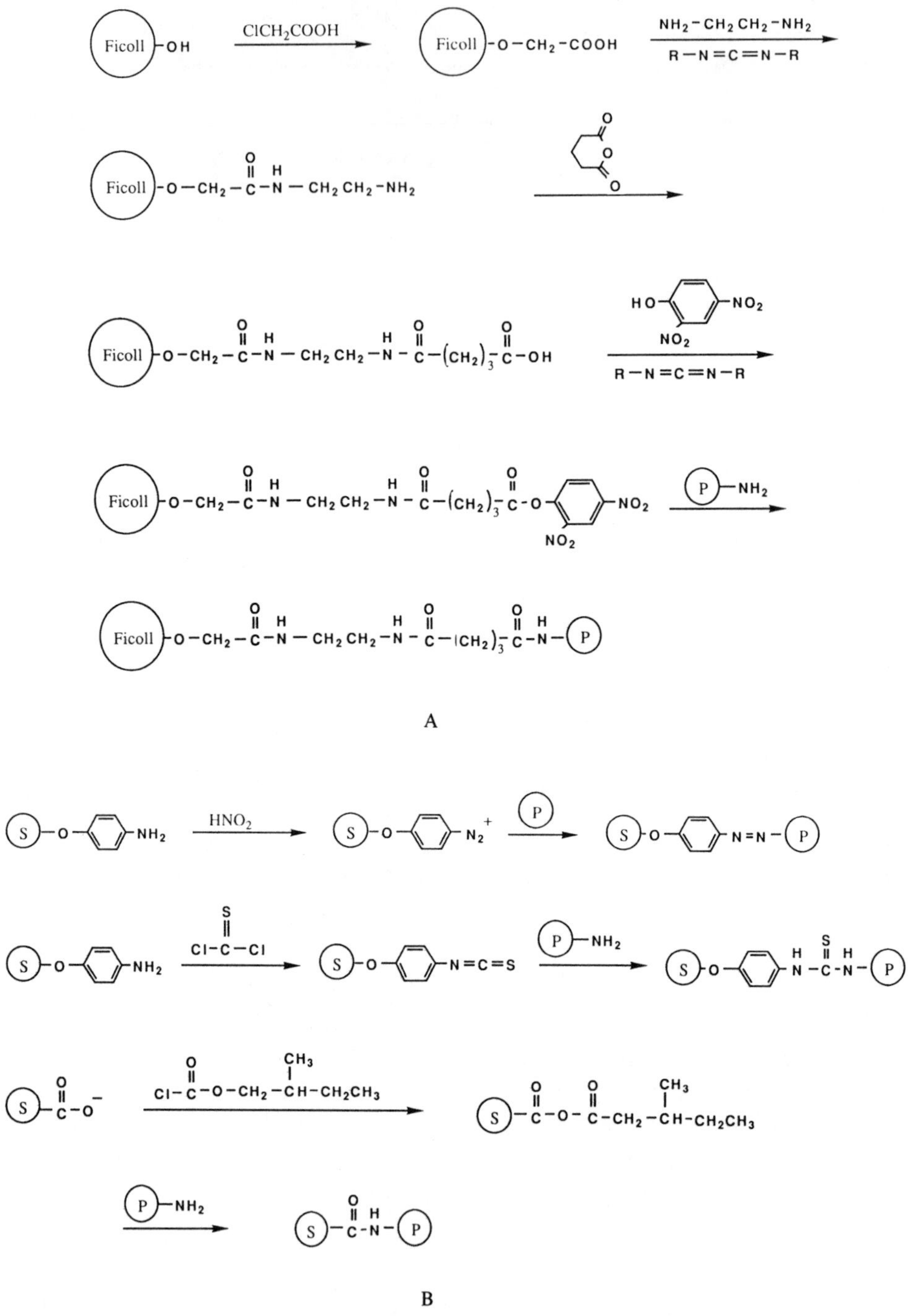

FIGURE 18.   Introduction of carbohydrate prosthetic group into proteins. (A) Modification of Ficoll for covalent attachment to proteins; (B) coupling of carbohydrates, S, to proteins through diazonium, thiocyanate and mixed anhydride.

reagents are absolutely specific toward a particular functional group. The lack of specificity has limited the usefulness of many reagents. However, the same reasons have also provided the basis of selectivity of a reagent. Since cross-reactivity of a reagent towards various groups is due to the diverse nucleophilicity of these groups as well as the effect of their

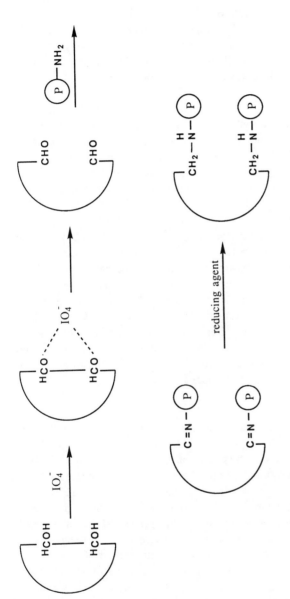

FIGURE 19. Reductive alkylation of amino groups with oxidized carbohydrates.

microenvironment, some reagents may react selectively faster with a particular group. For example, at neutral pH, N-ethylmaleimide reacts much faster with sulfhydryl groups than with amino groups. For other reagents, the stability of the product may play an important role. Cyanic acid, for instance, reacts with amino, sulfhydryl, imidazolyl, tyrosyl and carboxyl groups of proteins, but only the reaction with amino groups results in the formation of a stable product. Other adducts are readily reversible. Thus, under certain conditions, the relative reactivity of these reagents can be employed to selectively modify a particular group. Table 4 gives a summary of different classes of compounds and their reactivity toward the various functional groups found on proteins. These reactions will be discussed to illustrate the principle of chemical applications of cross-linking reagents. There are many excellent monographs and review articles that deal specifically with protein modifications.[6,70-76] Readers interested in this area are referred to these texts for detailed coverage.

A. SULFHYDRYL REAGENTS

The sulfhydryl moiety, with the thiolate ion as the active species, is the most reactive functional group in a protein. With a pK_a of about 8.6, the reactivity of the thiol is expected to increase with increasing pH, toward and above its pK_a. There are many reagents that react faster with the thiol than any other groups. The major classes of compounds that are particularly pertinent to the discussion of cross-linking reagents are presented here.

1. α-Haloacetyl Compounds

The reaction between α-haloacetates (for example, iodoacetate {1} [Figure 20]) and the sulfhydryl group has been discussed in Section III.B.2c. Their amide derivatives, α-haloacetamides {2}, follow the same trend of reactivity. These compounds have been used to carboxymethylate free thiols. They are not sulfhydryl specific. Other nucleophiles such as amines will also react. The reaction involves nucleophilic attack of the thiolate ion resulting in a displacement of the halide. The reactive haloacetyl moiety, $X-CH_2CO-$, has been incorporated into compounds for various purposes. For example, bromotrifluoroacetone {3} has been used for ^{19}F incorporation,[77] and N-chloroacetyliodotyramine {4} has been employed for the introduction of radioactive iodine[78] into proteins. Such α-haloacetyl group has also been incorporated into cross-linking reagents (see Chapters 4 and 5).

2. N-Maleimide Derivatives

Maleimides such as N-ethylmaleimide (NEM) {5} (Figure 21) are considered fairly specific to the sulfhydryl group, especially at pHs below 7 where other nucleophiles are protonated.[79-81] In acidic and near neutral solutions, the reaction rate with simple thiols is about 1000-fold faster than with the corresponding simple amines. Although the rate increases with pH, the reaction with the amino group also becomes significant at high pHs.[73] The other major competing reaction is the hydrolysis of maleimides to maleamic acids (Figure 21). However, at pH 7, the apparent rate of hydrolysis is only 3.2×10^{-4} min^{-1} in 0.1 M sodium phosphate buffer at 20°C which is too slow to interfere with the reaction with sulfhydryl groups.[82] The rate of decomposition becomes significant only at pH above neutrality.[83] At pH 7, it is estimated that the half-life of the reaction between minimolar concentrations of mercaptan and a maleimide is on the order of one second. Thiols undergo Michael reaction with maleimides to yield exclusively the adduct to the double bond (Figure 21). The resulting thioether bond is very stable and cannot be cleaved under physiological conditions. The quantitative reaction of maleimides with thiol groups forms the basis for a spectrophotometric assay of the latter. At pH 6.0, the loss in absorption of NEM at 300 nm can be used for the quantitative estimation of sulfhydryl groups on dilute solution in the presence of excess NEM.[84] However, the use of Ellman's reagent afford a greater sensitivity.

Many N-substituted maleimides have been synthesized for sulfhydryl group modification.

Table 4

EXAMPLES OF GROUP SPECIFIC REAGENTS

Reagent	Group specificity
α–Haloacetyl compounds (e.g. ICH_2COOH)	$-SH$, imidazole (−NH ring), $-S-CH_3$, $-NH_2$, aryl $-OH$
N-Maleimides (e.g. N-ethylmaleimide)	$-SH$, $-NH_2$
Mercurials (e.g. $ClHg\text{-}C_6H_4\text{-}COOH$)	$-SH$
Disulfides (e.g. NO_2/COOH–substituted phenyl $S-S$ phenyl NO_2/COOH)	$-SH$
Aryl Halides (e.g. F-, NO_2-, NO_2-substituted benzene)	$-SH$, $-NH_2$, aryl $-OH$, imidazole (−NH ring)
Acid anhydrides (e.g. succinic anhydride)	$-NH_2$, aryl $-OH$
Isocyanates (e.g. $HNCO$)	$-NH_2$
Isothiocyanates (e.g. phenyl–NCS)	$-NH_2$
Sulfonyl halides (e.g. $CH_3\text{-}C_6H_4\text{-}SO_2\text{-}Cl$)	$-NH_2$
Imidoesters (e.g. $CH_3 \cdot C(=NH_2^+Cl^-)\text{-}O\cdot CH_3$)	$-NH_2$
Diazoacetates (e.g. $N_2CH_2\text{-}C(=O)\text{-}N(H)\text{-}CH_2COOH$)	$-COOH$, $-SH$
Diazonium salts (e.g. phenyl–$N_2^+ Cl^-$)	aryl $-OH$, imidazole (−NH ring)
Dicarbonyl compound (e.g. 1,2-cyclohexanedione)	$-NH-C(=NH)-NH_2$

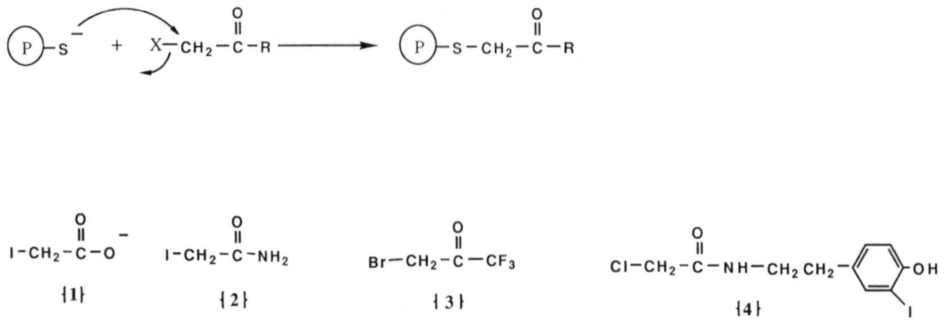

FIGURE 20.   Reaction of haloacetyl compounds with thiol. The equation shows nucleophilic substitution of the halide. Some representative compounds are: {1}: Iodoacetate; {2}: iodoacetamide; {3}: bromotrifluoroacetone; {4}: N-chloroacetyliodotyramine.

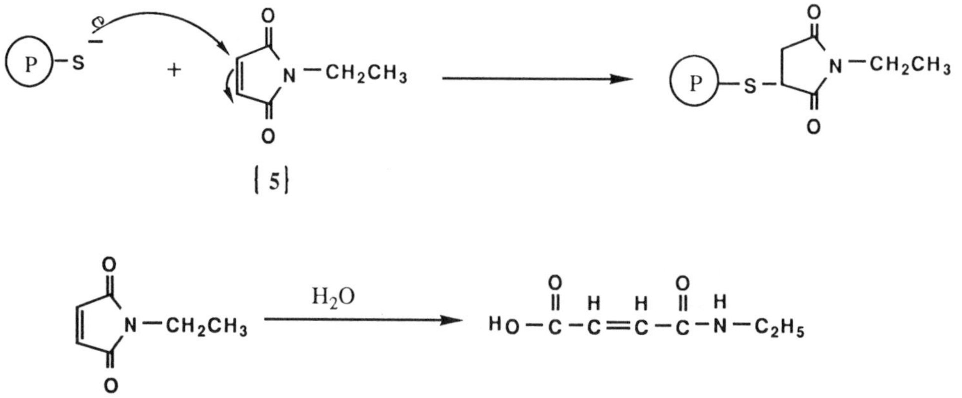

FIGURE 21.   Reaction of maleimides with thiol and the hydrolysis of maleimides. {5}: N-Ethylmaleimide (NEM).

The maleimide group has also been incorporated into cross-linking reagents to selectively react with the thiol group (see Chapter 4 and 5).

3. Mercurial Compounds

Mercurials are the most specific of all sulfhydryl reagents. Organomercurials such as p-chloromercuribenzoate (PCMB) {6} (converted to p-hydroxymercuribenzoate (PHMB) {7} in aqueous solution) (Figure 22), have been widely used.[84] The optimum rate of reaction is usually around pH 5, although the reaction rate for different mercurials and different sulfhydryl groups within one or different proteins can vary greatly. While the thiolate ion increases with pH, a faster rate of reaction does not result due to the competition of OH$^-$ for the reagent.

The affinity of mercurials for sulfhydryl groups is very strong. The dissociation constants of thiol-mercurial complexes is on the order of 10^{-20} which should allow specific modification of these groups in proteins. However, removal of mercurials from this complex is possible by competitive displacement with high concentrations of low-molecular weight mercaptans. Because of the specificity of the reaction at low pH towards the thiol group, the mercurial ion has been incorporated into other reagents.[85]

4. Disulfide Reagents

Disulfide interchange occurs when sulfhydryl groups react with disulfides as denoted earlier in Figure 5. The most commonly used disulfide in the modification of protein sulfhy-

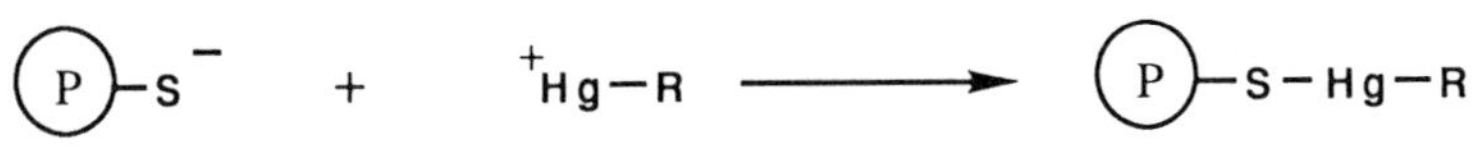

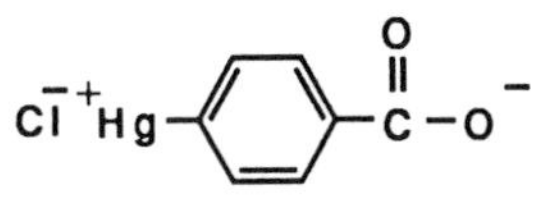

FIGURE 22.   Reaction of mercurials with thiol. {6}: *p*-Chloromercuribenzoate (PCMB); {7}: *p*-hydroxymercuribenzoate (PHMB).

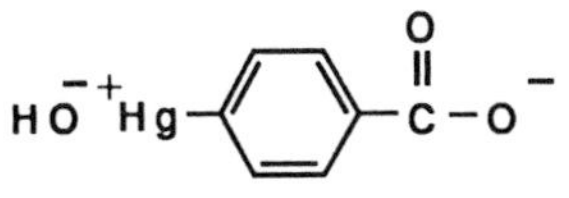

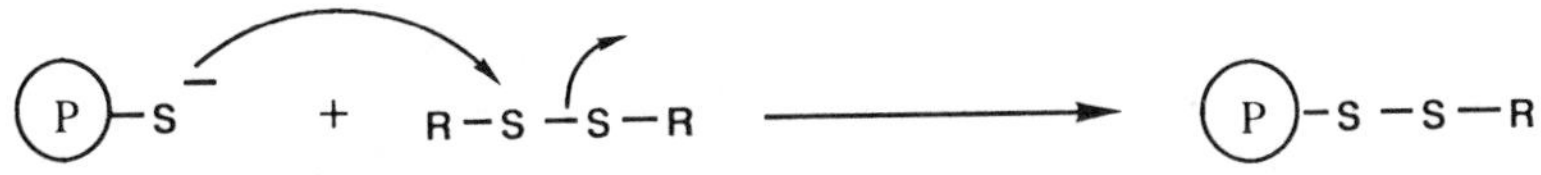

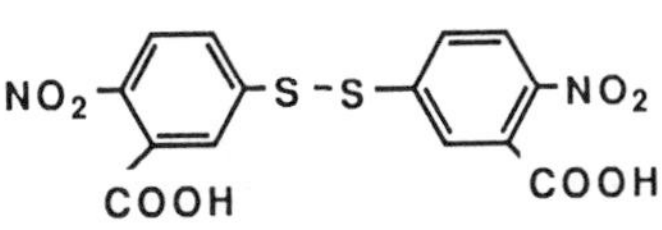

FIGURE 23.   Modification of sulfhydryl groups with disulfide compounds. {8}: Ellman's reagent, 5,5-dithiobis-(2-nitrobenzoic acid), (DTNB); {9}: 4,4′-dithiodipyridine; {10}: methyl-3-nitro-2-pyridyl disulfide; {11}: methyl-2-pyridyl disulfide; {12}: 2-mercaptoethanol; {13}: dithiothreitol (DTT).

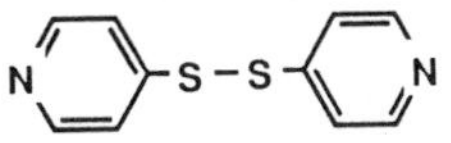

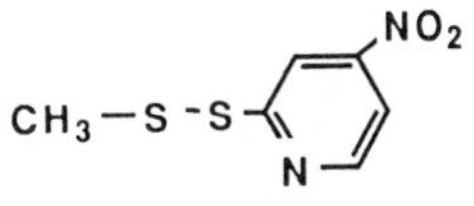

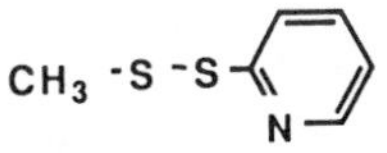

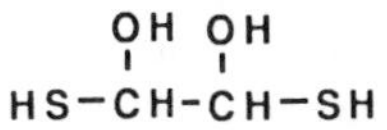

dryl groups is Ellman's reagent, 5,5′-dithiobis-(2-nitrobenzoic acid) (DTNB) {8} (Figure 23). Other disulfides, such as 4,4′-dithiodipyridine {9}, methyl-3-nitro-2-pyridyl disulfide {10} and methyl-2-pyridyl disulfide {11} (Figure 23), have also been synthesized.[86] The protein disulfides formed are readily reversible on incubation with another free mercaptan such as 2-mercaptoethanol {12} or dithiothreitol (DTT) {13} (Figure 23). As discussed in Section III.B.1, protein disulfides are easily reduced to free sulfhydryl groups. After reduction, the reduced thiols may react with other sulfhydryl reagents as mentioned above.

B. AMINO GROUP SPECIFIC REAGENTS

The amino group is another strong nucleophile in the protein. However, because of its abundance and omnipresence in proteins, it is the most important target for chemical mod-

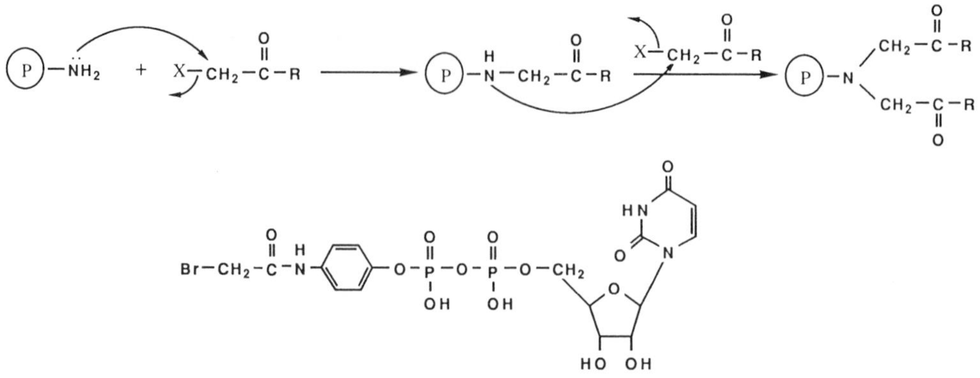

[14]

FIGURE 24.   Alkylation of amino groups. Reaction with haloacetyl compounds is shown with possible dialkylation. {14}: Bromoacetylaminophenyl uridylyl pyrophosphate.

ification, particularly in cases where cysteine residues are absent. Since the protonated species is not reactive, the rate increases with increasing pH as the free amine is formed. Due to the relatively high pK_a of the ammonium ion, most of the reagents that react with the amino group also react with other functionalities. However, many stable acylated products are formed only with the amino groups, providing the basis of selectivity. The most common reactions of amines are alkylation and acylation reactions. Some of the important reagents are described below.

1. Alkylating Agents
a. α-Haloacetyl Compounds

Compared with sulfhydryl groups, amino groups of proteins react with haloacetates and haloacetamides very slowly and only at high pHs, where they are unprotonated. Even then, the reaction rate is only one-hundredth of that of the thiol. The reaction is generally observed as a byproduct of extensive *S*-carboxymethylation and dialkylation may occur during the process (Figure 24). However, in proteins where there is no free sulfhydryl group or where the thiol is buried and is unreactive, the reaction with the amino group becomes important. It is under these circumstances that alkyl halides containing the reactive structure of haloacetyl moiety, -COCH$_2$X, react with the amino and other groups of proteins. These compounds have been used as affinity labels.[87] Wong and Frey[88] have synthesized a UDP-galactose affinity analog {14} (Figure 24) and have shown that it reacts with the nitrogen of adenine ring of NAD in UDP-galactose 4-epimerase. Many cross-linking reagents also contain such reactive functional group (see Chapter 4 and 5).

b. N-Maleimide Derivatives

Like haloacetates, the reaction of maleimides with amino group becomes significant only at alkaline pHs. Because of their greater reactivity with sulfhydryl groups, this reaction is useful only for proteins containing no cysteine residues.[89] Amino groups can conceivably react with maleimides in either one of the two ways (Figure 25). The amine nitrogen can add to the double bond of the maleimide ring (Michael reaction analogous to the addition of SH),[90] or undergo acylation by addition to the carbonyl group followed by ring opening.[81] Regardless of the reaction pathway, the reaction with amines is significant only above neutrality.

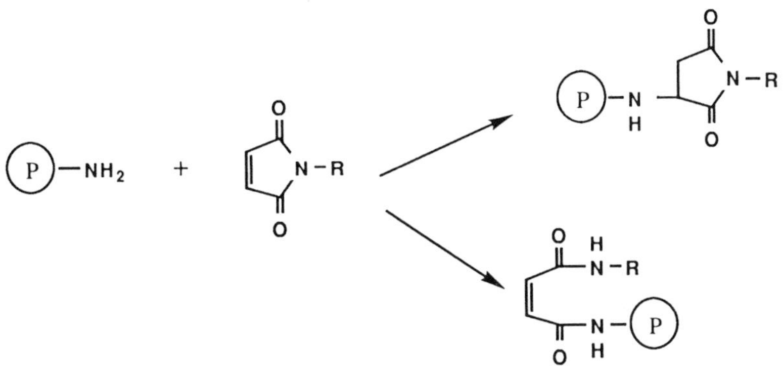

FIGURE 25. Reaction of *N*-maleimides with protein amino groups.

c. Aryl Halides

Aryl halides are usually regarded as amino-group reagents. Dinitrofluorobenzene {15} (Figure 26), for example, has been used to identify the N-terminal amino acid residues of polypeptides and proteins, and trinitrobenzenesulfonate {16} (Figure 26) has been used to quantitate amino groups in proteins because of their bright yellow-colored products. The reaction involves bimolecular nucleophilic substitution at the halogen bonded carbon and is accelerated at alkaline pH since only the unprotonated amino groups react. The reactivity of the aryl halides depends on their leaving group. Fluoro-compounds are the most reactive followed by chloro- or bromo- derivatives in the order: $F > CL^- Br > SO_3H$.[91] Trinitroaryl compounds are more reactive than the corresponding dinitro-compounds.

While aryl halides also react readily with other nucleophiles such as thiolate, phenolate, and imidazole, the products of these reactions are unstable at alkaline pH. In the case of sulfhydryl groups, the reaction is reversible in the presence of excess β-mercaptoethanol (Figure 23).[92]

Several reactive nitrohaloaromatic compounds have been used as cross-linking reagents. These will be further discussed in Chapters 4 and 5.

d. Aldehydes and Ketones

Carbonyl compounds such as aliphatic ketones and aldehydes react very readily and reversibly with amino groups of proteins to form Schiff bases.[93] As usual, the reaction is pH-dependent with the best result obtainable near pH 9. The adducts so formed can be stabilized with sodium borohydride or sodium cyanoborohydride, amine boranes such as dimethylamine, trimethylamine, *t*-butylamine, morpholine and pyridine boranes in a process known as reductive alkylation.[68,69] Reduction is best carried out at 0°C in 0.2 *M* borate buffer, pH 9.0. The concentration of sodium borohydride used for stabilization of the Schiff base is usually low so as not to reduce disulfide bonds, in which case, sodium cyanoborohydride may be a better choice. Pyridine borane has been reported to be an even superior reducing agent for reductive alkylation and is less toxic.[69] Under the mild, slightly alkaline conditions required for the extensive alkylation of amino groups, other protein side-chains do not give stable derivatives, constituting a high selectivity for the amino group.

Practically any carbonyl compounds can be used to alkylate amino groups of proteins. Reductive methylation of bovine pancreatic ribonuclease, ovomucoid, chymotrypsin, and other proteins has been carried out with formaldehyde {17} (Figure 27).[93] Pyridoxal phosphate (PLP) {18} has also been used to label lysine residues at or near pyridoxal phosphate binding sites of proteins and PLP binding sites in enzymes.[94,95] While some simple dialdehydes, notably glutaraldehyde, have been used for cross-linking of proteins (see Chapter 4), the

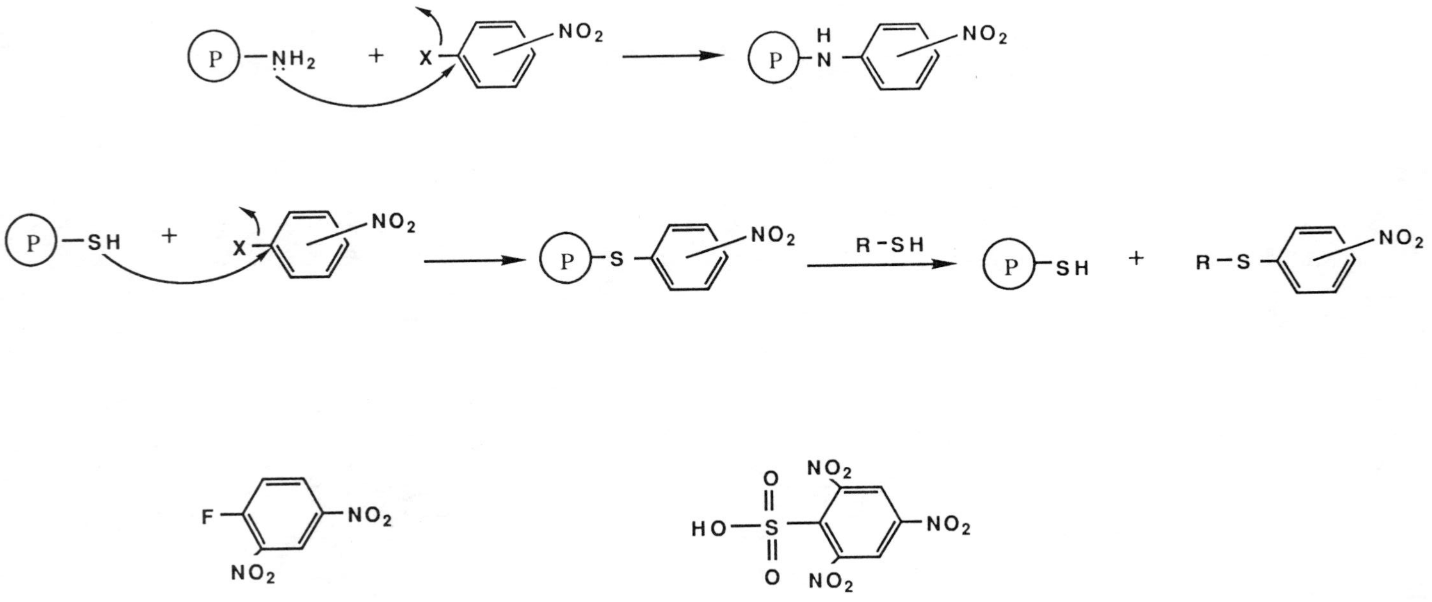

{15} {16}

FIGURE 26. Arylation of amino groups. Reaction with aryl halides is shown. The reaction with thiol is reversible. {**15**}: 2,4-Dinitrofluorobenzene; {**16**}: trinitrobenzenesulfonate.

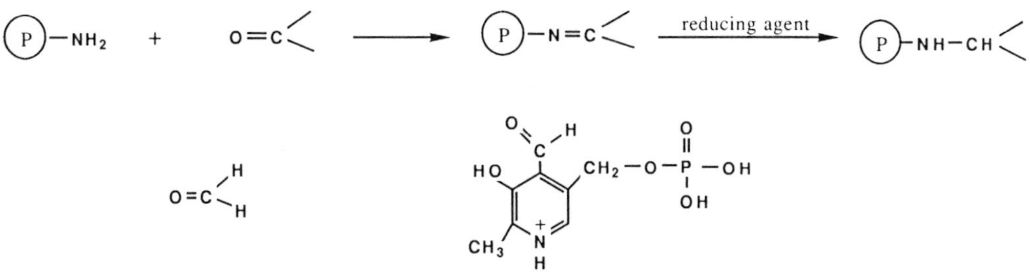

FIGURE 27.   Reaction of amino groups with aldehydes and ketones. Schiff base formed is reversible but can be stabilized with reducing agents. {**17**}: Formaldehyde; {**18**}: pyridoxal phosphate.

dialdehydes produced upon periodate cleavage of carbohydrate as discussed in Section III.B.4 provide an important and versatile means for protein conjugation. Examples of such reactions will be discussed in Chapter 10.

2. Acylating Agents

Acylating agents are compounds that contain an activated acyl group where the nucleophile attacks at the carbonyl carbon displacing a leaving group (Figure 28). The rate of acylation of a nucleophile depends on its pK_a, which, as discussed above, determines the nucleophilicity of a group. Other factors such as steric and proximity effects of other residues on a protein (microenvironment) also influence its reactivity. Since water is a nucleophile, hydrolysis of the acylating agent may be an important side reaction, consuming considerable portions of the reagent. Thus, even using large excess of the reagent, complete acylation of a protein may be difficult to achieve.

All the nucleophiles in proteins are susceptible to acylation. Tyrosine phenolic groups are generally less reactive than amino groups partly because of their higher pK_as and partly because they are generally shielded in proteins. Another important difference is that acylation of tyrosine residues is easily reversible. Hydroxylamine at neutral pH or simply mild alkaline conditions are sufficient to deacylate acylated tyrosines. Similar ease of reversibility of acylated products of other nucleophiles, such as sulfhydryl, imidazolyl and carboxyl, makes most acylating agents amino group selective. Thus, succinic anhydride (Figure 6) has been used to convert amino groups to carboxylic acids.

In addition to acid anhydrides, there are many acylating agents. Of particular importance are the isocyanates, isothiocyanates, imidoesters, acid halides, N-hydroxysuccinimidyl and other activated esters. Isocyanates and isothiocyanates react with amino, sulfhydryl, imidazolyl, tyrosyl and carboxyl groups of proteins. However, only amino groups yield stable products (Figure 28 A and B). Isocyanates are generally more reactive than isothiocyanates. Such difference in reactivity is reflected by the rates of reaction of these groups in 2-isocyanato-4-isothiocyanatotolulene, which contains both isocyanate and isothiocyanate functionalities.[96]

Imidoesters are the most specific reagents for amino groups among the acylating compounds. In mild alkaline pHs (between 7 to 10), imidoesters react only with amines to form imidoamides, the so-called amidines (Figure 28C). The products, like the amino groups they replace, carry a positive charge at physiological pH. Amidination, therefore, retains the net charges of the protein, minimizing the effect of charge on protein conformation.[97] For this reason, imidoesters are favorably incorporated into cross-linking reagents (see Chapters 4 and 5). Amidines are stable in acidic and neutral solutions, but slowly hydrolyze at high pHs. To preserve the cross-linked conjugates, it is necessary to lower the pH of the solution.

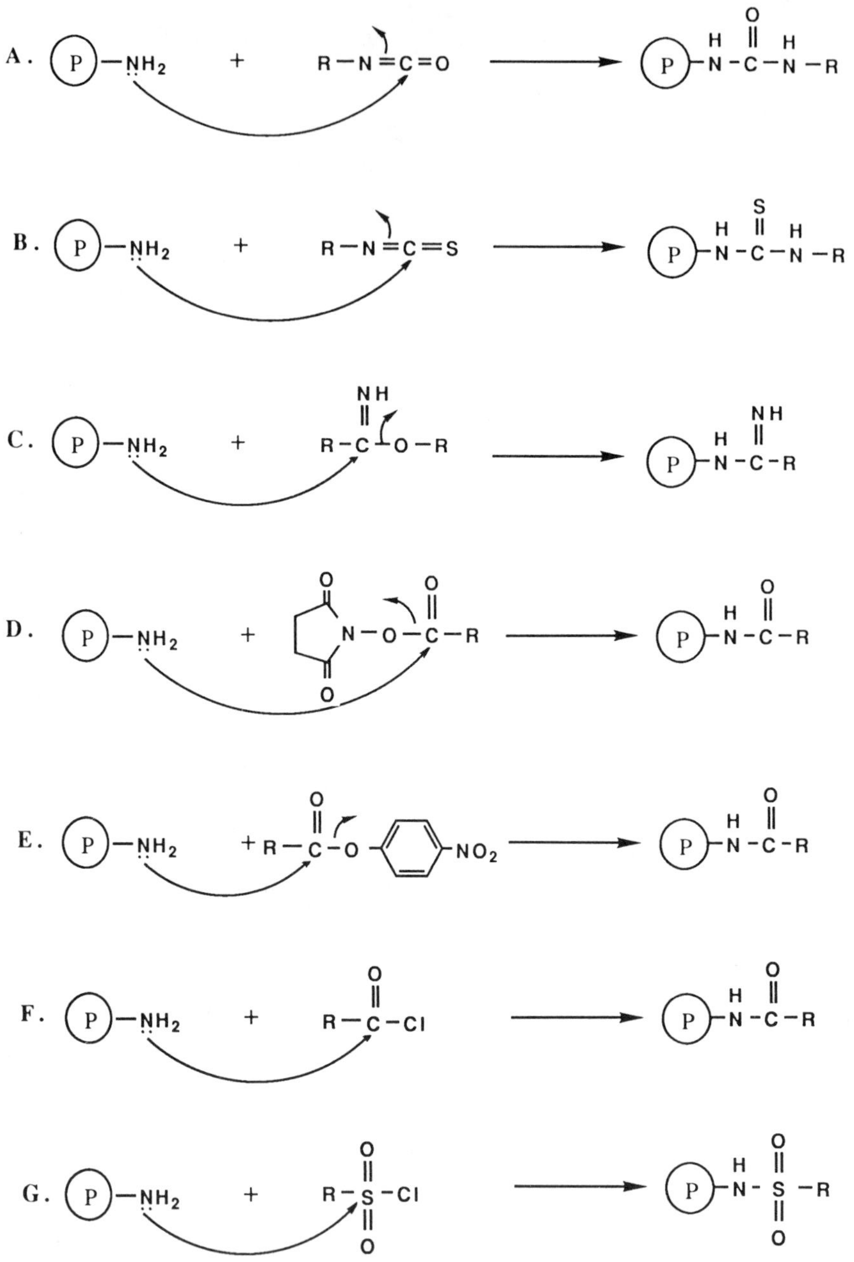

FIGURE 28. Acylation reactions of amino groups. Reaction with: (A) Isocyanate; (B) isothiocyanate; (C) imidoesters; (D) N-hydroxysuccinimidyl ester; (E) p-nitrophenyl ester, (F) acyl chloride; and (G) sulfonyl chloride.

Similar to imidoesters, there are many other activated esters that react with amino groups to form stable amide bonds. Common examples include N-hydroxysuccinimidyl esters and p-nitrophenyl esters (Figure 28 D and E). These compounds contain stable leaving group entities, N-hydroxysuccinimide and p-nitrophenol, respectively, and are reactive toward nucleophilic substitution. Such groups have been used frequently in the construction of cross-linking reagents.

Like acid anhydrides, acyl halides are very reactive compounds and react with all the

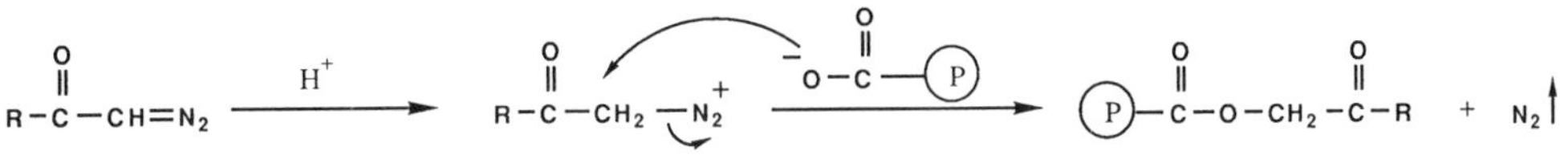

FIGURE 29. Reaction of diazoacetate esters and amides with carboxyl groups. A protonated intermediate is indicated.

nucleophiles of a protein. Reaction with amino groups results in the formation of stable amide bonds (Figure 28F). The adducts with other nucleophiles, for example, O-acyltyrosine, are unstable and hydrolyze under moderately alkaline conditions. The major competitive reaction is hydrolysis, since water molecules are present in high concentrations. Sulfonyl halides react with proteins in much the same way as acyl halides (Figure 28G). They are, however, more stable and therefore react more slowly. The reaction products are also more stable. Even S-sulfonylcysteine, O-sulfonyltyrosine, and imidazolesulfonylhistidine are stable in neutral solutions, unlike the corresponding acyl compounds. Thus, the reaction of sulfonyl chlorides and fluorides with proteins is not group specific. Many sulfonyl fluorides have been found to react with serine hydroxyl groups at the active sites of proteases.[98]

C. REAGENTS DIRECTED TOWARDS CARBOXYL GROUPS
1. Diazoacetate Esters and Diazoacetamides

Like diazomethane, which has been used to esterify carboxyl groups,[99] diazoacetate esters and diazoacetamides react with high specificity with carboxyl groups of proteins under mild acid conditions. The reaction is thought to involve protonation of the diazoacetyl group followed by nucleophilic displacement of nitrogen gas (Figure 29) and occurs optimally at a pH near 5.[100] At this pH, alkylation of sulfhydryl groups is an important side reaction. At higher pHs, other nucleophiles also react. Diazoacetate and diazoacetamides are highly reactive compounds and react rapidly with water and many simple inorganic anions which reduce their efficiency in reaction with proteins. At lower pHs, hydrolysis of the reagent is even more significant.

2. Carbodiimides

The most important chemical modification reactions of carboxyl groups utilize the carbodiimide-mediated process. In the presence of an amine, carbodiimides promote the formation of an amide bond in two steps. In the initial reaction, the carboxyl group adds to the carbodiimide to form an O-acylisourea intermediate. Subsequent reaction of the intermediate with an amine yields the corresponding amide (Figure 30).[57] The reactive intermediate also undergoes hydrolysis slowly and may react with other nucleophiles to form different carboxylated derivatives. With proteins, the optimum pH of the reaction is about 5.[57] Carbodiimides react not only with carboxylic acid but also with alcohols, amines, water and other nucleophiles. O-arylisourea is formed with phenolic group of tyrosine and S-cysteinylisourea with cysteine. These reactions may cause inactivation of the modified protein and decrease the efficiency of coupling.

Practically any carbodiimides can be used to facilitate the amide bond formation. There are several water soluble derivatives that are most useful for this application. The most distinct are 1-cyclohexyl-3-(2-morpholinyl-4-ethyl) carbodiimide (CMC) {19} and 1-ethyl-3-(3-dimethylaminopropyl) carbodiimide (EDC) {20} (Figure 30). The use of these compounds to convert a carboxyl group to a cationic amine has been discussed in Section III.B.2.e. In the absence of an added amine, cross-linking between proteins may also occur.[101] These reactions will be discussed in Chapter 6.

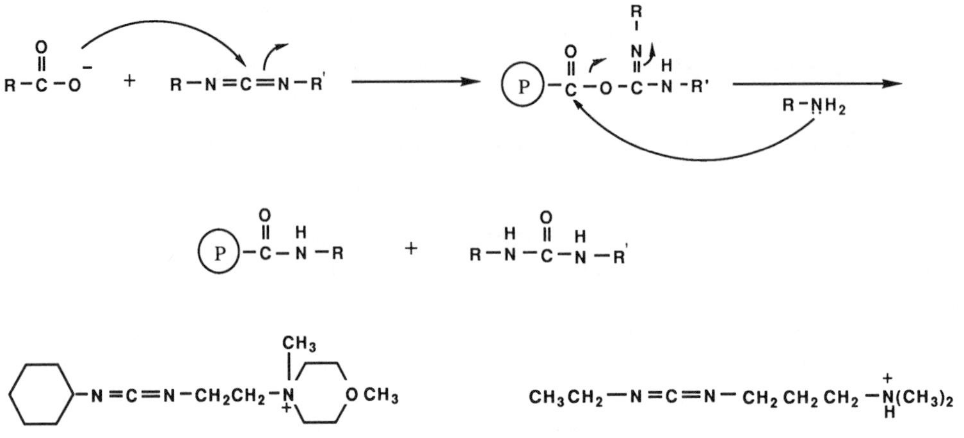

[19]   [20]

FIGURE 30. Carbodiimide-mediate modification of carboxyl groups. The reaction mechanism is shown. Also shown are the structures for 1-cyclohexyl-3-(2-morpholinyl-4-ethyl)carbodiimide (CMC) {**19**}; and 1-ethyl-3-(3-dimethylaminopropyl)carbodiimide (EDC) {**20**}.

D. TYROSINE SELECTIVE REAGENTS
1. Acylating Agents

The phenolate ion of tyrosine reacts similar to amino groups toward acylating agents. However, the tyrosyl group is generally perceived as having a lower reactivity. This is not because the phenolate ion has lower nucleophilicity, but because tyrosine residues are usually buried in proteins inaccessible for reactions due to their hydrophobicity. Such steric hindrance usually gives the amino groups the leading edge of selectivity. In addition, many O-acyl-tyrosine products are unstable, even at neutral pH, as has been noted above. The most important acylating reagent for the phenolic moiety of tyrosine is N-acetylimidazole (Figure 31A). Although it reacts with both amino and tyrosyl groups of proteins, it is more selective for tyrosine.[102] Such selectivity enables it to be used to determine the distribution of free and buried tyrosine residues in a protein. As with other acylating agents, the O-acetylphenol is reversible at high pHs or in the presence of hydroxylamine.[103]

2. Electrophilic Reagents

Tyrosine, histidine, and other aromatic residues of proteins are rich in electrons. These residues undergo electrophilic substitution reactions at the aromatic ring. The most important and useful electrophiles for reaction with tyrosine and histidine in proteins are diazonium compounds.[104] These reagents can be easily obtained by treating aromatic amines with nitrous acid at low temperature. Since diazonium ions are generally rather unstable, they are usually prepared fresh before use. The reaction with tyrosine involves an electrophilic attack of the diazonium ion at ortho-position of the phenol ring displacing a proton without disrupting its aromaticity (Figure 31B). Bisazotization of tyrosine is possible in the presence of excess reagent. A similar mechanism takes place at the imidazole moiety of histidine. The rate of the reaction increases with increasing alkalinity, with an optimal pH near 9. Other protein components such as lysine, tryptophan, cysteine, and arginine residues react very slowly, such that diazonium reagents can be regarded as tyrosine selective.

E. ARGININE SPECIFIC REAGENTS

A predominant reaction of the guanidinyl moiety of arginine residues is with 1,2-dicarbonyl reagents.[105,106] Commonly used vicinal diketones include glyoxal {**21**}, phenyl-

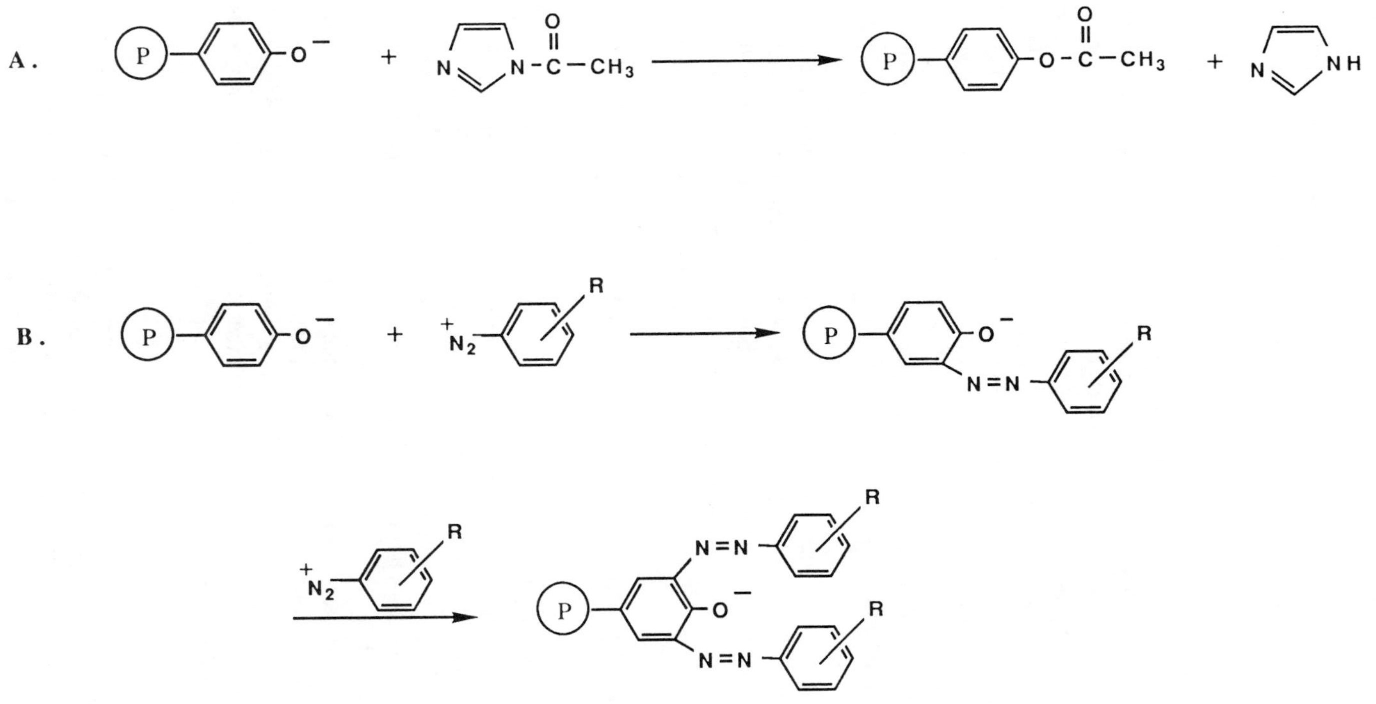

FIGURE 31. Modification of phenolic moiety of tyrosine residues in proteins. (A) Reaction with N-acetylimidazole; (B) reaction with diazonium salts.

glyoxal {**22**}, 2,3-butanedione {**23**} and 1,2-cyclohexanedione {**24**} (Figure 32). Under mild alkaline conditions, these compounds condense with the guanidinyl group in an initial reaction very similar to the Schiff base formation, which undergoes further rearrangement to different products with different compounds (Figure 32). With phenylglyoxal, a trimeric adduct {**25**} may be formed.[107] Borate ion may stabilize the adduct as in the case for 1,2-cyclohexanedione.[105]

The principal side reaction of modification of arginine with dicarbonyls is the reaction with lysine residues which gives unknown products. However, under slightly alkaline conditions, the reaction with arginine predominates.

F. HISTIDINE SELECTIVE REAGENTS

While a number of alkylating and acylating agents react with the imidazolyl moiety of histidines as referred to earlier, the rate of these reactions is generally inferior to other nucleophiles. Even with α-haloacetate, *N*-carboxymethylation is generally slow in comparison with sulfhydryl groups. However, when such reactive α-halocarbonyl group is incorporated into affinity labels, specific reaction may be achieved. For example, *p*-toluenesulfonylphenylalaninechloromethyl ketone (TPCK) {**26**} (Figure 33) specifically alkylates an active-center histidine in α-chymotrypsin. Similarly, *p*-toluenesulfonyllysinechloromethyl ketone (TLCK) {**27**} (Figure 33) reacts specifically with an histidine of trypsin. Such active-site-directed affinity labels may be useful in the design of active-site cross-linking reagents. Besides α-haloacetyl groups, other alkylating agents are not as reactive towards histidine.

With acylating reagents, histidine forms acylated products that are generally unstable and may undergo spontaneous hydrolysis. The most important acylating agent that has been commonly used for the modification of histidines is diethylypyrocarbonate or ethoxyformic anhydride (Figure 33).[108] At pH 4, this reagent shows good selectivity for histidine resulting in *N*-ethoxyformulation of one of the imidazole nitrogens. Amino groups are the major competitor of the reaction. The acylated imidazole is reversed at alkaline pH, resulting in the recovery of histidine. Deacylation can be achieved at neutral pH very rapidly with hydroxylamine.

G. METHIONINE ALKYLATING REAGENTS

The major chemical modification reactions of methionine are oxidation and alkylation, although cyanogen bromide has been used extensively to cleave peptide bonds formed with methionine residues. Oxidation of methionine to methionine sulfoxide can be achieved with hydrogen peroxide, but the reaction has no implication to protein cross-linking and therefore will not be further discussed.

Alkylation of methionine, for example with α-haloacids, is independent of pH and is most selective at pH 3 or less where other groups are protonated.[109,110] However, because methionine is often situated in the hydrophobic interior of proteins, some disruption of the tertiary structure is often a prerequisite for rapid reaction. For the same reasons, cross-linking reagents are generally not designed with methionine in mind.

H. TRYPTOPHAN SPECIFIC REAGENTS

Due to its hydrophobicity, tryptophan residues are generally buried in the interior of proteins. When exposed, they can be modified with *N*-bromosuccinimide (NBS) {**28**}[111] and 2-hydroxy-5-nitrobenzyl bromide (HNBB) {**29**} (Figure 34).[112] Whereas NBS selectively cleaves tryptophanyl peptide bonds in peptides and proteins, HNBB forms an adduct with tryptophan residue. However, practical use of these reagents in protein cross-linking has not been found.

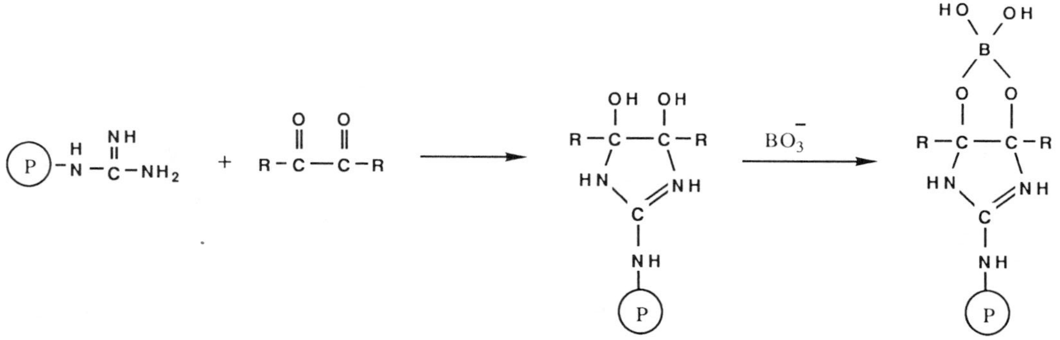

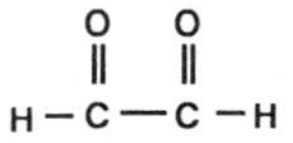

{21}

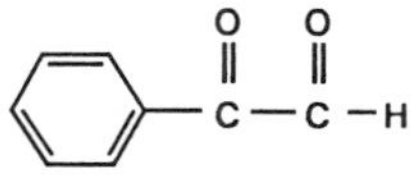

{22}

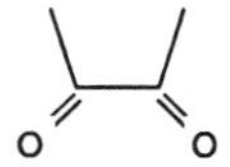

{23}

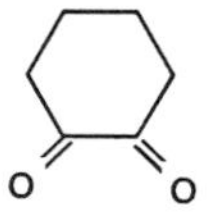

{24}

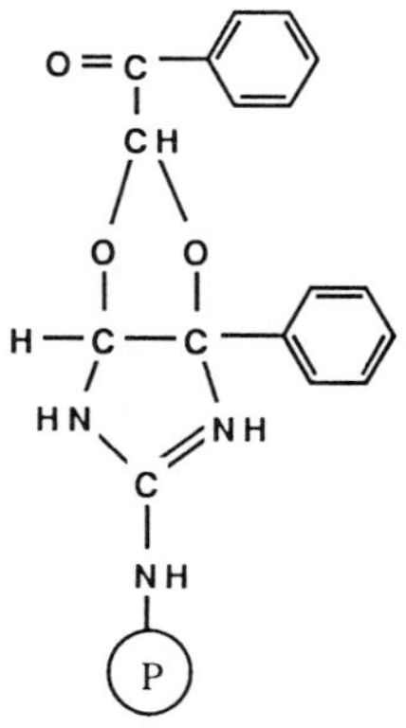

{25}

FIGURE 32. Modification of arginine residues with dicarbonyl compounds. The general reaction is indicated. Some common reagents are {**21**}: Glyoxal; {**22**}: phenylglyoxal; {**23**}: 2,3-butanedione; {**24**}: 1,2-cyclohexanedione. Phenylglyoxal may form a trimeric adduct {**25**}.

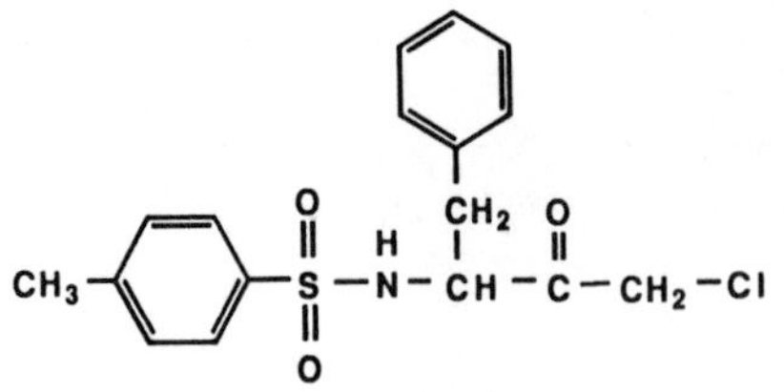

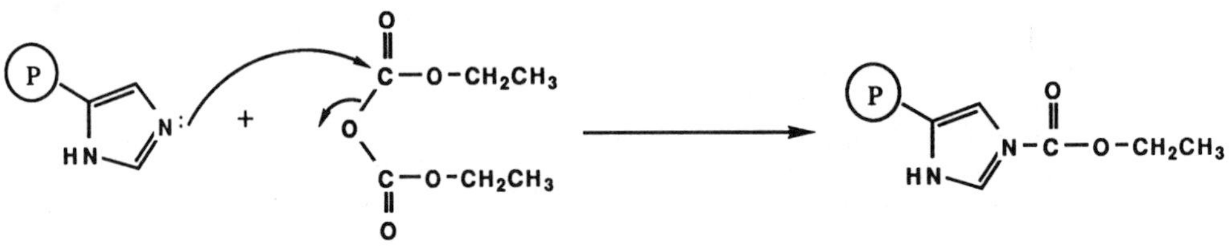

FIGURE 33. Modification with histidine residues. Active-site-directed affinity labels are *p*-toluenesulfonylphenylalanine chloromethylketone {**26**} and *p*-toluenesulfonyllysinechloromethylketone {**27**}. Acylation with diethylpyrocarbonate is shown.

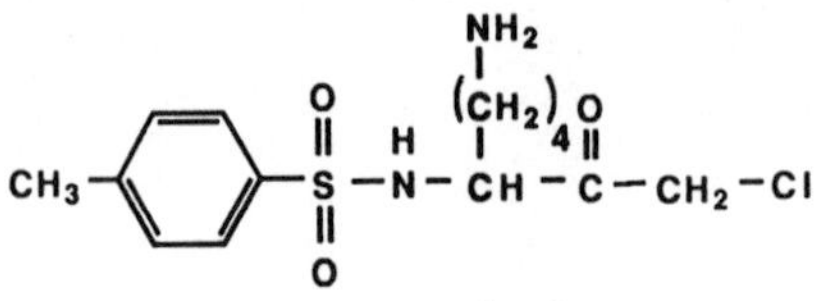

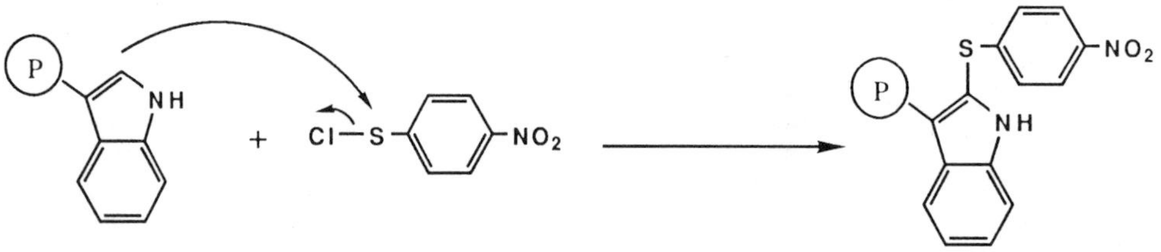

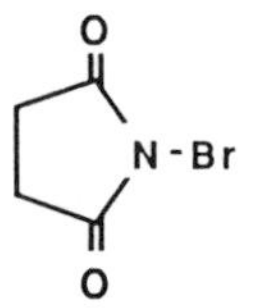

FIGURE 34. Modification of tryptophan residues. The reaction with *p*-nitrophenylsulfenyl chloride is shown. Other reagents are *N*-bromosuccinimide (NBS) {**28**} and 2-hydroxy-2-nitrobenzyl bromide (HNBB) {**29**}.

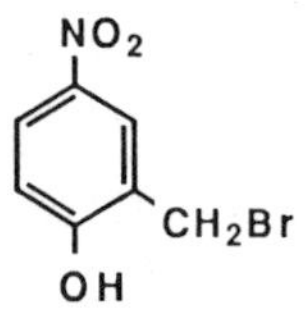

A distinct reagent, *p*-nitrophenylsulfenyl chloride, has been used for the modification of the indolyl moiety, giving rise to 2-thioether derivative (Figure 34).[113] The reaction is selective for tryptophan and cysteine residues and has been incorporated into a bifunctional cross-linking reagent.[114]

I. SERINE MODIFYING REAGENTS

As discussed above, alkyl alcohols are generally inert and are normally not subject to chemical modification in aqueous solutions unless highly reactive reagents are used. In organic solvents, hydroxyl groups of solid matrix can react with various reagents for the conjugation of proteins. These reactions will be discussed in Chapter 12. In proteins, hydroxyl groups of serines and threonines undergo modification only under certain circumstances where the hydroxyl group is activated by neighboring groups such as that at the active site of chymotrypsin and other serine proteases. Many reactive reagents such as diisopropyl-fluorophosphate, phenylmethylsulfonylfluoride and other arylsulfonyl fluorides have been found to react with the active-site serine.[98] But because of the strong competitive reaction of hydrolysis, these groups are generally not targets for chemical modification.

REFERENCES

1. **Fujioka, M., Takata, Y., Konishi, K., and Ogawa, H.,** Function and reactivity of sulfhydryl groups of rat liver glycine methyltransferase, *Biochemistry,* 26, 5696, 1987.
2. **Duggleby, K. G. and Kaplan, H.,** A competitive labeling method for the determination of the chemical properties of solitary functional groups in proteins, *Biochemistry,* 14, 5168, 1975.
3. **Schewale, J. G. and Brew, K.,** Effects of Fe^{3+} binding on the microenvironments of individual amino groups in human serum transferrin as determined by different kinetic labeling, *J. Biol. Chem.,* 257, 9406, 1982.
4. **Chothia, C.,** Principles that determine the structure of proteins, *Annu. Rev. Biochem.,* 53, 537, 1984.
5. **Kornfield, R. and Kornfield, S.,** Assembly of asparagine-linked oligosaccharides, *Annu. Rev. Biochem.,* 54, 631, 1985.
6. **Means, G. E. and Feeney, R. E.,** *Chemical Modification of Proteins,* Holden-Day, San Francisco, 1971.
7. **Loudon, J. D. and Shulman, N.,** Mobility of groups in chloronitrodiphenyl sulfones, *J. Chem. Soc.,* 1941, 722, 1941.
8. **Suhr, H.,** Effect of the leaving group on the velocity of nucleophilic aromatic substitutions, *Berichte,* 97, 3268, 1964.
9. **Bunnett, J. F.,** Nucleophilic reactivity, *Annu. Rev. Phys. Chem.,* 14, 271, 1963.
10. **Edwards, J. O. and Pearson, R. G.,** The factors determining nucleophilic reactivities, *J. Chem. Soc.,* 84, 26, 1962.
11. **Jencks, W. P.,** *Catalysis in Chemistry and Enzymology,* Dover Ed., Dover Publications, New York, 1987.
12. **Bruice, T. C. and Lapinski, R.,** Imidazole catalysis. IV. The reaction of general bases with *p*-nitrophenyl acetate in aqueous solution, *J. Am. Chem. Soc.,* 80, 2265, 1958.
13. **Freedman, R. B. and Radda, G. K.,** The reaction of 2,4,6-trinitrobenzenesulfonic acid with amino acids, peptides, and proteins, *Biochem. J.,* 108, 383, 1968.
14. **Hudson, R. F.,** Nucleophilic reactivity, in *Chemical Reactivity and Reaction Paths,* Klopman, G., Ed., John Wiley & Sons, New York, 1974, chap. 5.
15. **Stark, G. R.,** Reactions of cyanate with functional groups of proteins. IV. Inertness of aliphatic hydroxyl groups. Formation of carbamyl- and acylhydantoins, *Biochemistry,* 4, 2363, 1965.
16. **Mooz, E. D.,** Data on the naturally occurring amino acids, in *Handbook of Biochemistry and Molecular Biology,* Vol. 1, Fasman, G. D., Ed., 3rd ed., 1975, 111.
17. **Cantor, C. R. and Schimmel, P. R.,** *Physical Chemistry, Part I. The Conformation of Biological Macromolecules,* W. H. Freeman, San Francisco, 1980.
18. **Berliner, E.,** Kinetics of the iodination of phenol, *J. Am. Chem. Soc.,* 73, 4307, 1951.
19. **Mayberry, W. E. and Bertoli, D. A.,** Kinetics of the iodination. II. General base catalysis in the iodination of *N*-acetyl-L-tyrosine and *N*-acetyl-3-iodo-L-tyrosine, *J. Org. Chem.,* 30, 2029, 1965.
20. **Lundblad, R. L. and Noyes, C. M.,** *Chemical Reagents for Protein Modification,* Vol. 1, CRC Press, Boca Raton, FL, 1984, 16.
21. **Tanford, C. and Hauenstein, J. D.,** Hydrogen-ion equilibriums of ribonuclease, *J. Am. Chem. Soc.,* 78, 5287, 1956.
22. **Haschemeyer, R. H. and Haschemeyer, A. E. V.,** *Proteins: A Guide to Study by Physical and Chemical Methods,* Wiley-Interscience, New York, 1973.
23. **Kraut, J.,** Serine proteases: Structure and mechanism of catalysis, *Annu. Rev. Biochem.,* 46, 331, 1977.
24. **Blow, D. M., Birktoft, J. J., and Hartley, B. S.,** Role of a buried acid group in the mechanism of action of chymotrypsin, *Nature (London),* 221, 337, 1969.
25. **White, F. H., Jr.,** Regeneration of enzymic activity by air oxidation of reduced ribonuclease with observations of thiolation during reduction with thioglycolate, *J. Biol. Chem.,* 235, 383, 1960.
26. **Zahler, W. L. and Cleland, W. W.,** A specific and sensitive assay for disulfides, *J. Biol. Chem.,* 243, 716, 1968.
27. **Konigsberg, W.,** Reduction of disulfide bonds in proteins with dithiothreitol, *Methods Enzymol.,* 25B, 185, 1972.

28. **Ishikawa, E., Hashida, S., Kohno, T., and Tanaka, K.,** Methods for enzyme-labeling of antigens, antibodies and their fragments, in *Nonisotopic Immunoassays*, Ngo, T. T., Ed., Plenum Press, New York, 1988, 45.

29. **Imagawa, M., Hashida, S., Ishikawa, E., and Freytag, J. W.,** Preparation of a monomeric 2,4-dinitrophenyl Fab'-b-galactosidase conjugate for immunoenzymometric assays, *J. Biochem.*, 96, 1727, 1984.

30. **Butler, P. J. G., Harris, J. I., Hartley, B. S., and Leberman, R.,** Use of maleic anhydride for the reversible blocking of amino groups in polypeptide chains, *Biochem. J.*, 112, 679, 1969.

31. **Gounaris, A. D. and Perlman, G. E.,** Succinylation of pepsinogen, *J. Biol. Chem.*, 242, 2739, 1967.

32. **Klotz, I. M.,** Succinylation, *Methods Enzymol.*, 11, 576, 1967.

33. **Klapper, M. H. and Klotz, I. M.,** Hybridization of chemically modified proteins, *Methods Enzymol.*, 25, 536, 1972.

34. **White, F. H., Jr.,** Thiolation, *Methods Enzymol.*, 25, 541, 1972.

35. **Benesch, R. and Benesch, R. E.,** Thiolation of proteins, *Proc. Natl. Acad. Sci. U.S.A.*, 44, 848, 1958.

36. **Benesch, R. and Benesch, R. E.,** Formation of peptide bonds by aminolysis of homocysteine thiolactones, *J. Am. Chem. Soc.*, 78, 1597, 1956.

37. **Klotz, I. M. and Heiney, R. E.,** Introduction of sulfhydryl groups into proteins using acetyl-mercaptosuccinic anhydride, *Arch. Biochem. Biophys.*, 96, 605, 1962.

38. **Perham, R. N. and Thomas, J. O.,** Reaction of tobacco mosaic virus with a thiol-containing imido ester and a possible application to X-ray diffraction analysis, *J. Mol. Biol.*, 62, 415, 1971.

39. **Traut, R. R., Bollen, A., Sun, T. T., Hershey, J. W. B., Sundberg, J., and Pierce, L. R.,** Methyl 4-mercaptobutyrimidate as a cleavable crosslinking reagent and its application to the *Escherichia coli* 30s ribosome, *Biochemistry*, 12, 3266, 1973.

40. **Kenny, J. W., Sommer, A., and Traut, R. R.,** Cross-linking studies on the 50S ribosomal subunit of *Escherichia coli* with methyl 4-mercaptobutyrimidate, *J. Biol. Chem.*, 250, 9434, 1975.

41. **King, T. P., Li, Y., and Kochoumian, L.,** Preparation of protein conjugates via intermolecular disulfide bond formation, *Biochemistry*, 17, 1499, 1978.

42. **Jue, R., Lambert, J. M., Pierce, L. R., Traut, R. R.,** Addition of sulfhydryl groups of *Escherichia coli* ribosomes by protein modification with 2-iminothiolane (methyl 4-mercaptobutyrimidate), *Biochemistry*, 17, 5399, 1978.

43. **Schramm, H. J. and Dulffer, T.,** The use of 2-iminothiolane as a protein crosslinking reagent, *Hoppe-Seyler's Z. Physiol. Chem.*, 358, 137, 1977.

44. **Carlsson, J., Drevin, H., and Axen, R.,** Protein thiolation and reversible protein-protein conjugation. N-Succinimidyl 3-(2-pyridyldithio)propionate, a new heterobifunctional reagent, *Biochem. J.*, 173, 723, 1978.

45. **Lomant, A. J. and Fairbanks, G.,** Chemical probes of extended biological structures: synthesis and properties of the cleavable protein cross-linking reagent [^{35}S]-dithiobis(succinimidyl propionate), *J. Mol. Biol.*, 104, 243, 1976.

46. **Duncan, R. J. S., Weston, P. D., and Wrigglesworth, R.,** A new reagent which may be used to introduce sulfhydryl groups into proteins, and its use in the preparation of conjugates for immunoassay, *Anal. Biochem.*, 132, 68, 1983.

47. **Fuji, N., Akaji, K., Hayashi, Y., and Yajima, H.,** Studies on peptides. CXXV. 3-(3-p-methoxybenzylthiopropionyl)-thiazolidine-2-thione and its analogs as reagents for the introduction of the mercapto group into peptides and proteins, *Chem. Pharm. Bull.*, 33, 362, 1985.

48. **Jou, Y. H. and Bankert, R. B.,** Coupling of protein antigens to erythrocytes through disulfide bond formation: preparation of stable and sensitive target cells for immune hemolysis, *Proc. Natl. Acad. Sci. U.S.A.*, 78, 2493, 1981.

49. **Fujii, N., Hayashi, Y., Katakura, S., Akaji, K., Yajma, H., Inouye, A., and Segawa, T.,** Studies on peptides. CXXVIII. Application of new heterobifunctional crosslinking reagents for the preparation of neurokinin (A and B)-BSA (bovine serum albumin) conjugates, *Int. J. Pept. Protein Res.*, 26, 121, 1985.

50. **Yajima, H. and Fuji, N.,** Studies on peptides. 103. Chemical synthesis of a crystalline protein with full enzymatic activity of ribonuclease A, *J. Am. Chem. Soc.*, 103, 5867, 1981.

51. **Kato, K., Hamaguchi, Y., Fukui, H., and Ishikawa, I.,** Enzyme-linked immunoassay. I. Novel method for synthesis of the insulin-β-D-galactosidase conjugate and its applicability for insulin assay, *J. Biochem.*, 78, 235, 1975.

52. **Ishikawa, E., Imagawa, M., Hashida, S., Yoshitake, S., Hamaguchi, Y., and Ueno, T.,** Enzyme-labeling of antibodies and their fragments for enzyme immunoassay and immunohistochemical staining, *J. Immunoassay*, 4, 209, 1983.

53. **Gurd, F. R. N.,** Carboxymethylation, *Methods Enzymol.*, 11, 532, 1967.

54. **Raftery, M. A. and Cole, R. D.,** Tryptic cleavage at cysteinyl peptide bonds, *Biochem. Biophys. Res. Commun.*, 10, 967, 1963.

55. **Lindley, H.,** A new synthetic substrate for trypsin and its application to the determination of the amino-acid sequence of proteins, *Nature*, 178, 647, 1956.

56. **Wang, S. S. and Carpenter, F. H.,** Kinetic studies at high pH of the trypsin-catalyzed hydrolysis of N-alpha-benzoyl derivatives of L-arginamide, L-lysinamide and S-2-aminoethyl-L-cyteinamide and related compounds, *J. Biol. Chem.,* 243, 3702, 1968.

57. **Kurzer, F. and Douraghi-Zadeh, K.,** Advances in the chemistry of carbodiimides, *Chem. Rev.,* 67, 107, 1967.

58. **Ebert, C., Ebert, G., and Knipp, H.,** On the introduction of disulfide crosslinks into fibrous proteins and bovine serum albumin, in *Protein Cross-Linking: Nutritional and Medical Consequences,* Friedman, M., Ed., Plenum Press, New York, 1977, 235.

59. **Vincent, J. P., Lazdunski, M., and Delaage, M.,** Use of tetranitromethane as a nitration reagent. Reaction of phenol sidechains in bovine and porcine trypsinogens and trypsins, *Eur. J. Biochem.,* 12, 250, 1970.

60. **Sokolovsky, M., Riordan, J. F., and Vallee, B. L.,** Conversion of 3-nitrotyrosine to 3-aminotyrosine in peptides and proteins, *Biochem. Biophys. Res. Commun.,* 27, 20, 1967.

61. **McBroom, C. R., Samanen, C. H., and Goldstein, I. J.,** Carbohydrate antigens: Coupling of carbohydrates to proteins by diazonium and phenylisothiocyanate reactions, *Methods Enzymol.,* 22, 212, 1976.

62. **Gopalakrishnan, P. V., Zimmerman, U. J., and Karush, F.,** Labeling of antilactose antibody, *Methods Enzymol.,* 46, 516, 1977.

63. **Cuatrecasas, P. and Wilchek, M.,** Staphylococcal nuclease, *Methods Enzymol.,* 46, 516, 1977.

64. **Plotz, P. H. and Rifai, A.,** Stable, soluble, model immune complexes made with a versatile multivalent affinity-labeling antigen, *Biochemistry,* 21, 301, 1982.

65. **Ashwell, G.,** Carbohydrate antigens: coupling of carbohydrates to proteins by a mixed anhydride reaction, *Methods Enzymol.,* 22, 219, 1976.

66. **Lee, R. T., Wong, T.-C., Lee, R., Yue, L., and Lee, Y. C.,** Efficient coupling of glycopeptides to proteins with a heterobifunctional reagent, *Biochemistry,* 28, 1856, 1989.

67. **Bobbitt, J. M.,** Periodate oxidation of carbohydrates, *Adv. Carbohyd. Chem.,* 11, 1, 1956.

68. **Dottavio-Martin, D. and Ravel, J. M.,** Radiolabeling of proteins by reductive alkylation with [^{14}C]-formaldehyde and sodium cyanoborohydride, *Analyt. Biochem.,* 87, 562, 1978.

69. **Cabacungan, J. C., Ahmed, A. I., and Feeney, R. E.,** Amine boranes as alternative reducing agents for reductive alkylation of proteins, *Anal. Biochem.,* 124, 272, 1982.

70. **Lundblad, R. L. and Noyes, C. M.,** *Chemical Reagents for Protein Modification,* Vols. 1 and 2, CRC Press, Boca Raton, FL, 1984.

71. **Eyzaguirre, J., Ed.,** *Chemical Modification of Enzymes: Active Site Studies,* John Wiley & Sons, New York, 1987.

72. **Glazer, A. N., Delange, R. J., and Sigman, D. S.,** *Chemical Modification of Proteins,* North-Holland/Elsevier, Amsterdam, 1975.

73. **Glazer, A. N.,** The chemical modification of proteins by group-specific and site specific reagents, in *The Proteins,* 3rd ed., Neurath, H. and Hill, R. L., Eds., Academic Press, New York, 1976, chap. 2.

74. **Hirs, C. H. W. and Timasheff, S. N., Eds.,** Enzyme structure, *Methods Enzymol.,* 91, Section VII, 1983.

75. **Hirs, C. H. W., Ed.,** Enzyme structure, *Methods Enzymol.,* 11, 988, 1967.

76. **Hirs, C. H. W. and Timasheff, S. N., Eds.,** Enzyme structure (part B), *Methods Enzymol.,* 25, 1972.

77. **Huestis, W. H. and Raftery, M. A.,** A study of cooperative interactions in hemoglobin using fluorine nuclear magnetic resonance, *Biochemistry,* 11, 1648, 1972.

78. **Holowka, D.,** N-chloroacetyl-[^{125}I]-iodotyramine: an alkylating agent with high specific activity, *Analyt. Biochem.,* 117, 390, 1981.

79. **Gorin, G., Martin, P. A., and Doughty, G.,** Kinetics of the reaction of N-ethylmaleimide with cysteine and some congeners, *Arch. Biochem. Biophys.,* 115, 593, 1966.

80. **Gregory, J. D.,** The stability of N-ethylmaleimide and its reaction with sulfhydryl groups, *J. Am. Chem. Soc.,* 77, 3922, 1955.

81. **Smyth, D. G., Blumenfeld, O. O., and Konigsberg, W.,** Reaction of N-ethylmaleimide with peptides and amino acids, *Biochem. J.,* 91, 589, 1964.

82. **Heitz, J. R., Anderson, C. D., and Anderson, B. M.,** Inactivation of yeast alcohol dehydrogenase by N-alkylmaleimides, *Arch Biochem. Biophys.,* 127, 627, 1968.

83. **Gregory, J. D.,** The stability of N-ethylmaleimide and its reaction with sulfhydryl groups, *J. Am. Chem. Soc.,* 77, 392, 1955.

84. **Benesch, R. and Benesch, R. E.,** Determination of -SH groups in proteins, *Methods Biochem. Anal.,* 10, 43, 1962.

85. **Stefanini, S., Chiancone, E., McMurray, C. H., and Antonini, E.,** Dissociation human hemoglobin by different organomercurials, *Arch. Biochem. Biophys.,* 151, 28, 1972.

86. **Kimura, T., Matsueda, R., Nakagawa, Y., and Kaiser, E. T.,** New reagent for the introduction of the thiomethyl group at sulfhydryl residue of proteins with concomitant spectrophotometric titration of the sulfhydryl: methyl-3-nitro-2-pyridyl disulfide and methyl 2-pyridyl disulfide, *Analyt. Biochem.,* 122, 274, 1982.

87. **Baker, B. R.,** Design of active-site-directed irreversible enzyme inhibitors: the organic chemistry of the enzymic active site, John Wiley & Sons, New York, 1967.
88. **Wong, Y-H H. and Frey, P. A.,** Uridine diphosphate galactose 4-epimerase. Alkylation of enzyme-bound diphosphopyridine nucleotide by *p*-(bromoacetamido)phenyl uridyl pyrophosphate, an active-site-directed irreversible inhibitor, *Biochemistry,* 24, 5337, 1979.
89. **Brewer, C. F. and Riehm, J. P.,** Evidence for possible nonspecific reactions between *N*-ethylmaleimide and proteins, *Anal. Biochem.,* 18, 248, 1967.
90. **Smyth, D. G., Nagamatsu, A., and Fruton, J. S.,** Reactions of *N*-ethylmaleimide, *J. Am. Chem. Soc.,* 82, 4600, 1960.
91. **Eisen, H. N., Belman, S., and Carsten, M. E.,** The reaction of 2,4-dinitrobenzenesulfonic acid with free amino groups of proteins, *J. Am. Chem. Soc.,* 75, 4583, 1953.
92. **Shaltiel, S.,** Thiolysis of some dinitrophenyl derivatives of amino acids, *Biochem. Biophys. Res. Commun.,* 29, 178, 1967.
93. **Means, G. E. and Feeney, R. E.,** Reductive alkylation of amino groups in proteins, *Biochemistry,* 7, 2192, 1968.
94. **Schirich, L. G. and Mason, M.,** A study of the properties of a homogeneous enzyme preparation and of the nature of its interaction with substrates and pyridoxal 5-phosphate, *J. Biol. Chem.,* 238, 1032, 1963.
95. **Klein, S. M. and Sagers, R. D.,** Effect of borohydride reduction on the pyridoxal phosphate-containing glycine decarboxylase from *Peptococcus glycinophilus, J. Biol. Chem.,* 242, 301, 1967.
96. **Engvall, E. and Perlmann, P.,** Enzyme-linked immunosorbent assay, II. Quantitative assay of protein antigen, immunoglobulin G, by means of enzyme-labeled antigen and antibody-coated tubes, *Immunochemistry,* 8, 871, 1971.
97. **Wofsy, L. and Singer, S. J.,** Effects of the amidination reaction on antibody activity and on the physical properties of some proteins, *Biochemistry,* 2, 104, 1963.
98. **Wong, S. S., Quiggle, K., Triplett, C., and Berliner, L. J.,** Spin-labeled sulfonyl fluorides as active site of protease structure: II. Spin label synthesis and enzyme inhibition, *J. Biol. Chem.,* 249, 1678, 1974.
99. **Herriott, R. M.,** Reactions of native proteins with chemical reagents, *Adv. Protein Chem.,* 3, 169, 1947.
100. **Riehm, J. P. and Scheraga, H. A.,** Structural studies of ribonuclease. XVII. A reactive carboxyl group in ribonuclease, *Biochemistry,* 4, 772, 1965.
101. **Goodfriend, T. L., Levine, L., and Fasman, G. D.,** Antibodies to bradykinin and angiotensin: a use of carbodimides in immunology, *Science,* 144, 1344, 1964.
102. **Riordan, J. F., Wacker, W. E. C., and Valee, B. L.,** *N*-Acetylimidazole: a reagent for determination of "free" tyrosyl residues of proteins, *Biochemistry,* 4, 1758, 1965.
103. **Smyth, D. G.,** Acetylation of amino and tyrosine hydroxyl groups: preparation of inhibitors of oxytocin with no intrinsic activity on the isolated uterus, *J. Biol. Chem.,* 242, 1592, 1967.
104. **Riordan, J. F. and Vallee, B. L.,** Diazonium salts as specific reagents and probes of protein conformation, *Methods Enzymol.,* 25, 521, 1972.
105. **Pathy, L. and Smith, E. L.,** Reversible modification of arginine residues: application to sequence studies by restriction of tryptic hydrolysis to lysine residues, *J. Biol. Chem.,* 250, 557, 1975.
106. **Yankeelov, J. A., Jr.,** Modification of arginine by diketones, *Methods Enzymol.,* 25, 566, 1972.
107. **Takahashi, K.,** The reaction of phenylglyoxal with arginine residues in proteins, *J. Biol. Chem.,* 243, 6167, 1968.
108. **Miles, E. W.,** Modification of histidyl residues in proteins by diethylpyrocarbonate, *Methods Enzymol.,* 47, 431, 1977.
109. **Marks, R. H. L. and Miller, R. D.,** Chemical modification of methionine residues in azurin, *Biochem. Biophys. Res. Commun.,* 88, 661, 1979.
110. **Gundlach, H. G., Moore, S., and Stein, W. H.,** The reaction of iodoacetate with methionine, *J. Biol. Chem.,* 234, 1761, 1959.
111. **Spande, T. F. and Witkop, B.,** Determination of the tryptophan content of protein with *N*-bromosuccinimide, *Methods Enzymol.,* 11, 498, 1967.
112. **Loudon, G. M. and Koshland, D. E., Jr.,** The chemistry of a reporter group: 2-hydroxy-5-nitrobenzylbromide, *J. Biol. Chem.,* 245, 2247, 1970.
113. **Fontana, A. and Scoffone, E.,** Sulfenyl halides as modifying reagents for polypeptides and proteins, *Methods Enzymol.,* 25, 482, 1972.
114. **Demoliou, C. D. and Epand, R. M.,** Synthesis and characterization of a heterobifunctional photoaffinity reagent for modification of tryptophan residues and its application to the preparation of a photoreactive glycogan derivative, *Biochemistry,* 19, 4539, 1980.

Chapter 3

CHOICE AND DESIGN OF CROSS-LINKING REAGENTS

I. INTRODUCTION

The covalent bond formation between two components, either biological molecules or otherwise, can be catalyzed by enzymes, caused by activating agents, or facilitated by bifunctional reagents containing two reactive groups. The last strategy seems to be the most common and logical approach. By using two group-specific reagents linked through a spacer, bifunctional cross-linkers will react with specific functional groups within a molecule or between two different molecules resulting in a bond between these two components. The reactive groups in a cross-linker can be identical or different, providing a diversity of reagents that can bring about covalent bonding between any chemical species, either intramolecularly or intermolecularly.

The choice and design of a bifunctional reagent depends on its specific application. In each instance, the molecular species to be cross-linked may be different. These moieties may be located in disparate environments and the cross-linked products may require a certain configuration. Some of the conditions and requirements to be considered are

1. Reaction specificity towards a particular group, e.g., amino, sulfhydryl, carboxyl, guanidinyl, imidazolyl, and other amino acid side chains. If there is a special functional group on the protein to which another molecule will be linked, the cross-linking reagent must be specific to that group.
2. Hydrophobicity and hydrophilicity of the reagent. A protein in a hydrophobic environment may require a hydrophobic reagent to reach. For example, membrane permeability of a reagent may be necessary for labeling intramembranous proteins.
3. Cleavability of the reagent. It may be desirable in some cases to separate cross-linked proteins for identification of the components. In this case, the use of cleavable reagents will enable the cross-linking to be reversed.
4. Size and geometry of the reagent. The length of bridge between the reactive groups of cross-linkers may be used to measure the two cross-linked groups for intramolecular topology studies. The cross-linkers can also serve as spacers between two conjugated proteins.
5. Photosensitivity of the reagent. A photoactivatable reagent is essential for photoaffinity labeling studies. This is particularly important when one of the species is unknown, for example, the protein receptor of a hormone. Photoaffinity labels have been used to detect and identify membrane acceptors.
6. Presence of tracer (or reporter) group. Sometimes it may be necessary to follow a cross-linking reaction or detect molecular conformation after conjugation. In this case, a tracer or reporter group attached to the cross-linking reagent, for example, a fluorescent probe, spin label, or radioactivity, will facilitate the process.

Basically, cross-linking is essentially a chemical modification process. The reactive groups are located at the two ends of the reagent with a connecting backbone where various desirable functionalities can be introduced (Figure 1). A simple cross-linker can be made by connecting any of the group directed reagents discussed in Chapter 2 with a carbon chain.

The overall structure of the molecule can be constructed in different ways. For example, Chong and Hodges[1] have designed N-(4-azidobenzoylglycyl)-S-(2-thiopyridyl)-cysteine for the study of biological interactions (Figure 2). The rationale is as follows. The two ends of

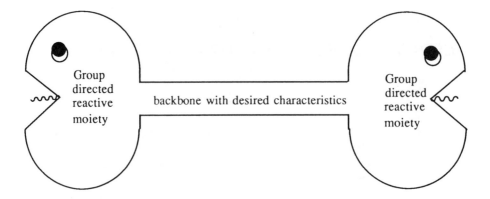

FIGURE 1.   General structure of a bifunctional reagent. The two group-specific reagents are connected by a bridge of different characteristics.

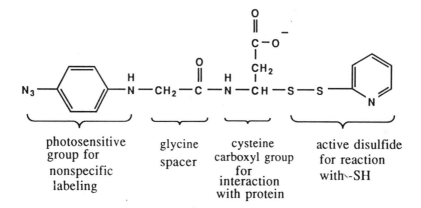

FIGURE 2.   Rationale for the design of *N*-(4-azidobenzoylglycyl)-*S*-(2-thiopyridyl)-cysteine.

the compound function as anchors that will react with proteins. One end contains a pyridinyl disulfide moiety that will react with a thiol through disulfide interchange because the displaced product, thiolpyridine, is a good leaving group. The other end contains a photoactivatable group that will nonspecifically insert into C-H or N-H bonds. Joining the two reactive ends are glycine and cysteine residues which contains functionalities that can interact with the protein.

The specificity of the cross-linking reagent for a protein depends on the relative reactivity of the amino acid side chains toward the reactive ends of the two-headed compounds. Many of these groups are derived from the reagents discussed in the last chapter. Since most of the chemical modifications are nucleophilic reactions with the amino acid side chains acting as the nucleophile, the modifying agent must contain certain characteristics for the nucleophilic attack. The design of cross-linking reagents involves the incorporation of these various special groups.

II. NUCLEOPHILIC REACTIONS

A. THE BASIC REACTION

In the nucleophilic substitution reactions, the amino acid side-chains with a lone pair of electrons attack the electrophilic centers of the substrate. The rate of the reaction depends on the nucleophilicity of the attacking reagent and the reactivity of the substrate.[2] In Chapter 2, the relative nucleophilicity of the amino acid side-chains has been discussed. The reactivity of the substrate, on the other hand, depends on its electrophilicity as well as its leaving group. This will be elaborated below.

1. Electrophilicity of the Substrate

The electrophilic centers are atoms deficient of electrons. This deficiency of electrons is generally induced by bonding more electronegative atoms or positive charges directly or next to the atom which in most cases is carbon. The electrons of the carbon atom will be drawn to the more electronegative atom inducing a slight positive charge. The examples in Figure 3 show that oxygen is probably the most commonly used electronegative atom. The arrows in the examples indicate the attraction of the electrons to the more electronegative atoms. These slightly positively charged centers are the targets of nucleophilic attack. The greater the positivity, the greater will be its electrophilicity.

Depending on whether the positive center is located on an alkyl or an acyl carbon, the nucleophile will be alkylated or acylated, respectively. These reagents can therefore be broadly classified into alkylating and acylating agents. Thus {1} and {2} are alkylating agents, while {3}, {4} and {5} (Figure 3) are acylating agents. Compound {6} is an arylating agent, a special alkylating reagent with an aromatic ring.

2. Leaving Group Reactivity

In the nucleophilic substitution reactions, the nucleophile attacks the electrophilic center and finally displaces a leaving group. The capability of a leaving group to leave affects the reactivity of the substrate toward the nucleophile.[3,4] The more stable the leaving group is as a free entity, the easier it will come off. Thus, the best leaving group is probably N_2 from aliphatic diazonium ions, $R-N_2^+$. Nitrogen can be quantitatively evolved from such diazonium compounds. Usually the capability of a leaving group to leave is inversely proportional to its basicity. The best leaving groups are usually the weakest bases. In this context, iodide is the best leaving group among the halide ions and fluoride the poorest. Also, OH^- and RO^- do not ordinarily leave since they are strong bases. They can, however, be converted to leaving groups after protonation, i.e., $-OH_2^+$ and $-O(H^+)R$. This may explain why alcohols and esters are stable at neutral pHs and usually undergo acid catalyzed reactions. In general, the approximate order of leaving ability for groups attached to saturated carbons is: $R-N_2^+$ > R-OTs (tosyl), $R-OSO_2OR$ > R-I > R-Br > $R-OH_2^+$ > R-Cl > $R-O(H^+)R$ > $R-ONO_2$ > $R-NR'_3^+$ > R-OCOR'. For groups bound to the carbonyl carbons the order is: R-CO-Cl (acyl halides) > R-CO-OCOR' (anhydrides) > R-CO-OAr (aryl esters) > R-CO-OH (carboxylic acids) > R-CO-OR (alkyl esters) > R-CO-NH2 (amides).

This list will provide a general guide to the reactivity of the reagents. The better the leaving group it possesses, the greater will be the reactivity of the reagent.

Some of the common leaving groups that have been incorporated in bifunctional reagents are N-hydroxysuccinamide, methanol, iodine and thiopyridine. These will be further illustrated in the next two sections.

B. ALKYLATION

The attack of a nucleophile at an alkyl or aryl carbon results in alkylation of the neucleophile. As discussed in Chapter 2, the best nucleophilic amino acid side-chain is the thiolate anion.[5] Thus, alkylation of the thiol is the most prevalent of all bifunctional reagents. Alkylation of the amino, thioether, imidazolyl, phenolate, and indolyl moieties will also occur under certain conditions as discussed in Chapter 2. In some instances, such as imidoesters, the amino group is highly preferred. In order to provide nucleophilic substitution reactions, good leaving groups must be introduced which should contain electron withdrawing

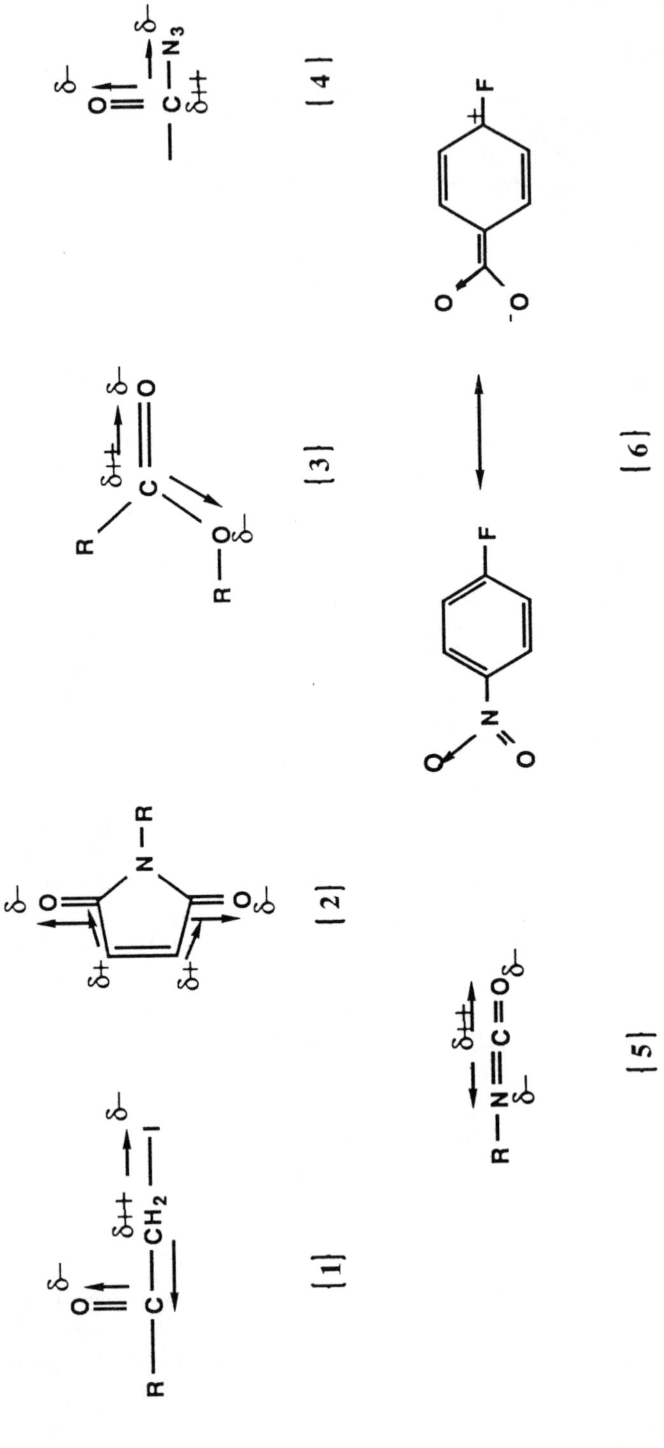

FIGURE 3. Induction of electrophilic centers. Electrons are drawn to the more electronegative atom creating a slight positivity at the carbon atom.

atoms to induce a positive charge at the desired center of attack. Some of such alkylating groups commonly incurred in cross-linking reagents are shown in Table 1.

The haloacetate and maleimide cross-linkers, such as dibromoacetone and bismaleimi-dohexane, react fastest with the thiol and are considered thiol specific.[6,7] Under conditions of higher alkaline pH, excess reagent and prolonged reaction time, other nucleophiles (i.e., amino, thioether, imidazolyl, phenolate) will also react. In the case of haloacetyl group, the halogen is the leaving group and the leaving ability follows the order: I > Br > Cl > F, which is the order of reactivity for the haloacetyl compounds. The homobifunctional cross-linker, dibromoacetone, reacts with two thiol nucleophiles, linking them together (Figure 4A). In the case of the bismaleimide derivative, the sulfur nucleophile attacks the double bond in a Michael addition reaction, again resulting a linkage between the two species (Figure 4B). In this reaction, there is no formal leaving group. The electrons in the double bond is shifted to stabilize the product.

Although usually considered an amino group reagent, aryl halides also readily react with the thiol, phenolate and imidazolyl groups in proteins. A simple example of the cross-linking reagent derived from such a functionality is 1,5-difluoro-2,4-dinitrobenzene.[8] At neutral pH, the reagent reacts preferentially with sulfhydryl groups. At higher pHs, it also modifies amino groups and other nucleophiles. The strong electronegative nitro groups create a positive center at the carbon where the fluorine atom is bonded (Figure 3). Nucleophilic attack at the positive carbon eliminates the fluorine which itself is also very electronegative (Figure 4C). The adduct formed with the thiol is reversible in the presence of an excess free thiol, for example 2-mercaptoethanol, which displaces the bonded nucleophile (See Chapter 2).[9]

The synthesis of a bifunctional reagent containing the diazoacetyl group has been reported for 1,1-*bis*(diazoacetyl)-2-phenylethane.[10] It has been shown to react with carboxyl groups of pepsin.[10,11] During the reaction, the diazoacetyl group is protonated and N_2 is eliminated. Thus, a low pH of about 5 is preferred for the reaction (Figure 4D). Other nucleophiles presumably can also react.

C. ACYLATION

In acylation, as in alkylation, the activated acyl compounds contain good leaving groups and are attacked by all amino acid nucleophiles. However, acylation primarily yields amino and phenolate derivatives. As discussed in Chapter 2, the adducts of sulfhydryl and imidazolyl groups are unstable, being rapidly hydrolyzed in aqueous solutions. Since tyrosine residues are generally buried in proteins and are not accessible for modification, the acylation reagents are generally regarded as amine specific.

Acylating agents can be easily derived from either carboxylic or sulfonic acids. In fact, any of the aliphatic or aromatic dicarboxylic acids or disulfonic acids can be activated to provide bifunctional acylating reagents capable of reacting under mild conditions to form covalent bonds with proteins. The activating groups are generally good leaving groups with electronegative atoms to further withdraw electrons from the carbonyl carbon such as *p*-nitrophenol and halides. Some commonly used leaving groups in bifunctional acylating agents are shown in Table 2. The products of these leaving entities are also shown. Of the carboxylic acid esters, the most commonly employed leaving groups are the succinimidyl derivatives and the imidoesters, although *p*-nitrophenol derivatives have also been used. *N*-Hydroxysuccinimidyl esters are popular because they can be easily synthesized, are highly activated, and react with amino groups under mild conditions (pH 7 to 9). The synthesis of these compounds is achieved, as for other esters such as *p*-nitrophenyl ester, by coupling *N*-hydroxysuccinimide to dicarboxylic acids using dicyclohexylcarbodiimide (Figure 5).[12] Another advantage of the hydroxysuccinimide functionality over the older hydroxylphthal-imide is the water solubility of the product of the reaction.[12] The *N*-hydroxysuccinimide product has been further sulfonylated to render it even more water soluble.

Table 1

COMMON ALKYLATING GROUPS THAT HAVE BEEN INCORPORATED
INTO CROSS-LINKING REAGENTS

Alkylating Groups	Examples of Cross-linking Reagents

Haloacetyl

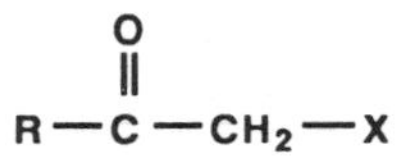

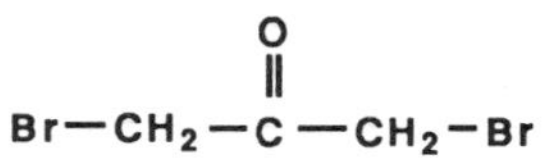

$$R-\overset{\overset{\displaystyle O}{\|}}{C}-CH_2-X$$

$$Br-CH_2-\overset{\overset{\displaystyle O}{\|}}{C}-CH_2-Br$$

N-Maleimido

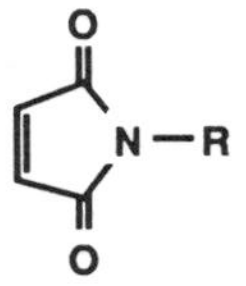

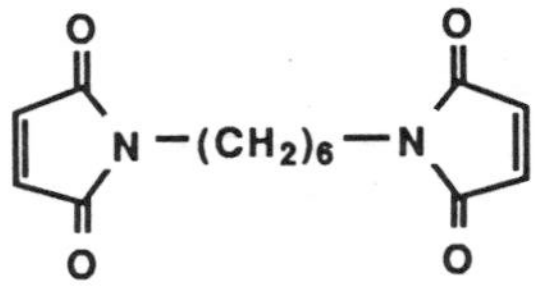

Aryl halide

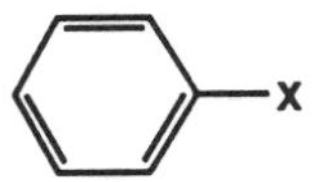

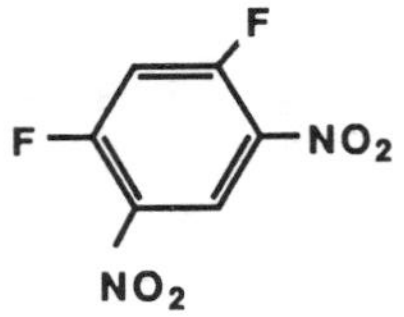

Diazoacetyl

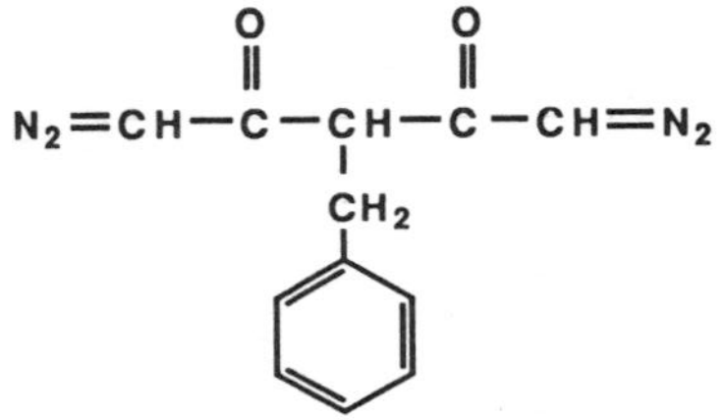

$$R-\overset{\overset{\displaystyle O}{\|}}{C}-CH=N_2$$

$$N_2=CH-\overset{\overset{\displaystyle O}{\|}}{C}-CH-\overset{\overset{\displaystyle O}{\|}}{C}-CH=N_2$$

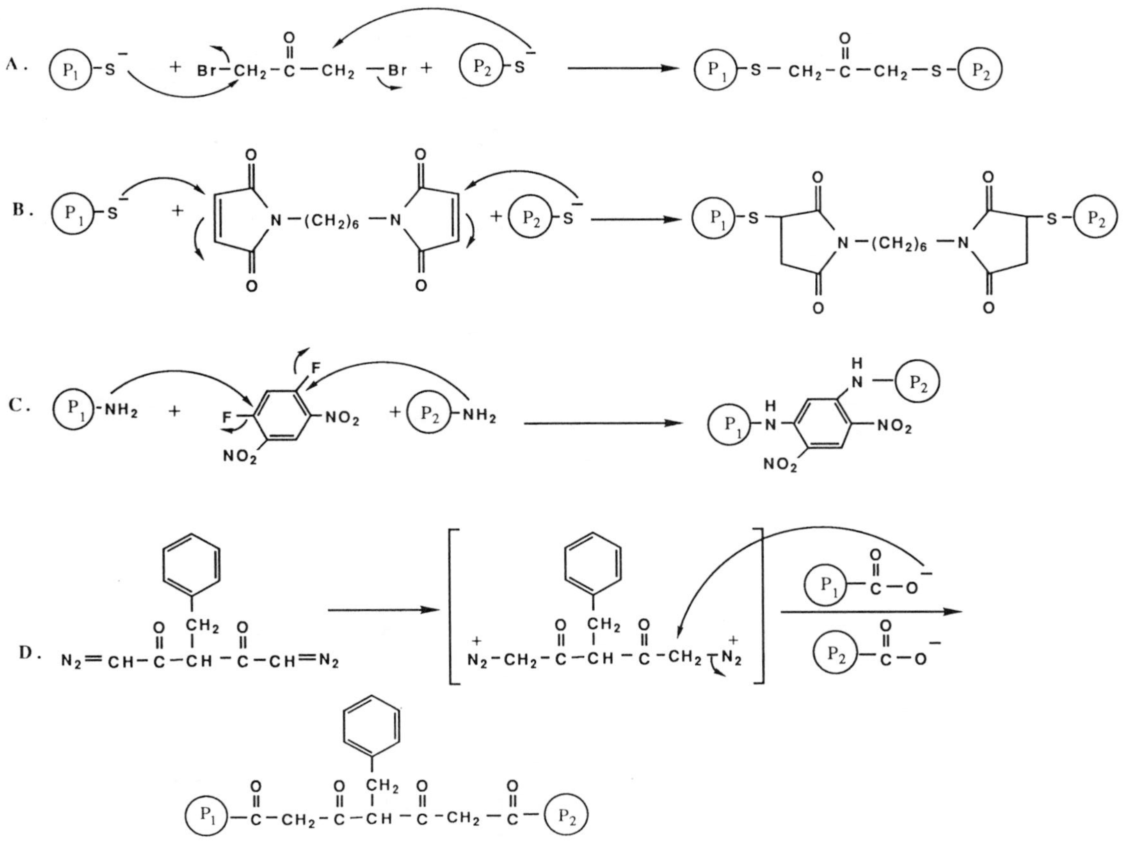

FIGURE 4.   Alkylation of the cross-linking reaction. Attack of the protein nucleophiles at the electrophilic center cross-links the proteins.

Table 2

ACTIVATED ACYL GROUPS IN BIFUNCTIONAL REAGENTS

Activated Functionality		Leaving Group	
Succinimidyl		N-hydroxysuccinimide	
Sulfosuccinimidyl		N-hydroxysuccinimide	
Imidoester		methanol	
Carbonyl chloride		chloride ion	
Azidocarbonyl		azide ion	
Sulfonyl chloride		chloride ion	
p-Nitrophenyl		p-nitrophenol	
isothiocyanate			
isocyanate			

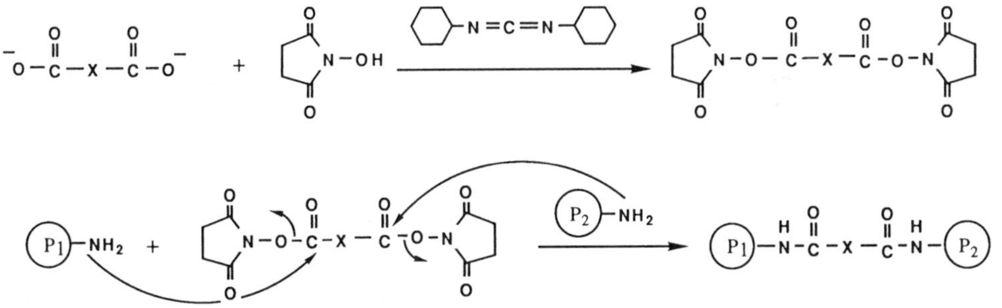

FIGURE 5. Synthesis of *bis-N*-hydroxysuccinimidyl esters and their cross-linking reaction with protein amino groups.

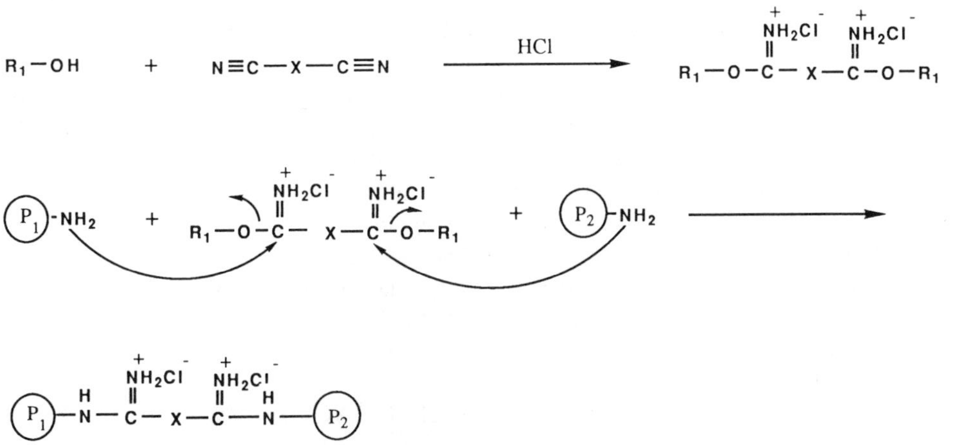

FIGURE 6. Synthesis of *bis*-imidoesters and their cross-linking reaction with protein amino groups.

In the case of imidoesters, these compounds are very soluble in water and react under mild conditions (pH 7 to 10) with a high degree of specificity with amino groups. As discussed in Chapter 2, the side reactions with thiol, phenolate, carboxyl, imidazolyl, and guanidinyl groups are negligible. *Bis*-imidoesters can be readily synthesized by a variety of procedures. The most common is the Pinner method where a primary alcohol is allowed to react with a nitrile in the presence of dry HCl (Figure 6).[13-15] These compounds are usually isolated as hydrochloride salts. After reaction of the imidoesters with amino groups, the resulting amidines are quite stable and resistant to acid hydrolysis. Total amino acid hydrolysis can be performed on the modified proteins for quantitation.

Acid halides are very reactive compounds with the halides as good leaving groups. Although these compounds can be synthesized easily in the absence of water, acyl halides are rarely used in the design of cross-linkers because of their instability in aqueous solutions. Similarly, acylazides are reactive by virtue of the good leaving azido group. However, bisacylazides have been used to cross-link proteins.[16] As discussed in Chapter 2, sulfonyl halides are more stable and have been incorporated into cross-linking reagents, such as phenol-2,4-di(sulfonyl chloride) (Figure 7). These compounds react with all the nucleophiles of the protein and are generally not group specific.

The isocyanates and isothiocyanates do not contain a leaving entity. During the nucleophilic attack at the electrophilic carbon, the electrons are shifted to the nitrogen to pick up a proton. With nitrogen as the nucleophile, isocyanates form stable urea derivatives and

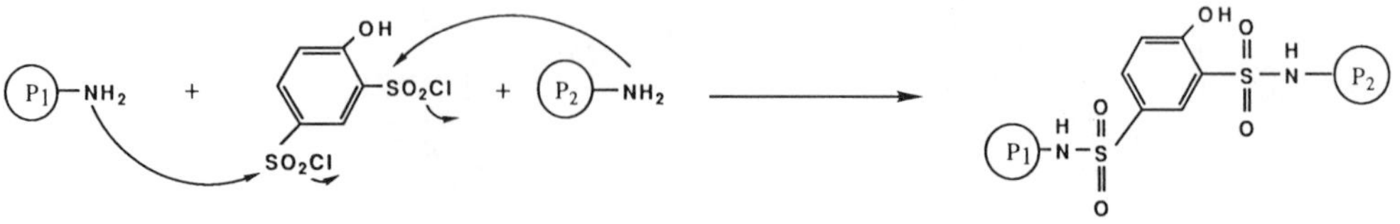

FIGURE 7.   Cross-linking with a di-sulfonyl chloride: phenol-2,4-di(sulfonyl chloride).

isothiocyanates form thiourea derivatives. As pointed out in Chapter 2, these compounds are considered amino group specific. Aryl di-isocyanates and di-isothiocyanates cross-linking reagents can be easily prepared from the corresponding diamines on reaction with phosgene or thiophosgene (Figure 8).[17] Condensation with amino groups of proteins will result in cross-linking.

III. ELECTROPHILIC REACTIONS

As the aromatic benzene and imidazole rings are relatively rich in electrons, the most important electrophilic reaction in protein modification is the reaction of diazonium salts with tyrosines and histidines. Such diazonium salts, particularly aryl diazonium compounds, may be incorporated into cross-linking reagents. An example is *bis*diazobenzidine which has been used to cross-link papain[18,19] and other proteins.[20] The aryl diazonium compounds are easily prepared from the corresponding aryl amines and nitrous acid which is generated *in situ* from sodium nitrite and hydrochloric acid. Diazotization of diaminobenzidine will yield *bis*diazobenzidine as a cross-linker (Figure 9). Thus cross-linking reagents may be designed with a phenylamine at one end of the bridge.

IV. GROUP-DIRECTED REACTIONS

There are relatively few reactive moieties that are specific for certain amino acid side-chains aside from their nucleophilicity. Any of the group-directed compounds narrated in the last chapter may be incorporated into bifunctional reagents. And a few have been successfully implemented. These are mercurials and disulfides for thiols, carbonyls for reductive alkylation with amines and vicinal dicarbonyls for arginine. The incorporation of these reactive groups into cross-linking reagents are illustrated below.

A. DISULFIDE REAGENTS
The specific thiol-disulfide interchange reaction is illustrated by the cross-linkers derived from pyridyl disulfide. The thiopyridine is an excellent leaving group, being stabilized by several resonance forms. An example of such a cross-linking reagent is *N*-succinimidyl 3-(2-pyridyldithio)propionate (Figure 10A).[21] The reagent reacts with an amino group of a protein at the *N*-succinimidyl ester bond and a thiol of another protein at the disulfide bond. The reaction with the thiol involves a direct displacement of the thiopyridine, generating a new disulfide bond (Figure 10).

Thiophthalimides and compounds containing -S-SO$_2$- are also thiol-specific reagents resulting in the formation of disulfide bonds.[22] The phthalimide and sulfinic acid provide excellent leaving groups and, at the same time, activate the thiol in the compound. Attack of a protein sulfhydryl group at the sulfur atom eliminates the leaving group and a disulfide bond is formed as illustrated in Figure 10 B and C for the cross-linking reagents, *N*-(4-azidophenylthio)phthalimide and *N,N'*-(4-azido-2-nitrophenyl)-cystamine dioxide, respectively. The disulfide bonds formed can be cleaved by reducing agents providing a reversible conjugate which will be discussed below.

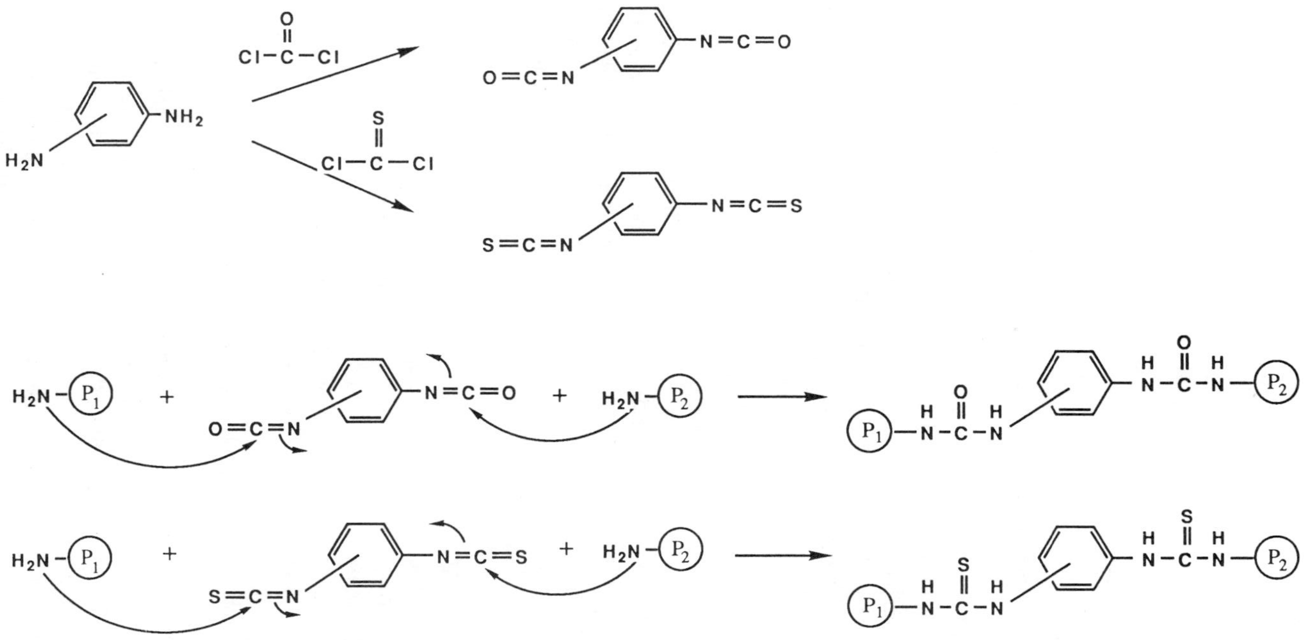

FIGURE 8. Synthesis of di-isocyanates and di-isothiocyanates and their cross-linking reaction with protein amino groups.

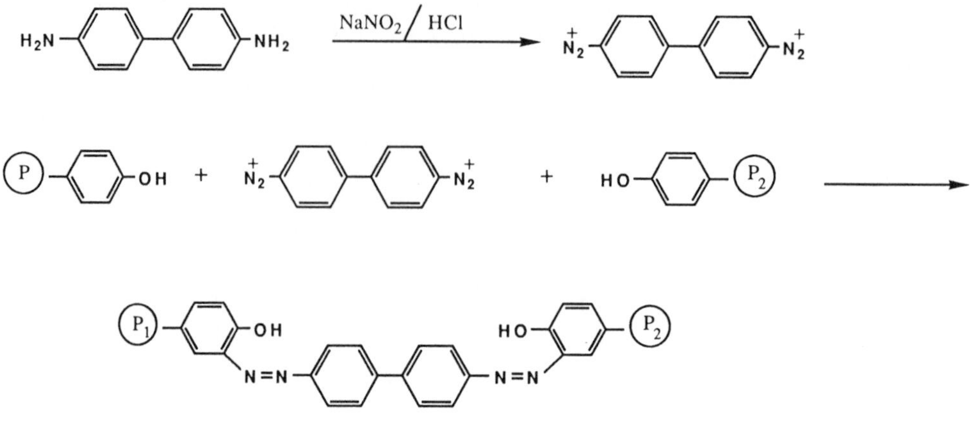

FIGURE 9.   Diazotization of diaminobenzidine and the cross-linking reaction of bisdiazobenzidine.

B. MERCURIAL REAGENTS

While mercuric ion itself can reversibly cross-link two sulfhydryl groups,[23,24] insertion of the ion into protein disulfide bonds makes this reagent undesirable for general conjugation.[25,26] However, mercurial derivatives have been synthesized where the mercurial ion is monovalent. One example is 3,6-*bis*(mercurimethyl)dioxane (Figure 11). It has been used to prepare mercaptalbumin dimers.[27]

C. REDUCTIVE ALKYLATION

As has been discussed in Chapter 2, reductive alkylation is a special reaction for the amino group. The process involves the formation of a Schiff base followed by reduction with a reducing agent. It should be noted that either the amino group or the carbonyl group may be contributed by the protein. Aldehydes may be derived from the carbohydrate moieties of glycoproteins by periodate treatment. In this case, glycoproteins may be cross-linked with diamines. On the other hand, proteins may be cross-linked through the amino groups by means of dialdehydes.[28] Examples of such dialdehydes can be found in Chapter 4.

D. VICINAL DICARBONYL REAGENTS

Vicinal dicarbonyl reagents are specific for the modification of guanidinyl group of arginine. Incorporation of such groups into bifunctional reagents will cross-link arginine moieties. Such a bifunctional derivative of phenylglyoxal, azidophenyl glyoxal (Figure 12), has been synthesized to study enzymes that contain an arginine at the active site.[29] This compound has also been used for cross-linking RNA to proteins.[30]

V. NONSPECIFIC REACTIONS

Nonspecific reactions involve reagents that react indiscriminately with the amino acid side chains. One class of such compounds that have evolved for biochemical studies are photoaffinity labels. The chemistry of these compounds have been reviewed by Knowles and Bayley.[31-33] Readers who are not familiar with this technique are encouraged to consult these publications for detailed treatment. Basically, photoaffinity labels are compounds that would generate reactive intermediates such as carbenes and nitrenes, on irradiation. These highly reactive species react immediately with their surrounding environment. Figure 13 outlines the principle reactions for nitrenes. Carbenes undergo similar reactions. The major reaction of interest is the insertion into C-H and N-H bonds resulting in a covalent bond with the compound. Common nitrene precursors are arylazides which have absorbance in

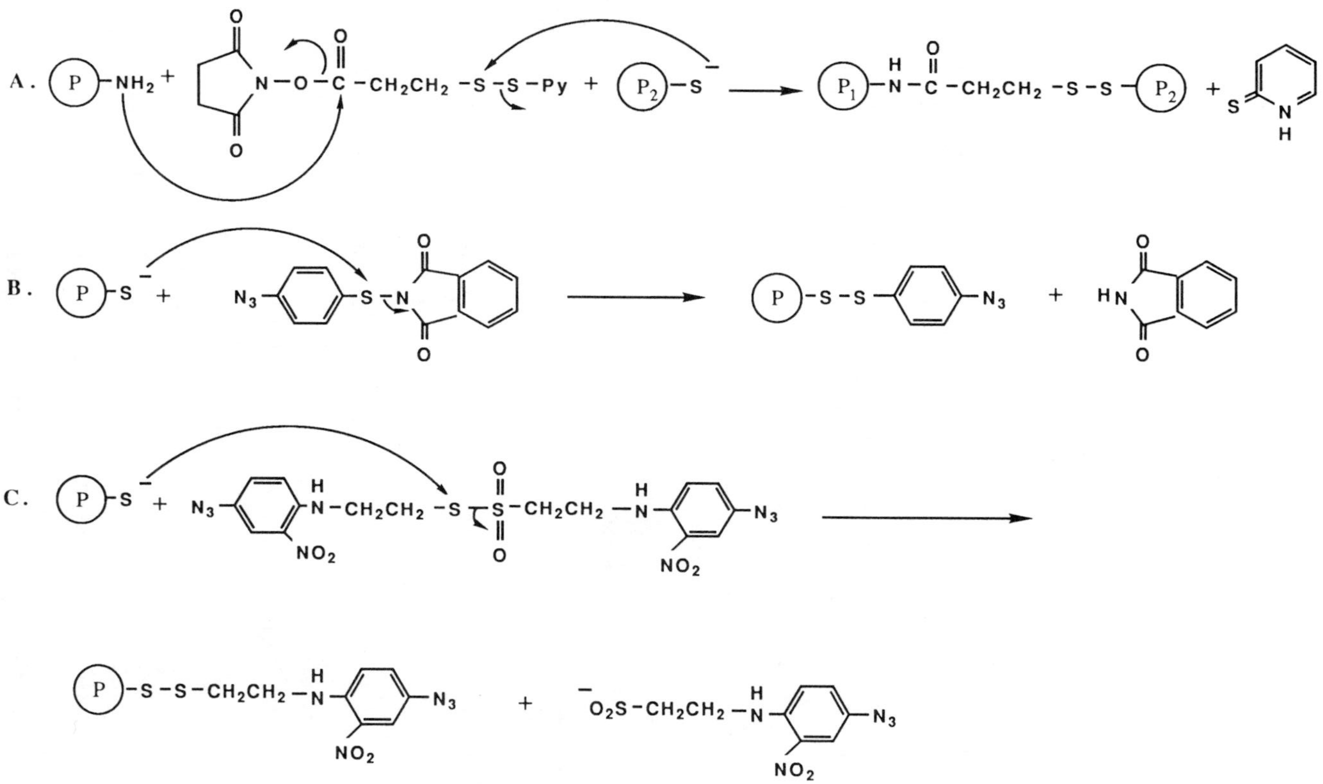

FIGURE 10. Disulfide forming cross-linkers: (A) cross-linking reaction with N-succinimidyl 3-(2-pyridyldithio)propionate; (B) reaction with N-(4-azidophenylthio)phthalimide; (C) reaction with N,N'-(4-azido-2-nitrophenyl)cystamine dioxide.

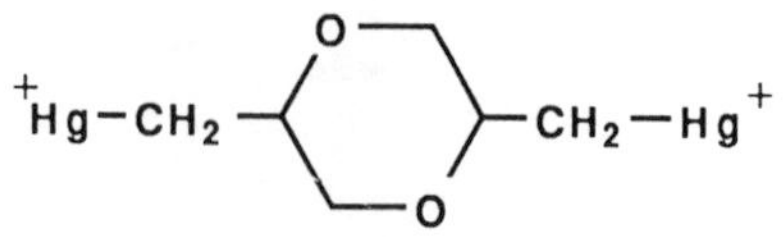

FIGURE 11.   Mercury derived thiol-specific reagent: 3,6-*bis*(mercurimethyl)dioxane.

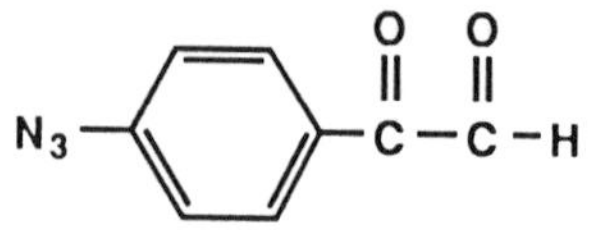

FIGURE 12.   An arginine specific dicarbonyl compound, azidophenyl glyoxal.

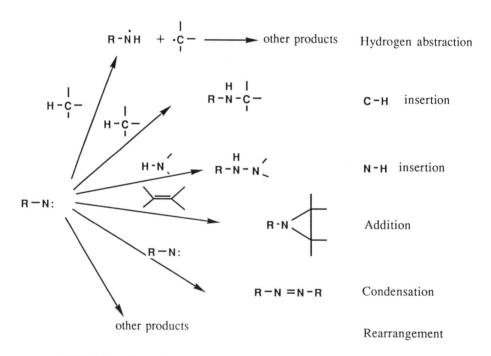

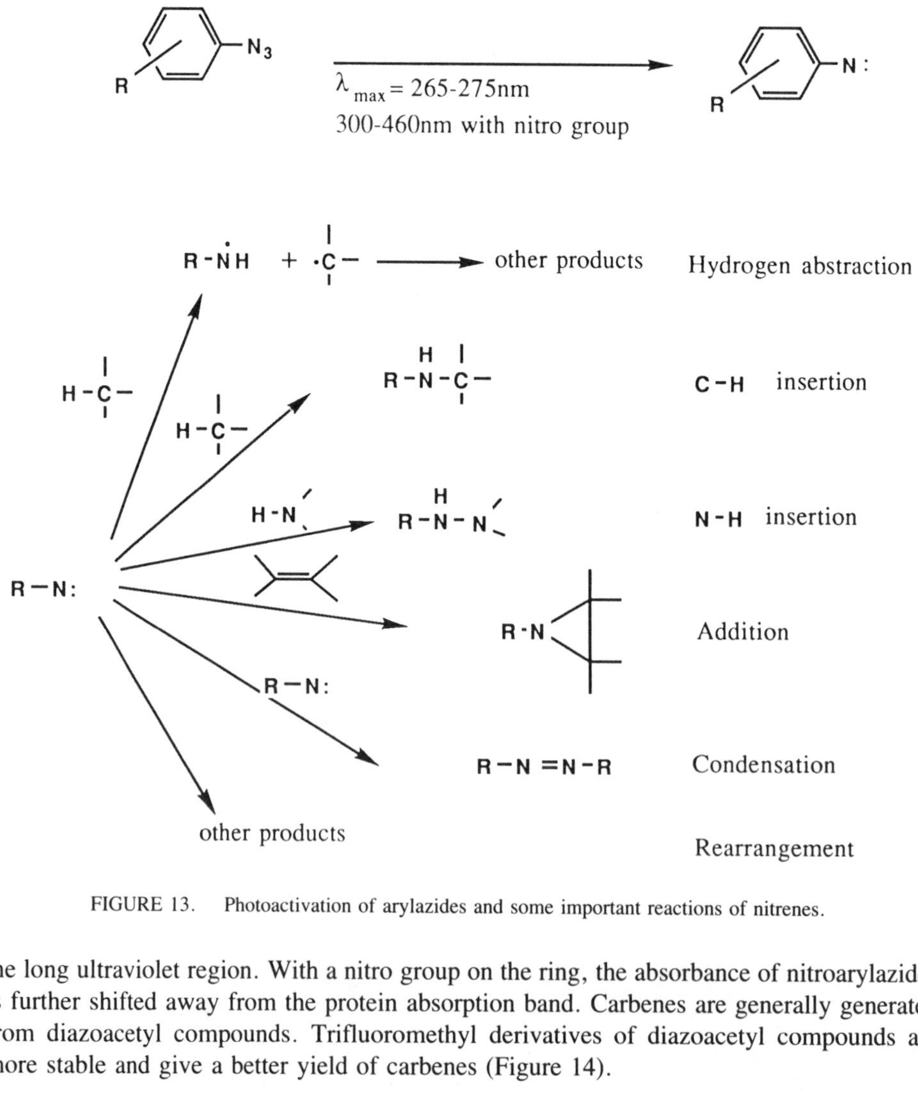

FIGURE 13.   Photoactivation of arylazides and some important reactions of nitrenes.

the long ultraviolet region. With a nitro group on the ring, the absorbance of nitroarylazides is further shifted away from the protein absorption band. Carbenes are generally generated from diazoacetyl compounds. Trifluoromethyl derivatives of diazoacetyl compounds are more stable and give a better yield of carbenes (Figure 14).

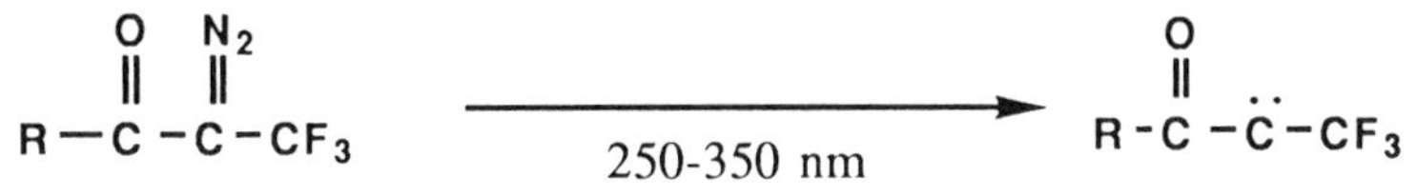

FIGURE 14. Photochemical generation of carbenes.

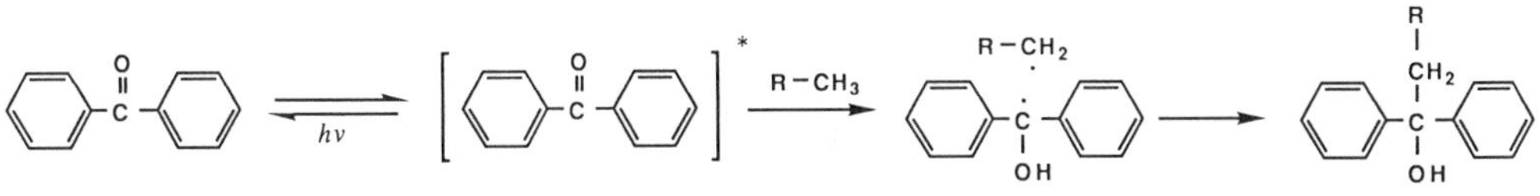

FIGURE 15. Photochemical reactions of benzophenone.

Another chemical entity that is photoactivatable is derived from benzophenone. Benzophenones and acetophenones, for example, are activated to the triplet state when irradiated with ultraviolet light. The excited state can either decay back to the starting material or react irreversibly with the nearby molecules as shown in Figure 15.[34] Together with arylazides and diazoacetyl compounds, these groups may be incorporated into bifunctional reagents for nonspecific cross-linking of proteins. Many reagents are synthesized for this purpose.[35] By far, however, azidophenyl derivatives are the most commonly used. Examples of these series of compounds can be found in Chapter 5.

VI. CLEAVAGE REAGENTS

In the bridge that joins the bifunctional reactive groups, various functionalities may be incorporated. The inclusion of cleavable bonds enables the cross-linked proteins to be separated for analysis. The advantages include confirmation of the cross-linking reactions, establishment of the amino acids modified, monitoring the effect of cross-linking, and identification of the cross-linked species. A number of cleavable bonds may be employed for this purpose. These include disulfide bonds,[36-38] mercurial group,[39] vicinal glycol,[40] azo,[41] sulfone,[42-44] ester[45,46] and thioester[47] linkages as shown in Table 3 and will be briefly discussed below.

A. DISULFIDE BOND

As mentioned above, disulfide bonds can be cleaved by any reducing agent, such as free mercaptans,[48] sodium borohydride,[49] sodium phosphorothioate,[50] and sulfite.[51,52] Cleavage is generally achieved under mild conditions with 2-mercaptoethanol, dithiothreitol or dithioerythritol at concentrations between 10 to 100 mM. The reaction is a simple two-step thiol-disulfide interchange as discussed in Chapter 2 (see Section III.B.1).

In the cross-linking reagents, the disulfide can be intrinsically built into the molecule as in Table 3 or generated during the modification reaction. The latter reactions have been demonstrated in Figure 10. These reagents must be used under nonreducing conditions and in the complete absence of free thiols to prevent disulfide-thiol interchange.

B. MERCURIAL GROUP

The reaction of mercurial compounds with sulfhydryl groups is reversible in the presence of free thiols.[39] Thus, mercurial cross-linked proteins may be separated in the presence of 2-mercaptoethanol or dithiothreitol. The reaction is similar to thiol-disulfide interchange. Because of the very small dissociation constant of thiol-mercurial complexes, relatively high concentrations of free thiol may be required for complete dissociation of the cross-linked proteins.

Table 3

CLEAVABLE GROUPS IN BIFUNCTIONAL REAGENTS

Cross-linked Product	Example of Cross-linker	Cleavage Condition	Cleaved Products	Reference
P_1—S—S—P_2	$CH_3 \cdot O \cdot C(=O) \cdot (CH_2)_2$—S—S—$(CH_2)_2 \cdot C(=O) \cdot O \cdot CH_3$ (with $\overset{+}{N}H_2Cl^-$ imidate groups)	reducing agent e.g. mercaptoethanol	P_1—SH + P_2—SH	36—38 48—52
P_1—S—Hg—X—Hg—S—P_2	$Hg—CH_2$ + $CH_2—\overset{+}{Hg}$ (dioxane mercurial)	free thiol e.g. mercaptoethanol	P_1—SH + P_2—SH	39
P_1—S—CH—CH—S—P_2 (OH, OH)	N—O—C(=O)—CH—CH—C(=O)—O—N (bis-NHS tartrate diester)	periodate	P_1—CHO + P_2—CHO	40 53,54
P_1—N=N—P_2	N_3—(phenyl)—N=N—(phenyl)—C(=O)—N(H)—$(CH_2)_2$—C(=O)—O—N	dithionite	P_1—NH$_2$ + P_2—NH$_2$	41
P_1—S(=O)(=O)—P_2	F—(phenyl, NO_2)—S(=O)(=O)—(phenyl, NO_2)—F	base	P_1—OH + P_2—SO$_3^-$	42—44

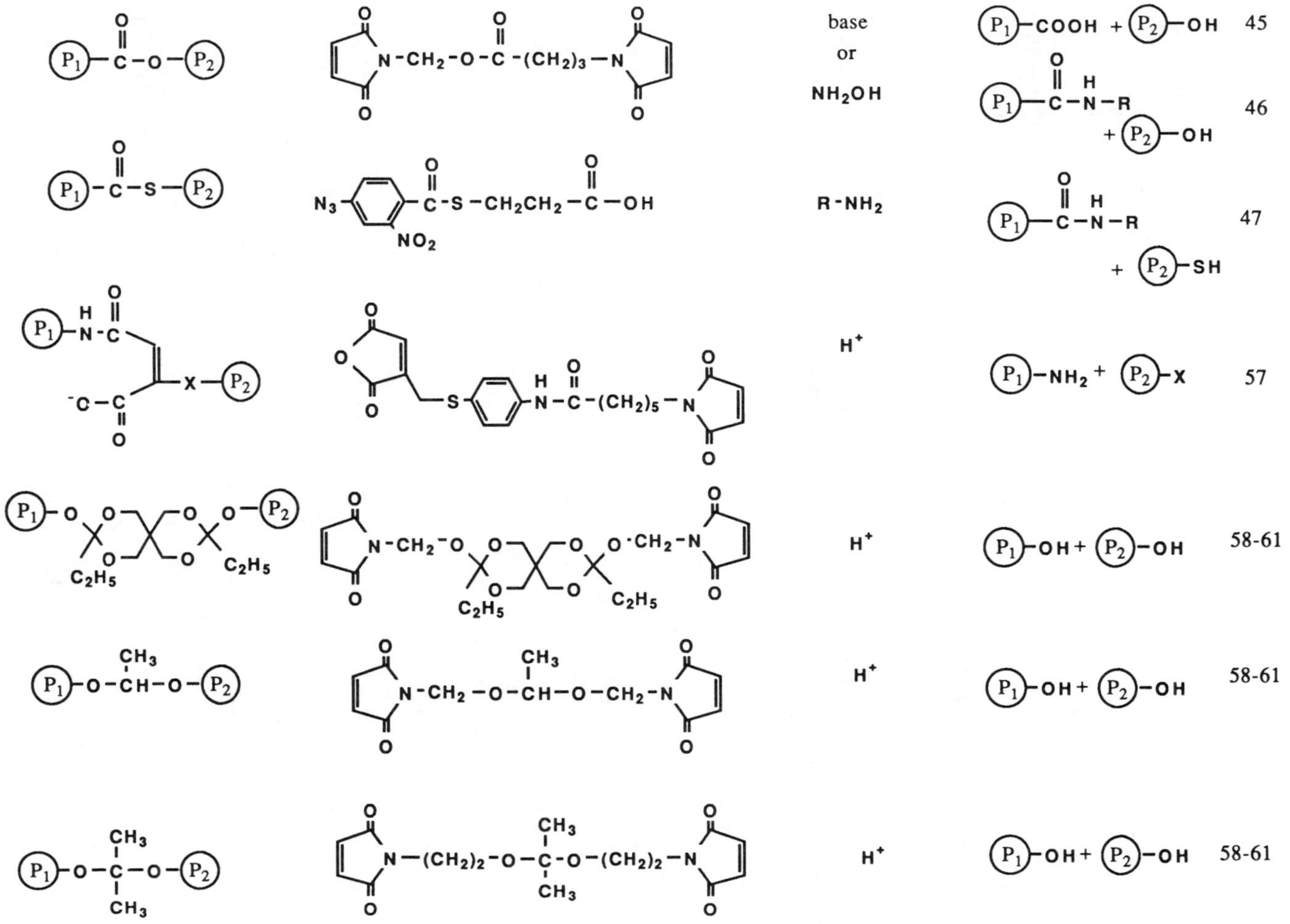

$P_1-COOH + P_2-OH$ 45

$P_1-\overset{O}{\overset{\|}{C}}-N-R \;+\; P_2-OH$ 46

$P_1-\overset{O}{\overset{\|}{C}}-N-R \;+\; P_2-SH$ 47

$P_1-NH_2^+ \;\; P_2-X$ 57

$P_1-OH + P_2-OH$ 58-61

$P_1-OH + P_2-OH$ 58-61

$P_1-OH + P_2-OH$ 58-61

base or NH_2OH

$R-NH_2$

H^+

H^+

H^+

H^+

C. VICINAL GLYCOL BOND

As in carbohydrates in glycoproteins (discussed in Chapter 2), vicinal glycol bonds can be cleaved with sodium periodate converting the diols into dialdehydes.[53,54] Although the rate of cleavage is lower than disulfide bonds, the oxidation reaction takes place mildly at neutral pH and ambient temperatures, for instance, 15 mM sodium periodate at pH 7.5 and 25°C for 4 to 5 h. Proteins coupled by reagents containing such a vicinal glycol linkage are therefore reversible. Disuccinimidyl tartarate shown in Table 3 is such a reagent. The cleavage reaction of cross-linked proteins produces aldehydes. It should be reminded that the carbohydrate moieties of cross-linked glycoproteins are also susceptible to oxidation. In addition, the reactivity of the aldehydes formed during the oxidation reaction may cause Schiff base formation with the amino groups on the proteins, rendering potential complications.

D. AZO LINKAGE

Azo linkages can be cleaved with 0.1 M sodium dithionite at pH 8.0 for 25 min.[41] The reduction produces two free amines. An example of such a cleavable reagent is N-[4-(p-azidophenylazo)-benzoyl]-3-aminohexyl-N'-oxysuccinimide ester shown in Table 3.

E. SULFONE LINKAGE

Sulfone linkages provide another means for cleavage. The bond is easily hydrolyzed with base at pH 11 to 12 for 2 h at 37°C to the sulfonic acid.[42-44] 4,4'-Difluoro-3,3'-dinitrophenyl-sulfone shown in Table 3 is an example of a cleavable cross-linking reagent containing such a group.

F. ESTER BOND

Theoretically esters can be hydrolyzed under acidic or basic conditions. Generally this is done with 0.1 N NaOH at ambient temperature.[45] The products of the hydrolysis are a carboxylic acid and an alcohol. The ester bond may also be cleaved by incubating with 1 M hydroxylamine at pH 7.5 to 8.5 and 25 to 37°C for 3 to 6 h.[46] This reaction produces an amide and an alcohol. However, the use of hydroxylamine is incompatible with proteins containing Asn-Gly bonds, which are known to be attacked by this chemical.[55] A bifunctional dimaleimide with such a cleavable ester bond, maleimidomethyl-3-maleimido propionate, synthesized by Sato and Nakao[45] is shown in Table 3.

G. THIOESTER BOND

As in ester bonds, thioesters can be hydrolyzed in acid or base. They are preferably cleaved with an amine which displaces the sulfide. A cross-linking reagent which forms a cleavable thioester on reaction with sulfhydryl groups of proteins is 3-(4-azido-2-nitrobenzoylseleno)propionic acid shown in Table 3.

H. MALEYLAMIDE LINKAGE

As indicated in Chapter 2, maleylation of amino groups with maleic anhydride is reversible. With 2-methylmaleic anhydride and 2,3-dimethylmaleic anhydride, the acyl derivatives are even more easily hydrolyzed under acidic conditions.[56] Blättler et al.[57] have incorporated the 2-methylmaleic anhydride in the synthesis of an acid labile heterobifunctional reagent. On reaction with an amino group, the maleylamide formed is stable above neutral pH, but hydrolyzes readily at mildly acidic pH (pH 4 to 5). The reaction is shown in Figure 16.

I. ACETALS, KETALS AND ORTHO ESTERS

Acetals, ketals, and ortho esters have long been established as acid labile and base stable.[58,59] Of these, ortho esters are the most susceptible to acid-catalyzed hydrolysis,

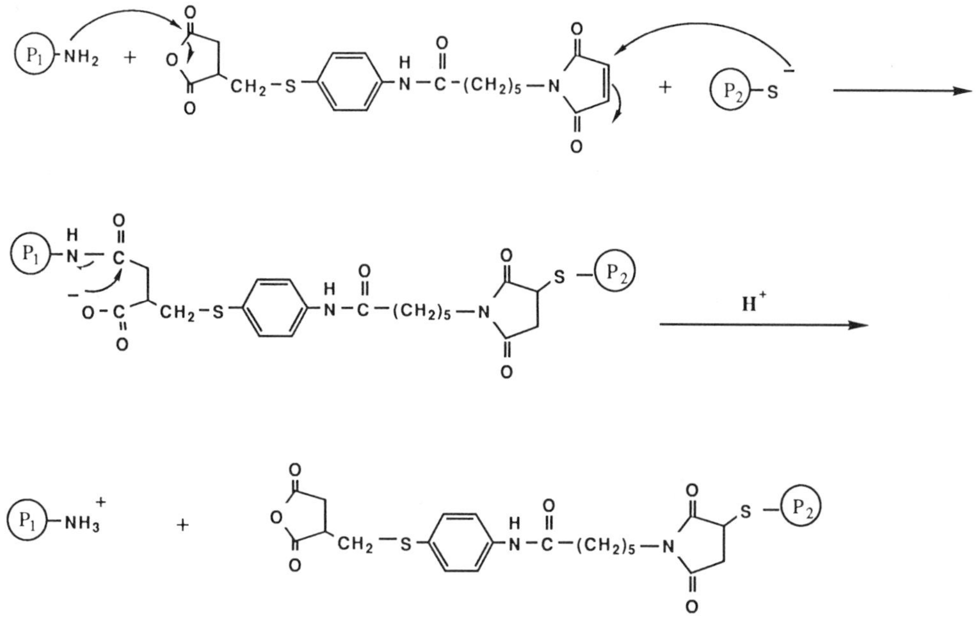

FIGURE 16. Reaction of a maleimido derivative of 2-methylmaleic anhydride.

followed by ketals and acetals. Ketals undergo faster hydrolysis than acetals because of enhanced stabilization of the intermediate carbonium ion by an additional alkyl group.[60] Among the ketals, acyclic ketals of acetone are the most acid labile. Recently, Srinivasachar and Neville[61] have incorporated these functionalities in the synthesis of acid-cleavable protein cross-linking reagents, bismaleimide ortho ester, and acetal and ketal cross-linkers, as shown in Table 3. These compounds are stable toward base and can be stored and manipulated at pH 8 to 9. They can be hydrolyzed at pH 5.5 with half-lives less than an hour for ortho esters and ketals.[61]

VII. LABELED CROSS-LINKING REAGENTS

The bridge joining the two ends of functional groups for cross-linking can not only be designed to contain cleavable functionalities as described above, but also can be constructed to include different architects to satisfy various requirements. For example, the cross-linking functional groups can be spaced between a desired distance using different lengths of carbon chains. This is represented by the *bis*-imidoester series shown in Table 4.[62] In addition, different reporter groups can also be incorporated as well as functional groups usable for radioactive labeling.

A. REPORTER GROUPS

There are many different reporter groups that are useful for the study of a protein subunit and protein environment interactions[63-65] as well as for monitoring the extent of chemical modification.[41,66,67] The common reporter groups are UV-VIS absorption chromophors,[68,69] fluorescent probes,[63] and spin labels.[64] These entities can be introduced into the backbone of the cross-linking reagents. Some of the examples that illustrate the possibilities are shown in Figure 17. These compounds have different applications. The spin label {7}, for example, has been used for membrane studies.[64,65]

Table 4

IMIDOESTER CROSS-LINKERS OF DIFFERENT CHAIN LENGTHS

Regent	Maximum Distance Between Cross-linked Groups (Å)

$$\overset{+}{N}H_2\overset{-}{Cl} \quad \overset{+}{N}H_2\overset{-}{Cl}$$
$$CH_3-O-\overset{\|}{C}-CH_2-\overset{\|}{C}-O-CH_3 \qquad\qquad 5$$

$-(CH_2)_2-$ 6

$-(CH_2)_4-$ 9

$-(CH_2)_5-$ 10

$-(CH_2)_6-$ 11

$-(CH_2)_8-$ 14

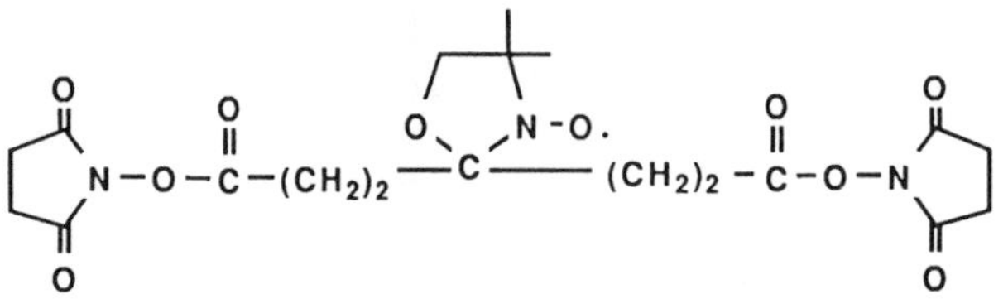

{7} 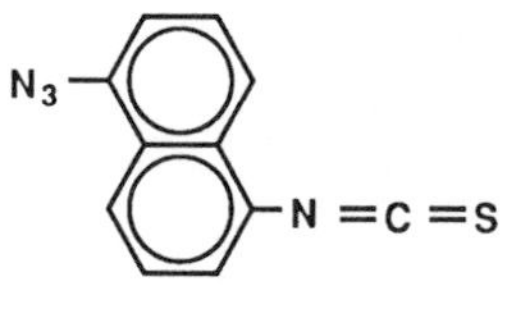   {8}

FIGURE 17.   Bifunctional cross-linking reagents containing reporter groups: {7}: a spin label and {8}: a fluorescent probe.

B. INTRODUCTION OF RADIOACTIVE ISOTOPES

Although radioactive cross-linking reagents can be synthesized de novo from compounds labeled with ^3H or ^{14}C, the ability to radiolabel an already-made cross-linker in a simple one-step process is of great appeal. Iodination has generally been used for such a purpose, particularly for proteins.[69,70] There are several reagents available for the introduction of ^{125}I. These include chloramine T,[71,72] iodine monochloride,[73] iodogen,[74] and lactoperoxidase.[75] The phenolic side chain of tyrosine provides a convenient major site of substitution. The incorporation of such hydroxyphenyl groups into cross-linking reagent has also been used for this purpose. An example, *N*-(4-azidocarbonyl-3-hydroxyphenyl)maleimide is shown in Figure 18. The iodine can be introduced readily with chloramine T.[76]

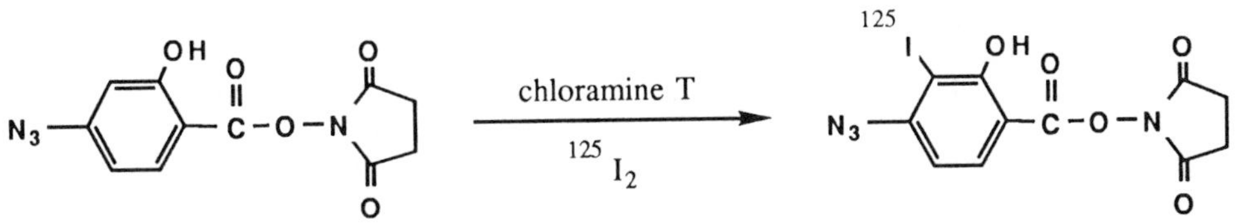

FIGURE 18. Introduction of radioactive isotope, ^{125}I, into a bifunctional reagent containing a hydroxyphenyl group, N-(4-azidocarbonyl-3-hydroxyphenyl)maleimide.

VIII. HYDROPHOBICITY AND HYDROPHILICITY OF REAGENTS

Bifunctional cross-linking reagents vary in their water solubility. The degree of water solubility may be enhanced by incorporating hydrophilic moieties into the molecules. For example the hydrophobicity of N-hydroxysuccinimides can be increased by adding a sulfonate group to the succinimide ring.[77] The resulting molecule with increased solubility in water has, however, decreased membrane permeability. Similarly, introducing a sulfonate group into the bridge increases the polarity of the compound and decreases the membrane solubility.[78,79] The hydrophobicity and hydrophilicity of the reagents therefore determine the depth of probe into the membrane. 4-Azidoiodobenzene, for example, was able to penetrate into the membrane and was able to cross-link phospholipid and proteins.[80] As shown in Table 5, other means of changing the hydrophobicity and hydrophilicity of the compounds are possible.[81-84] These include introduction of ether-oxygen groups, hydroxyl group, ester and amide bonds, and other water soluble functionalities into the bridging spacer.

Not only does increase in hydrophilicity of the reagent increase its solubility in water, the change of hydrophobicity also affects the mode and extent of intermolecular cross-linking. This is vividly demonstrated by Fasold et al.[84] who compared the efficiency of cross-linking of hemoglobulin using tartryl*bis*glycinazide and *bis*(ureido)prolylazobenzene (see Chapter 4 for structures). These two compounds have a similar span of 30 Å but the former is distinctly more hydrophilic than the latter. The results showed that the tartryl derivative gave a higher percentage of interchain bonds. It can be reasoned that the hydrophilic reagents tend to stick out from the surface of the modified protein to react with a second molecule, whereas the hydrophobic reagent, in contrast, may tend to stick to the hydrophobic portions of the protein favoring internal cross-linking. In addition to decreased efficiency of intermolecular cross-linking, hydrophobic reagents also have a higher tendency to denature proteins.[84]

Table 5

MODIFICATION OF HYDROPHOBICITY AND HYDROPHILICITY

More Hydrophobic   More Hydrophilic

REFERENCES

1. **Chong, P. C. S. and Hodges, R. S.,** A new heterobifunctional cross-linking reagent for the study of biological interactions between proteins. I. Design, synthesis and characterization, *J. Biol. Chem.*, 256, 5064, 1981.

2. **March, J.,** *Advanced Organic Chemistry: Reactions, Mechanism and Structure,* John Wiley & Sons, New York, 1985.

3. **Loudon, J. D. and Schulman, N.,** Mobility of groups in chloronitrodiphenyl sulfones, *J. Chem. Soc.,* 722, 1941.

4. **Suhr, H.,** Effects of the leaving group on the velocity of nucleophilic aromatic substitutions, *Berichte,* 97, 3268, 1964.

5. **Edwards, J. O. and Pearson, R. G.,** The factors determining nucleophilic reactivities, *J. Am. Chem. Soc.,* 84, 26, 1962.

6. **Gurd, F. R. N.,** Carboxymethylation, *Methods Enzymol.,* 11, 532, 1967.

7. **Gorin, G., Martin, P. A., and Doughty, G.,** Kinetics of the reaction of *N*-ethylmaleimide with cysteine and some congeners, *Arch. Biochim. Biophys.,* 115, 539, 1966.

8. **Zahn, H. and Meinhoffer, J.,** Reactions of 1,5-difluoro-2,4-dinitrobenzene with insulin, *Makromol. Chem.,* 26, 153, 1958.

9. **Shaltiel, S.,** Thiolysis of some dinitrophenyl derivatives of amino acids, *Biochem. Biophys. Res. Commun.,* 29, 178, 1967.

10. **Husain, S. S., Ferguson, J. B., and Fruton, J. S.,** Bifunctional inhibitors of pepsin, *Proc. Natl. Acad. Sci. U.S.A.,* 68, 2765, 1971.

11. **Lundblad, R. L. and Stein, W. H.,** On the reaction of diazoacetyl compounds with pepsin, *J. Biol. Chem.,* 244, 154, 1969.

12. **Anderson, G. W., Zimmerman, J. E., and Callahan, E. M.,** The use of esters of N-hydroxysuccinimide in peptide synthesis, *J. Am. Chem. Soc.,* 86, 1839, 1964.

13. **McElvain, S. M. and Schroeder, J. P.,** Ortho esters and related compounds from malono- and succinonitriles, *J. Am. Chem. Soc.,* 71, 40, 1949.

14. **Peters, K. and Richards, F. M.,** Chemical cross-linking: reagents and problems in studies of membrane structure, *Annu. Rev. Biochem.,* 46, 523, 1977.

15. **Hunter, M. J. and Ludwig, M. L.,** Amidination, *Methods Enzymol.,* 25, 585, 1972.

16. **Lutter, L. C., Ortanderl, F., and Fasold, H.,** Use of a new series of cleavable proteins-crosslinkers on the *Escherichia coli* ribosome, *FEBS Lett.,* 48, 288, 1974.

17. **Rifai, A. and Wong, S. S.,** Preparation of phosphorylcholine-conjugated antigens, *J. Immunol. Methods,* 94, 25, 1986.

18. **Silman, I. H., Albu-Weissenberg, M., and Katchalski, E.,** Some water-insoluble papain derivatives, *Biopolymers,* 4, 441, 1966.

19. **Silman, I. H. and Katchalski, E.,** Water-insoluble derivatives of enzymes, antigens, and antibodies, *Annu. Rev. Biochem.,* 35, 873, 1966.

20. **Gordon, J., Rose, B., and Sehon, A. H.,** Detection of "non-precipitating" antibodies in sera of individuals allergic to ragweed pollen by an *in vitro* method, *J. Exp. Med.,* 108, 37, 1958.

21. **Carlsson, J., Drevin, H., and Axén, R.,** Protein thiolation and reversible protein-protein conjugation. N-Succinimidyl 3-(2-pyridyldithio)propionate, a new heterobifunctional reagent, *Biochem. J.,* 173, 723, 1978.

22. **Ji, T. H.,** The application of chemical crosslinking for studies on cell membranes and the identification of surface reporters, *Biochem. Biophys. Acta,* 559, 39, 1979.

23. **Hughes, W. L.,** Albumin fraction isolated from human plasma as a crystalline mercuric salt, *J. Am. Chem. Soc.,* 69, 1836, 1947.

24. **Jovin, T. M., Englund, P. T., and Kornberg, A.,** Enzymatic synthesis of deoxyribonucleic acid. XXVII. Chemical modification of deoxyribonucleic acid polymerase, *J. Biol. Chem.,* 244, 3009, 1969.

25. **Arnon, R. and Shapira, E.,** Crystalline papain derivative containing an intramolecular mercury bridge, *J. Biol. Chem.,* 244, 1033, 1969.

26. **Sperling, R., Burstein, Y., and Steinberg, I. Z.,** Selective reduction and mercuration of cystine IV-V in bovine pancreatic ribonuclease, *Biochemistry,* 8, 3810, 1969.

27. **Kay, C. M. and Edsall, J. T.,** Dimerization of mercaptalbumin in the presence of mercurials. III. Bovine mercaptalbumin in water and in concentrated urea solutions, *Arch. Biochem. Biophys.,* 65, 354, 1956.

28. **Means, G. E. and Feeney, R. E.,** Reductive alkylation of amino groups in proteins, *Biochemistry,* 7, 2192, 1968.

29. **Ngo, T. T., Yam, C. F., Lenhoff, H. M., and Ivy, J.,** *p*-Azidophenylglyoxal: a heterobifunctional photoactivable cross-linking reagent selective for arginyl residues, *J. Biol. Chem.,* 256, 11313, 1981.

30. **Politz, S. M., Noller, H. F., McWhiter, P. D.,** Ribonucleic acid-protein cross-linking in *Escherichia coli* ribosomes: (4-azidophenyl)glyoxal, a novel heterobifunctional reagent, *Biochemistry,* 20, 372, 1981.

31. **Knowles, J. R.,** Photogenerated reagents for biological receptor-site labeling, *Acc. Chem. Res.,* 5, 155, 1972.

32. **Bayley, H. and Knowles, J. R.,** Photoaffinity labeling, *Methods Enzymol.,* 46, 69, 1977.

33. **Bayley, H.,** Photogenerated reagents in biochemistry and molecular biology, in *Laboratory Techniques in Biochemistry and Molecular Biology,* Vol. 12, Work, T. S. and Burdon, R. H., Eds., Elsevier, Amsterdam, 1984, 208.

34. **Walling, C. and Gibian, M. J.,** Hydrogen abstraction reactions by the triplet states of ketones, *J. Am. Chem. Soc.,* 87, 3361, 1965.

35. **Schafer, H.-J.,** Divalent azido-ATP analog for photoaffinity cross-linking of F_1 subunits, *Methods Enzymol.,* 126, 649, 1986.

36. **Traut, R. R., Bollen, A., Sun, T. T., Hershey, J. W. B., Sundberg, J., Pierce, L. R.,** Methyl-4-mercaptobutyrimidate as a cleavable cross-linking reagent and its application to the *Escherichia coli* 30S ribosome, *Biochemistry,* 12, 3266, 1973.

37. **Sun, T. T., Bollen, A., Kahan, L., Traut, R. R.,** Topography of ribosomal proteins of the *E. coli* 30S subunit as studied with the reversible cross-linking reagent methyl-4-mercaptobutyrimidate, *Biochemistry,* 13, 2334, 1974.

38. **Bragg, P. D. and Hou, C.,** Subunit composition, function, and spatial arrangement in the Ca^{2+}- and Mg^{2+}-activated adenosine triphosphatase of *Escherichia coli* and *Salmonella typhimurium, Arch. Biochem. Biophys.,* 167, 311, 1975.

39. **Webb, J. L.,** Enzyme and metabolic inhibitors, Vol. 2, Academic Press, New York, 1966, 729.
40. **Smith, R. J., Capaldi, R. A., Muchmore, D., and Dahlquist, F.,** Cross-linking of ubiquinone-cyto-chrome-c-reductase with periodate-cleavable bifunctional reagent, *Biochemistry,* 18, 3719, 1978.
41. **Jaffe, C. L., Lis, H., and Sharon, N.,** New cleavable photoreactive heterobifunctional cross-linking reagents for studying membrane organization, *Biochemistry,* 19, 4423, 1980.
42. **Wold, F.,** Reaction of bovine serum albumin with the bifunctional reagent *p,p'*-difluoro-*m,m'*-dinitrodi-phenylsulfone, *J. Biol. Chem.,* 236, 106, 1961.
43. **Wold, F.,** Bifunctional reagents, *Methods Enzymol.,* 25, 623, 1972.
44. **Zahling, D. A., Watson, A., and Bach, F. H.,** Mapping of lymphocyte surface polypeptide antigens by chemical cross-linking with BSOCOES, *J. Immunol.,* 124, 913, 1980.
45. **Sato, S. and Nakao, M.,** Cross-linking of intact erythrocyte membrane with a newly synthesized cleavable bifunctional reagent, *J. Biochem.,* 90, 1177, 1981.
46. **Abdella, P. M., Smith, P. K., and Royer, G. P.,** A new cleavable reagent for cross-linking and reusable immobilization of proteins, *Biochem. Biophys. Res. Commun.,* 87, 734, 1979.
47. **Friebel, K., Huth, H., Jany, K. D., and Trummer, W. E.,** Semireversible cross-linking synthesis and application of a novel heterobifunctional reagent, *Z. Physiol. Chem.,* 362, 421, 1981.
48. **Cleland, W. W.,** Dithiothreitol: a new protective reagent for SH groups, *Biochemistry,* 3, 480, 1964.
49. **Stockmayer, W. H. D., Rice, D. W., and Stephenson, C. C.,** Thermodynamic properties of sodium borohydride and aqueous borohydride ion, *J. Am. Chem. Soc.,* 77, 1980, 1955.
50. **Neumann, H., Goldberger, R. F., and Sela, M.,** Interaction of phosphorothioate with disulfide bonds of ribonuclease and lysozyme, *J. Biol. Chem.,* 239, 1536, 1964.
51. **Cecil, R. and McPhee, J. R.,** Kinetic study of the reactions on some disulfides with sodium sulfite, *Biochem. J.,* 60, 496, 1955.
52. **McPhee, J. R.,** Further studies on the reactions of disulfides with sodium sulfite, *Biochem. J.,* 64, 22, 1956.
53. **Bobbitt, J. M.,** Periodate oxidation of carbohydrates, *Adv. Carbohyd. Chem.,* 11, 1, 1956.
54. **Coggins, J. R., Hooper, E. A., and Perham, R. N.,** Use of dimethyl subermidate and novel periodate-cleavable *bis*(imidoesters) to study the quaternary structure of the pyruvate dehydrogenase multienzyme complex of *Escherichia coli, Biochemistry,* 15, 2527, 1976.
55. **Bornstein, P. and Balian, G.,** Cleavage of Asn-Gly bonds with hydroxylamine, *Methods Enzymol.,* 47, 132, 1977.
56. **Dixon, H. B. F. and Perham, R. N.,** Reversible blocking of amine groups with citraconic anhydride, *Biochem. J.,* 109, 312, 1968.
57. **Blättler, Kuenzi, B. S., Lambert, J. M., and Senter, P. D.,** New heterobifunctional protein cross-linking reagent that forms an acid-labile link, *Biochemistry,* 24, 1517, 1985.
58. **Cordes, E. H.,** Mechanism and catalysis for the hydrolysis of acetals, ketals, and ortho esters, *Prog. Phys. Org. Chem.,* 4, 1, 1967.
59. **Cordes, E. H. and Bull, H. G.,** Mechanism and catalysis for hydrolysis of acetals, ketals and orthoesters, *Chem. Rev.,* 74, 581, 1974.
60. **Cordes, E. H.,** in *The Chemistry of Carboxylic Acids and Esters,* Patai, S., Ed., Interscience Publishers, New York, 1979, 623.
61. **Srinivasachar, K. and Neville, D. M., Jr.,** New protein cross-linking reagents that are cleaved by mild acid, *Biochemistry,* 28, 2501, 1989.
62. **Ji, T. H.,** Cross-linking of glycolipids in erythrocyte ghost membrane, *J. Biol. Chem.,* 249, 7841, 1974.
63. **Gonzalez-Ros, J. M., Farach, M. C., and Martinez-Carrion, M.,** Ligand-induced effects at regions of acetylcholine receptor accessible to membrane lipids, *Biochemistry,* 22, 3807, 1983.
64. **Gaffney, B. J., Willingham, G. L., and Schepp, R. S.,** Synthesis and membrane interactions of spin label bifunctional reagents, *Biochemistry,* 22, 881, 1983.
65. **Willingham, G. L. and Gaffney, B. J.,** Reactions of spin-label cross-linking reagents with RBC proteins, *Biochemistry,* 22, 892, 1983.
66. **Ueno, T., Hikita, S., Muno, D., Sato, E., Kanaoka, Y., and Sekine, T.,** New fluorogenic, photoactivable heterobifunctional crosslinking thiol reagents, *Anal. Biochem.,* 140, 63, 1984.
67. **Maassen, J. A.,** Cross-linking of ribosomal proteins by 4-(6-formyl-3-azidophenoxy)butyrimidate, a het-erobifunctional, cleavable cross-linker, *Biochemistry,* 18, 1288, 1979.
68. **Sigrist, H., Allegrini, P. R., Kempf, C., Schnippering, C., and Zahler, P.,** 5-Isothiocyanato-1-na-phthalene azide and *p*-azidophenylisothiocyanate. Synthesis and application in hydrophobic heterobifunc-tional photoactive cross-linking of membrane proteins, *Eur. J. Biochem.,* 125, 197, 1982.
69. **Hughes, W. L.,** The chemistry of iodination, *Ann. N.Y. Acad. Sci.,* 70, 3, 1957.
70. **Seevers, R. H. and Counsell, R. E.,** Radioiodination techniques for small organic molecules, *Chem. Rev.,* 82, 575, 1982.
71. **Hunter, W. M. and Greenwood, F. C.,** Radioimmunoelectrophoretic assay for human growth hormone, *Nature (London),* 194, 495, 1962.

√ 72. **Greewood, F. C., Hunter, W. M., and Glover, J. S.,** The preparation of ^{131}I-labeled human growth hormone of high specific radioactivity, *Biochem. J.,* 89, 114, 1963.

73. **McFarlane, A. S.,** Efficient trace-labeling of proteins with iodine, *Nature (London),* 182, 53, 1958.

74. **Fraker, P. J. and Speck, J. C., Jr.,** Protein and cell membrane iodinations with a sparingly soluble chloroamide, 1,3,4,6-tetrachloro-3a,6a-diphenylglycouril, *Biochem. Biophys. Res. Commun.,* 80, 849, 1978.

75. **Marchalonis, J. J.,** Enzymic method for the trace iodination of immunoglobulins and other proteins, *Biochem. J.,* 113, 299, 1969.

76. **Ji, T. H. and Ji, I.,** Macromolecular photoaffinity labeling with radioactive photoactivable heterobifunctional reagents, *Analyt. Biochem.,* 121, 286, 1982.

77. **Staros, J. V.,** N-Hydroxysulfosuccinimide active esters: *bis(N*-hydroxysulfosuccinimide) esters of two dicarboxylic acids are hydrophilic, membrane-impermeant, protein cross-linkers, *Biochemistry,* 21, 3950, 1982.

78. **Staros, J. V., Morgan, D. G., and Appling, D. R.,** A membrane-impermeant, cleavable cross-linker: dimers of human erythrocyte band 3 subunits cross-linked at the extra-cytoplasmic face, *J. Biol. Chem.,* 256, 5890, 1981.

79. **Staros, J. V.,** Membrane impermeant cross-linking reagents: probes of the structure and dynamics of membrane proteins, *Acc. Chem. Res.,* 12, 435, 1988.

80. **Harris, R. and Findlay, J. B.,** Investigation of the organization of the major proteins in bovine myelin membranes. Use of chemical probes and bifunctional cross-linking reagents, *Biochim. Biophys. Acta,* 732, 75, 1983.

81. **Tesser, G. I., De Hoog-Declerck, R. A. O. M. M., and Westerhuis, L. W.,** New bifunctional reagents for the conjugation of a protein with an amine compound in water, *Hoppe-Seyler's Z. Physiol. Chem.,* 356, 1625, 1975.

82. **Staros, J. V., Lee, W. T., and Conrad, D. H.,** Membrane impermeant crosslinking reagent application to studies of the cell surface receptor for IgE, *Methods Enzymol.,* 150, 503, 1988.

83. **Schramm, H. J. and Dülffer, T.,** Synthesis and application of cleavable and hydrophilic crosslinking reagents, in *Protein Crosslinking: Biochemical and Molecular Aspects,* Friedman, M., Ed., Plenum Press, New York, 1976, 197.

84. **Fasold, H., Bäumert, H., and Fink, G.,** Comparison of hydrophobic and strongly hydrophilic cleavable cross-linking reagents in intermolecular bond formation in aggregates of proteins or protein-RNA, in *Protein Cross-Linking: Biochemical and Molecular Aspects,* Friedman, M., Ed., Plenum Press, New York, 1976, 207.

Chapter 4

HOMOBIFUNCTIONAL CROSS-LINKING REAGENTS

I. INTRODUCTION

Homobifunctional reagents are compounds that contain two reactive functional groups which will react with the same amino acid side-chain. In almost all cases, these two reactive groups are identical. These reagents have been shown to induce cross-linking intramolecularly between two locales within a macromolecule and intermolecularly between two molecules. Intramolecular cross-linking has been used to increase the stability of proteins against thermal and mechanical denaturation, to determine the distances between two reactive groups, and to detect different conformational states. Intermolecular cross-linking has been employed to characterize the nature and extent of protein-protein interactions, to couple two or more different proteins, and to determine the interacting protein neighbors. These reagents are also useful for the identification of receptors within membranes.[1-4] A variety of reactive groups have been incorporated into these compounds. These include group selective functionalities such as those presented in the last two chapters for thiols, amines, carboxylates, and phenols. Nonspecific photosensitive functional groups have also been synthesized. In addition, various cleavable bonds as discussed in the last chapter have been incorporated into these chemicals. In this chapter, the homobifunctional reagents will be presented according to their selectivity toward a certain amino acid side chain. It should be clear from the previous chapters that there are no absolutely group-specific reagents. For alkylating and acylating agents, the nucleophiles encompasses the sulfhydryl, amino, imidazolyl, phenolate, carboxylate, and hydroxyl groups. The conditions and the reactants, as discussed in Chapter 2, determine the selectivity of the reaction.

It may be noted that the nomenclature of these bifunctional reagents is inconsistent with that recommended by the International Union of Pure and Applied Chemistry. Since these compounds are generally complex and complicated, systemic names are usually long and difficult. The common names used in the literature are adopted here and throughout this book. Those who are interested in the systemic names are referred to the Chemical Abstracts.

II. AMINO GROUP DIRECTED CROSS-LINKERS

The homobifunctional cross-linking reagents that have been designed to react with amino groups of proteins are listed in Table 1. The imidates and N-succinimidyl derivatives are generally considered most selective for amino groups followed by aryl halides. Many of the alkylating and acylating agents which seemingly react preferentially with amino groups at alkaline pHs because of their abundance, also react with other nucleophiles.

A. BIS-IMIDOESTERS

Imidoesters readily react with amino groups eliminating an alcohol to form amidines (Chapter 3, Figure 6). Although cross-linking of proteins normally requires reactions at both ends of a bifunctional reagent, monofunctional imidoesters have been shown to induce protein cross-linking. Ethyl acetimidate, **II** (Table 1), for example, causes coupling of the subunits of Na^+,K^+-ATPase.[5] The mechanism of such a reaction is discussed in detail by Peters and Richards[1] and is shown in Figure 1. Primary amines attack imidates nucleophilically to produce an intermediate that breaks down to amidine at high pHs or to a new imidate at low pHs. The new imidate can react with another amino group, thus cross-linking the two components. In addition to ethyl acetimidate, methyl acetimidate **I** has also been shown to cross-link erythrocyte membrane proteins.[6]

Table 1

AMINO GROUP DIRECTED HOMOBIFUNCTIONAL CROSS-LINKING REAGENTS

Name (Abbreviation)	Structure	Cleavability/ Condition	Reference
A. BISIMIDOESTERS (BISIMIDATES)			
I. Methyl acetimidate · HCl (MA)	$CH_3-O-\overset{\overset{+NH_2Cl^-}{\|\|}}{C}-CH_2CH_3$	—	6
II. Ethyl acetimidate · HCl (EA)	$CH_3CH_2-O-\overset{\overset{+NH_2Cl^-}{\|\|}}{C}-CH_2CH_3$	—	5
III. Diethyl malonimidate 2HCl (DEM)	$CH_3CH_2-O-\overset{\overset{+NH_2Cl^-}{\|\|}}{C}-CH_2-\overset{\overset{+NH_2Cl^-}{\|\|}}{C}-O-CH_2CH_3$	—	9 - 13, 17
IV. Dimethyl malonimidate · 2HCl (DMM)	$CH_3-O-\overset{\overset{+NH_2Cl^-}{\|\|}}{C}-CH_2-\overset{\overset{+NH_2Cl^-}{\|\|}}{C}-O-CH_3$	—	13 - 15, 17
V. Dimethyl succinimidate · 2HCl (DMSC)	$CH_3-O-\overset{\overset{+NH_2Cl^-}{\|\|}}{C}-(CH_2)_2-\overset{\overset{+NH_2Cl^-}{\|\|}}{C}-O-CH_3$	—	13 - 15, 17

No.	Name	Structure	Refs.
VI.	Dimethyl glutarimidate · 2HCl (DMG)	$CH_3—O—C(=\overset{+}{N}H_2Cl^-)—(CH_2)_3—C(=\overset{+}{N}H_2Cl^-)—O—CH_3$	13-17
VII.	Dimethyl adipimidate · 2HCl (DMA)	$CH_3—O—C(=\overset{+}{N}H_2Cl^-)—(CH_2)_4—C(=\overset{+}{N}H_2Cl^-)—O—CH_3$	13 - 15, 17, 42
VIII.	Dimethyl pimelimidate · 2HCl (DMP)	$CH_3—O—C(=\overset{+}{N}H_2Cl^-)—(CH_2)_5—C(=\overset{+}{N}H_2Cl^-)—O—CH_3$	13-17, 40
IX.	Dimethyl suberimidate · 2HCl (DMS)	$CH_3—O—C(=\overset{+}{N}H_2Cl^-)—(CH_2)_6—C(=\overset{+}{N}H_2Cl^-)—O—CH_3$	7, 13 - 17, 41
X.	Dimethyl azelaimidate · 2HCl	$CH_3—O—C(=\overset{+}{N}H_2Cl^-)—(CH_2)_7—C(=\overset{+}{N}H_2Cl^-)—O—CH_3$	13 - 17
XI.	Dimethyl sebacimidate · 2HCl	$CH_3—O—C(=\overset{+}{N}H_2Cl^-)—(CH_2)_8—C(=\overset{+}{N}H_2Cl^-)—O—CH_3$	13 - 17
XII.	Dimethyl dodecimidate · 2HCl	$CH_3—O—C(=\overset{+}{N}H_2Cl^-)—(CH_2)_{10}—C(=\overset{+}{N}H_2Cl^-)—O—CH_3$	17

Table 1 (continued)

AMINO GROUP DIRECTED HOMOBIFUNCTIONAL CROSS-LINKING REAGENTS

Name (Abbreviation)	Structure	Cleavability/ Condition	Reference
A. BISIMIDOESTERS (BISIMIDATES)			
XIII. Dimethyl 3,3'-oxydipropion-imidate · 2HCl (DODP)	$CH_3-O-\overset{\overset{+}{N}H_2Cl^-}{\overset{\|}{C}}-(CH_2)_2-O-(CH_2)_2-\overset{\overset{+}{N}H_2Cl^-}{\overset{\|}{C}}-O-CH_3$	—	18
XIV. Dimethyl 3,3'-(methylenedioxy)-dipropionimidate · 2HCl (DMDP)	$CH_3-O-\overset{\overset{+}{N}H_2Cl^-}{\overset{\|}{C}}-(CH_2)_2-O-CH_2-O-(CH_2)_2-\overset{\overset{+}{N}H_2Cl^-}{\overset{\|}{C}}-O-CH_3$	—	18
XV. Dimethyl 3,3'-(dimethylenedioxy)-dipropionimidate · 2HCl (DDDP)	$CH_3-O-\overset{\overset{+}{N}H_2Cl^-}{\overset{\|}{C}}-(CH_2)_2-O-(CH_2)_2-O-(CH_2)_2-\overset{\overset{+}{N}H_2Cl^-}{\overset{\|}{C}}-O-CH_3$	—	18
XVI. Dimethyl 3,3'-(tetramethylenedioxy)-dipropionimidate · 2HCl (DTDP)	$CH_3-O-\overset{\overset{+}{N}H_2Cl^-}{\overset{\|}{C}}-(CH_2)_2-O-(CH_2)_4-O-(CH_2)_2-\overset{\overset{+}{N}H_2Cl^-}{\overset{\|}{C}}-O-CH_3$	—	18
XVII. Dimethyl 3,3'-(diethyletherdioxy)-dipropionimidate · 2HCl	$CH_3-O-\overset{\overset{+}{N}H_2Cl^-}{\overset{\|}{C}}-(CH_2)_2-O-(CH_2)_2-O-(CH_2)_2\,\overset{\overset{+}{N}H_2Cl^-}{\overset{\|}{C}}-O-CH_3$	—	18

XVIII. Diisethionyl 3,3'-dithiobis propionimidate · 2HCl	$HO_3S-(CH_2)_2-O-C(=NH_2^+Cl^-)-(CH_2)_2-S-S-(CH_2)_2-C(=NH_2^+Cl^-)-O-(CH_2)_2-SO_3H$	+/thiol	23
XIX. Dimethyl 3,3'-dithiobispropionimidate · 2HCl (DTBP)	$CH_3-O-C(=NH_2^+Cl^-)-(CH_2)_2-S-S-(CH_2)_2-C(=NH_2^+Cl^-)-O-CH_3$	+/thiol	24,25
XX. Dimethyl 4,4'-dithiobisbutyrimidate · 2HCl (DTBB)	$CH_3-O-C(=NH_2^+Cl^-)-(CH_2)_3-S-S-(CH_2)_3-C(=NH_2^+Cl^-)-O-CH_3$	+/thiol	26
XXI. Dimethyl 5,5'-dithiobisvalerimidate · 2HCl (DTBV)	$CH_3-O-C(=NH_2^+Cl^-)-(CH_2)_4-S-S-(CH_2)_4-C(=NH_2^+Cl^-)-O-CH_3$	+/thiol	26
XXII. Dimethyl 7,7'-dithiobisenanthimidate · 2HCl (DTBE)	$CH_3-O-C(=NH_2^+Cl^-)-(CH_2)_6-S-S-(CH_2)_6-C(=NH_2^+Cl^-)-O-CH_3$	+/thiol	26
XXIII. Dimethyl 3,3'-(dithiodimethylenediosy)-bisimidate · 2HCl	$CH_3-O-C(=NH_2^+Cl^-)-(CH_2)_2-O-(CH_2)_2-S-S-(CH_2)_2-O-(CH_2)_2-C(=NH_2^+Cl^-)-O-CH_3$	+/thiol	18

Table 1 (continued)

AMINO GROUP DIRECTED HOMOBIFUNCTIONAL CROSS-LINKING REAGENTS

Name (Abbreviation)	Structure	Cleavability/Condition	Reference
A. BISIMIDOESTERS (BISIMIDATES)			
XXIV. Dimethyl 3,3'-(dithiodimethylene diamido)-bispropionimidate· 2HCl	$CH_3-O-C(=\overset{+}{N}H_2\,\overset{-}{C}l)-(CH_2)_2-\overset{H}{N}-C(=O)-(CH_2)_2-S-S-(CH_2)_2-C(=O)-\overset{H}{N}-(CH_2)_2-C(=\overset{+}{N}H_2\,\overset{-}{C}l)-O-CH_3$	+/thiol	18
XXV. N,N'-bis(2-carboximido-methyl)tartarimide dimethyl ester · 2HCl (CMTD); Tartryldi(methyl-2-aminoacetimidate) · 2HCl (TDAA)	$CH_3-O-C(=\overset{+}{N}H_2\,\overset{-}{C}l)-CH_2-\overset{H}{N}-C(=O)-\underset{OH}{CH}-\underset{OH}{CH}-C(=O)-\overset{H}{N}-CH_2-C(=\overset{+}{N}H_2\,\overset{-}{C}l)-O-CH_3$	+/periodate	27
XXVI. N,N'-bis(2-carboximido-ethyl)tartarimide dimethyl ester · 2HCl (CETD); Tartryldi(methyl-3-aminopropionimidate) · 2HCl	$CH_3-O-C(=\overset{+}{N}H_2\,\overset{-}{C}l)-(CH_2)_2-\overset{H}{N}-C(=O)-\underset{OH}{CH}-\underset{OH}{CH}-C(=O)-\overset{H}{N}-(CH_2)_2-C(=\overset{+}{N}H_2\,\overset{-}{C}l)-O-CH_3$	+/periodate	27
XXVII. 3,4,5,6-tetrahydroxy-suberimidate · 2HCl (THS)	$CH_3-O-C(=\overset{+}{N}H_2\,\overset{-}{C}l)-CH_2-\underset{OH}{CH}-\underset{OH}{CH}-\underset{OH}{CH}-\underset{OH}{CH}-CH_2-C(=\overset{+}{N}H_2\,\overset{-}{C}l)-O-CH_3$	+/periodate	28,29

XXXVIII. Dimethyl 3,3'-(N-2,4-dinitrophenyl)-bispropionimidate · 2HCl

$$CH_3-O-C(=\overset{+}{N}H_2\;\bar{Cl})-(CH_2)_2-N-(CH_2)_2-C(=\overset{+}{N}H_2\;\bar{Cl})-O-CH_3$$

(N-aryl = 2,4-dinitrophenyl: ring bearing NO_2 and NO_2)

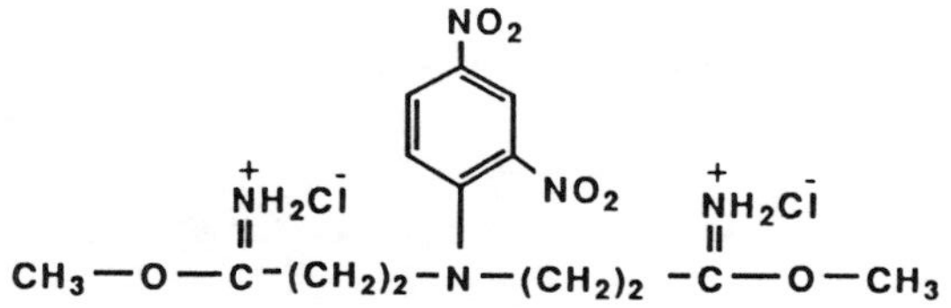

— 30

XXIX. Dimethyl 3,3'-(N-2,4-dinitro-5-carboxyphenyl)-bispropionimidate · 2HCl

$$CH_3-O-C(=\overset{+}{N}H_2\;\bar{Cl})-(CH_2)_2-N-(CH_2)_2-C(=\overset{+}{N}H_2\;\bar{Cl})-O-CH_3$$

(N-aryl = 2,4-dinitro-5-carboxyphenyl: ring bearing NO_2, NO_2, COO^-)

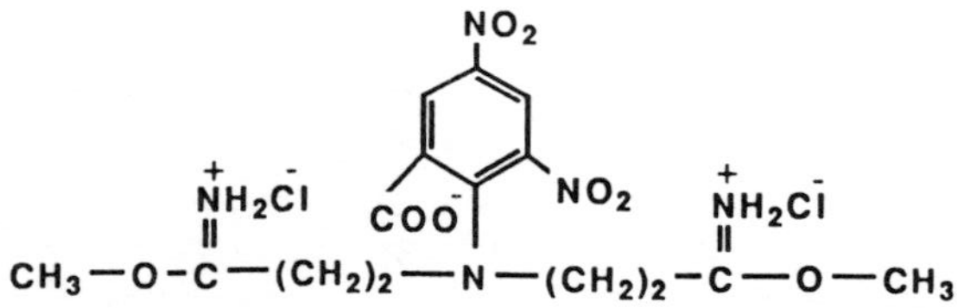

— 30

XXX. Dimethyl 3,3'-[N-(5-(N,N-dimethylamino)naphthyl)sulfonyl]bispropionimidate · 2HCl

$$CH_3-O-C(=\overset{+}{N}H_2\;\bar{Cl})-(CH_2)_2-N-(CH_2)_2-C(=\overset{+}{N}H_2\;\bar{Cl})-O-CH_3$$

(N-substituent = $O=S=O$ linked to naphthyl bearing $N(CH_3)CH_3$)

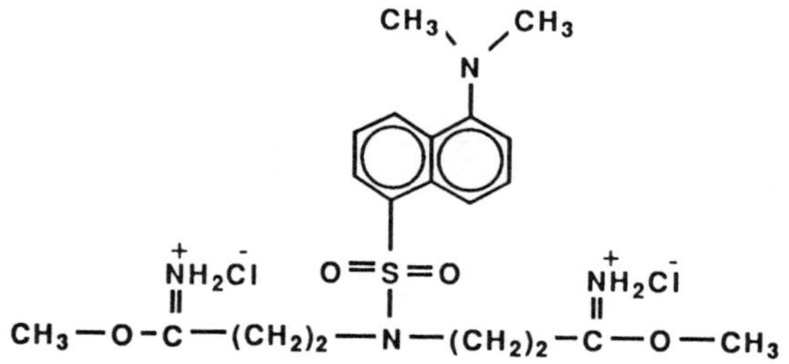

— 31

B. BIS-N-SUCCINIMIDYL DERIVATIVES

XXXI. Disuccinimidylsuberate (DSS); N-hydroxysuccinimidylsuberate (NHS-SA)

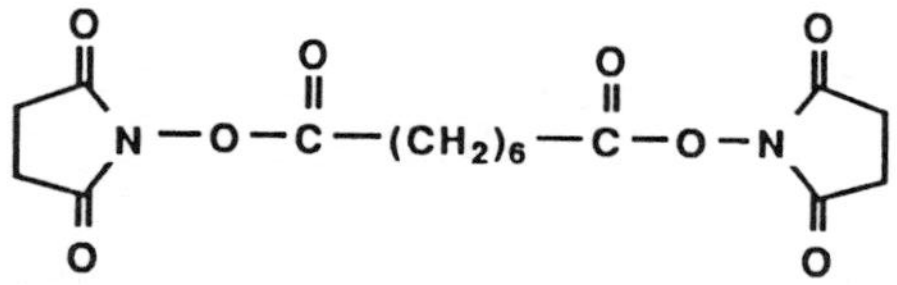

$$N-O-C(=O)-(CH_2)_6-C(=O)-O-N$$

(each terminus = N-succinimidyl, succinimide rings with $O=$ and $=O$)

— 56

Table 1 (continued)

AMINO GROUP DIRECTED HOMOBIFUNCTIONAL CROSS-LINKING REAGENTS

Name (Abbreviation)	Structure	Cleavability/ Condition	Reference
B. BIS-N-SUCCINIMIDYL DERIVATIVES			
XXXII. Bis(sulfosuccinimidyl)-suberate (BSSS)	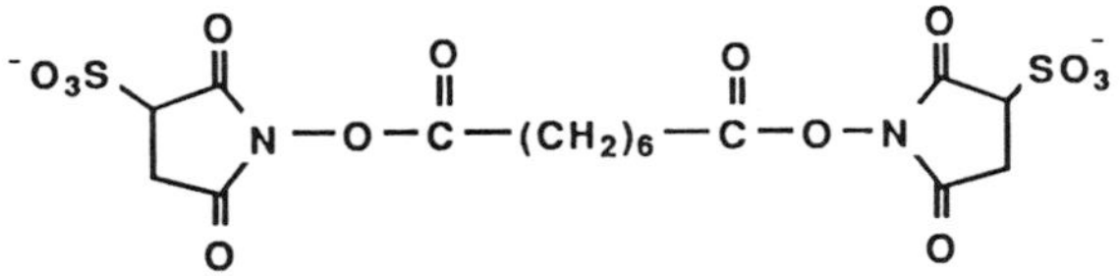	—	57,58
XXXIII. Succinate bis-(N-hydroxy-succinimide ester)	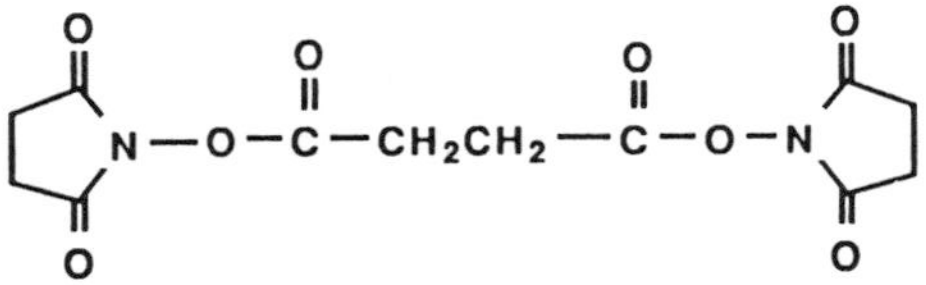	—	59,60
Spin-labeled bis-(N-hydroxysuccinimide ester)	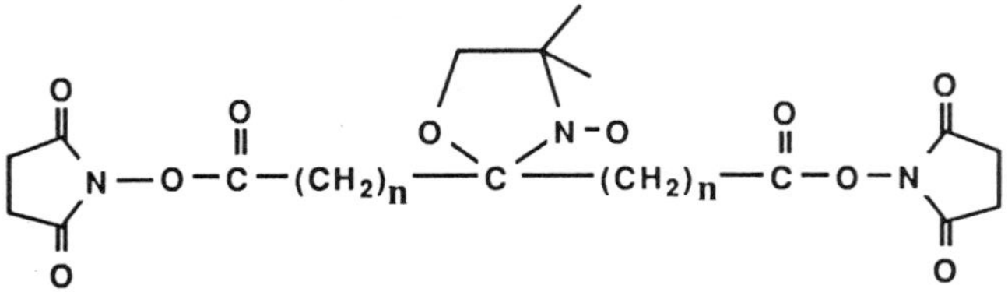		
XXXIV. n = 6 : 3-Oxy-2,2-bis[6-((N-succinimidyloxy)carbonyl)hexanyl]-4,4-dimethyloxazolidine		—	61,62
XXXV. n = 8 : 3-Oxy-2,2-bis[8-((N-succinimidyloxy)carbonyl)octanyl]-4,4-dimethyloxazolidine		—	61

Other spin-labeled bis-(N-hycroxysuccinimide esters)

XXXVI.	n = 3	3-Oxy-2,2-bis[6-(((3-((N-succinimidyloxy)carbonyl)propyl)amino)-carbonyl)hexanyl]-4,4-dimethyloxazolidine	— 61
XXXVII.	n = 5	3-Oxy-2,2-bis[6-(((5-((N-succinimidyloxy)carbonyl)pentanyl)amino)-carbonyl)hexanyl]-4,4-dimethyloxazolidine	— 61
XXXVIII.		Bis(sulfo-N-succinimidyl)-doxyl-2-spiro-5'-azelate (BSSDA)	— 63
XXXIX.		Bis(sulfo-N-succinimidyl)-doxyl-2-spiro-4'-pimelate (BSSDP)	— 64,65

Table 1 (continued)

AMINO GROUP DIRECTED HOMOBIFUNCTIONAL CROSS-LINKING REAGENTS

Name (Abbreviation)	Structure	Cleavability/ Condition	Reference

B. BIS-N-SUCCINIMIDYL DERIVATIVES

XL. Ethyleneglycol bis-(succinimidysuccinate) (EGS)

$$N-O-C-(CH_2)_2C-O-(CH_2)_2-O-C(CH_2)_2-C-O-N$$

+/NH₂OH → +/NH_2OH 66,67

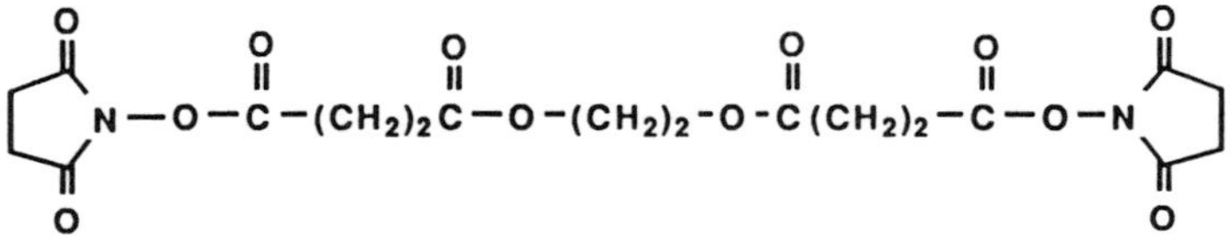

XLI. Bis[2-(succinimido-oxycarbonyl)ethyl]sulfone (BSES)

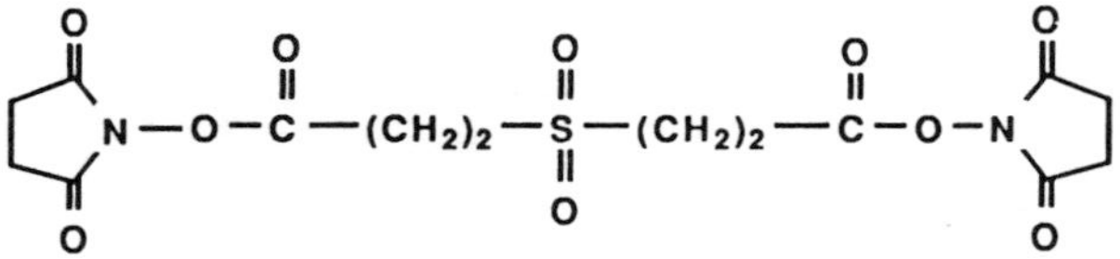

$$N-O-C-(CH_2)_2-S-(CH_2)_2-C-O-N$$

+/base   68

XLII. Bis[2-(succinimido-oxycarbonyloxy)ethyl]-sulfone (BSOCOES)

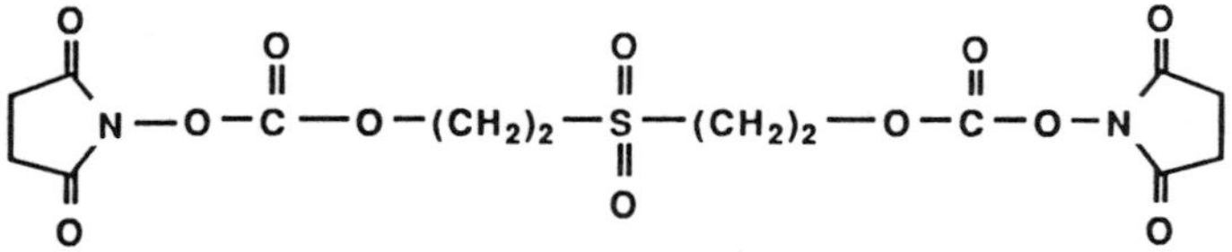

$$N-O-C-O-(CH_2)_2-S-(CH_2)_2-O-C-O-N$$

+/base   1,68

XLIII. 3,3'-dithiobis-(succinimidylpropionate) (DTSP, DSP); Lomant's reagent

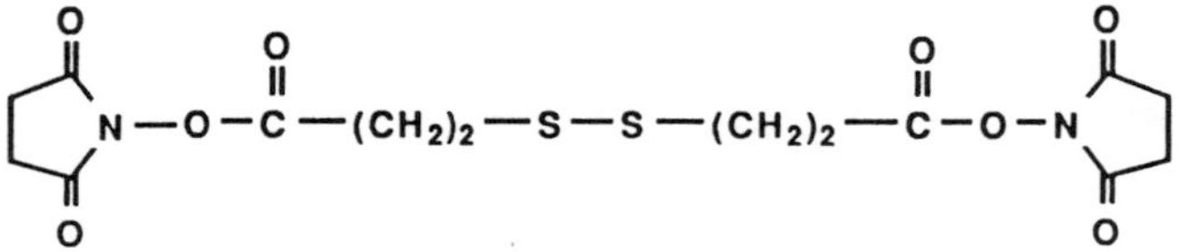

$$N-O-C-(CH_2)_2-S-S-(CH_2)_2-C-O-N$$

+/thiol   69

XLIV. 2,2'-dithiobis-(succinimidylpropionate) (2,2'-DSP)

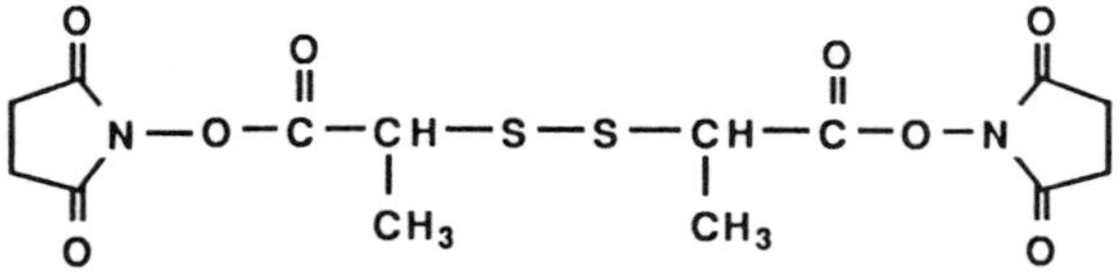

$$N-O-C-CH-S-S-CH-C-O-N$$

+/thiol   70

XLV. 3,3'-dithiobis(sulfo-succinimidyl propionate) (DTSSP)

+/thiol 57,71

XLVI. Disuccinimidyl-(N,N'-diacetylhomocystine)

+/thiol 72

XLVII. Disuccinimidyl tartarate (DST)

+/periodate 73

XLVIII. N,N'·bis(3-succinimidyloxy-carbonylpropyl)tartaramide (BSOPT)

+/periodate 73

C. BIFUNCTIONAL ARYL HALIDES

XLIX. 1,5-difluoro-2,4-dinitrobenzene (FFDNB, DFBN, DFDNB)

— 95,96,105, 107-110, 115

Table 1 (continued)

AMINO GROUP DIRECTED HOMOBIFUNCTIONAL CROSS-LINKING REAGENTS

Name (Abbreviation)	Structure	Cleavability/ Condition	Reference
C. BIFUNCTIONAL ARYL HALIDES			
L. 1,5-dichloro-2,4-dinitrobenzene		–	97
LI. 1,5-dibromo-2,4-dinitrobenzene		–	97
LII. Bis(3,5-dibromosalicyl) fumarate		+/base	98,113
LIII. Bis(3,5-dibromosalicyl) succinate		+/base	98,99
LIV. 4,4'-difluoro-3,3'-dinitro-diphenylsulfone; Bis(3-nitro-4-fluorophenyl)sulfone		+/base	21,100, 101103, 104

D. BIFUNCTIONAL ACYLATING AGENTS

1. *Diisocyanates and diisothiocyanates*

LV. 1,6-hexamethylene diisocyanate (HMDI)

$O=C=N-(CH_2)_6-N=C=O$ — 117,118, 132

LVI. 1,3-dicyanatobenzene — 119

LVII. 1,4-dicyanatobenzene — 119

LVIII. Toluene-2,4-diisocyanate — 117,119, 120,122

LIX. Toluene-2-isocyanate-4-isothiocyanate; 2-cyanato-4-isothiocyanatotoluene — 119

LX. Xylene diisocyanate — 117,119

LXI. Benzidine diisocyanate (BDI) — 119,122

LXII. 2,2'-dimethoxybenzidine diisocyanate — 119

Table 1 (continued)

AMINO GROUP DIRECTED HOMOBIFUNCTIONAL CROSS-LINKING REAGENTS

D. BIFUNCTIONAL ACYLATING AGENTS

Name (Abbreviation)	Structure	Cleavability/Condition	Reference
LXIII. Diphenylmethane-4,4'-diisocyanate	$O{=}C{=}N$—⬡—CH_2—⬡—$N{=}C{=}O$	—	122
LXIV. 3-methoxydiphenylmethane-4,4'-diisocyanate	$O{=}C{=}N$—⬡—CH_2—⬡(OCH$_3$)—$N{=}C{=}O$	—	117,119, 120
LXV. Dicyclohexylmethane-4,4'-diisocyanate	$O{=}C{=}N$—⬡—CH_2—⬡—$N{=}C{=}O$	—	122
LXVI. Hexahydrobiphenyl-4,4'-diisocyanate	$O{=}C{=}N$—⬡—⬡—$N{=}C{=}O$	—	122
LXVII. 2,2'-dicarboxy-4,4'-azophenyldiisocyanate	$O{=}C{=}N$—⬡(COOH)—$N{=}N$—⬡(COOH)—$N{=}C{=}O$	+/dithionite	28,117,119 120,131
LXVIII. 2,2'-dicarboxy-4,4'-azophenyldithioiso-cyanate	$S{=}C{=}N$—⬡(COOH)—$N{=}N$—⬡(COOH)—$N{=}C{=}S$	+/dithionite	28,117, 119,120

LXIX. 2,2'-dicarboxy-6,6'-azophenyldiisocyanate — +/dithionite — 123

LXX. Bis-*p*-(2-carboxy-4'-azophenylisocyanate) — +/dithionite — 28,123

LXXI. Bis-p-(2-carboxy-4'-azophenyl-isothiocyanate) — +/dithionite — 28,123

LXXII. p-Phenylene-diisothiocyanate; 1,4-phenylene diisothiocyanate — — — 13

LXXIII. Diphenyl-4,4'-diisothiocyanato-2,2'-disulfonic acid — — — 124

LXXIV. 4,4'-diisothiocyanato-2,2'-disulfonic acid stilbene (DIDS) — — — 125

LXXV. 4,4'-diisothiocyanato-dihydrostilbene-2,2'-disulfonic acid — — — 126,127

Table 1 (continued)

AMINO GROUP DIRECTED HOMOBIFUNCTIONAL CROSS-LINKING REAGENTS

Name (Abbreviation)	Structure	Cleavability/Condition	Reference
D. BIFUNCTIONAL ACYLATING AGENTS			
2. Bifunctional sulfonyl halides			
LXXVI. Phenol-2,4-disulfonyl-chloride		—	134,135
LXXVII. α–Naphthol-2,4-disulfonyl chloride		—	134–136
LXXVIII. Naphthalene-1,5-disulfonyl chloride		—	122
3. Bis-nitrophenol esters			
Bis-(p-nitrophenyl ester) of carboxylic acids			
LXXIX. n = 4 Bis-(p-nitrophenyl) adipate		—	137,138

LXXX. n = 5 Bis-(p-nitrophenyl) pimelate — 138

LXXXI. n = 8 Bis-(p-nitrophenyl) suberate — 138

LXXXII. Carbonyl bis(L-methionine p-nitrophenyl ester) (CBMNPE) +/CNBr 139,140

4. Bifunctional acylazides

LXXXIII. Tartryl diazide (TDA) +/periodate 29,144

Other tartryl diazides +/periodate 29,144

LXXXIV. n = 1 Tartryl di(glycylazide) (TDGA)

LXXXV. n = 2 Tartryl di(β–alanylazide) (TDAA)

LXXXVI. n = 3 Tartryl di(γ-aminobutyrylazide) (TDBA)

LXXXVII. n = 4 Tartryl di(δ–aminovalerylazide) (TDVA)

LXXXVIII. n = 5 Tartryl di(ε–aminocaproylazide) (TDCA)

Table 1 (continued)

AMINO GROUP DIRECTED HOMOBIFUNCTIONAL CROSS-LINKING REAGENTS

Name (Abbreviation)	Structure	Cleavability/ Condition	Reference

D. **BIFUNCTIONAL ACYLATING AGENTS**

LXXXIX. p-Bis-(ureido)azido-
oligoprolylylazobenzene
(PAPA)

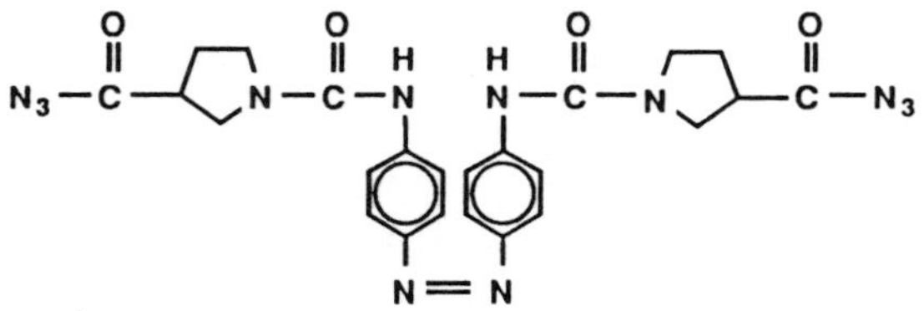

+/dithionite   143

XC. Trimesyl-tris-β-alanylazide
(TTA)

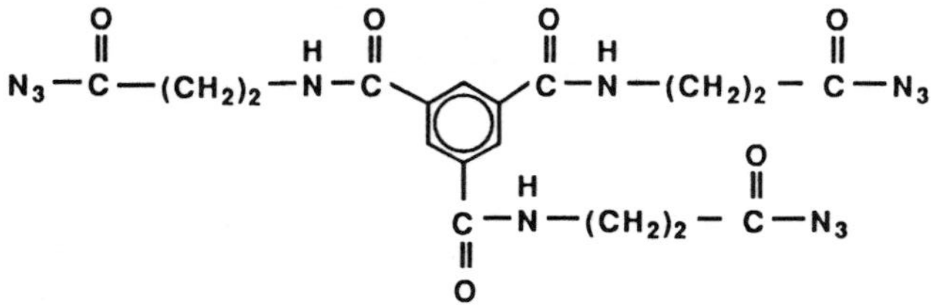

—   143

E. **DIALDEHYDES**

XCI. Glyoxal

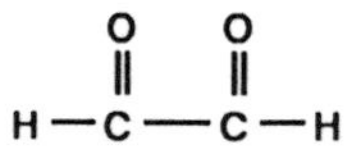

—   145-146

XCII. Malondialdehyde
(MDA)

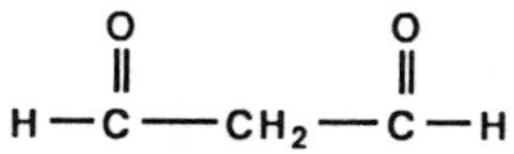

—   146,153

	Structure	Ref.
XCIII. Succinialdehyde	$H-C-(CH_2)_2-C-H$ (dialdehyde, both carbonyls)	146
XCIV. Adipaldehyde	$H-C-(CH_2)_4-C-H$	146,147
XCV. α-Hydroxyadipaldehyde	$H-C-CH-(CH_2)_3-C-H$ (with OH)	149-151
XCVI. Glutaraldehyde	$H-C-(CH_2)_3-C-H$	145-148,151 155-169
XCVII. 3-Methylglutaraldehyde	$H-C-CH_2-CH-CH_2-C-H$ (with CH_3)	149-151
XCVIII. 2-Methoxy-2,4-dimethyl-glutaraldehyde	$H-C-C-CH_2-CH-C-H$ (with OCH_3, CH_3, CH_3)	149-151
XCIX. o-Phthalaldehyde	benzene ring with two CHO groups	152

Table 1 (continued)

AMINO GROUP DIRECTED HOMOBIFUNCTIONAL CROSS-LINKING REAGENTS

Name (Abbreviation)	Structure	Cleavability/Condition	Reference
E. DIALDEHYDES			
P^1,P^2-bis (5'-pyridoxal)-polyphosphate		$+/H^+$, OH^-	170,171
C. P^1,P^2-bis(5'-pyridoxal)diphosphate (Bis-PLP)	$R = O$		
CI. P^1,P^3-bis(5'-pyridoxal)triphosphate		$+/H^+$, OH^-	171
CII. Bis(5'-pyridoxalpyrophospho)methane		$+/H^+$, OH^-	171
CIII. Bis(5'-pyridoxalpyrophospho)1,6-fructose		$+/H^+$, OH^-	171

Compound	Structure		Ref.
CIV. Bis(5′-pyridoxalpyrophospho)-2,3-glycerate		+/H⁺, OH⁻	171
CV. Formaldehyde	$H-\overset{O}{\overset{\|}{C}}-H$	–	151, 172-176
F. DIKETONES			
CVI. 2,5-Hexanedione	$CH_3-\overset{O}{\overset{\|}{C}}-CH_2-CH_2-\overset{O}{\overset{\|}{C}}-CH_3$	–	177
CVII. 3,4-Dimethyl-2,5-hexanedione	$CH_3-\overset{O}{\overset{\|}{C}}-\overset{CH_3}{\overset{\|}{CH}}-\overset{CH_3}{\overset{\|}{CH}}-\overset{O}{\overset{\|}{C}}-CH_3$	–	187
G. OTHER AMINO GROUP DIRECTED REAGENTS			
CVIII. Benzoquinone		–	179
CIX. 2-Iminothiolane		+/thiol	181
CX. Erythreitolbiscarbonate (EBC)		+/periodate	72

R =

$$\overset{O}{\overset{\|}{-O-P-}}\overset{COOH}{\overset{\|}{CH}}\overset{O}{\overset{\|}{-CH_2-O-P-}}\overset{O}{\overset{\|}{O-P-O-}}$$

Table 1 (continued)

AMINO GROUP DIRECTED HOMOBIFUNCTIONAL CROSS-LINKING REAGENTS

Name (Abbreviation)	Structure	Cleavability/ Condition	Reference
G. OTHER AMINO GROUP DIRECTED REAGENTS			
CXI. Mucobromic acid		—	187
CXII. Mucochloric acid		—	187
CXIII. Ethylchloroformate		—	188
CXIV. p-nitrophenylchloroformate		—	188

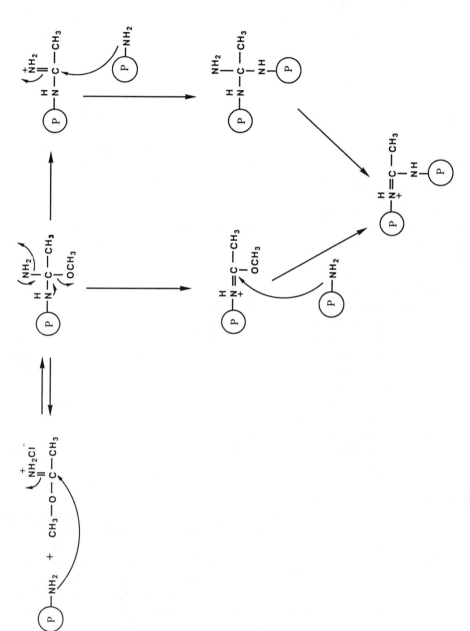

FIGURE 1. Cross-linking of proteins by monofunctional imidoester, methyl acetimidate.

Because of the ease of synthesis,[1,7,8] a large number of *bis*-imidoester reagents with different spacer groups are available as shown in Table 1. A series of compounds, **III** to **XII**, with different chain lengths from 5 to 14 Å have been synthesized.[9-17] To increase the hydrophilicity of the bridge, ether-oxygen-groups have been incorporated into the spacer as in **XIII** to **XVII** which also increase the chain span to 17 Å.[18] In general, these *bis*-imidoesters are water soluble and react with amino groups with a high degree of specificity under mild conditions (pH 7 to 10) to form amidine derivatives.[8] The reaction is favored at alkaline pHs to offset the major competitive hydrolysis at neutral pH.[19,20] In order to maximize the reaction, increment additions of the reagent are recommended.[2] The side reactions with other groups are negligible.[1]

A potential advantage of imidoesters over other amino group directed cross-linkers is the retention of positive charge on the original amino group. The products are normally stable to mild acid hydrolysis[21] but are reported to be susceptible to hydrolysis by hydrazine[22] or ammonia.[20]

Several cleavable *bis*-imidoesters, compounds **XVIII** to **XXIV**, containing the disulfide bond have been synthesized.[18,23-26] These compounds can be cleaved in the presence of thiol-containing reagents. Another series of imidoesters, **XXV** to **XXVII**, containing vicinal diols cleavable by periodate, have also been used.[27-29]

Compounds **XXVIII** to **XXX** are bridge-labeled reagents.[30,31] The colored dinitrophenyl group introduced into the spacer was used for studies on the tertiary structures of proteins (**XXVIII** and **XXIX**).[30] Reagent **XXX** is fluorescent and was demonstrated to cross-link lactate dehydrogenase.[31]

It is impossible to cite all of the proteins that have been cross-linked using these *bis*-imidoesters. The homologous series with different chain lengths has been used for studying α-crystallin.[17] Other reagents have been used to study proteins in erythrocytes,[2] synaptosomal membranes,[32] Mengo virus capsids,[33] rous sarcoma virus,[34] ion-channels,[35] and ribosomes.[36] Many enzymes such as glycogen phosphorylase b,[37] transhydrogenase,[38] phosphorylase-kinase,[39] lactose synthase,[40] phosphofructokinase,[41] ribonuclease,[42] aldose,[7] glyceraldehyde-3-phosphate dehydrogenase,[7] tryptophan synthetase B protein,[7] L-arabinose isomerase,[7] (Na,K)ATPase,[43] monooxygenases[44,45] and asparate transcarbamylase[7] have been cross-linked as well as a series of proteins, for example, nucleosomal histones,[46] tubulin,[47] IgE,[48] alpha-tropomyosin,[49] low density lipoprotein,[50] concanavalin A,[51] hemoglobin,[52] and simian virus induced antigen.[53] These reagents have also been used to introduce cross-links between immunoglobulins and other proteins,[11] bovine leukemia virus RNA and proteins,[54] and lipid and proteins in retrovirus.[55]

B. *BIS*-SUCCINIMIDYL DERIVATIVES

Compounds **XXXI** to **XLVIII**, are cross-linkers activated by *N*-hydroxysuccinimide ester.[56-73] These compounds are easily synthesized by condensing *N*-hydroxysuccinimide with the corresponding dicarboxylic acids in the presence of dicyclohexylcarbodiimide (Chapter 3, Figure 5).[74] They react preferentially with amino groups eliminating *N*-hydroxysuccinimide as the leaving group.[75] The reaction is complete within 10 to 20 min at pH 6 to 9.[60,75,76] Since unprotonated amino groups are the reactive species, increase in pH would increase the rate of the reaction. However, rapid hydrolysis at alkaline pHs effectively competes against the reaction. For example, the half-life of hydrolysis at pH 7.5 and 8.6 is 4 to 5 h and less than 10 min, respectively.[69] Although the imidazolyl group of histidine competes with amines, the product is unstable and is readily hydrolyzed. In this manner imidazolyl groups accelerate the hydrolysis.[75] *N*-hydroxysuccinimide esters, however, are stable for months under anhydrous conditions.

These *bis*-succinimidyl cross-linkers are generally sparingly soluble in water. Increase in hydrophilicity has been achieved by the introduction of the sulfonate group into the succinimidyl ring as shown in compounds **XXXII, XXXVIII, XXXIX** and **XLV.** Incorporation of hydrogen bonding atoms such as ester oxygens in **XL** and hydroxyl groups in **XLVII** and **XLVIII** also increases the solubility. These reagents may be dissolved initially in an organic solvent such as acetone or dimethyl sulfoxide before adding to the protein solution. A tenfold molar excess is generally sufficient to modify all accessible amino groups. The product of the reaction with an amino group is an amide and therefore resists mild acid hydrolysis.

Bis-succinimidyl esters of various chain lengths have been used in studies of the spatial arrangements of muscle proteins.[77] Theoretically any homologous series may be synthesized from the available dicarboxylic acids. Inclusion of a spin label in the bridge as in compounds **XXXIV** to **XXXIX** has enabled these reagents to be used for the study of membranes.[61-65] Deuterium and ^{15}N isotopes have been introduced into reagent **XXXIX** for studies of human erythrocytes.[65] Cleavable *bis*-succinimidyl cross-linkers have also been synthesized. Reagent **XL** with ester bonds are susceptible to cleavage by hydroxylamine. Compounds with sulfone linkage, **XLI** and **XLII,** are cleavable in base. Free mercaptans can be used to cleave disulfide bonds in compounds **XLIII** to **XLVI.** The vicinal hydroxyl groups in **XLVII** and **XLVIII** are cleavable by periodate. These compounds have been used to study prothrombin and thrombin,[78] thrombin receptor,[79] platelet proteins,[80] membrane proteins,[81] transhydrogenase,[82] P450 and P450 reductase,[83] ubiquinone-cytochrome c reductase,[84] vasoactive intestinal peptide receptor,[85] lymphotoxin receptors,[86] galanin receptor,[87] gonadotropin-releasing hormone,[88] F_1-adenosine triphosphatase,[70] virus induced antigen,[53] and viral proteins.[33,89] Other reagents have been used to investigate von Willebrand factor,[90] chorionic gonadotropin receptor,[91] epidermal growth factor-induced receptor[92,93] and asialoorosomucoid receptor.[94]

B. BIFUNCTIONAL ARYL HALIDES

Although bifunctional aryl halides, **XLIX** to **LIV,** react preferentially with amino and tyrosine phenolic groups, they also react with thiol and imidazolyl groups as mentioned in Chapter 2.[95-100] The electron withdrawing carboxyl and nitro groups on the benzene ring activate the halogen for nucleophilic substitution (see Chapter 3, Figure 3). However, rapid reaction requires relatively high pH values. Fluoro compounds are most reactive, followed by chloro-, and then bromo-derivatives. The reaction can be followed by the release of hydrogen halide using a pH stat, or by the characteristic visible and UV spectra of the product. Unfortunately, all the aryl halides are insoluble in water. They can be added to an aqueous protein solution after first dissolved in acetone, dioxane, alcohol, or other water miscible organic solvents. The lysine and tyrosine adducts are resistant to acid hydrolysis, but alkaline hydrolysis will liberate the reacted amino acids.[101,102] Of all the aryl halides, only the salicyl, **LII** and **LIII,** and the sulfone derivative, **LIV,** are cleavable by base or by Ni-catalytic reduction.[21,98-104]

The reactivities of the two halogens on the phenyl ring are different. The first halogen generally reacts faster than the second.[21] For example, the replacement of one of the fluorine atoms of DFDNB (**XLIX**) decreased the reactivity of the remaining fluorine atom.

A number of proteins have been cross-linked with the bifunctional aryl halides. Examples are collagen,[103] bovine serum albumin,[100,104] aspartate aminotransferase,[105] (Na,K)-ATPase,[106] thymidylate synthetase,[107] silk,[101] wool,[97] insulin,[108] ribonuclease,[109-111] erythrocyte membranes,[112] hemoglobin,[98,99,113] and retrovirus proteins.[89] These compounds have also been used to form covalently linked conjugates of several pairs of proteins, for example, IgG and ferritin,[114] IgG and horse radish peroxidase,[115] and coupling of antigens to erythrocytes.[116]

D. BIFUNCTIONAL ACYLATING AGENTS
1. Di-isocyanates and Di-isothiocyanates

Of the di-isocyanates and di-isothiocyanates cross-linkers, **LV** to **LXXV,** the majority are aromatic derivatives.[117-127] These reagents generally react with amino groups to form stable urea and thiourea derivatives, respectively (see Chapter 3, Figure 8). Their reactions with sulfhydryl, imidazolyl and phenolic groups give relatively unstable bonds which undergo spontaneous hydrolysis.[13,21] The isocyanates are not stable in aqueous solutions. Hydrolysis to form the amine is significant. The half-life of aliphatic isocyanates at pH 7.6 is less than 2 min.[21] Aromatic isocyanates are somewhat more reactive than the aliphatic ones. There is also a differential reactivity between the two functional groups in asymmetric compounds. In compound **LIX,** the isocyanate is more reactive than the isothiocyanate. Also in compound **LXIV,** the 4'-isocyanate is more reactive than the hindered 4-isocyanate group.[21] Most of the compounds are insoluble in aqueous solutions; only those with carboxyl and sulfonyl substitutions are water soluble. Compounds **LXVII** to **LXXI** containing the azo bond are reversible, being cleaved in the presence of dithionite. The stilbene derivative, **LXXIV,** is fluorescent and has many applications.[128-130] Its reduced analog, dihyrostibene (**LXXV**), however, is nonfluorescent.

These bifunctional di-isocyanate and di-isothiocyanate reagents having a wide variety of reactivity, size and solubility, have been used for different purposes. They have been used to study myoglobin,[131] chymotrypsin,[117,132] ribonuclease,[117] elastase,[132] collagen,[133] and bovine serum albumin.[134] They have been used to cross-link different protein molecules, for example, IgG and bovine serum albumin,[119] IgG and ferritin,[119] IgG and ovalbumin.[122]

2. Bifunctional Sulfonyl Halides

Aromatic sulfonyl chlorides, **LXXVI** to **LXXVIII,** react with amino groups to form sulfonamide derivatives with chloride as the leaving group (see Chapter 3, Figure 7).[134-136] The sulfonamide linkage can be cleaved with HBr in glacial acetic acid without breaking the peptide bonds.[21] These reagents are quite insoluble in water and hydrolyze rapidly in aqueous solutions. They have been used to cross-link lysozyme[134,135] insulin[136] and ovalbumin.[122]

3. *bis*-Nitrophenyl Esters

Compounds **LXXIX** to **LXXXII** are activated as dinitrophenyl esters that react with amino groups most rapidly, although their specificity is not very high.[137-140] The reaction involves nucleophilic attack at the ester carbonyl carbon displacing nitrophenol (Figure 2). The liberated nitrophenol has a distinct yellow color above pH 7 where the phenolic group is deprotonated. Absorbance at 410 nm may be used to quantitate the reaction. Other nucleophiles also react to form acylated derivatives. The most important competitive reaction in aqueous solutions is hydrolysis. These compounds can be easily synthesized by coupling nitrophenol to the corresponding dicarboxylic acids and, like the succinimidyl esters, a large number of diphenyl esters may be prepared.[141] Of particular interest is compound **LXXXII** synthesized by Busse and Carpenter.[140] This reagent is a derivative of methionine and therefore is cleavable by CNBr. It has been used to cross-link insulin. Other reagents have been used to cross-link silk fibroin, wool and collagen.[142]

4. Bifunctional Acylazides

Bifunctional acylazides, like other acylating reagents, readily react with amino groups to produce amide bonds. These compounds were originally designed for photoaffinity labeling but reacted readily in the dark as acylating agents because of the good leaving azido group. The tartryl diazides of different lengths, **LXXXIII** to **LXXXVIII,** were developed by Lutter et al.[29,144] They contain vicinal hydroxyl groups and thus can be cleaved by mild treatment with periodate. These compounds have been used to cross-link various ribosomal proteins[29] and erythrocyte proteins.[144]

The acyl azide derivatives of *p-bis*-(ureido)oligoprolylazobenzene, PAPA (**LXXXIX**), and trimesyl-*tris*-β-alanylazide, TTA (**XC**), are synthesized and purified as the hydrazide.

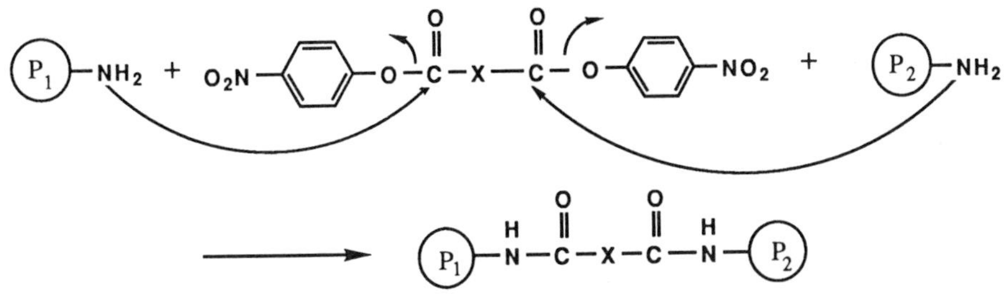

FIGURE 2.   Cross-linking reaction of *bis*-nitrophenyl esters.

Prior to protein cross-linking, the hydrazide is converted to the acylazide by sodium nitrite and 1 *N* HCl.[143] These compounds have been tested for their cross-linking ability on hemoglobin. PAPA has also been shown to cross-link ferritin subunits. TTA is a trifunctional reagent and has been used to link three-stranded collagen model oligopeptides.

E. DIALDEHYDES

Several simple dialdehydes such as glyoxal, malondialdehyde, succinialdehyde, adipaldehyde, glutaraldehyde and phthalaldehyde (**XCI** to **XCIX**) have been shown to form protein cross-links.[145-156] The most extensively used reagent is glutaraldehyde. Conceivably the reaction should proceed through a Schiff base. However, the product formed is irreversible. Glutaraldehyde has been found to form polymers in solution. At acidic pH, the polymers are cyclic hemiacetal. At neutral or slightly alkaline pHs at which cross-linking is carried out, the α,β-unsaturated aldehyde polymers are formed which increases in length as pH is raised.[154,155] Presumably it is the unsaturated polymer that cross-links the amino groups of proteins as illustrated in Figure 3.[147,156] The interaction of the Schiff base with adjacent double bonds provide the stability towards hydrolysis. With excess amino groups, nucleophilic addition of the ethylenic double bond is possible.[1]

Because of its different polymeric forms, the distance between two cross-linked groups cannot be estimated. Glutaraldehyde has been used extensively to cross-link and stabilize protein crystals[156-159] as well as proteins in solution.[160,161] Hemoglobin,[162-164] phosphofructokinase,[165] lactate dehydrogenase,[166] and insulin[167] have been studied with glutaraldehyde. Glutaraldehyde has also been used to form enzyme-immunoglobulin conjugates.[168,169]

Other glutaraldehyde derivatives have also been found to cross-link proteins. The two glutaraldehyde derivatives, 3-methylglutaraldehyde (**XCVII**) and 3-methoxy-2,4-dimethylglutaraldehyde (**XCVIII**) have been used in tanning and to cross-link albumin and casein.[149-151] *o*-Phthaladehyde (**XCIX**), a fluorogenic aromatic dialdehyde, has been used to react with primary amines for their detection.[152] Pyridoxal phosphate which has been used as an monofunctional reagent to modify lysine residues at the phosphate binding sites has been synthesized into various dimeric forms (**C** to **CIV**). These compounds contain the pyrophosphate bond which may be hydrolyzed in acid or base. They have been used to cross-link glycogen phosphorylase b and hemoglobin.[170,171]

Formaldehyde (**CV**), strictly not a dialdehyde but capable of reacting bifunctionally, is the simplest of all cross-linking reagents.[172-176] In concentrated aqueous solution (formalin), it exists as a series of low-molecular-weight polymers.[22] In dilute solutions, it reverts to the monomeric form.[174] Cross-linking reaction involves the attack of an amino group to form a quaternary ammonium salt, which loses a molecule of water to produce an immonium cation (Figure 4). This strongly electrophilic cation then reacts with a number of nucleophiles in the protein producing a methylene-bridged cross-link.[175] In addition to amines, it reacts with sulfhydryl, phenolic, imidazolyl, indolyl and guanidinyl groups. Methylene bridges between lysine-tyrosine have been isolated from formaldehyde-treated tetanus and diphtheria toxins.[176]

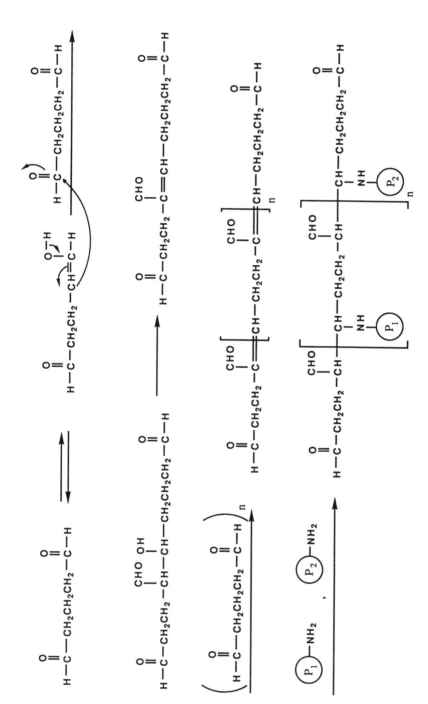

FIGURE 3.   Polymerization of glutaraldehyde and its cross-linking of proteins.

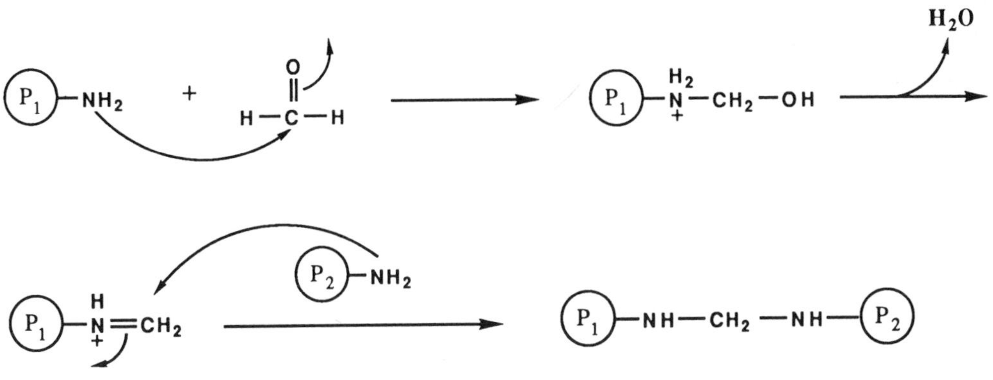

FIGURE 4.   Mechanism of protein cross-linking by formaldehyde.

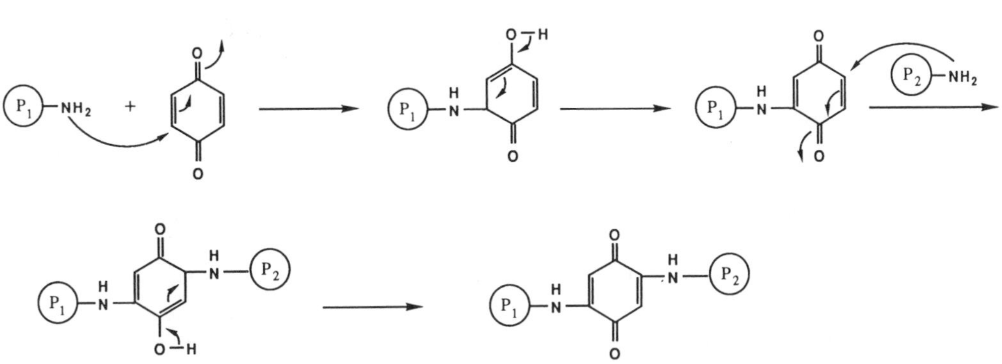

FIGURE 5.   Cross-linking of proteins with benzoquinone.

F. DIKETONES

Ketones, like aldehydes, are capable of forming Schiff bases with amines. 2,5-Hexanedione, **CVI**, for example, can react with the amino group of lysine and cyclize to form a pyrrole which subsequently leads to cross-linking of proteins.[177] Another derivative, 3,4-dimethyl-2,5-hexanedione (**CVII**), behaves similarly.[178]

G. OTHER AMINO GROUP REACTING CROSS-LINKING REAGENTS

There are several other reagents that do not readily fit into the above categories of amino group directed cross-linkers but have been implicated to react with amino groups. Among these, benzoquinones such as *p*-benzoquinone (**CVIII**) have been used in protein-protein and enzyme antibody coupling.[179,180] The mechanism of the reaction follows a two-step pathway as shown in Figure 5. First the nucleophile is added to the benzoquinone to form an adduct which undergoes oxidation to generate substituted benzoquinone. A second nucleophile is added in the subsequent step to yield a cross-linked product.

2-Iminothiolane, **CIX,** which has been discussed in Chapter 2 as a thiolating agent has also been used as a cross-linker.[181] In this context, the first step in the cross-linking process is amidination of amino groups of proteins to introduce free thiols. The second step consists of the formation of disulfide bonds between the incorporated thiols in the presence of an oxidizing agent such as hydrogen peroxide. In this manner, the monofunctional reagent has formed cross-links between progesterone-receptor subunits,[182] ribosomal proteins,[183,184] and ribosome proteins and 23S RNA.[185]

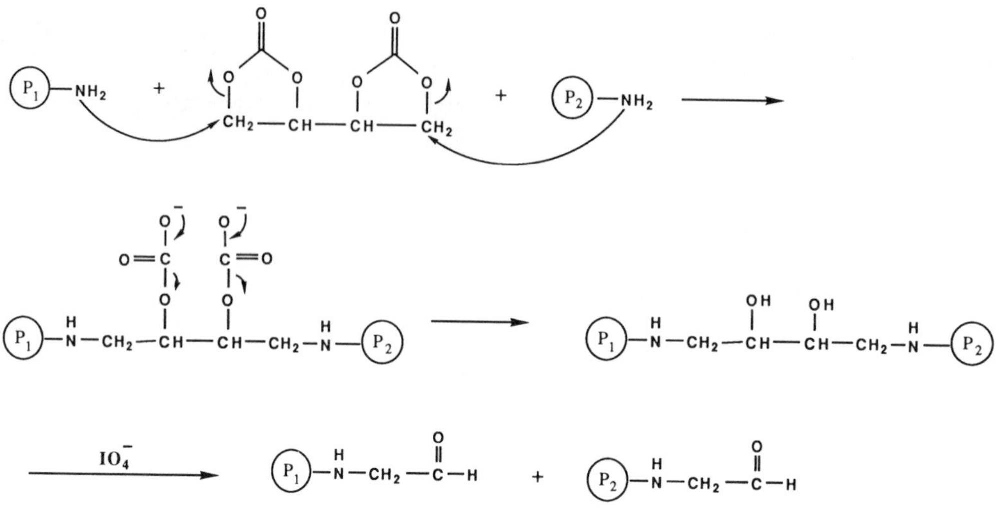

FIGURE 6.   Reaction mechanism of erythritolbiscarbonate.

Erythritolbiscarbonate, **CX,** is a bifunctional reagent that reacts with amino groups at the pH ranges 7.5 to 9.5 to yield *bis*-carbonate. After decarboxylation to form vicinal diols, the cross-linked product may be cleaved by periodate.[72,186] The reaction is shown in Figure 6.

Mucobromic and mucochloric acids, **CXI** and **CXII,** have been shown to cross-link gelatin.[187] These compounds contain a carboxyl and an aldehyde group which can exist in a cyclic lactone. In this form, amino groups may attack the carbonyl carbon to form an amide bond. Schiff base formation of the aldehyde with another amino group completes the cross-linking reaction. Michael addition of a nucleophile to the double bond is another possible mechanism for the cross-linking reaction (Figure 7).

Ethylchloroformate, **CXIII,** and *p*-nitrophenylchloroformate, **CXIV,** are another pair of similar reagents that have been found to cross-link proteins.[188] These compounds join two amino groups through a carbonyl group. The speculated reaction is shown in Figure 8. It is possible that these reagents may function as a zero-length cross-linker which will be discussed in Chapter 6.

Multidiazonium compounds derived from poly-condensates of 4,4'-diamino-diphenyl-methane-3,3'-dicarboxylic acid and mixed poly-condensates of 4,4'-diaminodiphenylme-thane-3,3'-dicarboxylic acid, and 4-nitro-2-aminobenzoic acid are found to cross-link bovine serum albumin, ribonuclease, kallikrein inactivator and human chorionic gonadotropin.[347] Various other compounds of the intermediate metabolites are also potential cross-linkers. Homogentisic acid, for example, has been shown to cross-link collagen preparations.[348]

III. SULFHYDRYL GROUP DIRECTED CROSS-LINKERS

The sulfhydryl cross-linking reagents consist of derivatives of mercury, maleimide, disulfide, halomethylketone, and other alkylating agents as listed in Table 2. Of these reagents, only the mercurial derivatives, disulfide- and disulfide-forming compounds are thiol specific. *N*-maleimido derivatives, although considered as sulfhydryl group specific, may react with other nucleophiles. This is also true for the other alkylating agents. In the absence of a free sulfhydryl group, other amino acid nucleophiles, such as the amino group, become important and would be the subject of cross-linking. Since the thiol is the best nucleophile at neutral pH, all the alkylating agents are listed under Table 2 as thiol directing.

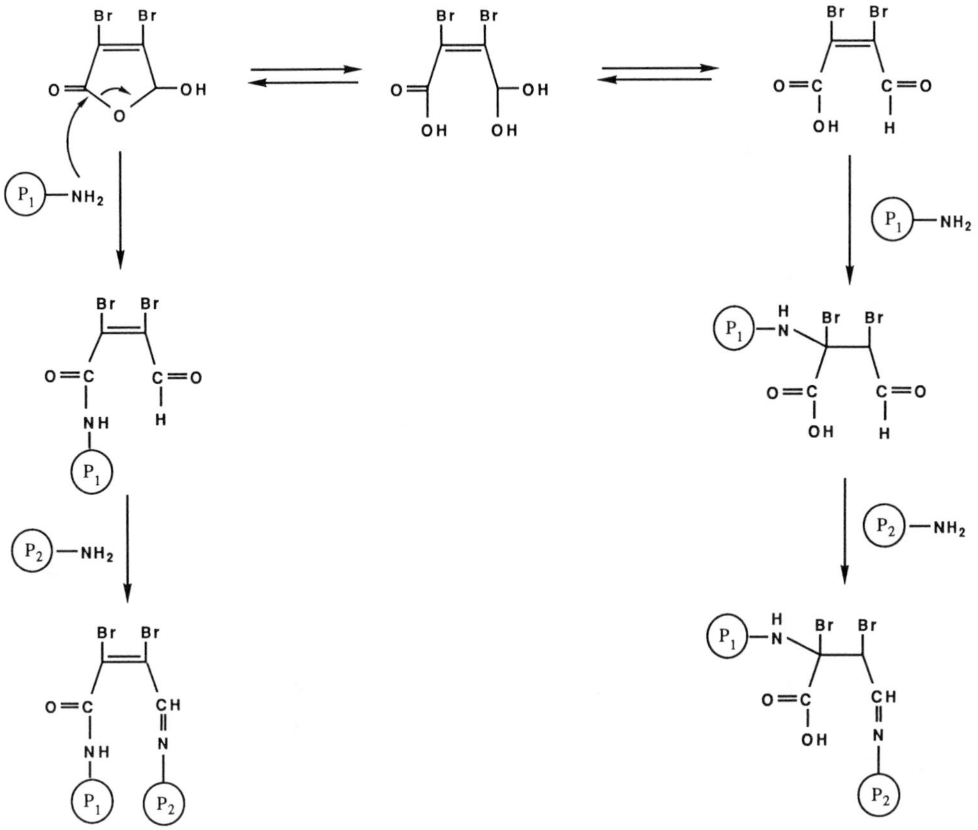

FIGURE 7. Possible cross-linking reaction of mucobromic acid.

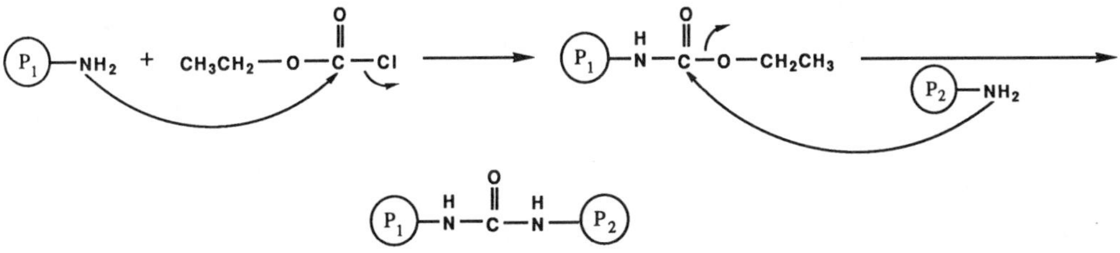

FIGURE 8. Speculation on the cross-linking reaction of ethylchloroformate.

A. MERCURIAL REAGENTS

Mercuric ion, **CXV**, reacts reversibly with sulfhydryl groups.[189-192] The first thiol group reacts very fast followed by a slower reaction with the second. Cross-linking occurs when the ratio Hg^{2+}/-SH is 0.5 or less. Intramolecular cross-linking has been prepared with reduced papain[191] and pancreatic ribonuclease.[192] Intermolecular cross-links have been shown with mercaptalbumin[189] and *E. coli* DNA polymerase.[190]

3,6-*bis*-(Mercurimethyl)dioxane with counter ions of acetate (**CXVI**), chloride (**CXVII**), or nitrate (**CXVIII**), can be prepared from allyl alcohol and the corresponding mercury (II) salt.[193] These compounds react similarly to but faster than the mercuric ion. They also introduce a linkage distance of 15 Å vs. about 5 Å for mercuric ion. Inter- and intramolecular cross-links between bovine serum albumin,[193-195] ovalbumin[196] and ribonuclease[196,197] have been introduced with these compounds.

Another mercurial derivative, 1,4-*bis*(bromomercuri)butane (**CXIX**), was synthesized by Vas and Csanády[198] by reacting $HgBr_2$ and 1,4-dibromobutane using the appropriate

Table 2

SULFHYDRYL GROUP DIRECTED HOMOBIFUNCTIONAL CROSS-LINKERS

Name (Abbreviation)	Structure	Cleavability/ Condition	Reference
A. MERCURICAL REAGENT			
CXV. Mercuric ion	Hg^{++}	+/thiol	189-192
3,6-Bis(mercurimethyl) dioxane derivatives		+/thiol	193-197
CXVI. X= CH_3-COO^- : 3,6-Bis(acetoxymercurimethyl)dioxane			
CXVII. X= Cl^- : 3,6-Bis(chloromercurimethyl)dioxane			
CXVIII. X= NO_2^- : 3,6-Bis(nitromercurimethyl)dioxane			
CXIX. 1,4-Bis(bromomercuri)butane	$Br\ \overset{-}{Hg}-(CH_2)_4-\overset{+}{Hg}\ Br$	+/thiol	193-197
B. DISULFIDE FORMING REAGENTS			
CXX. 3-Oxy-2,2-bis[[((2-((3-carboxy-4-nitrophenyl)dithio)ethyl)-amino)carbonyl]hexanyl]-4,4-dimethyloxazolidine		+/thiol	61, 62

Polymethylenebis(methanethiosulfonate)

$$H_3C-\overset{\overset{O}{\|}}{\underset{\underset{O}{\|}}{S}}-S-(CH_2)_n-S-\overset{\overset{O}{\|}}{\underset{\underset{O}{\|}}{S}}-CH_3$$

CXXI. n=5 : 5,5'-Pentamethylenebis(methanethiosulfonate)	+/thiol	200
CXXII. n=6 : 6,6'-Hexamethylenebis(methanethiosulfonate)	+/thiol	200
CXXIII. n=8 : 8,8'-Octamethylenebis(methanethiosulfonate)	+/thiol	200, 201
CXXIV. n=10 : 10,10'-Decamethylenebis(methanethiosulfonate)	+/thiol	200
CXXV. n=12 : 12,12'-Dodecamethylenebis(methanethiosulfonate)	+/thiol	200
CXXVI. Crabescein	+/thiol	252

C. BISMALEIMIDES

CXXVII. N,N'-Methylenebismaleimide	—	203

Table 2 (continued)

SULFHYDRYL GROUP DIRECTED HOMOBIFUNCTIONAL CROSS-LINKERS

Name (Abbreviation)	Structure	Cleavability/Condition	Reference
C. BISMALEIMIDES			
CXXVIII. N,N'-Trimethylenebismaleimide		—	204-206
CXXIX. N,N'-Hexamethylenebismaleimide; Bis(N-maleimido)-1,6-hexane (BMH)		—	206, 207
CXXX. N,N'-Octamethylenebismaleimide; Bis(N-maleimido)-1,8-octane (BMO)		—	208
CXXXI. N,N'-Dodecamethylenebismaleimide; Bis(N-maleimido)-1,12-dodecane (BMD)		—	208
CXXXII. Bis(N-maleimidomethyl)ether		—	203, 209 210

204

204, 211

204, 206

208

212, 213

205

—

—

—

—

—

+/dithionite

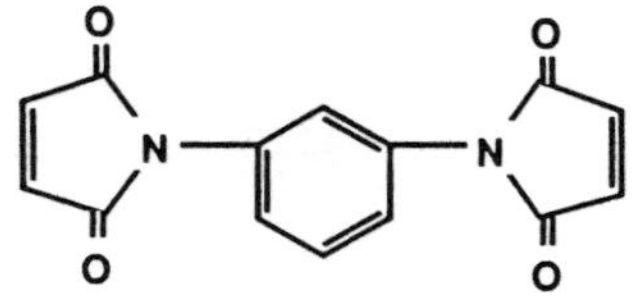

CXXXIII. N,N'-(1,3-Phenylene)bismaleimide

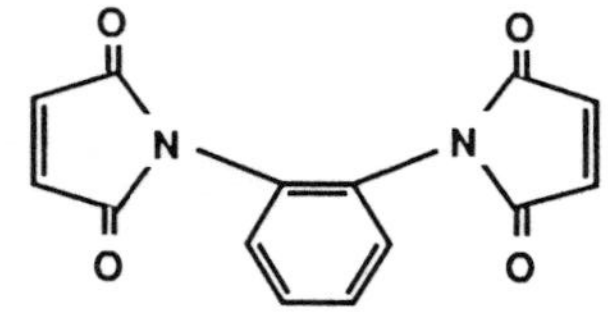

CXXXIV. N,N'-(1,2-Phenylene)-bismaleimide

CXXXV. N,N'-(1,4-Phenylene)-bismaleimide

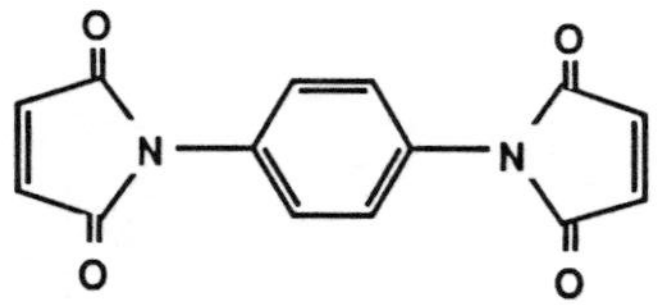

CXXXVI. Bis(N-maleimido)-4,4'-bibenzyl (BMB)

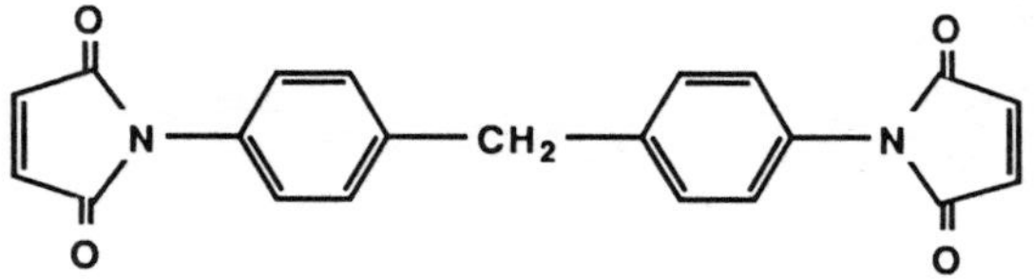

CXXXVII. Naphthalene-1,5-dimaleimide (NDM)

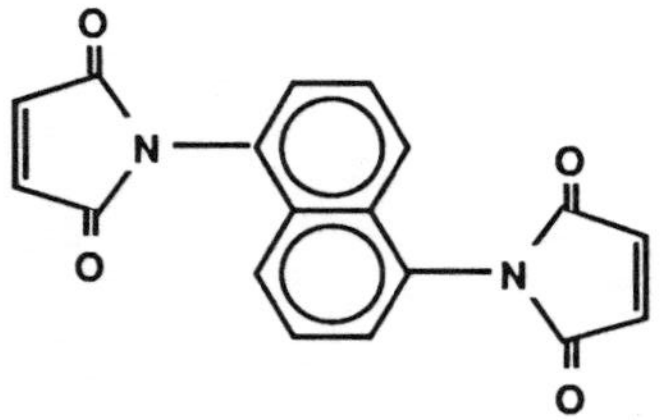

CXXXVIII. Azophenyldimaleimide

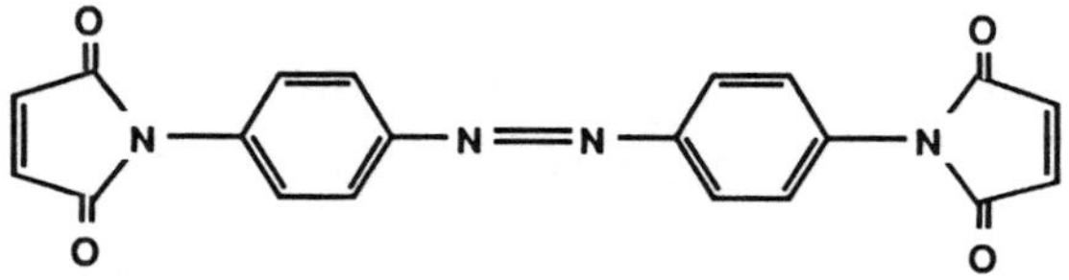

Table 2 (continued)

SULFHYDRYL GROUP DIRECTED HOMOBIFUNCTIONAL CROSS-LINKERS

Name (Abbreviation)	Structure	Cleavability/ Condition	Reference

C. **BISMALEIMIDES**

CXXXIX. 4,4'-Dimaleimidostilbene	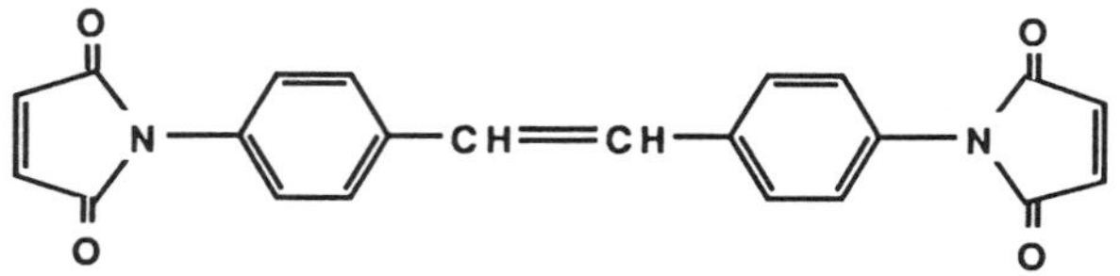	—	214
CXL. 4,4'-Dimaleimidylstilbene-2,2'-disulfonic acid (DMSDS)	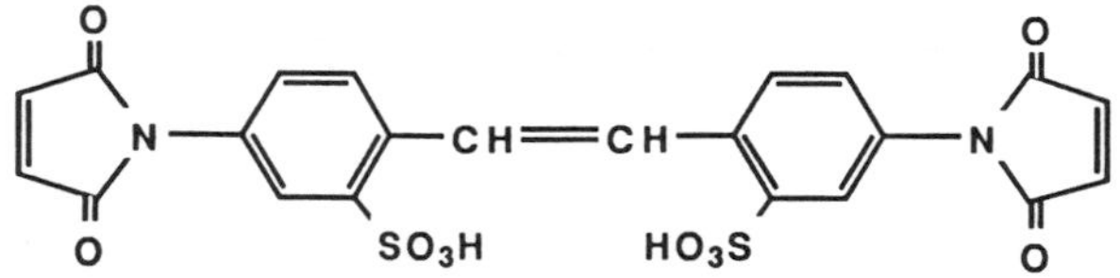	—	214
CXLI. Maleimidomethyl-3-maleimido-propionate (MMP)	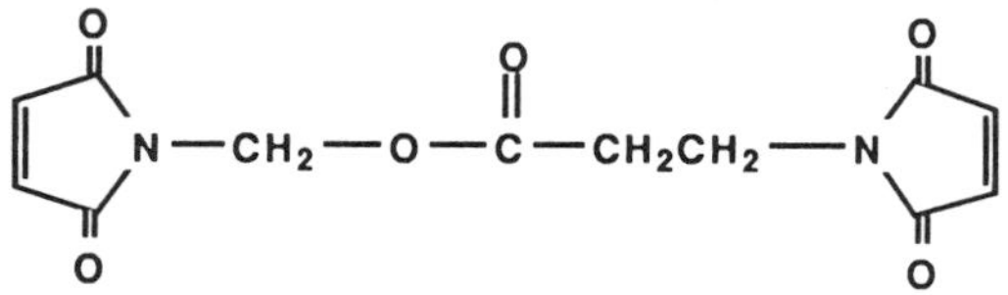	+/base	215
CXLII. 2,2-Bis(maleimidoethoxy)-propane	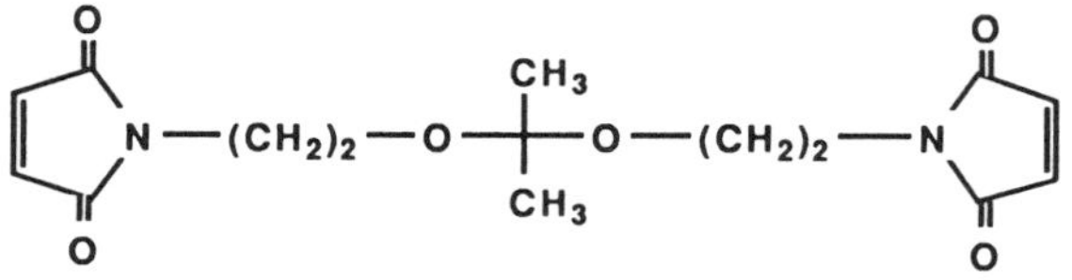	+/H$^+$	216

CXLIII. 2,2-Bis(maleimidomethoxy)-propane   +/H$^+$ 216

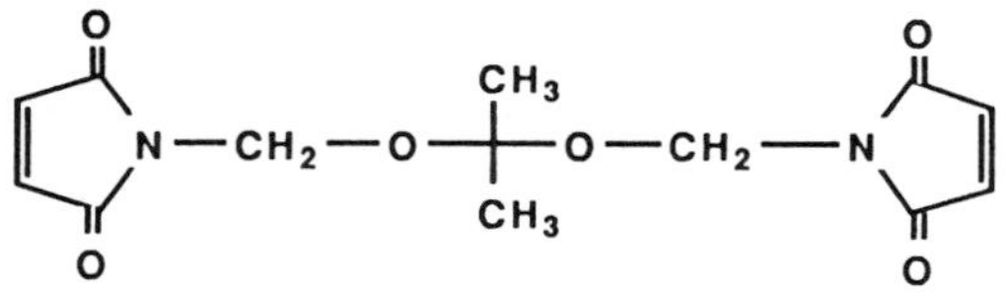

CXLIV. 1,1'-[[(3,9-Diethyl-2,4,8,10-tetraoxaspiro[5.5]undecane-3,9-diyl}bis(oxymethylene)]-bis-1H-pyrrole-2,5-dione   +/H$^+$ 216

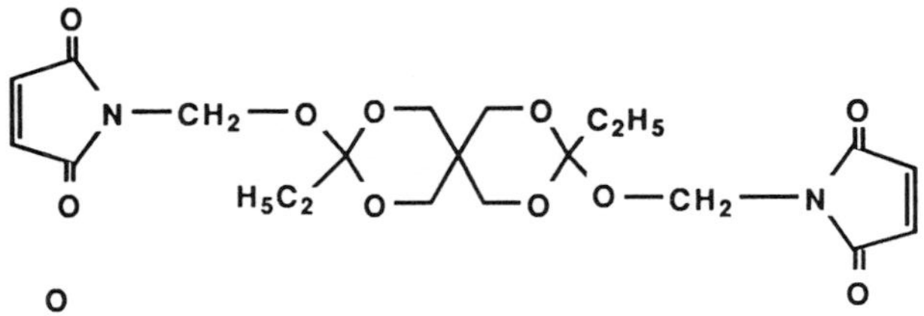

CXLV. 1,1'-[[(3,9-Diethyl-2,4,8,10-tetraoxaspiro[5.5]undecane-3,9-diyl}bis(oxy-2,1-ethanediyl)]-bis-1H-pyrrole-2,5-dione   +/H$^+$ 216

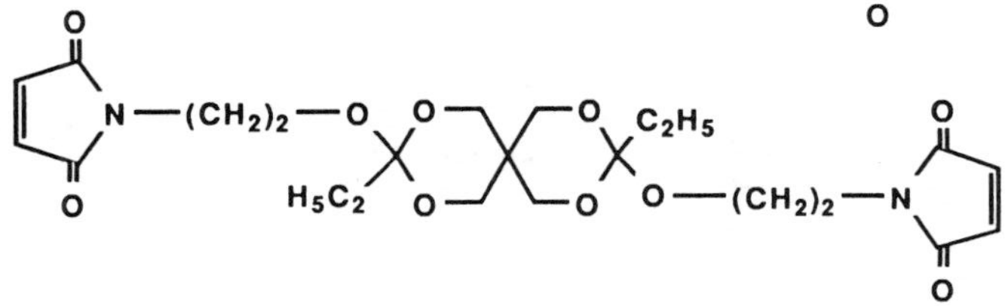

D. ALKYLATING AGENTS

1. Bio-haloacetyl derivatives

CXLVI. 1,3-Dibromoacetone   –   225

N,N'-Bis(iodoacetyl)poly-methylenediamine   –   226

CXLVII. n=2 : N,N'-Bis(iodoacetyl)ethylenediamine; N,N'-ethylene-bis(iodoacetamide)

CXLVIII. n=6 : N,N'-Bis(iodoacetyl)hexamethylenediamine; N,N'-hexamethylene-bis(iodoacetamide)

CXLV. n=11 : N,N'-Bis(iodoacetyl)-undecamethylene-diamine; N,N'-undecamethylene-bis(iodoacetamide)

Table 2 (continued)

SULFHYDRYL GROUP DIRECTED HOMOBIFUNCTIONAL CROSS-LINKERS

Name (Abbreviation)	Structure	Cleavability/ Condition	Reference
D. ALKYLATING AGENTS			
CL. N,N′-Di(bromoacetyl)phenyl-hydrazine		—	227
CLI. 1,2-Di(bromoacetyl)amino-3-phenylhydrazine		—	227
CLII. γ-(2,4-Dinitrophenyl)-α–bromoacetyl-L-diaminobutyric acid bromoacetyl-hydrazide(DIBAB)		—	228

229 —

CLIII. Bis-[α-bromoacetyl-ε-(2,4-dinitrophenyl)-lysylprolyl]-ethylenediamine

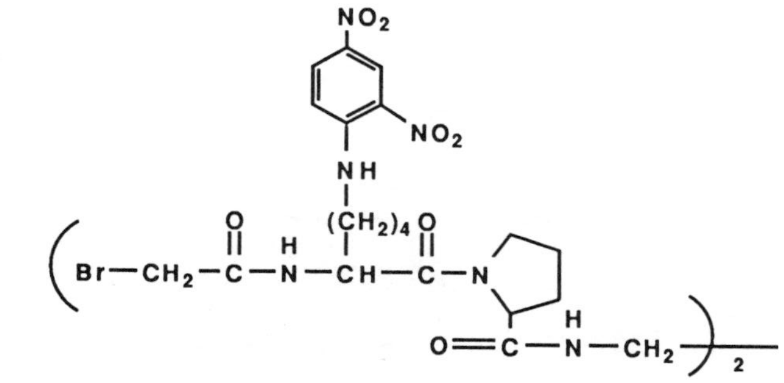

232 +/dithionite

CLIV. 2,2'-Dicarboxy-4,4'-diiodo-acetamidoazobenzene

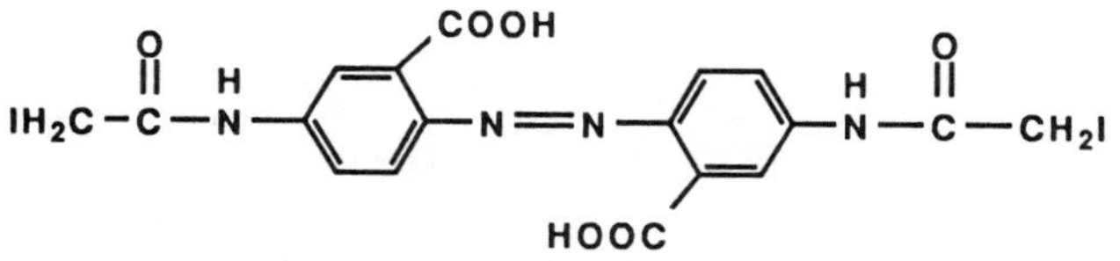

232 +/dithionite

CLV. 2,2'-Dicarboxy-4,4'-dibromo-acetamidoazobenzene

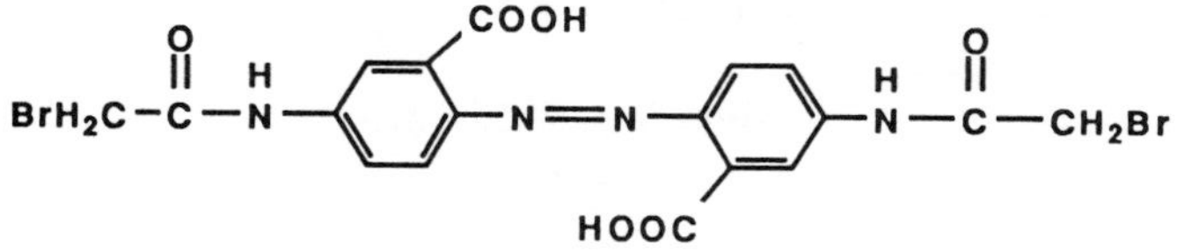

230 +/thiol

CLVI. N,N'-Bis(α-iodoacetyl)-2,2'-dithiobisethylamine (DIDBE)

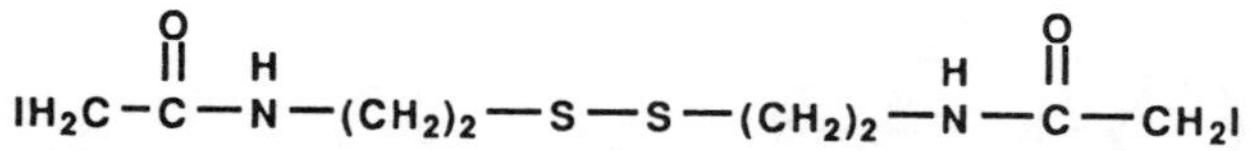

231 —

CLVII. 4,5'-Di[[(iodoacetyl)-amino]-methyl]fluorescein

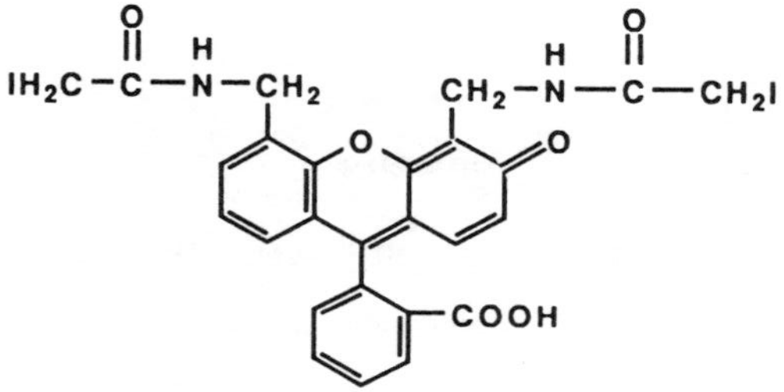

Table 2 (continued)

SULFHYDRYL GROUP DIRECTED HOMOBIFUNCTIONAL CROSS-LINKERS

Name (Abbreviation)	Structure	Cleavability/ Condition	Reference
D. ALKYLATING AGENTS			
CLVIII. p-Bis(ureido)-(1-iodoacet-amido-2-ethylamino)oligo-prolylazobenzene		+/dithionite	28, 143
2. Di-alkyl halides			
CLIX. α,α'-Dibromo-p-xylene sulfonic acid		—	233
CLX. α,α'-Diiodo-p-xylene sulfonic acid		—	233
CLXI. Di(2-chloroethyl)sulfide	$Cl-CH_2CH_2-S-CH_2CH_2-Cl$	—	235, 238
CLXII. Di(2-chloroethyl)sulfone		+/dithionate	241
CLXIII. Di(2-chloroethyl)methylamine		—	236

CLXIV. Tri(2-chloroethyl)amine (TCEA)

$N(CH_2CH_2Cl)_3$

— 242

CLXV. N,N-Bis(β-bromoethyl)benzyl-amine

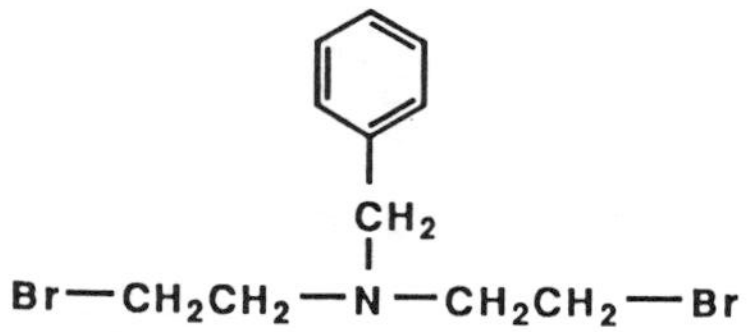

$Br-CH_2CH_2-N-CH_2CH_2-Br$ (with benzyl CH_2-phenyl group)

— 227

CLXVI. Di(2-chloroethyl)-p-methoxy-phenylamine

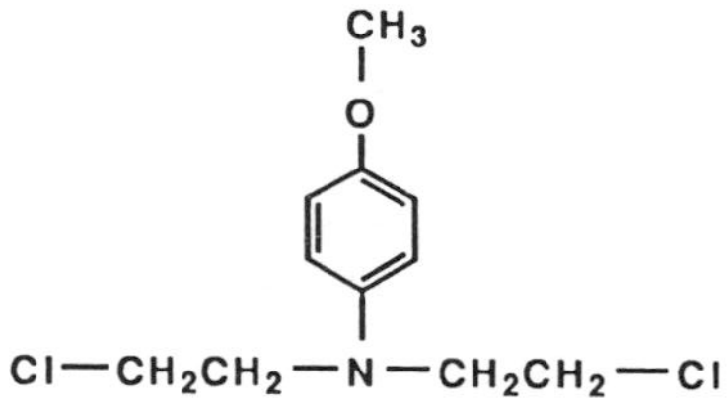

CH_3-O-(phenyl)$-N-CH_2CH_2-Cl$, $Cl-CH_2CH_2-$

— 72

Nitrosoureas

$$R-N-C-N-R$$
with N–O and C=O

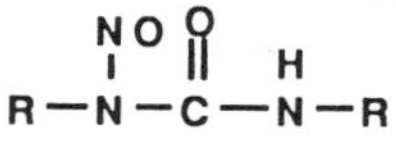

CLXVII. R= CH_2CH_2Cl : 1,3-Bis(2-chloroethyl)-1-nitrosourea

— 243

CLXVIII. R= (cyclohexyl)–OH : 1,3-Bis(trans-4-hydroxyhexyl)-1-nitrosourea

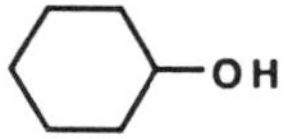

— 246

3. *s-Triazines*

CLXIX. 2,4-Dichloro-6-methoxy-s-triazine

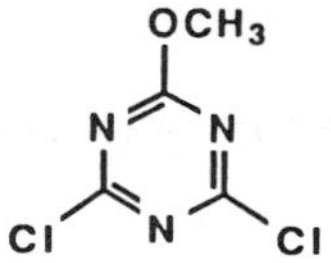

(triazine ring with OCH_3, Cl, Cl)

— 247

Table 2 (continued)

SULFHYDRYL GROUP DIRECTED HOMOBIFUNCTIONAL CROSS-LINKERS

Name (Abbreviation)	Structure	Cleavability/ Condition	Reference
D. ALKYLATING AGENTS			
CLXX. 2,4,6-Trichloro-s-triazine; Cyanuric chloride		—	250
CLXXI. 2,4-Dichloro-6-(3'-methyl-4-aminoanilino)-s-triazine		—	133, 249
CLXXII. 2,4-Dichloro-6-amino-s-triazine		—	133, 249
CLXXIII. 2,4-Dichloro-6-(sulfonic-acid)-s-triazine; 2,4-Dichloro-s-triazin-2-yl-6-sulfonic acid		—	133, 249

123, 249

251-253

254

—

—

—

CLXXIV. 2,4-Dichloro-6-(5'-sulfonic-acid-naphthaleneamino)-s-triazine; 5-[(4,6-dichloro-s-triazin-2-yl)amino]naphthalene-1-sulfonic acid

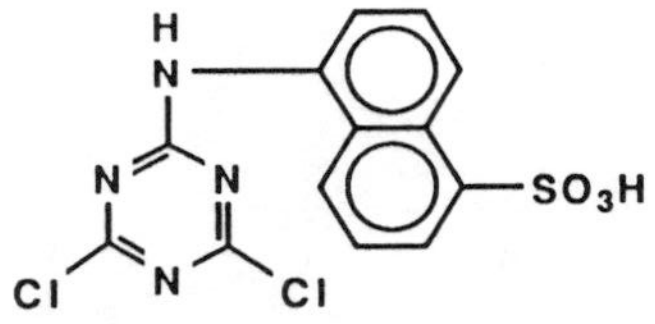

CLXXV. 5-[(4,6-Dichloro-s-triazin-2-yl)amino]fluorescein (5-DTAF)

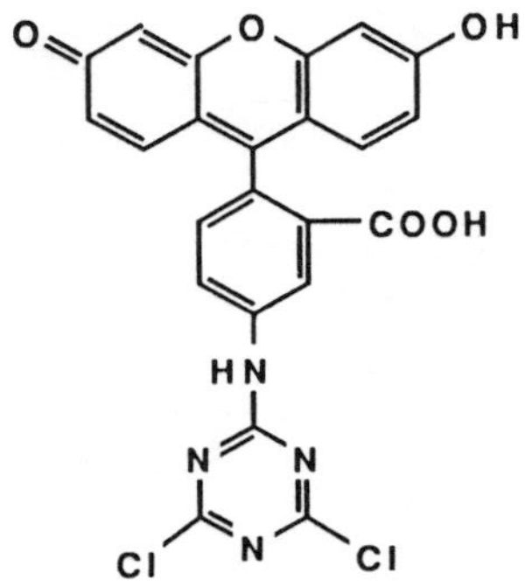

CLXXVI. 5-(4,6-Dichloro-s-trizin-2-yl)-aminofluorescein diacetate

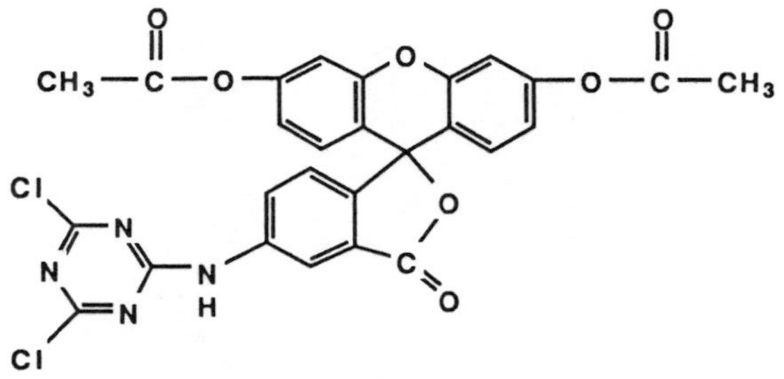

Table 2 (continued)

SULFHYDRYL GROUP DIRECTED HOMOBIFUNCTIONAL CROSS-LINKERS

Name (Abbreviation)	Structure	Cleavability/ Condition	Reference
D. ALKYLATING AGENTS			
4. Aziridines			
CLXXVII. 2,4,6-Tri(ethyleneimino)-s-triazine		—	234, 235
CLXXVIII. N,N'-Ethyleneiminoyl-1,6-diaminohexane		—	234, 235
CLXXIX. Tri{1-(2-methylaziridenyl)}-phosphine oxide		—	234, 235 255, 256

5. Bis-epoxides

CLXXX. 1,2:3,4-Diepoxybutane $+/IO_4^-$ 257, 258

CLXXXI. 1,2:5,6-Diepoxyhexane — 259

CLXXXII. Bis(2,3-epoxypropyl)ether — 258

CLXXXIII. 1,4-Butadioldiglycidoxy-ether — 260

CLXXXIV. 3,4-Isopropylidene-1,2:5,6-dianhydromannitol — 254

6. Other sulfhydryl reacting cross-linkers

CLXXXV. Divinyl sulfone +/dithionate 261, 262

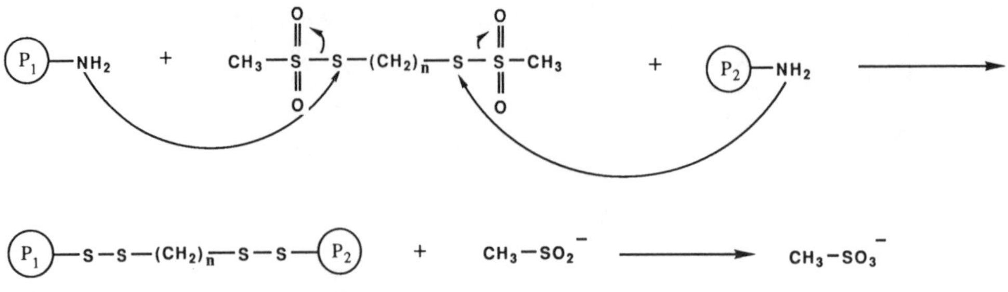

FIGURE 9.   Cross-linking reaction of polymethylenebis(methanethiosulfonate).

Grignard reagent. The reagent was found to inhibit 3-phosphoglycerate kinase cross-linking two fast-reacting thiol groups.

B. DISULFIDE FORMING REAGENTS

Disulfide-thiol interchange is the basis of cross-linking caused by the spin labeled compound **CXX**.[61,62] The reaction is easily reversible in the presence of excess free mercaptans. Radioactive analogs labeled with ^{35}S and ^{14}C have also been synthesized.[62] These compounds have been used to cross-link hemoglobin[62] and incorporate spin-labels into membranes similar to spin-labels **XXXIV** to **XXXIX**.

Another series of compounds that cross-link thiol groups by the formation of disulfide bonds are disulfide dioxide derivatives.[199] The polymethylene *bis*-methane thiosulfonate reagents, **CXXI** to **CXXV**, have been developed by Bloxham and Sharma for studying pairs of thiol groups in lactate dehydrogenase, pyruvate kinase, phosphofructokinase and glyceraldehyde-3-phosphate dehydrogenase.[200,201] These series of compounds are synthesized by reacting dibromoalkanes and potassium methane thiosulfonate in refluxing methanol. Radioactive analogs have also been synthesized. The reaction with thiols involves disulfide formation and elimination of methanesulfinate which is subsequently oxidized to methane sulfonic acid (Figure 9). Although cleavable by thiols, the cross-linked peptides are stable to BrCN and tryptic digestion. This allows the identification of the modified residue with routine procedures.

Crabescein, **CXXVI**, is a special compound in this category. It is a fluorescent derivative of fluorescein containing two free sulfhydryls which have been shown to add across disulfide bonds of reduced antibody.[252] It was used to measure the rotational correlation time of IgG.

With the introduction of extrinsic sulfhydryl groups into protein molecules as discussed in Chapter 2, the scope of disulfide forming cross-linking reagents can be further broadened. For example, intrinsic protein sulfhydryl and extrinsic thiol may be cross-linked.[1]

C. BISMALEIMIDES

The most commonly used sulfhydryl reagents are probably the *N*-substituted bismaleimide derivatives.[22,202-227] They are generally synthesized from the corresponding amine with maleic anhydride and either acetic anhydride[217] or dicyclohexyl carbodiimide.[218] The cross-linking reagents listed in Table 2, **CXXVII** to **CXLV**, react readily at pH 7 to 8 with thiol groups to form sulfides through Michael addition. They also react at a much slower rate with amino and imidazolyl groups. At pH 7, for example, the reaction with simple thiols is about 1,000-fold faster than the corresponding amines.[219] The characteristic absorbance change in the 300 nm region associated with the reaction provides a convenient method for monitoring the reaction.[220] Since these reagents are generally insoluble in water, they can be added to the aqueous protein solution as solids or after dissolving in water-miscible organic solvents. These compounds are stable at low pHs but are susceptible to hydrolysis at high pHs. Maleimido groups attached to benzene rings are also labile at neutral pH.[221,222]

The maleimide cross-linkers have been used for cross-linking various proteins, such as wool,[204,207] myosin,[214] bovine serum albumin,[204,206,207] hemoglobin,[203] tryptophan synthetase[208,210] and membrane components.[215]

Few compounds of this class are reversible. The azo bond of **CXXXVIII** can be cleaved by dithionite and the ester bond of **CXLI** is hydrolyzed by base.[215] The acetal, ketal, and ortho ester bonds of **CXLII** to **CXLV** are susceptible to hydrolysis at acidic pHs.[216]

Compounds **CXXXVII, CXXXIX** and **CXL** are fluorogenic. Naphthalene dimaleimide, **CXXXVII**, is fluorescent and has been used to study myosin subfragment 1.[212,223,224] The stilbene compounds, **CXXXIX** and **CXL,** have a low fluorescence until both maleimides have reacted. Maximum absorbance is at about 316 nm. After reaction with thiol, the emission maximum occurs at about 385 to 430 nm.[128-130]

D. BIFUNCTIONAL ALKYLATING AGENTS

The bifunctional alkylating agents are attacked by any of the nucleophiles mentioned earlier in Chapter 2. Since the thiol is the most potent nucleophile at neutral pH, these reagents are listed in Table 2 as sulfhydryl selective reagents. It should be borne in mind that they are not thiol specific and amino groups may be the target of modification, particularly at higher pHs.

1. *bis*-Haloacetyl Derivatives

The haloacetyl containing cross-linkers, **CXLVI** to **CLVIII**, react primarily with sulfhydryl, imidazolyl and amino groups.[225-232] With ficin and stem-bromelain, for example, 1,3-dibromoacetone (**CXLVI**) cross-links histidine and cysteine residues at the active site.[225] At neutral pH, the reagents can be considered as thiol specific. At higher pHs, increased reactions with amino groups become prominent. The reaction with proteins can be readily followed by the production of halogen acids in a pH-stat. Compounds **CLIV** and **CLV** also have maximum absorbance at 370 nm which can be used to monitor the reaction.[21] They are also reversible. Quantitative cleavage can be achieved by reduction of the azo group with dithionite.[232] Similarly compound **CLVI** is cleavable through thio-disulfide exchange with mercaptans.[230] The fluorescein derivative, **CLVII**, is fluorescent with emission at about 552 nm when excited at 496 nm and may be used to follow the extent of modification and structural studies.

Various proteins have been cross-linked by these *bis*-haloacetylketones for structural studies. These include *H*-meromyosin,[21,232] chymotrypsin,[227] aldolase,[226,230] IgG,[229] and tubulin.[230]

2. *bis*-Alkyl Halides

Benzyl halides, **CLIX** and **CLX,** are activated by the benzene ring through resonance and react similarly as haloacetyl ketones.[233] They have been used to intramolecularly cross-link lysozyme.

The halogen atoms beta to sulfur and nitrogen as in *S*- and *N*-mustards, respectively, are readily replaced by nucleophiles. Alkylation of proteins is favored at neutral pH, although reactivity increases with increasing pH.[234-238] Only one of the compounds in this series is cleavable. The sulfone derivative, **CLXII,** may be hydrolyzed under basic conditions. These nitrogen and sulfur mustards have been used to cross-link collagen,[239] bacteriophage,[240] wool fibers,[72] hemoglobin,[234-236] serum albumin,[234-236,240] keratine, insulin, gelatin, pepsin, egg albumin, hexokinase, protein components of tobacco mosaic virus, chymotrypsinogen,[234,235] chymotrypsin,[227] serum globulin, fibrinogen,[234,235,241] and ovalbumin.[236]

It may be of interest to note that the nitrogen mustard, TCEA (**CLXIV**), is a trifunctional cross-linking reagent with a span of 5 Å. Its potential to cross-link three nucleophiles has been applied to investigate myosin ATPase.[242]

Within this category of compounds are the derivatives of nitrosourea, for example 1,3-*bis*(2-chloroethyl)-1-nitrosourea, **CLXVII**.[243] These compounds have been used as tumor therapeutic agents and have been found to cross-link proteins and nucleic acids.[243-245] Although the exact mechanism of cross-linking is not understood, it has been postulated that these compounds decompose to alkylating carbonium ions as well as organic isocyanates under physiological conditions.[244] Thus, derivatives different from haloalkyls such as **CLXVIII** are effective cross-linkers.[246] However, these compounds have not been used *in vitro* as bifunctional reagents.

3. *s*-Triazines

The chlorine atoms of the *s*-triazine series, **CLXIX** to **CLXXVI**, are very reactive towards nucleophiles,[72] including hydroxyl groups of carbohydrates.[133,247-249] Because of their reactivity, they are rapidly hydrolyzed in aqueous solutions resulting in poor cross-linking. The rates of reaction of the chlorine atoms of cyanuric chloride, **CLXX**, are markedly different. Displacement of the first chlorine proceeds very fast even in the cold with a half-life of 30 s. The second chlorine reacts comparatively slower with a half-life of 30 min at 40°C. The last chlorine is displaced very slowly requiring hours of incubation at elevated temperatures. These triazines have been used to cross-link collagen[133,249] and conjugate antigens to erythrocytes.[259] Fluorescein derivatives, **CLXXV** and **CLXXVI**, are fluorescent compounds.[251-254] 5-DTAF (**CLXXV**) has been used to cross-link glycoproteins and to study the rotational correlation time after intramolecular cross-linking of IgG.

4. Aziridines

While **CLXXVII** is a derivative of triazine, it also belongs to the group of aziridines. Aziridines containing a strained three-membered heterocyclic nitrogen ring are highly reactive. Nucleophiles react with these compounds, **CLXXVII** to **CLXXIX**, by ring opening.[234,235,255,256] Compounds **CLXXVII** and **CLXXIX**, like cyanuric chloride, have three potential reactive groups and may react as a trifunctional reagent. These aziridines effectively cross-link proteins such as albumin,[256] wool fibers, and γ-globulin.[255,256] as well as immobilize ligands to solid matrices (see Chapter 12). The major competitive reaction is hydrolysis.

5. bis-Epoxides (Bisoxiranes)

Like aziridines, epoxides also undergo ring opening reactions with nucleophiles, including hydroxyl groups. These *bis*-epoxides, **CLXXX** to **CLXXXIV**, have been used to cross-link proteins to RNA,[54,257] wool fibers[258,259] and bovine serum albumin as well as to activate matrices for immobilization of ligands which will be discussed in Chapter 12.[260,261] Of these reagents, diepoxybutane, **CLXXX**, will produce a vicinal diol linkage after reaction. It is therefore cleavable by sodium periodate. In addition to these compounds listed, a series of glycerol polyglycidyl ether, trimethylolpropane polyglycidyl ether and sorbitol polyglycidyl ether have been used to cross-link collagen preparations.[349]

6. Other Sulfhydryl Reacting Cross-Linkers

Divinylsulfone, **CLXXXV**, is an alkylating agent with nucleophilic addition to the double bond. It has been used to couple nucleophilic ligands to solid materials such as agarose.[261,262] At alkaline pHs (pH 11), divinylsulfone reacts with hydroxyl groups of agarose. Reactions also occur with thiols and amino groups at lower pHs and at somewhat higher rate. The product with hydroxyl functions is unstable above pH 9.0 and with amino group, unstable above pH 8.0. The scheme of reaction is shown in Figure 10.

IV. CARBOXYL GROUP DIRECTED CROSS-LINKING REAGENTS

There are relatively few carboxyl group directed homobifunctional cross-linking reagents. Carbodiimides which are generally regarded as zero-length cross-linkers (see Chapter

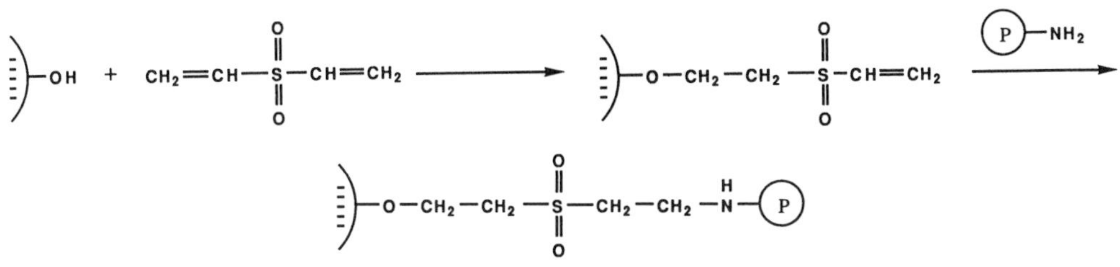

FIGURE 10. Reaction mechanism of divinylsulfone in the immobilization of proteins to agarose.

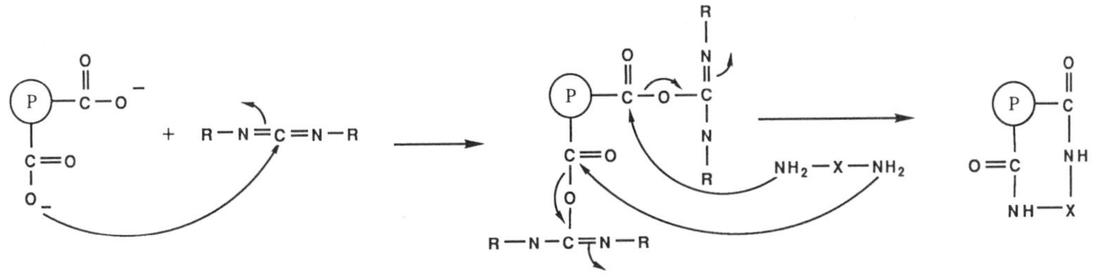

FIGURE 11. Cross-linking of proteins with carbodiimides and diamines.

6) can, in the presence of diamines such as ethyl diamine or cleavable cystamine, cross-link carboxyl groups as shown in Figure 11.[263] In this case, carbodiimides activate the carboxyl groups of proteins to form O-acylisoureas which are cross-linked by diamines. They are also capable of coupling heterobifunctional reagents which will be discussed in the next chapter. Theoretically any carbodiimides can be used, but water-soluble compounds such as those shown in Chapter 2 are the most common choices. Other carbodiimides will be discussed in Chapter 6.

The carboxylate ion, as a nucleophile, can effectively compete with other nucleophiles at neutral or slightly acidic pH, where the other nucleophiles such as amino, thiol, and phenolic groups, are present in less-reactive protonated forms. Thus, the carboxyl group can react with nitrogen and sulfur mustards and other reagents.[234-236,239,241] Similarly, the diazo compounds, **CLXXXVI** and **CLXXXVII**, shown in Table 3 have been used to modify carboxyl groups.[264-266] Bisdiazohexane, **CLXXXVII**, which is also reactive toward thiols, phenols, and other nucleophiles, have been used to cross-link wool and fibroin,[266] whereas compound **CLXXXVI** was used to cross-link pepsin intramolecularly.[264,265]

V. PHENOLATE AND IMIDAZOLYL GROUP DIRECTED CROSS-LINKING REAGENTS

Diazonium salts readily react with aromatic amino acid residues, tyrosine and histidine, by electrophilic substitution reactions. Proteins deficient in aromatic amino acid residues do not react.[267,268] *bis*-Diazonium compounds are easily prepared by treatment of aryl diamines with sodium nitrite in acidic conditions. The diazonium salts from the diamines listed in Table 3, **CLXXXVIII** to **CXCVII**, have been used to couple antigens,[271-280] for example, serum albumin,[271] insulin, and ovalbumin to erythrocytes,[281] conjugate antigens and antibodies to ferritin,[282,283] and prepare protein aggregates.[270,273,274,284] Poly-diazonium salts derived from compounds **CXCVI** and **CXCVII** have been used to prepare antigens and insoluble enzymes.[277,280] Of the diazonium compounds, the ones containing disulfide bonds, **CXCIV** and **CXCV**, are cleavable by mercaptans.

Table 3

OTHER GROUP DIRECTED HOMOBIFUNCTIONAL CROSS-LINKERS

Name (Abbreviation)	Structure	Cleavable/ condition	Reference

A. CARBOXYL GROUP DIRECTED REAGENTS

CLXXXVI. 1,1-Bis(diazoacetyl)-2-phenylethane		—	264, 265
CLXXXVII. Bisdiazohexane	$N_2 = CH - (CH_2)_4 - CH = N_2$	—	266

B. PHENOLATE AND IMIDAZOLYL GROUP DIRECTED REAGENTS - Bisdiazonium precursors

CLXXXVIII. p–Phenylenediamine		—	269, 270
CLXXXIX. Bis-benzidine		—	271-274

CXC. 3,3'-Dimethoxybenzidine; (o-Dianisidine)	—	275
CXCI. Benzidine-2,2'-disulfonic acid	—	275
CXCII. Benzidine-3,3'-disulfonic acid	—	275
CXCIII. 4,4'-Diaminodiphenylamine	—	269, 270
CXCIV. 4,4'-Diaminodiphenyldisulfide	+/thiol	24, 276
CXCV. 2,2'-Dinitro-4,4'-diamino-diphenyldisulfide	+/thiol	285

Table 3 (continued)

OTHER GROUP DIRECTED HOMOBIFUNCTIONAL CROSS-LINKERS

Name (Abbreviation)	Structure	Cleavable/condition	Reference
B. PHENOLATE AND IMIDAZOLYL GROUP DIRECTED REAGENTS - Bisdiazonium precursors			
CXCVI. Poly(4,4'-diaminodiphenyl-amine-3,3'-dicarboxylic acid)		—	277
CXCVII. Poly(p-amino-D,L-phenyl-alanyl-L-leucine)		—	278–280
CXCVIII. Potassium nitrosyl disulfonate		—	286, 287
CXCIX. Tetranitromethane	$C(NO_2)_4$	—	288

C. ARGININE REAGENT

CC. p-Phenylenediglyoxal	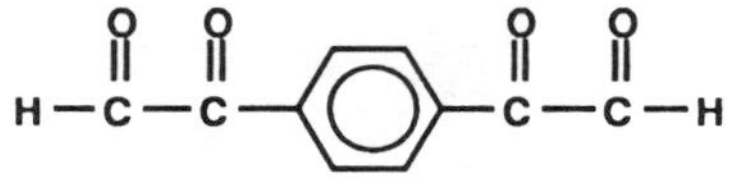	–	292, 293

D. MISCELLANEOUS REAGENTS

CCI. cis-Dichlorodiaminoplatinum(II) (cis-DDP)	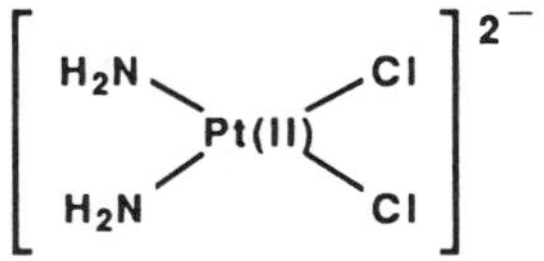	+/diethyldithio-carbamate	336-340
CCII. Adipic acid dihydrazide	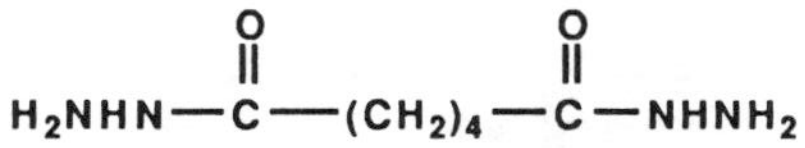	–	341
CCIII. N,N'-Bis(β-aminoethyl)-tartramide	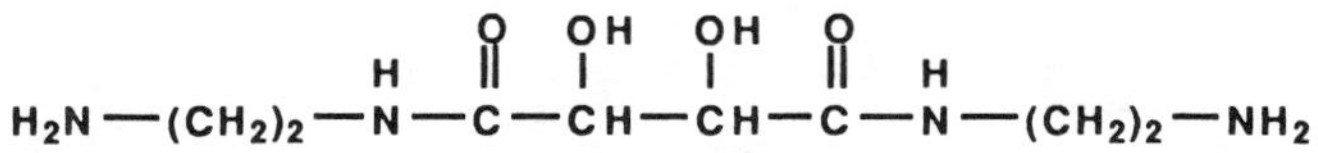	–	348

Potassium nitrosyldisulfonate, **CXCVIII,** has been shown to form cross-links in wool, silk, casein, insulin and collagen.[286,287] It is speculated that tyrosine residues are involved in the cross-linking formation. However, the reaction mechanism is not understood. It is possible that tryptophan residues may undergo oxidation and somehow take part in the cross-linking reaction.

Tetranitromethane, **CXCIX,** is another reagent that causes insolubilization of collagen, γ-globulin and carboxypeptidase A.[288] The reaction mechanism is unclear. Since tetranitromethane reacts with tyrosine residues, this amino acid may be involved.

VI. ARGININE RESIDUE DIRECTED CROSS-LINKER

As discussed in Chapter 2, vicinal diones react with the guanidinyl moiety of arginine. Such reactive functional group has been incorporated into *p*-phenylenediglyoxal, **CC,** to make a bifunctional cross-linking reagent that has been shown to react with arginine residues of proteins.[289,290] Such a reagent also cross-links guanosine bases in nucleic acids.

VII. NONDISCRIMINATORY PHOTOACTIVATABLE CROSS-LINKERS

Of the photoactivatable moieties that have been incorporated into cross-linking reagents, diazoalkanes have not been used in homobifunctional reagents, although it is found in a few heterobifunctional reagents which will be discussed in the next chapter. The more commonly used group is the azido function. Alkyl azides, however, have not been favorably used.[1,291] Acylazides as well as sulfonyl and phosphoryl azides are generally not used as photoactivatable reagents because of their nucleophilic reactivity in the dark.[29] They are therefore listed under Table 1. The only photoactivatable homobifunctional reagents are derived from aryl azides and are shown in Table 4.[292,293] These aryl azides are photolyzed at wavelengths 300 to 400 nm so that the biological component, be it proteins or nucleic acids, would not be damaged by photoirradiation.

The photoactivatable homobifunctional cross-linking reagents, **CCIV** to **CCVII,** shown in Table 4 are used mainly to label erythrocyte membrane proteins.[292,293] Compound **CCVII** was synthesized by Mikkelsen and Wallach[292] as a cleavable reagent in the present of thiols. It has been used to identify erythrocyte membranes. It should be noted that the reagent may be used as a heterobifunctional reagent by first reacting with a sulfhydryl group through the disulfide-thiol interchange reaction.

VIII. NONCOVALENT HOMOBIFUNCTIONAL CROSS-LINKING REAGENTS

While chemical cross-linking has the connotation that the reagents react to form covalent bonds, there are species that physically associate with proteins to form such a tight complex that it is essentially irreversible. The avidin-biotin complex is a well known example of such a system. Some immunoglobulins can also bind to antigens with dissociations in the order of fetomolar. Another class of proteins that have specific binding are the lectins. These molecules may be important under certain circumstances and may be useful for some applications.

A. AVIDIN AND STREPTAVIDIN

Avidin found in hen egg white is a glycoprotein with a molecular weight of 67,000 Da. It has four extraordinarily high affinity binding sites for biotin with a dissociation constant of approximately 10^{-15} *M*. The avidin-biotin interaction has been exploited in immunoassays,

Table 4

PHOTOACTIVATABLE HOMOBIFUNCTIONAL CROSS-LINKING REAGENTS

Name (Abbreviation)	Structure	Cleavable/ condition	Reference
CCIV. 4,4'-Diazidobiphenyl (DABP)		–	292
CCV. 1,5-Diazidonaphthalene (DAN)		–	292
CCVI. N,N'-Bis(p-azido-o-nitro-phenyl)-1,3-diamino-2-propanol		–	293
CCVII. 4,4'-Dithiobisphenylazide (DTPA, DTBPA)		+/thiol	293

in labeling techniques, and in purification of macromolecules.[294] Streptavidin is similar to avidin except that it is free of carbohydrates and is isolated from Streptomyces. Streptavidin has a lower isoelectric point of about 5.[295,296] It has been proved to reduced nonspecific binding in a number of studies.[297-299]

Since avidin and streptavidin are tetravalent, they can cross-link biotin containing proteins. In most immunochemical applications, biotin is artificially incorporated into various macromolecules. Avidin is then used to cross-link biotinylated molecules. In the bridged avidin-biotin (BRAB) method of immunoassay, both the immunoglobulin and the indicator enzyme is biotinylated and avidin is used to cross-link these species as shown in Figure 12A. Because of the multiple binding sites on avidin, more biotinylated enzyme can be bound to increase the intensity of the substrate color development. A similar cross-linking complex is seen in the avidin-biotin complex (ABC) system (Figure 12B). In addition to immunoassays, the avidin-biotin technology has been used in gene probes, protein-blotting and immunohistochemistry.[300]

B. ANTIGEN AND ANTIBODY AS CROSS-LINKERS

The basic unit (monomer) of immunoglobulins contains two Fab (antigen-binding) fragments and one Fc (crystallizable) fragment. These molecules are therefore bifunctional in that they can bind specifically two antigens, thus cross-linking them. With monoclonal antibodies, the dissociation constant of antibody-antigen complex could be as low as 10^{-15} M, making the complex essentially irreversible. There are five classes of immunoglobulins, designated IgG, IgA, IgM, IgD, and IgE. All the antibodies exist as monomers except IgM which is pentameric. In adult animals, the majority of serum immunoglobulin is of the IgG class. IgG is thus the most common reagent. Treatment of IgG with pepsin yields a F(ab)$'_2$ fragment which retains the two antigen binding sites. F(ab)$'_2$ can be purified and used as a cross-linking reagent.

Not only are homobifunctional antibodies used in various assays, particularly in agglutination precipitin assays, antibodies of different specificities have been cross-linked to yield heterobifunctional agents. Bode et al.[301] have prepared bispecific antibody by cross-linking antifibrin antibody and 2-iminothiolane-modified anti-tissue plasminogen activator (tPA) with N-succinimidyl-3-(2-pyridyldithio)propionate. Such bispecific antibody recognizes both fibrin and tPA and is able to conjugate these components with an apparent dissociation constant of 10^{-9} to 10^{-10} M. This application has extended immunoglobulins as homobifunctional cross-linkers to heterobifunctional cross-linkers. Bispecific monoclonal antibodies reactive to different molecules have also been constructed.[302-305]

Similarly, compounds that contain two antigenic determinants will cross-link antibody molecules. Simple molecules such as L-tyrosine-*bis*(*p*-azobenzenearsenate) (*Bis*-RAT) and dinitrophenyl-1-tyrosine-*p*-azobenzenearsenate (DNP-RAT) (Figure 13) have been shown to cross-link anti-RAT antibodies, anti-RAT and anti-DNP antibodies, respectively.[306] Similar cross-linking of the immunoglobulins is achieved by *bis*-azobenzenearsenate and by dinitrophenylazobenzenearsenate separated by various spacers. Other examples of multiepitopic molecules are hapten-conjugated antigens such as dinitrophenyl labeled bovine serum albumin.[307] In addition to proteins, polysaccharides have also been used as carriers. Ficoll, a polymer of fructose, has been covalently bonded with dinitrophenyl and phosphorylcholine.[308,309] Between 30 to 35 haptens can be conjugated per molecule of ficoll. Such molecules are able to cross-link immunoglobulins to form large immune complexes used for nephropathy studies.[310]

C. LECTINS

Lectins are sugar-binding proteins that agglutinate cells or precipitate carbohydrates or glycoconjugates. They are multivalent, possessing at least two sugar-binding sites which

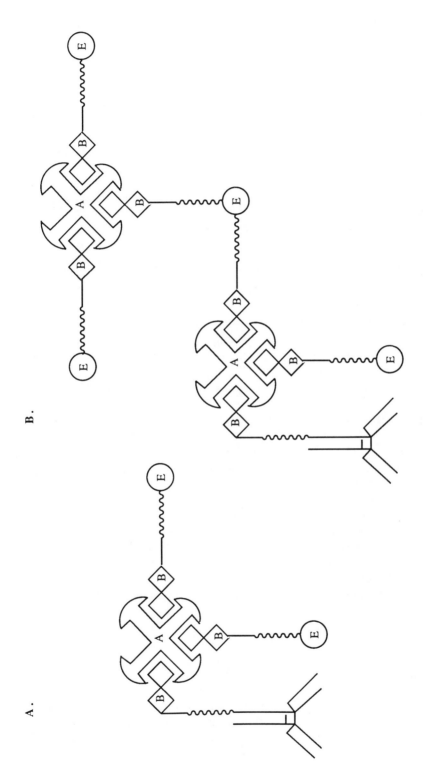

FIGURE 12. Use of avidin as cross-linker in immunoassays. (A) Bridged avidin-biotin complex; (B) avidin-biotin complex. In the figure B represents biotin, A, avidin and E, enzyme.

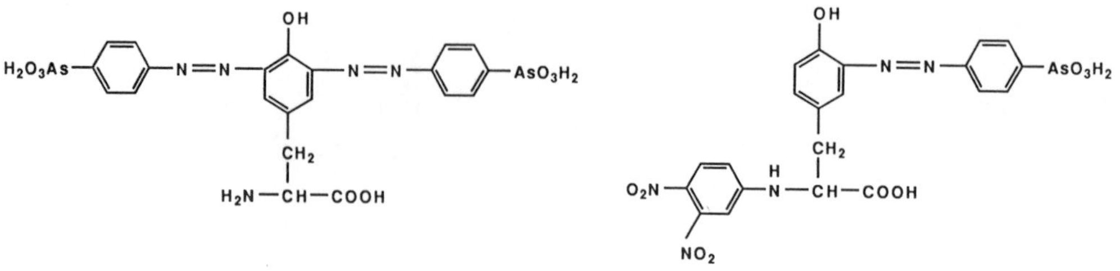

FIGURE 13.   Structure of bifunctional antigens: *Bis*-RAT, L-tyrosine-*bis*(*p*-azobenzenearsenate); DNP-RAT, din-itrophenyl-1-tyrosine-*p*-azobenzenearsenate.

enable them to agglutinate animal and plant cells and/or to precipitate polysaccharides, glycoproteins, peptidoglycans, teichoic acids, glycolipids, etc. Each lectin binds specifically to a certain sugar sequence in oligosaccharides and glycoconjugates. There are many extensive reviews, symposia proceedings as well as books that have been written on the properties, functions, and applications of lectins.[311-317] It is beyond the scope of this book to review all the literature. The interested reader is urged to consult the references listed. In this section, we will summarize the relevant literature related to the application of lectins in cross-linking.

There must be hundreds, if not thousands, of different lectins that have been detected in and isolated from various plants and animals, bacteria and viruses. The most celebrated example is probably concanavalin A (con A) purified from jack bean. Con A at pH 7 is composed of four subunits of $M_r = 26,500$ Da.[318,319] It is a metalloprotein, each subunit of which contains one Ca^{2+} and one Mn^{2+} which are required for carbohydrate-binding activity.[320] Con A would bind α-mannosyl, α-glucosyl and α-*N*-acetylglucosaminyl groups and would precipitate branched polysaccharides containing these sugar units as nonreducing termini.[321,322] Mannose, in its α-anomeric form, is the monosaccharide most complementary to the con A sugar binding site.[323-325] Being tetravalent, con A can cross-link and precipitate glycogen, yeast mannan, and various glycoproteins in a manner resembling antibodies. Such interactions have been used for structural analysis and isolation and resolution of glycoproteins.[326]

Like con A, many lectins have been used for the detection, fractionation, isolation and purification of glycoproteins, which do not necessarily employ their multivalent properties. Their cross-linking nature, however, is the basis in the Ouchterlony double-diffusion and the affinoimmunoelectrophoresis technique in which the lectin is used in place of antibodies for the detection of glycoproteins.[327,329] Other applications include blood typing[329,330] and the identification of microorganisms.[331-334] It seem that the carbohydrate specificity and the multivalent nature of the lectins can be further employed in cross-linking of components for enzyme immunoassays, for the study of immune complexes, and for colloidal gold staining techniques.[335]

IX. MISCELLANEOUS BIFUNCTIONAL REAGENTS

There are few additional homobifunctional cross-linking reagents with undefined group selectivity that are worth mentioning. cis-Dichlorodiammine platinum (II), **CCI** (Table 3) has been found to cross-link $α_2$-macroglobulin with one or more methionine residues at or near the receptor recognition site.[336-338] The cross-linking can be reversed by diethyldithio-carbamate which is a potent platinum chelator.[338] The reagent is also speculated to cross-link complementary strands of DNA.[339,340]

Adipic acid dihydrazide, **CCII** (Table 3), has been used to cross-link glycoproteins, acid phosphatase and invertase.[341] The carbohydrate moieties are first oxidized by sodium periodate to dialdehydes which are cross-linked through Schiff base formation with the hydrazine derivative. Reductive alkylation is then completed with reducing agents as discussed in Chapter 2. Other diamines of different chain lengths may also be used in this process.

In addition to chemical cross-linking, enzyme catalyzed cross-linking with monofunctional reagents has also been achieved. The ability of transglutaminase to catalyze the aminolysis of the γ-carboxamide group of peptide bound glutamine residues has been used to cross-link proteins.[342,343] Gorman and Folk[344] have used transglutaminase and a cleavable diamine, *N,N'-bis*(β-aminoethyl)tartramide (**CCIII**, Table 3) to cross-link guanidinated β-casein, which has the lysine protected to prevent ε-(γ-glutamyl)lysine cross-links. Other exogenous amines have been shown to compete effectively with ε-amino groups of lysine residues and are incorporated.[345] Thus, many other diamines can be used for this purpose.[346] This enzyme catalyzed reaction is analogous to carbodiimide mediated cross-linking, except that glutamine residues are involved instead of carboxyl groups.

REFERENCES

1. **Peters, K. and Richards, F. M.,** Chemical cross-linking: Reagents and problems in studies of membrane structure, *Annu. Rev. Biochem.,* 46, 523, 1977.
2. **Ji, T. H.,** The application of chemical crosslinking for studies on cell membranes and the identification of surface reporters, *Biochim. Biophys. Acta,* 559, 39, 1979.
3. **Patel, R. P. and Price, S.,** Derivatives of proteins. I. Polymerization of α-chymotrypsin by use of *N*-ethyl-5-phenylisoxazolium-3'-sulfonate, *Biopolymers,* 5, 583, 1967.
4. **Pilch, P. F. and Czech, M. P.,** Interaction of cross-linking agents with the insulin effector system of isolated fat cells. Covalent linkage of ^{125}I-insulin to a plasma membrane receptor protein of 140,000 daltons, *J. Biol. Chem.,* 254, 3375, 1979.
5. **Sweadner, K. J.,** Crosslinking and modification of Na,K-ATPase by ethyl acetimidate, *Biochem. Biophys. Res. Commun.,* 78, 962, 1977.
6. **Shaw, A. B. and Marinetti, G. V.,** Cross linking of erythrocyte membrane proteins and phospholipids by chemical probes, *Membrane Biochem.,* 3, 1, 1980.
7. **Davies, G. E. and Stark, G. R.,** Use of dimethyl suberimidate, a cross-linking reagent, in studying the subunit structure of oligomeric proteins, *Proc. Natl. Acad. Sci. U.S.A.,* 66, 651, 1970.
8. **Hunter, M. J. and Ludwig, M. L.,** Amidination, *Methods Enzymol.,* 25, 585, 1972.
9. **Wang, T.-W. and Kassell, B.,** The preparation of a chemically cross-linked complex of the basic pancreatic trypsin inhibitor with trypsin, *Biochemistry,* 13, 698, 1974.
10. **Bartholeyns, J. and Moore, S.,** Pancreatic ribonuclease: enzymic and physiological properties of a cross-linked dimer, *Science,* 186, 444, 1974.
11. **Dutton, A., Adam, M., and Singer, S. J.,** Bifunctional imidoesters as cross-linking reagents, *Biochem. Biophys. Res. Commun.,* 23, 730, 1966.
12. **Niehaus, Jr., W. G. and Wold, F.,** Cross-linking of erythrocyte membranes with dimethyl adipimidate, *Biochem. Biophys. Acta,* 196, 170, 1970.
13. **Hartman, F. C. and Wold, F.,** Bifunctional reagents. Cross-linking of pancreatic ribonuclease with a diimido ester, *J. Am. Chem. Soc.,* 88, 3890, 1966.
14. **Ji, T. H.,** Cross-linking of glycolipids in erythrocyte ghost membrane, *J. Biol. Chem.,* 249, 7841, 1974.
15. **Hucho, F., Mullner, H., and Sund, H.,** Investigation of the symmetry of oligomeric enzymes with bifunctional reagents, *Eur. J. Biochem.,* 59, 79, 1975.
16. **Tinberg, H. M., Nayudu, P. R. V., and Packer, L.,** Crosslinking of membranes: The effect of dimethylsuberimidate, a bifunctional alkylating agent, on mitochondrial electron transport and ATPase, *Arch. Biochem. Biophys.,* 172, 734, 1976.
17. **Siezen, R. J., Bindels, J. G., and Hoenders, H. J.,** The quaternary structure of bovine α-crystallin: Chemical crosslinking with bifunctional imidoesters, *Eur. J. Biochem.,* 107, 243, 1980.
18. **Schramm, H. J. and Dülffer, T.,** Synthesis and application of cleavable and hydrophilic cross-linking reagents, in *Protein Cross-Linking: Biochemical and Molecular Aspects,* Friedman, M., Ed., Plenum Press, New York, 1976, 197.

19. **Browne, D. T. and Kent, B. H.,** Formation of nonamidine products in the reaction of primary amines with imido esters, *Biochem. Biophys. Res. Commun.*, 67, 126, 1975.

20. **Hunter, M. J. and Ludwig, M. L.,** The reaction of imidoesters with proteins and related small molecules, *J. Am. Chem. Soc.*, 84, 3491, 1962.

21. **Wold, F.,** Bifunctional reagents, *Methods Enzymol.*, 25, 623, 1972.

22. **Means, G. E. and Feeney, R. E.,** *Chemical Modification of Proteins*, Holden-Day, San Francisco, 1971.

23. **Staros, J. V., Morgan, D. G., and Appling, D. R.,** A membrane-impermeant, cleavable cross-linker. Dimers of human erythrocyte band 3 subunits cross-linked at the extracytoplasmic membrane face, *J. Biol. Chem.*, 256, 5890, 1981.

24. **Wang, K. and Richards, F. M.,** Behavior of cleavable cross-linking reagents based on the disulfide group, *Isr. J. Chem.*, 12, 375, 1974.

25. **Ruoho, A., Bartett, P. A., Dutton, A., and Singer, S. J.,** Disulfide-bridge bifunctional imidoester as a reversible cross-linking reagent, *Biochem. Biophys. Res. Commun.*, 63, 417, 1975.

26. **Aizawa, S., Kurimoto, F., and Yokono, O.,** Crosslinking studies with different length dithiobisalkylimidates. (I). Solubilized erythrocyte spectrin, *Biochem. Biophys. Res. Commun.*, 75, 870, 1977.

27. **Coggins, J. R., Hooper, E. A., and Perham, R. N.,** Use of DMS and novel periodate-cleavable *bis*-imido-esters to study the quaternary structure of the pyruvate dehydrogenase multienzyme complex of *E. coli.*, *Biochemistry*, 15, 2527, 1976.

28. **Fasold, H., Baumert, H., and Fink, G.,** Comparison of hydrophobic and strongly hydrophilic cleavable cross-linking reagents in intermolecular by formation in aggregates of proteins or protein-RNA, in *Protein Cross-Linking: Biochemical and Molecular Aspects*, Friedman, M., Ed., Plenum Press, New York, 1976, 207.

29. **Lutter, L. C., Ortanderl, F., and Fasold, H.,** Use of a new series of cleavable protein-crossliners on the *Escherichia coli* ribosome, *FEBS Lett.*, 48, 288, 1974.

30. **Schramm, H. J.,** Synthese von farbigen Nitrilen, Dinitril und bifunktimellen Imidsoureestern, *Hoppe-Seyler's Z. Physiol. Chem.*, 348, 289, 1967.

31. **Schramm, H. J.,** The synthesis of mono- and bifunctional nitriles and imidoesters carrying a fluorescent group, *Hoppe-Seyler's Z. Physiol. Chem.*, 356, 1375, 1975.

32. **Smith, A. P. and Loh, H. H.,** Effect of bisimidate cross-linking reagents on synaptosomal plasma membrane, *Biochemistry*, 17, 1761, 1978.

33. **Hordern, J. S., Leonard, J. D., and Scraba, D. G.,** Structure of the Mengo virion, *Virology*, 97, 131, 1979.

34. **Gebhardt, A., Bosch, J. V., Ziemiecki, A., and Griis, R. R.,** Rous sarcoma virus p19 and gp35 can be chemically crosslinked to high molecular weight complexes: an insight into virus assembly, *J. Mol. Biol.*, 174, 297, 1984.

35. **Drews, G. and Rack, M.,** Modification of sodium and gating currents by amino group specific cross-linking and monofunctional reagents, *Biophys J.*, 54, 383, 1988.

36. **Uchiumi, T., Terao, K., and Ogata, K.,** Identification of neighboring protein pairs cross-linked with DTBP in rat liver 40 S ribosomal subunits, *J. Biochem.*, 90, 185, 1981.

37. **Hajdu, J., Dombradi, V., Bot, G., and Friedrich, P.,** Structural changes in glycogen phosphorylase as revealed by cross-linking with bifunctional di-imidates: phosphorylase-b, *Biochemistry*, 18, 4137, 1979.

38. **Anderson, W. M. and Fisher, R. R.,** The subunit structure of bovine heart mitochondrial transhydrogenase, *Biochim. Biophys. Acta*, 635, 194, 1981.

39. **Lambooy, P. K. and Steiner, R. F.,** The cross-linking of phosphorylase-kinase, *Arch. Biochem. Biophys.*, 213, 551, 1982.

40. **Brew, K., Shaper, J. H., Olsen, K. W., Trayer, I. P., and Hill, R. L.,** Cross-linking of components of lactose synthetase with dimethyl-pimelimidate, *J. Biol. Chem.*, 250, 1434, 1975.

41. **Lad, P. M. and Hammes, G. G.,** Physical and chemical properties of rabbit muscle phosphofructinase crosslinked with dimethyl suberimidate, *Biochemistry*, 13, 4530, 1974.

42. **Hartman, F. C. and Wold, F.,** Cross-linking of bovine pancreatic ribonuclease A with dimethyl adipimidate, *Biochemistry*, 6, 2439, 1967.

43. **De Pont, J. J. H. H. M.,** Reversible inactivation of (Na,K)ATPase by use of a cleavable bifunctional reagent, *Biochim. Biophys. Acta*, 567, 247, 1979.

44. **Baskin, L. S. and Yang, C. S.,** Cross-linking studies of monooxygenase enzymes, in *Microsomes, Drug Oxidations, and Chemical Carcinogenesis*, Vol. 1, Coon, M. J., and Conney, A. H., Estabrook, R. W., Gelboin, H. V., Gillette, J. R., and O'Brien, P. J., Eds., Academic Press, New York, 1980, 102.

45. **Baskin, L. S. and Yang, C. S.,** Identification of cross-linked cytochrome P-450 in rat liver microsomes by enzyme immunoassay, *Biochem. Biophys. Res. Commun.*, 108, 700, 1982.

46. **Suda, M. and Iwai, K.,** Identification of suberimidate cross-linking sites of four histone sequences in the H1-depleted chromatin. Histone arrangement in nucleosome core, *J. Biochem.*, 86, 1659, 1979.

47. **Galella, G. and Smith, D. B.,** The cross-linking of tubulin with imido-esters, *Can. J. Biochem.*, 60, 71, 1982.

48. **Kagey-Sobotka, A., Dembo, M., Goldstein, B., Metzger, H., and Lightenstein, L. M.,** Qualitative characteristics of histamine release from human basophils by covalently cross-linked Ig-E, *J. Immunol.*, 127, 2285, 1981.

49. **Ohara, O., Takanashi, S., and Ooi, T.,** Cross-linking study on tropomyosin, *J. Biochem.*, 87, 1795, 1980.

50. **Ikai, A. and Yanagita, Y.,** A cross-linking study of apo-low density lipoprotein, *J. Biochem.*, 88, 1359, 1980.

51. **Ji, T. H.,** A novel approach to the identification of surface receptors. The use of photosensitive hetero-bifunctional cross-linking reagent, *J. Biol. Chem.*, 252, 1566, 1977.

52. **Pennathur-Das, R., Heath, R. H., Mentzer, W. C., and Lubin, B. H.,** Modification of hemoglobin S with dimethyl adipimidate: Contribution of individual reacted subunits to changes in properties, *Biochim. Biophys. Acta*, 704, 389, 1982.

53. **Dietrich, J. B.,** Chemical crosslinking of different forms of the simion virus 40 large T antigen using bifunctional reagents, *FEBS Lett.*, 201, 311, 1986.

54. **Uckert, W., Wunderlich, V., Ghysdael, J., Portetelle, D., and Burny, A.,** Bovine leukemia virus (BLV): A structural model based on chemical cross-linking studies, *Virology*, 133, 386, 1984.

55. **Uckert, W. and Rudolph, M.,** Chemical crosslinking of lipids and proteins within mason-pfigen monkey and squirrel monkey type D retrovirus, *Arch. Geschwulstforsch*, 56, 107, 1986.

56. **Pilch, P. F. and Czech, M. P.,** Interaction of cross-linking agents with the insulin effector system of isolated fat cells. Covalent linkage of ^{125}I-insulin to a plasma membrane receptor protein of 140,000 daltons, *J. Biol. Chem.*, 254, 3375, 1979.

57. **Staros, J. V.,** *N*-Hydroxysulfosuccinimide active esters: *Bis(N*-hydroxysulfosuccinimide) esters of two carboxylic acids are hydrophilic, membrane-impermeant, protein cross-linkers, *Biochemistry*, 21, 3950, 1982.

58. **Giedroc, D. P., Puett, D., Ling, N., and Staros, J. V.,** Demonstration by covalent cross-linking of a specific interaction between β-endorphin and calmodulin, *J. Biol. Chem.*, 258, 16, 1983.

59. **Ryan, J. W., Day, A. R., Schultz, D. R., Ryan, U. S., Chung, A., Marlborough, D. I., and Dorer, F. E.,** Localization of angiotension converting enzyme (kininase II). I. Preparation of antibody-hemeoctapeptide conjugates, *Tissue Cell*, 8, 111, 1976.

60. **Lindsay, D. G.,** Intramolecular cross-linked insulin, *FEBS Lett.*, 21, 105, 1972.

61. **Gaffney, B. J., Willingham, G. L., and Schopp, R. S.,** Synthesis and membrane interactions of spin-label bifunctional reagent, *Biochemistry*, 22, 881, 1983.

62. **Willingham, G. L. and Gaffney, B. J.,** Reactions of spin-label cross-linking reagents with red blood cell proteins, *Biochemistry*, 22, 892, 1983.

63. **Anjaneyulu, P. S. R., Beth, A. H., Cobb, C. E., Juliao, S. F., Sweetman, B. J., and Staros, J. V.,** *Bis*(sulfo-*N*-succinimidyl)doxyl-2-spiro-5′-azelate: synthesis, characterization and reaction with the anion-exchange channel in intact human erythrocyte, *Biochemistry*, 28, 6583, 1989.

64. **Beth, A. H., Conturo, T. E., Venakataramu, S. D., and Staros, J. V.,** Dynamic and interactions of the anion channel in intact human erythrocytes: an electron paramagnetic resonance spectroscopic study employing a new membrane-impermeable bifunctional spin label, *Biochemistry*, 25, 3824, 1986.

65. **Anjaneyulu, P. S., Beth, A. H., Sweetmen, B. J., Faulkner, L. A., and Staros, J. V.,** Bis(sulfo-*N*-succinimidyl) [^{15}N,^{2}H16]doxyl-2-spiro-4′-pimelate, a stable isotope-substituted, membrane-impermeant bifunctional spin label for studies of the dynamics of membrane proteins: application to the anion-exchange channel in intact human erythrocytes, *Biochemistry*, 27, 6844, 1988.

66. **Baskin, L. S. and Yang, C. S.,** Cross-linking studies of cytochrome P-450 and reduced NADPH-cytochrome-P 450 reductase, *Biochemistry*, 19, 2260, 1980.

67. **Abdella, P. M., Smith, P. K., and Royer, G. P.,** A new cleavable reagent for cross-linking and reversible immobilization of proteins, *Biochem. Biophys. Res. Commun.*, 87, 734, 1979.

68. **Zarling, D. A., Watson, A., and Bach, F. H.,** Mapping of lymphocyte surface polypeptide antigens by chemical cross-linking with BSOCOES, *J. Immunol.*, 124, 913, 1980.

69. **Lomant, A. J. and Fairbanks, G.,** Chemical probes of extended biological structures: synthesis and properties of the cleavable protein cross-linking reagent [^{35}S]dithiobis(succinimidylpropionate), *J. Mol. Biol.*, 104, 243, 1976.

70. **Bragg, P. D. and Hou, C.,** Chemical cross-linking of α-subunit in the F_1 adenosine triphosphatase of *Escherichia coli*, *Arch. Biochem. Biophys.*, 244, 361, 1986.

71. **Staros, J. V., Kakkad, B. P.,** Crosslinking and chymotryptic digestion of the extracytoplasmic domain of the anion exchange channel in intact human erythrocytes, *J. Membr. Biol.*, 74, 247, 1983.

72. **Han, K.-K., Richard, C., and Delacourte, A.,** Chemical cross-links of proteins by using bifunctional reagents, *Int. J. Biochem.*, 16, 129, 1984.

73. **Smith, R. J., Capaldi, R. A., Muchmore, D., and Dahlquist, F.,** Cross-linking of ubiquinone-cytochrome-C-reductase with periodate-cleavable bifunctional reagent, *Biochemistry*, 17, 3719, 1978.

74. **Anderson, G. W., Zimmerman, J. E., and Callahan, F. M.,** Esters of *N*-hydroxysuccinimide in peptide synthesis, *J. Am. Chem. Soc.,* 86, 1839, 1964.
75. **Cuatrecasas, P. and Parikh, I.,** Adsorbents for affinity chromatography. Use of *N*-hydroxysuccinimide esters of agarose, *Biochemistry,* 11, 2291, 1972.
76. **Vanin, E. F. and Ji, T. H.,** Synthesis and application of cleavable photoactivable heterobifunctional reagents, *Biochemistry,* 20, 6754, 1981.
77. **Hill, M., Bechet, J. J., and D'Albis, A.,** Disuccinimidyl esters as bifunctional cross-linking reagents for proteins, *FEBS Letts.,* 102, 282, 1979.
78. **Travers, R. C., Noyes, C. M., Roberts, H. R., and Lundblad, R. L.,** Influence of metal ions on prothrombin self association, *J. Biol. Chem.,* 257, 10708, 1982.
79. **Takamatsu, T., Horne, M. K., III, and Gralnick, H. R.,** Identification of the thrombin receptor on human platelets by chemical crosslinking, *J. Clin. Invest.,* 77, 362, 1986.
80. **Davies, G. E. and Palek, J.,** Platelet protein organization analysis by treatment with membrane permeable cross-linking reagents, *Blood,* 59, 502, 1982.
81. **Wiemken, V., Theiler, R., and Bachofen, R.,** Lateral organization of proteins in chromatophore membrane of R. rubrum, studies by chemical cross-linking, *J. Bioenerg. Biomembr.,* 13, 181, 1981.
82. **Anderson, W. M. and Fisher, R. R.,** The subunit structure of bovine heart mitochondrial transhydrogenase, *Biochim. Biophys. Acta,* 635, 194, 1981.
83. **Baskin, L. S. and Yang, C. S.,** Cross-linking studies of cytochrome P-450 and reduced NADPH-cytochrome-P450 reductase, *Biochemistry,* 19, 2260, 1980.
84. **Smith, R. J., Capaldi, R. A., Muchmore, D., and Dahlquist, F.,** Cross-linking of uniquinone-cytochrome-c-reductase with periodate-cleavable bifunctional reagent, *Biochemistry,* 17, 1761, 1978.
85. **Laburthe, M., Breant, B., and Rouyen-Fessard, C.,** Molecular identification of receptors for vasoactive intestinal peptide in rat intestinal epithelium by covalent cross-linking: evidence for two classes of binding sites with different structural and functional properties, *Eur. J. Biochem.,* 139, 181, 1984.
86. **Stauber, G. and Aggarwal, B. B.,** Characterization and affinity cross-linking of receptors for human recombinant lymphotoxin (tumor necrosis factor-β) on a human histiocytic lymphoma cell line, U-937, *J. Biol. Chem.,* 264, 3573, 1989.
87. **Amiranoff, B., Lorinet, A. M., and Laburthe, M.,** Galanin receptor in the rat pancreatic beta-cell line Rin M-5F: molecular characterization by chemical cross-linking, *J. Biol. Chem.,* 264, 20714, 1989.
88. **Conn, P. M., Rogers, D. C., Stewart, J. M., Niedel, J., and Sheffield, T.,** Conversion of a gonadotropin-releasing hormone antagonist to an antagonist, *Nature (London),* 296, 653, 1982.
89. **Pinter, A. and Fleissner, E.,** Structural studies of retroviruses: characterization of oligomeric complexes of murine and feline leukemia virus envelope and core components formed upon cross-linking, *J. Virol.,* 30, 157, 1979.
90. **Andrews, R. K., Gorman, J. J., Booth, W. J., Corino, G. L., Castaldi, P. A., and Berndt, M. C.,** Cross-linking of a monomeric 39/34-kDa diapase fragment of von Willebrand factor (leu-480/val-481-gly-718) to the *N*-terminal region of the α-chain of membrane glycoprotein Ib on intact platelets with bis(sulfosuccinimidyl)suberate, *Biochemistry,* 28, 8326, 1989.
91. **Zhang, Q.-Y. and Menon, K. M. J.,** Characterization of rat leydig cell gonadotropin receptor structure by affinity cross-linking, *J. Biol. Chem.,* 263, 1002, 1988.
92. **Fanger, B. O., Stephans, J. E., and Staros, J. V.,** Trapping of epidermal growth factor-induced receptor dimers by chemical cross-linking, *FASEB J.,* 2, A1774, 1988.
93. **Fanger, B. O., Stephens, J. E., and Staros, J. V.,** High-yield trapping of EGF-induced receptor dimers by chemical cross-linking, *FASEB J.,* 3, 71, 1989.
94. **Herzig, M. C. and Weigel, P. H.,** Synthesis and characterization of *N*-hydroxysuccinimide ester chemical affinity derivatives of asialoorosomucoid that covalently cross-link to galactosyl receptors on isolated rat hepatocytes, *Biochemistry,* 28, 600, 1989.
95. **Zahn, H.,** Bridge reactions in amino acids and fiber proteins, *Angew. Chem.,* 67, 56, 1955.
96. **Grow, T. E. and Fried, M.,** Lipoprotein geometry. I. Spatial relations of human HDL (high density lipoproteins) apoproteins studied with a bifunctional reagent, *Biochem. Biophys. Res. Commun.,* 66, 352, 1975.
97. **Fraenkel-Conrat, H.,** The chemistry of proteins and peptides, *Annu. Rev. Biochem.,* 25, 318, 1956.
98. **Walder, J. A., Zaugg, R. H., Walder, R. Y., Steele, J. M., and Klotz, I. M.,** Diaspirins that cross-link β chains of hemoglobin: *bis*(3,5-dibromosalicyl)succinate and *bis*(3,5-dibromosalicyl)fumarate, *Biochemistry,* 18, 4265, 1979.
99. **Chatterjee, R., Welty, E. V., Walder, R. Y., Pruitt, S. L., Rogers, P. H., Arone, A., and Walder, J. A.,** Isolation and characterization of a new hemoglobin derivative cross-linked between α chains (lysine 99α$_1$ → lysine 99α$_2$), *J. Biol. Chem.,* 261, 9929, 1986.
100. **Wold, F.,** Reaction of bovine serum albumin with the bifunctional reagent *p,p'*-difluoro-*m,m'*-dinitrodiphenylsulfone, *J. Biol. Chem.,* 236, 106, 1961.

101. **Zhan, H., Zuber, H., Ditscher, W., Wegerle, D., and Meienhofer, J.,** Reactions of aromatic fluoro compounds with amino acids and proteins. XV. Reaction of p,p'-difluoro-m,m'-dinitrodiphenylsulfone with silk fibroin, *Chem. Ber.*, 89, 407, 1956.

102. **Mills, G. L.,** Identification of dinitrophenylamino acids, *Nature (London)*, 165, 403, 1950.

103. **Zhan, H. and Wegerle, D.,** Collagen. III. Reactions with *bis*(3-nitro-4-fluorophenyl)sulfone, *Kolloid-Z.*, 172, 29, 1960.

104. **Tawde, S., Ram, J. S., and Iyengar, M. R.,** Physicochemical and immunochemical studies on the reaction of bovine serum albumin with p,p'-difluoro-m,m'-dinitrophenylsulfone, *Arch. Biochem. Biophys.*, 100, 270, 1963.

105. **Deyev, S. M., Afanasenko, G. A., and Polyanovsky, O. L.,** Two steps modification of aspartate amino transferase with DFNB-cross-linking localization, *Biochim. Biophys. Acta*, 534, 358, 1978.

106. **Harris, W. E. and Stahl, W. L.,** Organizational of thiol groups of electric-eel electric organ Na, K ion-stimulated ATPase studied with bifunctional reagents, *Biochem. J.*, 185, 787, 1980.

107. **Munroe, W. A. and Dunlap, R. B.,** Chemical modification of *Lactobacillus casei* thymidylate synthetase with FDBN and FDNB, *Arch. Biochem. Biophys.*, 214, 742, 1982.

108. **Zahn, H., and Meienhofer, J.,** Reactions of 1,5-difluoro-2,4-dinitrobenzene with insulin. I. Synthesis of model compounds, *Makromol. Chem.*, 26, 153, 1958.

109. **Marfey, P. S., Nowak, H., Uziel, M., and Yphantis, D. A.,** Reaction of bovine pancreatic ribonuclease A with 1,5-difluoro-2,4-dinitrobenzene. I. Preparation of monomeric intramolecularly bridged derivatives, *J. Biol. Chem.*, 240, 3264, 1965.

110. **Marfey, P. S. and King, M. V.,** Chemical modification of ribonuclease A crystals. I. Reaction with 1,5-difluoro-2,4-dinitrobenzene, *Biochim. Biophys. Acta*, 105, 178, 1965.

111. **Cuatrecasas, P., Fuchs, S., and Anfinsen, C. B.,** Cross-linking of aminotyrosyl residues in the active site of Staphylococcal nuclease, *J. Biol. Chem.*, 244, 406, 1969.

112. **Berg, H. C., Diamond, J. M., and Marfey, P. S.,** Erythrocyte membrane: chemical modification, *Science*, 150, 64, 1965.

113. **Vandegriff, K. D., Medina, F., Marini, M. A., and Winslow, R. M.,** Equilibrium oxygen binding to human hemoglobin cross-linked between the α chains by *bis*(3,5-dibromosalicyl)fumarate, *J. Biol. Chem.*, 264, 17824, 1989.

114. **Ram, J. S., Tawde, S. S., Pierce, Jr., G. B., and Midgley, A. R., Jr.,** Preparation of antibody-ferritin conjugates for immuno electron microscopy, *J. Cell Biol.*, 17, 673, 1963.

115. **Modesto, R. R. and Pesce, A. J.,** The reaction of 4,4'-difluoro-3,3'-dinitrodiphenyl sulfone with γ-globulin and horse-radish peroxidase, *Biochim. Biophys. Acta*, 229, 384, 1971.

116. **Ling, N. R.,** Coupling of protein antigens to erythrocytes with difluoro-dinitrobenzene, *Immunology*, 4, 49, 1961.

117. **Ozawa H.,** Bridging reagent for protein. I. The reaction of diisocyantes with lysine and enzyme proteins, *J. Biochem. (Tokyo)*, 62, 419, 1967.

118. **Snyder, P. D., Jr., Wold, F., Bernlohr, R. W., Dullum, C., Desnick, R. J., Krivit, W., and Condie, R. M.,** Enzyme therapy. II. Purified human α-galactosidase A. Stabilization to heat and protease degradation by complexing with antibody and by chemical modification, *Biochem. Biophys. Acta*, 350, 432, 1974.

119. **Schick, A. F. and Singer, S. J.,** On the formation of covalent linkages between two protein molecules, *J. Biol. Chem.*, 236, 2477, 1961.

120. **Fasold, H.,** Synthese und reaktionen eines wasserlöslichen, spaltbaren, starren reagens zur verknüpfung von freien aminogrup-pen in proteinen, *Biochem. Z.*, 339, 482, 1964.

121. **Haimovich, J., Hurwitz, E., Novik, N., and Sela, M.,** Preparation of protein-bacteriophage conjugates and their use in detection of anti-protein antibodies, *Biochem. Biophys. Acta*, 207, 115, 1970.

122. **Borek, F. and Silverstein, A. M.,** Characterization and purification of ferritin-antibody globulin conjugates, *J. Immunol.*, 87, 555, 1961.

123. **Fasold, H. and Lusty, C. J.,** The application of azo dyes to identify reactive groups and determine distances in proteins, *7th Int. Congr. Biochem. Abstr.*, III, 1, 1967.

124. **Manecke, G. and Gunzel, G.,** Darstellung eines wasserunlöslichen, aktiven papains, *Naturwissenschaften*, 54, 647, 1967.

125. **Cabantchik, I. Z., Balshin, M., Breuer, W., and Rothstein, A.,** Pyridoxal phosphate. An anionic probe for protein amine groups exposed on the outer and inner surfaces of intact human red blood cells, *J. Biol. Chem.*, 250, 5130, 1975.

126. **Macara, I. G. and Cantley, L. C.,** Mechanism of anion exchange across the red cell membrane by band 3: interactions between stilbene sulfonate and NAP-taurine binding sites, *Biochemistry*, 20, 5695, 1981.

127. **Lepke, S. and Passow, H.,** Inverse effect of dansylation of red blood cell membrane on band 3 protein-mediated transport of sulate and chloride, *J. Physiol.*, 328, 27, 1982.

128. **Wells, J. A. and Yount, R. G.,** Chemical modification of myosin by active-site trapping of metal-nucleotides with thiol crosslinking reagents, *Methods Enzymol.*, 85, 93, 1982.

129. **Rao, A., Martin, P., Reithmeier, R. A. F., and Cantley, L. C.,** Location of the stilbenesulfonate binding sites of the human erythrocyte anion-exchange system by resonance energy transfer, *Biochemistry,* 18, 4505, 1979.

130. **Dissing, S., Jesaitis, A. J., and Fortes, P. A. G.,** Fluorescence labeling of the human erythrocyte anion transport system. Subunit structure studied with energy transfer, *Biochim. Biophys. Acta,* 553, 66, 1979.

131. **Fasold, H.,** Chemical investigation of the tertiary structure of proteins. I. Cross-linking of myoglobin, checking its molecular weight and native state, *Biochem. Z.,* 342, 288, 1965.

132. **Brown, W. E. and Wold, F.,** Alkyl isocyanates as active site-specific inhibitors of chymotrypsin and elastase, *Science,* 174, 608, 1971.

133. **Bowes, J. H. and Cater, C. W.,** Crosslinking of collagen, *J. Appl. Chem.,* 15, 296, 1965.

134. **Herzig, D. J., Rees, A. W., and Day, R. A.,** Bifunctional reagents and protein structure determination. The reaction of phenolic disulfonyl chlorides with lysozyme, *Biopolymers,* 2, 349, 1964.

135. **Herzig, D. J., and Rees, A. W., and Day, R. A.,** The use of cross-linking reagents to study the conformation of lysozyme, *Fed. Proc.,* 21, 410, 1962.

136. **Zahn, H. and Meienhoffer, J.,** Experiments with insulin, *Makromol. Chem.,* 26, 153, 1958.

137. **Brandenburg, D.,** Peptides. 87. Preparation of $N^{\alpha A1}$, $N^{\epsilon B29}$-adipoyl-insulin, an intramolecularly crosslinked derivative of beef insulin, *Hoppe-Seyler's Z. Physiol. Chem.,* 353, 869, 1972.

138. **Plotz, P. H.,** Bivalent affinity labeling haptens in the formation of model immune complexes, *Methods Enzymol.,* 46, 505, 1977.

139. **Busse, W. D. and Carpenter, F. H.,** Carbonylbis(L-methionine p-nitrophenyl ester). A new reagent for the reversible intramolecular cross-linking of insulin, *J. Am. Chem. Soc.,* 96, 5947, 1974.

140. **Busse, W. D. and Carpenter, F. H.,** Synthesis and properties of CBM-insulin, a pro-insulin analogue which is convertible to insulin by CNBr cleavage, *Biochemistry,* 15, 1649, 1976.

141. **Zahn, H. and Schade, F.,** Nitrophenyl esters, *Chem. Ber.,* 96, 1747, 1963.

142. **Zahn, H. and Schade, F.,** Chemische modifizierung von insulin, seidenfibroin, sehnenkollagen und wollkeratin mit nitrophenylestern, *Angew. Chem.,* 75, 377, 1963.

143. **Wetz, K., Fasold, H., and Meyer, C.,** Synthesis of long, hydrophilic, protein cross-linking reagents, *Anal. Biochem.,* 58, 347, 1974.

144. **Miyakawa, T., Takemoto, L. J., and Fox, C. F.,** Membrane permeability of bifunctional, amino site-specific, cross-linking reagents, *J. Supramol. Struct.,* 8, 303, 1978.

145. **Brooks, B. R. and Klamerth, O. L.,** Interaction of DNA with bifunctional aldehydes, *Eur. J. Biochem.,* 5, 178, 1968.

146. **Cater, C. W.,** The evaluation of aldehydes and other difunctional compounds as cross-linking agents for collagen, *J. Soc. Leather Trade Chem.,* 47, 259, 1963.

147. **Richard, F. M. and Knowles, J. R.,** Glutaraldehyde as a protein cross-linking reagent, *J. Mol. Biol.,* 37, 231, 1968.

148. **Josephs, R., Eisenberg, H., and Reisler, E.,** Some properties of cross-linked polymers of glutamic dehydrogenase, *Biochemistry,* 12, 4060, 1973.

149. **Fein, M. L. and Filachione, E. M.,** Tanning studies with aldehydes, *J. Am. Leather Chem. Assoc.,* 52, 17, 1957.

150. **Seligsberger, L. and Sadlier, C.,** New developments in tanning with aldehydes, *J. Am. Leather Chem. Assoc.,* 52, 2, 1957.

151. **Hopwood, D.,** Comparison of the crosslinking abilities of glutaraldehyde, formaldehyde, and α-hydroxyadipaldehyde with bovine serum albumin and casein, *Histochemie,* 17, 151, 1969.

152. **Benson, J. R. and Hare, P. E.,** o-Phthalaldehyde: fluorogenic detection of primary amines in the picomole range. Comparison with fluorescamine and ninhydrin, *Proc. Natl. Acad. Sci. U.S.A.,* 72, 619, 1975.

153. **Kergonou, J. F., Pennacino, I., Lafite, C., and Ducousso, R.,** Immunological relevance of molonic dialdehyde (MDA). III. Immuno-enzymatic determination of human immunoglobulin binding to MDA-crosslinked proteins, *Biochem. Int.,* 16, 845, 1988.

154. **Hardy, P. M., Nicholls, A. C., and Rydon, H. N.,** The nature of glutaraldehyde in aqueous solution, *Chem. Commun.,* 565, 1969.

155. **Hardy, P. M., Nicholls, A. C., and Rydon, H. N.,** The nature of the crosslinking of proteins by glutaraldehyde. Part I. Interaction of glutaraldehyde with the amino groups of 6-aminohexanoic acid and of α-N-acetyl-lysine, *J. Chem. Soc. Perkin Trans.,* 1, 958, 1976.

156. **Monsan, P., Puzo, G., and Mazarguil, H.,** Etude du mecanisme d'etablissement des liaisons glutaraldehyde proteins, *Biochimie,* 57, 1281, 1975.

157. **Quiocho, F. A. and Richards, F. M.,** Intermolecular cross linking of a protein in the crystalline state: carboxypeptidase A, *Proc. Natl. Acad. Sci. U.S.A.,* 52, 833, 1964.

158. **Bishop, W. H. and Richards, F. M.,** Isoelectric point of a protein in the crosslinked crystalline state: β-lactoglobulin, *J. Mol. Biol.,* 33, 415, 1968.

159. **Reeke, G. N., Hartsuck, J. A., Ludwig, M. L., Quiocho, F. A., Steitz, T. A., and Lipscomb, W. N.**, The structure of carboxypeptidase A. VI. Some results at 2.0 Å resolution, and the complex with glycyl-tyrosine at 2.8 Å resolution, *Proc. Natl. Acad. Sci. U.S.A.*, 58, 2220, 1967.

160. **Habeeb, A. F. S. A. and Hiramoto, R.**, Reaction of proteins with glutaraldehyde, *Arch. Biochem. Biophys.*, 126, 16, 1968.

161. **Schejter, A. and Bar-Eli, A.**, Preparation and properties of crosslinked water-insoluble catalase, *Arch. Biochem. Biophys.*, 136, 325, 1970.

162. **Tam, J. W. and Cheng, L. Y.**, Chemical cross-linking of hemoglobin-H, *Biochim. Biophys. Acta*, 580, 75, 1979.

163. **Scannon, P. J.**, Molecular modification of hemoglobin, *Crit. Care Med.*, 10, 261, 1982.

164. **Guillochon, D., Esclade, L., Remy, M. H., and Thomas, D.**, Studies on hemoglobin immobilized by cross-linking with glutaraldehyde, *Biochim. Biophys. Acta*, 670, 332, 1981.

165. **Cambou, B., Laurent, M., Hervagault, J. F., and Thomas, D.**, Modulation of phosphofructokinase behavior by chemical modification during the immobilization process, *Eur. J. Biochem.*, 121, 99, 1981.

166. **Hermann, R., Jaenicke, R., and Rudolph, R.**, Analysis of the reconstitution of oligomeric enzymes by cross-linking with glutaraldehyde, *Biochemistry*, 20, 5195, 1981.

167. **Alwan, S., Smith, H. J., Mahbouba, M., Evans, J. C., and Morgan, P. H.**, Cross-linking of insulin with glutaraldehyde to form macromolecules, *J. Pharm. Pharmac.*, 33, 323, 1981.

168. **Engvall, E., and Perlmann, P.**, Enzyme-linked immunosorbent assay (ELISA). Quantitative assay of immunoglobulin G, *Immunochemistry*, 8, 871, 1971.

169. **Avrameas, S. and Ternynck, T.**, Peroxidase-labeled antibody and Fab conjugates with enhanced intracellular penetration, *Immunochemistry*, 8, 1175, 1971.

170. **Shimomura, S. and Fukui, T.**, Characterization of the pyridoxal phosphate site in glycogen phosphorylase b from rabbit muscle, *Biochemistry*, 17, 5359, 1978.

171. **Benesch, R. E. and Kwong, S.**, *Bis*-pyridoxal polyphosphates: A new class of specific intramolecular crosslinking agents for hemoglobin, *Biochem. Biophys. Res. Commun.*, 156, 9, 1988.

172. **Fraenkel-Conrat, H. and Olcott, H. S.**, Reaction of formaldehyde with proteins. VI. Cross-linking of amino groups with phenol, imidazole, or indole groups, *J. Biol. Chem.*, 174, 827, 1948.

173. **Fraenkel-Conrat, H. and Olcott, H. S.**, Crosslinking between amino and primary amide or guanidyl groups, *J. Am. Chem. Soc.*, 70, 2673, 1948.

174. **French, D. and Edsall, J. T.**, Reactions of formaldehyde with amino acids and proteins, *Adv. Prot. Chem.*, 2, 277, 1945.

175. **Ji, T. H.**, Bifunctional reagents, *Methods Enzymol.*, 91, 580, 1983.

176. **Blass, J., Bizzini, B., and Raynaud, M.**, Mechanism of detoxication by formol, *Compt. Rend.*, 261, 1448, 1965.

177. **Sager, P. R.**, Cytoskeletal effects of acrylamide and 2,5-hexanedione: Selective aggregation of vimentin filaments, *Toxicol. Appl. Pharmacol.*, 97, 141, 1989.

178. **Graham, D. G., Szakal-Quin, G., Priest, J. W., and Anthony, D. C.**, *In vitro* evidence that covalent cross-linking of neurofilaments occur in γ-diketone neuropathy, *Proc. Natl. Acad. Sci. U.S.A.*, 81, 4979, 1984.

179. **Ternynck, T. and Avrameas, S.**, Conjugation of *p*-benzoquinone treated enzymes with antibodies and Fab fragments, *Immunochemistry*, 14, 767, 1977.

180. **Avrameas, S., Ternynck, T., and Guesdon, J.-L.**, Coupling of enzymes to antibodies and antigens, *Scand. J. Immunol.*, (Suppl. 7), 7, 1978.

181. **Traut, R. R., Bollen, A., Sun, T. T., Hershey, J. W. B., Sundberg, J., and Pierce, L. R.**, Methyl 4-mercaptobutyrimidate as a cleavable cross-linking reagent and its application to the *Escherichia coli* 30S ribosome, *Biochemistry*, 12, 3266, 1973.

182. **Birnbaumer, M. E., Schrader, W. T., and O'Malley, B. W.**, Chemical cross-linking of chick oviduct progesterone-receptor subunit by using a reversible bifunctional cross-linking reagent, *Biochem. J.*, 181, 201, 1079.

183. **Lambert, J. M., Boileau, G., Cover, J. A., and Traut, R. R.**, Cross-links between ribosomal proteins of 30S subunits in 70S tight couples and in 30S subunits, *Biochemistry*, 22, 3913, 1983.

184. **Walleczek, J., Redl, B., Stöffler-Meiliche, M., and Stöffler, G.**, Protein-protein cross-linking of the 50S ribosomal subunit of *Escherichia coli* using 2-iminothiolane: Identification of cross-links by immunoblotting techniques, *J. Biol. Chem.*, 264, 4231, 1989.

185. **Gulle, H., Hoppe, E., Osswald, M., Greuer, B., Brimacombe, R., and Stöffler, G.**, RNA-protein cross-linking in *Escherichia coli* 50S ribosomal subunits. Determination of sites on 23S RNA that are cross-linked to proteins L2, L4, L24 and L27 by treatment with 2-iminothiolane, *Nucleic Acids Res.*, 16, 815, 1988.

186. **Coggins, J. R., Hooper, E. A., and Perham, R. N.**, Use of DMS and novel periodate-cleavable *bis*-imido-esters to study the quaternary structure of the pyruvate dehydrogenase multienzyme complex of E. coli, *Biochemistry*, 15, 2527, 1976.

187. **Robinson, I. D.,** Role of crosslinking of gelatin in aqueous solutions, *J. Appl. Polymer Sci.,* 8, 1903, 1964.

188. **Avrameas, S. and Ternynck, T.,** Biologically active water-insoluble protein polymers. I. Their use for isolation of antigens and antibodies, *J. Biol. Chem.,* 242, 1651, 1967.

189. **Hughes, W. L.,** Albumin fraction isolated from human plasma as a crystalline mercuric salt, *J. Am. Chem. Soc.,* 69, 1836, 1947.

190. **Jovin, T. M., Englund, P. T., and Kornberg, A.,** Enzymatic synthesis of deoxyribonucleic acid, *J. Biol. Chem.,* 244, 3009, 1969.

191. **Arnon, R. and Shapira, E.,** Crystalline papain derivative containing an intramolecular mercury bridge, *J. Biol. Chem.,* 244, 1033, 1969.

192. **Sperling, R., Burstein, Y., and Steinberg, I. Z.,** Selective reduction and mercuration of cystine IV-V in bovine pancreatic ribonuclease, *Biochemistry,* 8, 3810, 1969.

193. **Edsall, J. T., Maybury, R. H., Simpson, R. B., and Straessle, R.,** Dimerization of serum mercaptalbumin in the presence of mercurials. II. Studies with a bifunctional organic mercurial, *J. Am. Chem. Soc.,* 76, 3131, 1954.

194. **Edelhoch, H., Katchalsk, E., Maybury, R. H., Hughes, Jr., W. L., and Edsall, J. T.,** Dimerization of serum mercaptalbumin in presence of mercurials. I. Kinetic and equilibrium studies with mercuric salts, *J. Am. Chem. Soc.,* 75, 5058, 1953.

195. **Kay, C. M. and Edsall, J. T.,** Dimerization of mercaptalbumin in the presence of mercurials. III. Bovine mercaptalbumin in water and in concentrasted urea solutions, *Arch. Biochem. Biophys.,* 65, 354, 1956.

196. **Singer, S. J., Fothergill, J. E., and Shainoff, J. R.,** A general method for the isolation of antibodies, *J. Am. Chem. Soc.,* 82, 565, 1960.

197. **Mandy, W. J., Rivers, M. M., and Nisonoff, A.,** Recombination of univalent subunits derived from rabbit antibody, *J. Biol. Chem.,* 236, 3221, 1961.

198. **Vas, M. and Csanády, G.,** The two fast-reacting thiols of 3-phosphoglycerate kinase are structurally juxtaposed: chemical modification with bifunctional reagents, *Eur. J. Biochem.,* 163, 365, 1987.

199. **Huang, C. K. and Richards, F. M.,** Reaction of a lipid-soluble, unsymmetrical, cleavable, cross-linking reagent with muscle aldolase and erythrocyte membrane proteins, *J. Biol. Chem.,* 252, 5514, 1977.

200. **Bloxham, D. P. and Sharma, R. P.,** The development of S,S'-polymethylenebis(methanethiosulfonates) as reversible cross-linking reagent for thiol groups and their use to form stable catalytically active cross-linked dimers with glyceraldehyde-3-phosphate dehydrogenase, *Biochem. J.,* 181, 355, 1979.

201. **Bloxham, D. P. and Cooper, G. K.,** Formation of a polymethylene bis(disulfide) intersubunit cross-link between cys-281 residues in rabbit muscle glyceraldehyde-3-phosphate dehydrogenase using octamethylene bis(methane[^{35}]thiosulfonate), *Biochemistry,* 21, 1807, 1982.

202. **Vallee, B. L. and Riordon, J. F.,** Chemical approaches to the properties of active sites of enzymes, *Annu. Rev. Biochem.,* 38, 733, 1969.

203. **Simon, S. R. and Konigsberg, W. H.,** Chemical modification of hemoglobins: a study of conformation restraint by internal bridging, *Proc. Natl. Acad. Sci. U.S.A.,* 56, 749, 1966.

204. **Moore, J. E. and Ward, W. H.,** Cross-linking of bovine plasma albumin with wool keratin, *J. Am. Chem. Soc.,* 78, 2414, 1956.

205. **Fasold, H., Gröschel-Stewart, U., and Turba, F.,** Azophenyl-dimaleimide als spaltbare peptidbrücken-bildende reagentien zwischen cysteinresten, *Biochem. Z.,* 337, 425, 1963.

206. **Zahn, H. and Lumper, L.,** Specificity of bifunctional sulfhydryl reagents and synthesis of a defined dimer of bovine serum albumin, *Hoppe-Seyler's Z. Physiol. Chem.,* 349, 485, 1968.

207. **Kovacic, P. and Hein, R. W.,** Cross-linking of polymers with dimaleimide, *J. Am. Chem. Soc.,* 81, 1187, 1959.

208. **Heilmann, H. D. and Holzner, M.,** The spatial organization of the active sites of the bifunctional oligomeric enzyme tryptophan synthetase: cross-linking by a novel method, *Biochem. Biophys. Res. Commun.,* 99, 1146, 1981.

209. **Tawney, P. O., Snyder, R. H., Conger, R. P., Leibbrand, K. A., Stiteler, C. H., and Williams, A. R.,** Maleimide and derivatives. II. Maleimide and *N*-methylmaleimide, *J. Org. Chem.,* 26, 15, 1961.

210. **Freedberg, W. B. and Hardman, J. K.,** Structural and functional roles of the cysteine residues in the α subunit of the *Escherichia coli* tryptophan synthetase, *J. Biol. Chem.,* 246, 1439, 1971.

211. **Chang, F. N. and Flaks, J. G.,** Specific crosslinking of *Escherichia coli* 30S ribosomal subunit, *J. Mol. Biol.,* 68, 177, 1972.

212. **Wells, J. A., Knoeber, C., Sheldon, M. C., Werber, M. M., and Yount, R. G.,** Cross-linking of myosin subfragment 1. Nucleotide-enhanced modification by a variety of bifunctional reagents, *J. Biol. Chem.,* 255, 11135, 1980.

213. **Moroney, J. V., Warncke, K., and McCarthy, R. E.,** The distance between thiol groups in the gamma subunit of coupling factor 1 influences the proteon permeability of thylakoid membranes, *J. Bioenerg. Biomembr.,* 14, 347, 1982.

214. **Chantler, P. and Bower, S. M.,** Cross-linking between translationally equivalent sites on the heads of myosin: relationship to energy transfer results between the same pair of sites, *J. Biol. Chem.,* 263, 938, 1988.

215. **Sato, S. and Nakao, M.,** Cross-linking of intact erythrocyte membrane with a newly synthesized cleavable bifunctional reagent, *J. Biochem.,* 90, 1177, 1981.

216. **Srinivasachar, K. and Neville, D. M. Jr.,** New protein cross-linking reagents that are cleaved by mild acid, *Biochemistry,* 28, 2501, 1989.

217. **Cava, M. P., Deana, A. A., Muth, K., and Mitchell, M. J.,** N-phenylmaleimide, *Org. Synth.,* 41, 93, 1961.

218. **Trommer, W. E. and Hendrick, M.,** Formation of maleimides by a new mild cyclization procedure, *Synthesis,* 8, 484, 1973.

219. **Brewer, C. F. and Riehm, J. P.,** Evidence for possible nonspecific reactions between N-ethylmaleimide and proteins, *Anal. Biochem.,* 18, 248, 1967.

220. **Riordan, J. F. and Vallee, B. L.,** Reactions with N-ethylmaleimide and p-mercuribenzoate, *Methods Enzymol.,* 25, 449, 1972.

221. **Hashida, S., Imagawa, M., Inoue, S., Ruan, K.-H., and Ishikawa, E.,** More useful maleimide compounds for the conjugation of Fab' to horseradish peroxidase through thiol groups in the hinge, *J. Appl. Chem.,* 6, 56, 1984.

222. **Yoshitake, S., Hamaguchi, Y., and Ishikawa, E.,** Efficient conjugation of rabbit Fab' with β-D-galactosidase from *Escherichia coli, Scand. J. Immunol.,* 10, 81, 1979.

223. **Perkins, W. J., Wells, J. A., and Young, R. G.,** Characterization of the properties of ethenoadenosine nucleotides bound or trapped at the active site myosin subfragment, *Biochemistry,* 23, 3994, 1984.

224. **Miller, L., Coopedge, J., and Reisler, E.,** The reactive SH1 and SH2 cysteines in myosin subfragment 1 are cross-linked at similar rates with reagents of different length, *Biochem. Biophys. Res. Commun.,* 106, 117, 1982.

225. **Husain, S. S. and Lowe, G.,** Evidence for histidine in the active sites for ficin and stem-bromelain, *Biochem. J.,* 110, 53, 1968.

226. **Ozawa, H.,** Bridging reagent for protein. II. The reaction of N,N'-polymethylenebis(iodoacetamide) with cysteine and rabbit muscle aldolase, *J. Biochem. (Tokyo),* 62, 531, 1967.

227. **Gundlash, H. G.,** Habilitationsschrift, University of Wurzburg, 1965, 44.

228. **Wilchek, M. and Givol, D.,** Affinity cross-linking of heavy and light chains, *Methods Enzymol.,* 46, 501, 1977.

229. **Segal, D. M. and Hurwitz, E.,** Dimers and trimers of immunoglobulin G covalently cross-linked with a bivalent affinity label, *Biochemistry,* 15, 5253, 1976.

230. **Luduena, R. F., Roach, M. C., Trcka, P. P., and Weintraub, S.,** Bioiodoacetyldithioethylamine: A reversible cross-linking reagent for protein sulfhydryl group, *Analyt. Biochem.,* 117, 76, 1982.

231. **Haugland, R. P.,** Handbook of fluorescent probes and research chemicals, Molecular Probes, Inc., Eugene, Oregon, 1990, 22.

232. **Fasold, H., Gröschel-Stewart, U., and Turba, F.,** Synthese and reaktionen eines wasserlöslichen, spaltbarenreagens zur verknupfung frier SH-gruppen in proteinen, *Biochem. Z.,* 339, 287, 1964.

233. **Hiremath, C. B. and Day, R. A.,** Introduction of covalent cross-linkages into lysozyme by reaction with α,α'-dibromo-p-xylenesulfonic acid, *J. Am. Chem. Soc.,* 86, 5027, 1964.

234. **Ross, W. C. J.,** The chemistry of cytotoxic alkylating agents, *Adv. Cancer Res.,* 1, 397, 1953.

235. **Alexander, P.,** The reactions of carcinogens with macromolecules, *Adv. Cancer Res.,* 2, 1, 1954.

236. **Burnop, V. C. E., Francis, G. E., Richards, D. E., and Wormall, A.,** Nitrogen-15-labeled nitrogen mustard. The combination of bis(2-chloroethyl)methylamine with proteins, *Biochem. J.,* 66, 504, 1957.

237. **Boursnell, J. C.,** Some reactions of mustard gas (β,β'-di-chlorodiethyl sulfide) with proteins, *Biochem. Soc. Symp.,* 2, 8, 1948.

238. **Philips, F. S.,** Recent contributions to the pharmacology of bis(2-haloethyl)amines and sulfides, *J. Pharmacol. Exp. Therap.,* 99, 281, 1950.

239. **Goodlad, G. A. J.,** Cross-linking of collagen by sulfur- and nitrogen-mustards, *Biochim. Biophys. Acta,* 25, 202, 1957.

240. **Brookes, P., and Lawley, P. D.,** Effects of alkylating agents on T_2 and T_4 bacteriophages, *Biochem. J.,* 89, 138, 1963.

241. **Berenblum, I. and Wormall, A.,** The immunological properties of proteins treated with β,β'-dichlorodiethyl sulfide (mustard gas) and β,β'-dichlorodiethyl sulfones, *Biochem. J.,* 33, 75, 1939.

242. **Hiratsuka, T.,** Cross-linking of three heavy-chain domains of myosin adenosine triphosphatase with a trifunctional alkylating agent, *Biochemistry,* 27, 4110, 1988.

243. **Aliosman, F., Caughlan, J., Gray, G. S.,** Diseased DNA intrastrand cross-linking and cytotoxicity induced in human brain tumor cells by 1,3-bis(2-chloroethyl)-1-nitrosourea after *in vitro* reaction with glutathione, *Cancer Res.,* 49, 5954, 1989.

244. **Prestayko, A. W., Baker, L. H., Crooke, S. T., Carter, S. K., and Schein, P. S.,** *Nitrosoureas. Current Status and New Developments,* Academic Press, New York, 1981, chap. 4.

245. **Ewig, R. A. G. and Kohn, K. W.,** DNA-protein cross-linking and DNA interstrand cross-linking by haloethylnitrosoureas in L1210 cells, *Cancer Res.,* 38, 3197, 1977.

246. **Ali-Osman, F., Giblin, J., Berger, M., Murphy, M. J., Jr., and Rosenblum, M. L.,** Chemical structure of carbamoylating groups and their relationship to bone marrow toxicity and antiglioma activity of bifunctionally alkylating and carbamoylating nitrosoureas, *Cancer Res.,* 45, 4185, 1985.

247. **Agarwal, K. L., Grudzinski, S., Kenner, G. W., Rogers, N. H., Sheppard, R. C., and McGuigan, J. E.,** Immunochemical differentiation between gastrin and related peptide hormones through a novel conjugation of peptides to proteins, *Experientia,* 27, 514, 1971.

248. **Nakane, P. K.,** Simultaneous localization of multiple tissue antigens using the peroxidase-labeled antibody method. A study on pituitary glands of the rat, *J. Histochem. Cytochem.,* 16, 557, 1968.

249. **Cater, C. W.,** The efficiency of dialdehydes and other compounds as cross-linking agents for collagen, *J. Soc. Leather Trades Chem.,* 49, 455, 1965.

250. **Avrameas, S., Taudou, B., and Chuilon, S.,** Glutaraldehyde, cyanuric chloride, and tetra-azotized *o*-dianisidine as coupling reagents in the passive hemogglutination test, *Immunochemistry,* 6, 67, 1969.

251. **Wadsworth, P., Sloboda, R. D.,** Modification of tubulin with the fluorochrome 5-(4,6-dichlorotriazin-2-yl)aminofluorescein and the interaction of the fluorescent protein with the isolated meiotic apparatus, *Biol. Bull.,* 166, 357, 1984.

252. **Mahoney, C. W. and Azzi, A.,** The synthesis of fluorescent chlorotriazinylaminofluorescein-concanavalin A and its use as a glycoprotein stain on sodium dodecyl sulfate/polyacrylamide gels, *Biochem. J.,* 243, 567, 1987.

253. **Packard, B., Edidin, M., and Komoriya, A.,** Site-directed labeling of a monoclonal antibody. Targeting to a disulfide bond, *Biochemistry,* 25, 2548, 1986.

254. **Haugland, R. P.,** *Handbook of Fluoresent Probes and Research Chemicals,* Molecular Probes, Inc., Eugene, Oregon, 1990, 40.

255. **Onoue, K., Yagi, Y., and Pressmann, D.,** Immunoadsorbents with high capacity, *Immunochemistry,* 2, 181, 1965.

256. **Likhite, V. and Sehon, A. H.,** Protein-protein conjugation, in *Methods in Immunology and Immunochemistry,* Williams, C. A. and Chase, M. W., Eds., Vol. 1, Academic Press, 1967, 150.

257. **Sköld, S.-E.,** Chemical crosslinking of elongation factor G to the 23SRNA in 70S ribosomes from *Escherichia coli, Nucleic Acid Res.,* 11, 4923, 1983.

258. **Kohn, K. W., Spears, C. L., and Doty, P.,** Interstrand cross-linking of DNA by nitrogen mustard, *J. Mol. Biol.,* 19, 266, 1966.

259. **Fearnley, C. and Speakman, J. B.,** Cross-linkage formation in keratin, *Nature (London),* 166, 743, 1950.

260. **Sundberg, L. and Porath, J.,** Preparation of adsorbents for biospecific affinity chromatography. I. Attachment of group-containing ligands to insoluble polymers by means of bifunctional oxiranes, *J. Chromatogr.,* 90, 87, 1974.

261. **Porath, J.,** General methods and coupling procedures, *Methods Enzymol.,* 34, 13, 1974.

262. **Porath, J. and Axén, R.,** Immobilization of enzymes to agar, agarose, and sephadex supports, *Methods Enzymol.,* 44, 19, 1976.

263. **Carraway, K. L. and Koshland, D. E., Jr.,** Carbodiimide modification of proteins, *Methods Enzymol.,* 25, 616, 1972.

264. **Husain, S. S., Ferguson, J. B., Fruton, J. S.,** Bifunctional inhibitor of pepsin, *Proc. Natl. Acad. Sci. U.S.A.,* 68, 2765, 1971.

265. **Lundblad, R. L. and Stein, W. H.,** On the reaction of diazoacetyl compounds with pepsin, *J. Biol. Chem.,* 244, 154, 1969.

266. **Zahn, H. and Waschka, O.,** Reaction of bisdiazohexane with amino acids, mercaptans, wool keratin, and silk fibroin, *Makromol. Chem.,* 18, 201, 1956.

267. **Gold, P., Jonson, J., and Freedman, S. O.,** The effect of the sequence of erythrocyte sensitization on the *bis*-diazotized benzidine hemagglutination reaction, *J. Allergy,* 37, 311, 1966.

268. **Stavitsky, A. B. and Arquilla, E. R.,** Studies of proteins and antibodies by specific hemagglutination and hemolysis of protein-conjugated erythrocytes, *Int. Arch. Allergy Appl. Immunol.,* 13, 1, 1958.

269. **Howard, A. N. and Wild, F.,** A two-stage method of cross-linking proteins suitable for use in serological techniques, *Br. J. Exp. Pathol.,* 38, 640, 1957.

270. **Olovnikov, A. M.,** Sensitization of erythrocytes by polycondensed proteins of immune serum and their use for determining antigen content, *Immunochemistry,* 4, 77, 1967.

271. **Gordon, J., Rose, B., and Schon, A. H.,** Detection of ''nonprecipitating'' antibodies of individuals allergic to ragweed pollen by an *in vitro* method, *J. Exp. Med.,* 108, 376, 1958.

272. **Arquilla, E. R. and Stavitsky, A. B.,** The production and identification of antibodies to insulin and their use in assaying insulin, *J. Clin. Invest.,* 35, 458, 1956.

273. **Ishizaka, K., and Ishizaka, T.,** Molecular basis of passive sensitization. II. The role of fragment III of γ-globulin and its disulfide bonds in passive sensitization and complement fixation, *J. Immunol.,* 93, 59, 1964.

274. **DeCarvalho, S., Lewis, A. J., Rand, H. J., and Uhrick, J. R.,** Immunochromatographic partition of soluble antigens on columns of insoluble diazo-gamma-globulins, *Nature (London),* 204, 265, 1964.

275. **Goldman, R., Kedam, O., Silman, I. H., Caplan, S. R., and Katchalski, E.,** Papain-collodion membranes. I. Preparation and properties, *Biochemistry,* 7, 486, 1968.

276. **Koch, N. and Haustein, D.,** Association of surface IgM with two membrane proteins on murine B lymphocytes detected by chemical crosslinking, *Mol. Immunol.,* 20, 33, 1983.

277. **Anderer, F. A. and Schlumberger, H. D.,** Antigenic properties of proteins cross-linked by multidiazonium compounds, *Immunochemistry,* 6, 1, 1969.

278. **Bar-Eli, A. and Katchalski, E.,** Preparation and properties of water-insoluble derivatives of trypsin, *J. Biol. Chem.,* 238, 1690, 1963.

279. **Cetra, J. J., Givol, D., Silman, H. I., and Katchalski, E.,** A two-stage cleavage of rabbit γ-globulin by a water-insoluble papain preparation followed by cysteine, *J. Biol. Chem.,* 236, 1720, 1961.

280. **Riesel, E. and Katchalski, E.,** Preparation and properties of water-insoluble derivatives of urease, *J. Biol. Chem.,* 239, 1521, 1964.

281. **Pressman, D., Campbell, D. H., and Pauling, L.,** The agglutination of intact azo-erythrocytes by antisera homologous to the attached groups, *J. Immunol.,* 44, 101, 1942.

282. **Borek, F.,** A new two-stage method for cross-linking proteins, *Nature (London),* 191, 1293, 1961.

283. **Kalnins, V. I., Stich, H. F., and Yohn, D. S.,** Electron microscopic localization of virus-associated antigens in human amnion cells (AV-3) infected with human adenovirus, type 12, *Virology,* 28, 751, 1966.

284. **Ishizaka, K., and Ishizaka, T.,** Biologic activity of aggregated γ-globulin. II. A study of various methods for aggregation and species differences, *J. Immunol.,* 85, 163, 1960.

285. **Demoliou, C. D. and Epand, R. M.,** Synthesis and characterization of a heterobifunctional photoaffinity reagent for modification of tryptophan residues and its application to the preparation of a photoreactive glucagon derivative, *Biochemistry,* 19, 4539, 1980.

286. **Earland, C. and Stell, J. P. C.,** Formation of new cross-linkages in proteins by oxidation of tyrosine residues with potassium nitrosyldisulfonate, *Polymer,* 7, 549, 1966.

287. **Consden, R. and Kirrane, J. A.,** Cross-linking in collagen by potassium nitrosyldisulfonate, *Nature (London),* 218, 957, 1968.

288. **Doyle, R. J., Bello, J., and Roholt, O. A.,** Probable protein cross-linking with tetranitromethane, *Biochim. Biophys. Acta,* 160, 274, 1968.

289. **Wagner, R. and Garrett, R. A.,** A new RNA-RNA cross-linking reagent and its application to ribosomal 5S RNA, *Nucleic Acid Res.,* 5, 4065, 1978.

290. **Hancock, J. and Wagner, R.,** A structural model of 5S RNA from E. coli based on intramolecular cross-linking evidence, *Nucleic Acid Res.,* 10, 1257, 1982.

291. **Das, M. and Fox, F.,** Chemical cross-linking in biology, *Annu. Rev. Biophys. Bioeng.,* 8, 165, 1979.

292. **Mikkelsen, R. B. and Wallach, D. F. H.,** Photoactivated cross-linking of protein within the erythrocyte membrane core, *J. Biol. Chem.,* 251, 7413, 1976.

293. **Guire, P.,** Stepwise thermophotochemical cross-linking agents for enzyme stabilization and immobilization, *Fed. Proc.,* 35, 1632, 1976.

294. **Bayer, E. A. and Wilchek, M.,** The use of avidin-biotin complex, *Methods Biochem. Anal.,* 26, 1, 1980.

295. **Chaiet, L. and Wolf, F. J.,** The properties of streptavidin, a biotin-binding protein produced by streptomycetes, *Arch. Biochem. Biophys.,* 106, 1, 1964.

296. **Weber, P. Cl., Ohlendore, D. H., Wendoloski, J. J., and Salemme, F. R.,** Structural origins of high-affinity biotin binding to streptavidin, *Science (London),* 243, 85, 1989.

297. **Hacuptle, M. T., Aubert, M. L., Djiane, J., and Krachenbuhl, J. P.,** Binding sites for lactogenic and somatogenic hormones from rabbit mammary gland and liver, *J. Biol. Chem.,* 258, 305, 1983.

298. **Gardner, L.,** Nonradioactive DNA labeling. Detection of specific DNA and RNA sequences on nitrocellulose and *in situ* hybridizations, *Biotechniques,* 1, 39, 1983.

299. **Hofmann, K., Wood, S. W., Brinton, C. C., Montibeller, J. A., and Finn, F. M.,** Iminobiotin affinity columns and their application to retrieval of streptavidin, *Proc. Natl. Acad. Sci. U.S.A.,* 77, 4666, 1980.

300. **Wilchek, M. and Bayer, E. A., Eds.,** Avidin-biotin technology, *Methods Enzymol.,* 184, 1990.

301. **Bode, C., Runge, M. S., Branscomb, E. E., Newell, J. B., Matsueda, G. R., and Haber, E.,** Antibody-directed fibrinolysis: An antibody specific for both fibrin and tissue plasminogen activator, *J. Biol. Chem.,* 264, 944, 1989.

302. **Kurokawa, T., Iwasa, S., and Kakinuma, A.,** Enhanced fibrinolysis by a bispecific monoclonal antibody reactive to fibrin and tissue plasminogen activator, *Bio/Technology,* 7, 1163, 1989.

303. **Staerz, U. D. and Bevan, M. J.,** Hybrid hybridoma producing a bispecific monoclonal antibody that can focus effector T-cell activity, *Proc. Natl. Acad. Sci. U.S.A.,* 83, 1453, 1986.

304. **Tada, H., Toyoda, Y., and Iwasa, S.,** Bispecific antibody producing hybrid hybridoma and its use in one-step immunoassays for human lymphotoxin, *Hybridoma*, 8, 73, 1989.

305. **Karawajew, L., Micheel, B., Behrsing, O., and Gaestel, M.,** Bispecific antibody-producing hybrid hybridoma selected by a fluorescence activated cell sorter, *J. Immunol. Methods*, 96, 265, 1987.

306. **Nitecki, D. E., Woods, V., and Goodman, J. W.,** Crosslinking of antibody molecules by bifunctional antigens, in *Protein Cross-Linking. Biochemical and Molecular Aspects,* Friedman, M., Ed., Plenum Press, New York, 1976, 139.

307. **Rifai, A.,** Experimental models for IgA-associated nephritis, *Kidney Int.,* 31, 1, 1987.

308. **Plotz, P. H. and Rifai, A.,** Stable, soluble, model immune complexes made with a versatile multivalent affinity-labeling antigen, *Biochemistry*, 21, 301, 1982.

309. **Rifai, A. and Wong, S. S.,** Preparation of phosphorylcholine-conjugate antigens, *J. Immunol. Methods*, 94, 25, 1986.

310. **Chen, A., Wong, S. S., and Rifai, A.,** Glomerular immune deposits in experimental IgA nephropathy: a continuum of circulating and *in situ* formed immune complexes, *Am. J. Pathol.,* 130, 216, 1988.

311. **Sharon, N. and Lis, H.,** *Lectins,* Chapman and Hall, New York, 1989.

312. **Bog-Hansen, T. C., Ed.,** *Lectins: Biology, Biochemistry, Clinical Biochemistry,* Vols. 1 to 5, De Gruyter, New York, 1981 to 1986.

313. **Lis, H. and Sharon, N.,** Lectins. Properties and applications to the study of complex carbohydrates in solution and on cell surfaces, in *Biology of Carbohydrates,* Ginsburg, V. and Robins, P. W., Eds., Vol. 2, Wiley, New York, 1984, 1.

314. **Etzler, M. E.,** Plant lectins: molecular and biological aspects, *Annu. Rev. Plant Physiol.,* 36, 209, 1985.

315. **Liener, I. E., Sharon, N., and Goldstein, I. J., Eds.,** *The Lectins: Properties, Functions and Applications in Biology and Medicine,* New York, Academic Press, 1986.

316. **Lis, H. and Sharon, N.,** Lectin as molecules and as tools, *Annu. Rev. Biochem.,* 55, 35, 1986.

317. **Osawa, T. and Tsuji, T.,** Fractionation and structural assessment of oligasaccharides and glycopeptides by use of immobilized lectins, *Annu. Rev. Biochem.,* 56, 21, 1987.

318. **Agrawal, B. B. L. and Goldstein, I. J.,** Protein-carbohydrate interaction. VI. Isolation of concanavalin A by specific adsorption on cross-linked dextran gels, *Biochim. Biophys. Acta,* 147, 262, 1967.

319. **McKenzie, G. H., Sawyer, W. H., and Nichol, L. W.,** The molecular weight and stability of concanavalin A, *Biochim. Biophys. Acta,* 263, 283, 1972.

320. **Agrawal, B. B. L. and Goldstein, I. J.,** Protein-carbohydrate interaction. XV. The role of bivalent cations in concanavalin A-polysaccharide interaction, *Can. J. Biochem.,* 46, 1147, 1968.

321. **Goldstein, I. J.,** Studies on the combining sites of concanavalin A, *Adv. Exp. Med. Biol.,* 55, 35, 1974.

322. **Goldstein, I. J. and Hayes, C. E.,** The lectins. Carbohydrate-binding proteins of plants and animals, *Adv. Carbohydr. Chem. Biochem.,* 35, 127, 1978.

323. **Goldstein, I. J., Iyer, R. N., Smith, E. E., and So, L. L.,** Protein-carbohydrate interaction. XX. The interaction of concanavalin A with sepharose and some of its derivatives, *Biochemistry*, 6, 2373, 1967.

324. **So, L. L. and Goldstein, I. J.,** Protein-carbohydrate interaction. XXI. Interaction of concanavalin A with o-fructans, *Carbohydr. Res.,* 10, 231, 1969.

325. **Poretz, R. D. and Goldstein, I. J.,** An examination of the topography of the saccharide binding sites of concanavalin A and of the forces involved in complexation, *Biochemistry*, 9, 2890, 1970.

326. **Dulaney, J. T.,** Binding interactions of glycoproteins with lectins, *Mol. Cell. Biochem.,* 21, 43, 1978.

327. **Goldstein, I. J.,** Use of concanavalin A for structural studies, *Methods Carbohydr. Chem.,* 6, 106, 1972.

328. **Bog-Hansen, T. C.,** Affinity electrophoresis with lectins for characterization of glycoproteins, in Proc. 3rd Int. Symp. Affinity Chromatography and Molecular Interaction, Egly, J. M., Ed., *INSERM Symp. Ser.,* 1979, 399.

329. **Watkins, W. M., Yates, A. D., and Greenwell, P.,** Blood group antigens and the enzymes involved in their synthesis. Past and present, *Biochem. Soc. Trans.,* 9, 186, 1981.

330. **Judd, W. J.,** The role of lectins in blood group serology, *CRC Crit. Rev. Clin. Lab. Sci.,* 12, 171, 1980.

331. **Yajko, D. M., Chu, A., and Hadley, W. K.,** Rapid confirmatory identification of *Neisseria gonorrhoeae* with lectins and chromatogenic substrates, *J. Clin. Microbiol.,* 19, 380, 1984.

332. **Raychowdhury, M. K., Goswami, R., and Chakrabarti, P.,** Agglutination of some bacterial cells by concanavalin A, *Indian J. Exp. Biol.,* 20, 748, 1982.

333. **DeLucca, A. J., II.,** Lectins grouping of *Bacillus thuringiensis* serovars, *Can J. Microbiol.,* 30, 1100, 1984.

334. **Graham, K., Keller, K., Ezzell, J., and Doyle, R.,** Enzyme-linked lectinosorbent assay (ELIA) for detecting *Bacillus anthracis, Eur. J. Clin. Microbiol.,* 3, 210, 1984.

335. **Roth, J.,** Application of lectin-gold complexes for electron microscopic localization of glycoconjugation on thin sections, *J. Histochem. Cytochem.,* 31, 987, 1983.

336. **Roche, P. A., Jensen, P. E. H., and Pizzo, S. V.,** Intersubunit cross-linking by *cis*-dichlorodiammine-platinum (II) stabilizes an α_2-macroglobulin "nascent" state: evidence that thiol ester bond cleavage correlates with receptor recognition site exposure, *Biochemistry*, 27, 759, 1988.

337. **Pizzo, S. V., Roche, P. A., Feldman, S. R., and Gonias, S. L.,** Further characterization of platinum-reactive component of the α_2-macroglobulin-receptor recognition site, *Biochem. J.,* 238, 217, 1986.
338. **Gonias, S. L., Oakley, A. C., Walther, P. J., and Pizzo, S. V.,** Effect of diethyldithiocarbamate and nine other nucleophiles on the intersubunit protein cross-linking and inactivation of purified human α_2-macroglobulin by *cis*-diamminedichloroplatinum (II), *Cancer Res.,* 44, 5764, 1984.
339. **Rosenberg, B., van Camp, L., Trosko, J. E., and Mansour, V. H.,** Platinum compounds: a new class of potent antitumor agents, *Nature (London),* 222, 385, 1969.
340. **Roberts, J. J. and Pascoe, J. M.,** Cross-linking of complementary strand of DNA in mammalian cells by antitumor platinum compounds, *Nature (London),* 235, 282, 1972.
341. **Kozulic, B., Barbaric, S., Ries, B., and Mildner, P.,** Study of the carbohydrate part of yeast acid phosphatase, *Biochem. Biophys. Res. Commun.,* 122, 1083, 1984.
342. **Folk, J. E. and Chung, S. I.,** Molecular and catalytic properties of transglutaminases, *Adv. Enzymol.,* 38, 109, 1973.
343. **Folk, J. E. and Finlayson, J. S.,** The ϵ-(γ-glutamyl)lysine crosslink and the catalytic role of transglutaminases, *Adv. Prot. Chem.,* 31, 1, 1977.
344. **Gorman, J. J. and Folk, J. E.,** Transglutaminase amine substrates for photochemical labeling and cleavable cross-linking of proteins, *J. Biol. Chem.,* 255, 1175, 1980.
345. **Gorman, J. J. and Folk, J. E.,** Structural features of glutamine substrates for human plasma factor XIIIa (activated blood coagulation factor XIII), *J. Biol. Chem.,* 255, 419, 1980.
346. **Folk, J. E.,** Transglutaminase, *Annu. Rev. Biochem.,* 49, 517, 1980.
347. **Anderer, F. A. and Schlumberger, H. D.,** Antigenic properties of proteins cross-linked by multidiazonium compounds, *Immunochemistry,* 6, 1, 1969.
348. **Milch, R. A.,** Viscometric hardening of gelatin sols in the presence of certain intermediary metabolites, *J. Surg. Res.,* 3, 254, 1963.
349. **Murayama, Y., Satoh, S., Oka, T., Imanishi, J., and Noishiki, Y.,** Reduction of the antigenicity and immunogenicity of xenografts by a new cross-linking reagent, *ASAIO Trans.,* 34, 546, 1988.

Chapter 5

HETEROBIFUNCTIONAL CROSS-LINKERS

I. INTRODUCTION

In contrast to homobifunctional reagents, heterobifunctional reagents contain two dissimilar functional groups of different specificities. These two reactive functionalities can be any combination of the conventional group-selective moieties. For example, one end of the cross-linker may be selective for an amino group while the other end directed to a sulfhydryl group. As categorized in Table 1, a variety of combinations is possible; thus, different amino acids in proteins can be linked together. It should be stressed again, however, that with few exceptions, no reagent is absolutely group specific. There is cross-reactivity between the various classes of heterobifunctional cross-linkers presented in Table 1. The classification is a guide for selectivity rather than specificity. By taking advantage of the differential reactivity of the two different functional groups, cross-linking can be controlled both selectively and sequentially.

In addition to the conventional group-specific functionalities, one end of the heterobifunctional cross-linker may be photosensitive as represented in Table 2. These photosensitive groups, as elaborated in Chapter 3, react indiscriminately on activation by irradiation. With one end of the cross-linker anchored to an amino acid residue, the photoreactive moiety can be used to probe its surrounding environment. The photosensitive reagents are most effectively used in macromolecular photoaffinity labeling, in which the reagent is first incorporated into a polypeptide ligand in the dark. The labeled ligand is then allowed to bind to specific receptors which are in turn cross-linked to the ligand on photolysis.[1,2] The use of this technique for the identification of receptors will be discussed in Chapter 9. The following will be a general discussion of the reactivities of these reagents.

II. GROUP SELECTIVE HETEROBIFUNCTIONAL CROSS-LINKING REAGENTS

A. AMINO AND SULFHYDRYL GROUP DIRECTED CROSS-LINKERS

This assembly of heterobifunctional cross-linking reagents contains two-headed reactive functionalities directed toward amino and sulfhydryl groups. Most of the amino group directed reactive functionalities are acylating agents, while those directed toward sulfhydryl groups are alkylating agents. One end of the compounds I to XXVI (Table 1) are activated by the hydroxysuccinimidyl ester moiety which will undergo nucleophilic substitution reaction, liberating N-hydroxysuccinimide.[1,3-16] Other activated esters include p-nitrophenyl ester, as in XXVII to XXXIII.[14-21] and imidoester, as in XXXV to XXXVIII.[23-25] Acyl azides, XXXIX,[26,27] acyl chlorides, XL to XLII,[28] and aryl halide, XXXIV,[22] are also directed toward the amino group. Since they are highly activated, they may also react with hydroxyl groups.[28] Haloketone, XLIII, and alkyl halide, XLIV, which react with various nucleophiles are also included as amino group selective.[27,29,30] These functionalities will, of course, react faster with the thiol and are intended as thiol selective in reagents XIX to XXVIII, XXXI, XXXII, XXXVI to XXXVIII and XLVI.[32] Other thiol-directed groups in these two-headed compounds are the maleimido moiety as in II to XVIII, XXIX, XXX, and XXXIX to XLV, disulfide bonds as in I, XXXIV and XXXV and other alkylating moieties as in XXXIII and XLVII to LIX.[33,34] Since maleimido group attached to aromatic ring are labile at neutral pH, the most stable compound is probably SMCC (VI and VII).[33,34] Nitrovinyl of XXXIII reacts with the thiol through Michael addition to the double bond

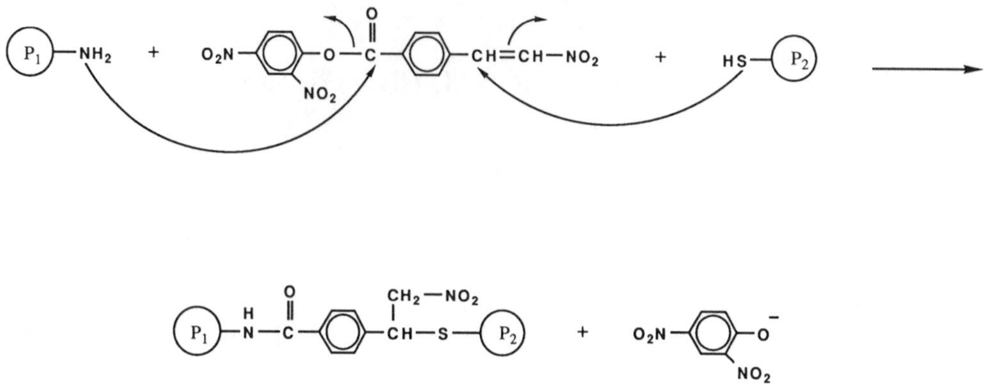

FIGURE 1.   Cross-linking reaction of 2,4-dinitrophenyl-*p*-(β-nitrovinyl)benzoate.

more readily than *N*-maleimide derivatives under acidic conditions.[21] The dinitrophenyl ester of the other end of the compound also reacts with amines at a much faster rate than *N*-hydroxysuccinimide esters under basic conditions. The reaction is shown in Figure 1. Cross-linking of sulfhydryl and amino groups is also achievable with epichlorohydrin (**XLVI**) although its main application is in the immobilization of proteins and carbohydrates.

Many of these two-headed compounds can react with different amino acid side chains. Ethyl haloacetimidates (**XXXVI** to **XXXVIII**), for example, have a broad reactivity. Although their imidate moieties react quite specifically with lysine residues through the amidination reaction, the haloacetamido group can react with an amino group or any nucleophile including histidine as shown by Diopoh and Olomucki on cross-linking studies with RNAse.[24,25] Since the thiol is a strong nucleophile, many of the reagents such as **XLIII** and **XLIV** may act as homobifunctional reagents cross-linking two sulfhydryl groups if the thiol group is in excess. On the other hand, in its absence, other nucleophiles will be cross-linked. In general, however, these reagents are used to cross-link proteins with known amino acid side chains in a sequence of reaction steps. Proteins which are known to contain free thiols will be reacted first with the compound. After modification, the free end of the bifunctional reagent is reacted with another protein with desired amino acid side chains. For example, to use **XLIII** as a heterobifunctional cross-linker, the maleimide end is first reacted with a protein containing a thiol group. The alkyl halide end is then made to react with an amino group in another protein.

There are several compounds in this category that are cleavable. Obviously, compounds that result in the formation of disulfide bonds, for example, **I** and **XXXV**, are cleavable in the presence of a free thiol. The sulfone-containing compounds **XXIII**, **XXIV**, **XXVII** and **XXVIII** are cleavable by dithionite. Compound **XLV** is cleavable under mildly acidic conditions.[31]

Among all these compounds, few are particularly worth noting. Compound **XXXIX** is photosensitive but undergoes nucleophilic substitution in the dark. This compound contains a phenolic ring and is, therefore, iodinatable by various iodinating agents (see Chapter 3), providing the possibility of introducing radioisotope ^{125}I for various applications. Compounds **XXV** and **XXVI** contain three functional groups, one succinimide ester and two alkyl halides. These compounds have the potential of cross-linking three sites in a protein or proteins. Although the reagents have not been employed under this context, useful information may be obtained through this application.

Of particular interest are the reagents **XLVII** to **LIX** which belong to a class of compounds referred to as equilibrium transfer alkylating cross-linkers (ETAC) described by Mitra and Lawton.[33,34] These compounds contain a good leaving group and an electron-withdrawing

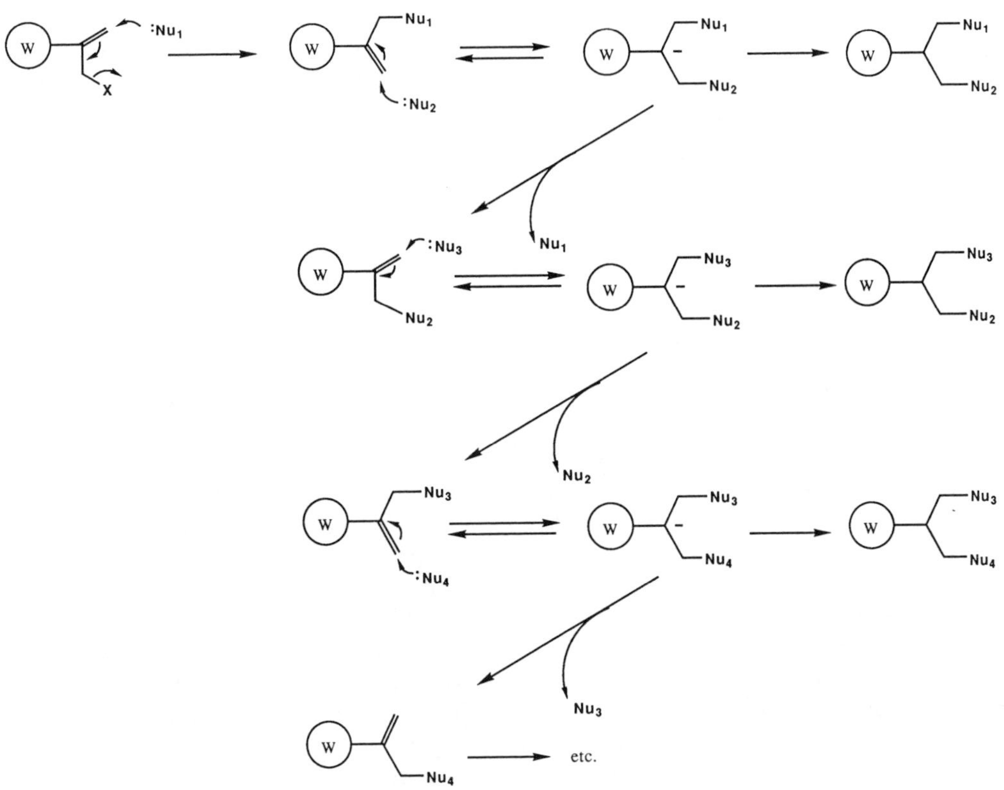

FIGURE 2. Mechanism of reaction of equilibrium transfer alkylating cross-linkers (ETAC). W designates an electron withdrawing group and X a good leaving roup. (Adapted from Libertore, F. A. et al., *Bioconjugate Chem.*, 1, 36, 1990).

group in resonance with a double bond. In the first described compounds (**XLVII** and **XLVIII**), *p*-nitrophenyl served as an electron-withdrawing group and 5-thio-2-nitrobenzoate and ammonium iodide, respectively, served as leaving groups. Protein nucleophilic residues undergo Michael addition at the double bond eliminating the leaving group to generate a new resonant double bond. A second Michael addition is then possible with another nucleophile on the same or different protein resulting in a cross-linkage. Alternatively, during the second Michael reaction process, the first added nucleophile may be eliminated reforming the double bond. A third nucleophile may then undergo similar Michael addition and the process continues. The reagent can be transferred from the initial site of protein attachment to other groups until the most thermodynamically stable bond is formed (Figure 2). With the protein side chains of lysine, tyrosine, glutamic and aspartic acids, the reagent will undergo Michael addition and Michael elimination indefinitely, until a thermodynamically stable bond is formed. A more stable bond is formed when thiol addition is encountered to form the thioether bridge. Since the first publication, several more compounds have been synthesized, for example **XLIX** to **LIX**.[34] Not only can these compounds be used to cross-link different groups heterobifunctionally or homobifunctionally, these compounds have also been used to introduce various reporter groups to reduced immunoglobulins.[123]

The amino and sulfhydryl directed heterobifunctional reagents have been extensively used to cross-link various proteins. SPDP (**I**) is probably the most popular of all. It has been used to conjugate antifibrin antibody to urokinase and TPA as well as in the preparation of bispecific antibodies.[37,38] It has also been used in the preparation of immunotoxins (see Chapter 11).[18,39,40] Many other compounds have also been frequently used in the preparation of enzyme-immunoglobulin conjugates.[41,42]

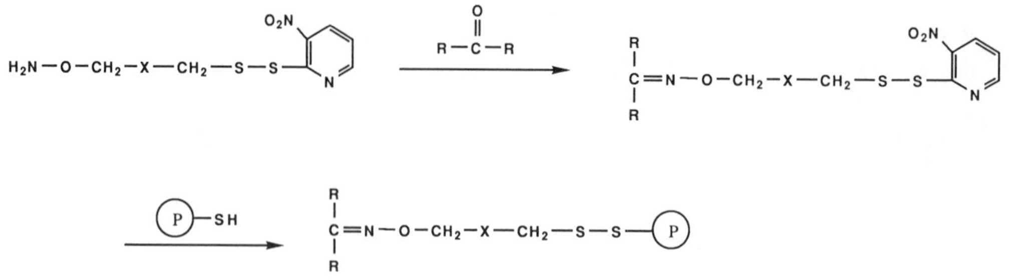

FIGURE 3.   Cross-linking carbonyl and sulfhydryl groups with aminooxypyridyldithio derivatives.

B. CROSS-LINKERS DIRECTED TOWARD CARBOXYL AND EITHER SULFHYDRYL OR AMINO GROUPS

As discussed in Chapter 3, compounds containing an diazoacetyl group are photosensitive. In the dark, however, they are reactive towards the carboxyl group at acidic pHs (see Chapter 2).[60] Therefore, in addition to being photoaffinity labels, the cross-linking reagents **LX** to **LXIII** are potential cross-linkers for carboxyl groups.[43-46] At the opposite end of the molecule, reagent **LX** is thiol specific because of the disulfide bond which will undergo thiol-disulfide interchange with sulfhydryl groups. While compound **LXI** is an alkylating agent, **LXII** and **LXIII** are acylating agents. They will cross-link carboxyl group with any other nucleophiles. In the presence of sulfhydryl groups, **LXI** will be thiol directing. In its absence, amino groups and other nucleophiles will react. This is true for **LXII** and **LXIII** except that the acylating agent will form more stable products with amino groups. Compound **LXI** was shown to react with a cysteine at the active site of pepsin,[44] while **LXII** was used to photolabel the active site of chymotrypsin.[45]

C. CARBONYL AND SULFHYDRYL GROUP DIRECTED CROSS-LINKERS

Compounds **LXIV** and **LXV** are analogous compounds containing a disulfide bond and a free alkoxylamino group which confer the reactivity toward the sulfhydryl and carbonyl groups, respectively. Through thiol-disulfide interchange, protein sulfhydryls will form new disulfide bonds with these compounds eliminating aminothiopyridine.[47] The alkyoxylamino moiety of these compounds reacts readily with ketones and aldehydes to produce stable alkoxime as shown in Figure 3. These compounds have been used to cross-link adriamycin and thiolated antibody.[47] The alkoxylamino group is also able to form alkoxime with di-aldehydes formed from glycoproteins on periodate treatment.

D. MISCELLANEOUS HETEROBIFUNCTIONAL CROSS-LINKERS

In addition to the above category of compounds, there are heterobifunctional reagents that cross-link various nucleophiles. The Cyssor reagent, **LXVI,** has been found to cross-link antibodies at pH 8.0 without cleavage of the protein, which was found to occur at cysteine residues under acidic conditions.[48] The molecule is essentially an alkylating agent. Nucleophiles attack at both the quinone ring and the bromoacetyl group as shown in Figure 4. Although the nucleophiles in this particular reaction have not been identified, it is speculated that carboxyl groups of aspartate and glutamate, and the indolyl ring of tryptophan may serve as nucleophiles in addition to sulfhydryls and amino groups.

Compounds **LXVII** to **LXIX** contain an aldehyde group which will form Schiff base with amino groups.[49-53] The succinimidyl ester of **LXVII** and the imidoester of **LXVIII** are also reactive toward the amino group, making these two compounds function like a homobifunctional reagent. Compound **LXVIII,** for example, cross-links lysine 51 and lysine 29 of the ribosomal protein L7/L12.[50] The double bond of acrolein, **LXIX,** is subject to Michael addition with nucleophiles and has been shown to cross-link collagen.[53]

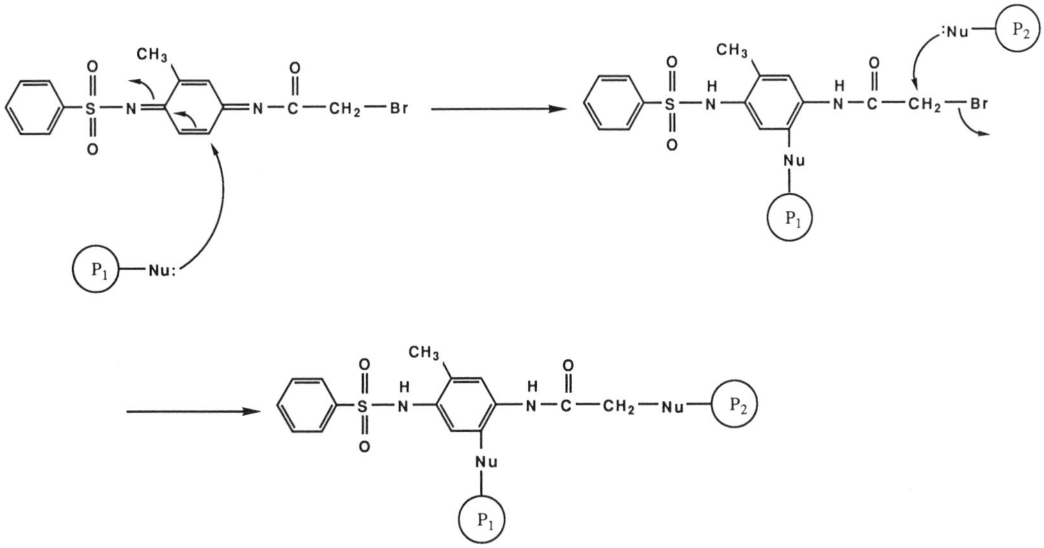

FIGURE 4.   Mechanism of cross-linking by 2-methyl-N^1-benzenesulfonyl-N^4-bromoacetylquinonediimide.

III. PHOTOSENSITIVE HETEROBIFUNCTIONAL CROSS-LINKING REAGENTS

By far, photosensitive heterobifunctional cross-linkers represent the largest portion of the heterobifunctional reagents. Because these functionalities are inert until they are photolyzed, these reagents are first linked to the protein in the dark through a group directed reaction as presented in Table 2. The labeled protein is then irradiated to activate the photosensitive group which reacts indiscriminatively with its environment as discussed in Chapter 3. For example, 4-fluoro-3-nitrophenylazide is first reacted with the amino group of a protein. The labeled protein is then irradiated to generate the reactive nitrene which reacts immediately with its surrounding environment. Cross-linking occurs only after this reaction has been completed.

The photosensitive labels are generally classified according to the active species they produce. Nitrenes are generated from azides whereas carbenes are derived from diazo compounds.[54] Only a few carbene generating diazo reagents are synthesized. The cross-linking reagents, **CXI** to **CXIV** and **CXXIII** are the only carbene precursors developed.[43,45,46,56-59] This is probably because of the ability of carbenes to undergo a variety of reactions including the very efficient reaction with water. In addition, the parent diazoacetyl compounds are generally unstable, particularly at low pH and are reactive towards nucleophiles such as carboxyl groups.[60]

Azido derivatives constitute the majority of the photoactivatable cross-linking agents. Three types of azides have been synthesized: the aryl, alkyl and acyl azides. Alkylazides, however, are not used in cross-linking for several reasons. First, they have absorption maxima in the UV region at which irradiation may damage proteins, nucleic acids and other components. Second, the alkylnitrene intermediates readily undergo rearrangement to form inactive imines. Last but not least, alkylazides are reactive and may undergo nucleophilic displacement reactions. For the same reasons, acylazides are generally used as acylating agents.[61] Only arylazides have been extensively used in photoactivatable cross-linkers.[62-95,97-106,111-117] Aryl azides have a low activation energy and can be photolyzed in the long UV region.[22,118] The presence of electron-withdrawing substituents such as nitro- and hydroxyl-groups further increases the wavelength of absorption into the 300 nm region.[60]

Table 1

GROUP SELECTIVE HETEROBIFUNCTIONAL CROSS-LINKERS

Name (Abbreviation)	Structure	Cleavable/condition	Reference
A. AMINO AND SULFHYDRYL GROUP DIRECTED BIFUNCTIONAL REAGENTS			
I. N-Succinimidyl 3-(2-pyridyldithio) propionate (SPDP)		+/thiol	3
II. N-Succinimidyl maleimidoacetate (AMAS)		—	4, 5
III. N-Succinimidyl 3-maleimido-propionate (BMPS)		—	4, 5
IV. N-Succinimidyl 4-maleimidobutyrate		—	1

4, 5

6

7, 8

9

7, 8

V. N-Succinimidyl 6-maleimidocaproate;
N-Succinimidyl 6-maleimidylhexanoate
(SMH)

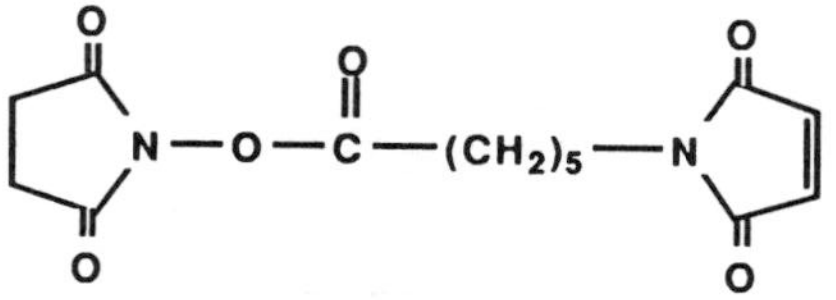

VI. N-Succinimidyl 4-(N-maleimido-
methyl)cyclohexane-1-carboxylate
(SMCC)

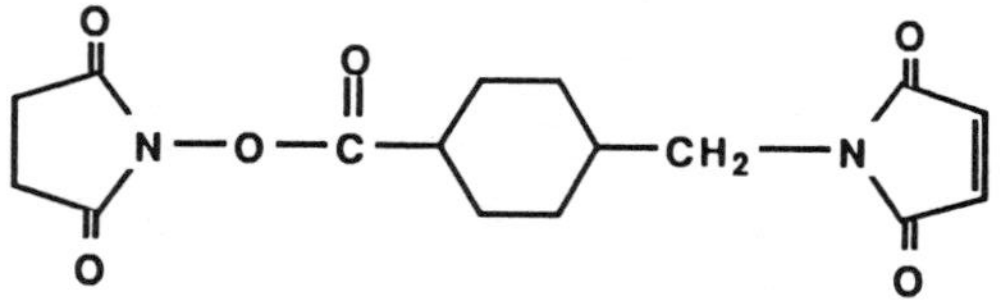

VII. N-Sulfosuccinimidyl 4-(N-maleimido-
methyl)cyclohexane-1-carboxylate
(Sulfo-SMCC)

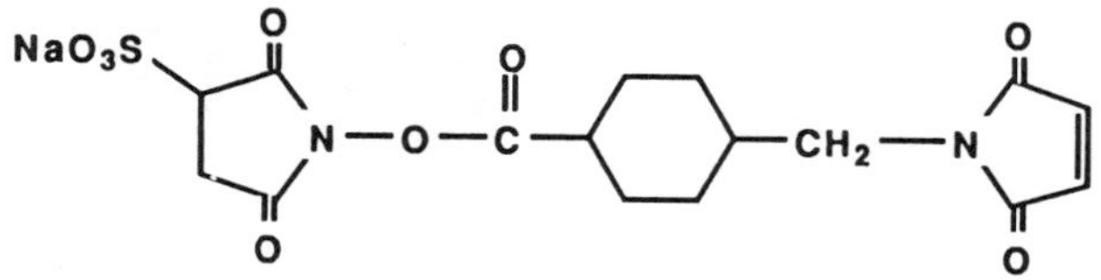

VIII. N-Succinimidyl 4-(p-maleimidophenyl)butyrate (SMPB)

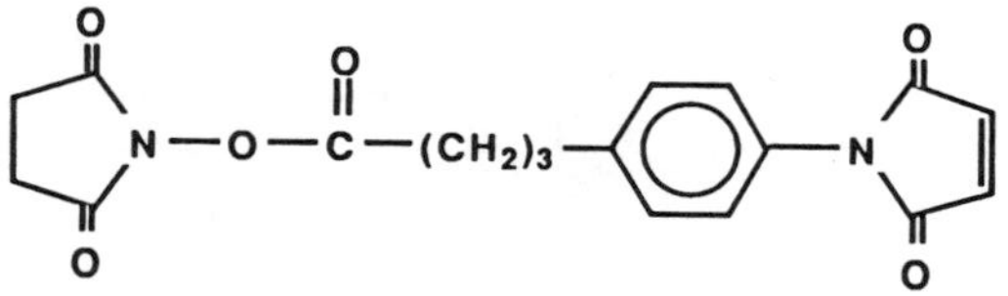

IX. N-Sulfosuccinimidyl 4-(p-maleimidophenyl)butyrate (Sulfo-SMPB)

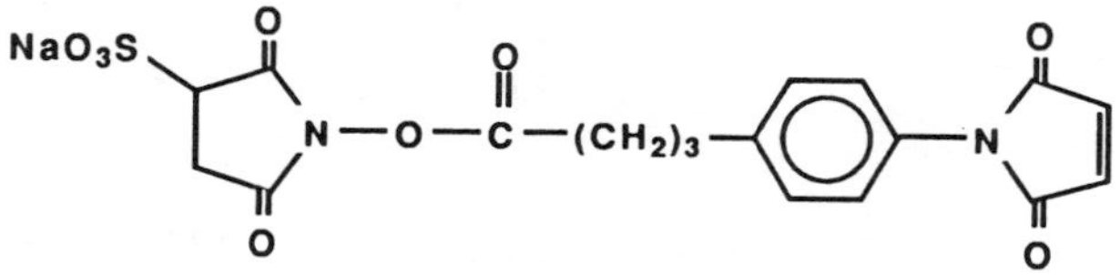

Table 1 (continued)

GROUP SELECTIVE HETEROBIFUNCTIONAL CROSS-LINKERS

A. AMINO AND SULFHYDRYL GROUP DIRECTED BIFUNCTIONAL REAGENTS

Name (Abbreviation)	Structure	Cleavable/condition	Reference
X. N-Succinimidyl o-maleimidobenzoate		—	4
XI. N-Succinimidyl m-maleimidobenzoate (SMB) ; m-Maleimidobenzoyl-N-hydroxysuccinimide ester (MBS)		—	4
XII. N-Sulfosuccinimidyl m-maleimidobenzoate ; m-Maleimidobenzoyl-sulfosuccinimide ester (Sulfo-MBS)		—	7, 8

XIII. N-Succinimidyl p-maleimido-benzoate

XIV. N-Succinimidyl 4-maleimido-3-methoxybenzoate

XV. N-Succinimidyl 5-maleimido-2-methoxybenzoate

XVI. N-Succinimidyl 3-maleimido-4-methoxybenzoate

Table 1 (continued)

GROUP SELECTIVE HETEROBIFUNCTIONAL CROSS-LINKERS

Name (Abbreviation)	Structure	Cleavable/ condition	Reference

A. AMINO AND SULFHYDRYL GROUP DIRECTED BIFUNCTIONAL REAGENTS

Name (Abbreviation)	Cleavable/ condition	Reference
XVII. N-Succinimidyl 3-maleimido-4-(N,N-dimethyl)aminobenzoate	—	4
XVIII. Maleimidoethoxy{p-(N-succinimidylpropionato)-phenoxy}ethane	+/H$^+$	10
XIX. N-Succinimidyl 4-{(N-iodo-acetyl)amino}benzoate (SIAB)	—	11

XX. N-Sulfosuccinimidyl 4-[(N-iodo-acetyl)amino]benzoate (Sulfo-SIAB) — 11

XXI. N-Succinimidyliodoacetate — 12

XXII. N-Succinimidylbromoacetate — 13

XXIII. N-Succinimidyl 3-(2-bromo-3-oxobutane-1-sulfonyl)propionate +/dithionate 14

XXIV. N-Succinimidyl 3-(4-bromo-3-oxobutane-1-sulfonyl)propionate +/dithionate 15

Table 1 (continued)

GROUP SELECTIVE HETEROBIFUNCTIONAL CROSS-LINKERS

Name (Abbreviation)	Structure	Cleavable/ condition	Reference

A. AMINO AND SULFHYDRYL GROUP DIRECTED BIFUNCTIONAL REAGENTS

Name (Abbreviation)	Structure	Cleavable/ condition	Reference
XXV. N-Succinimidyl 2,3-dibromopro-pionate		—	16
XXVI. N-Succinimidy 4-[{N,N-bis(2-chloroethyl)}aminophenyl-butyrate; Chlorambucil-N-hydroxysuccinimide ester		—	16
XXVII. p-Nitrophenyl 3-(2-bromo-3-oxobutane-1-sulfonyl)pro-pionate		+/dithionate	14
XXVIII. p-Nitrophenyl-3-(4-bromo-3-oxobutane-1-sulfonyl)pro-pionate		+/dithionate	15

17

18

19

20

21

—

—

—

—

—

XXIX. p-Nitrophenyl 6-maleimido-caproate

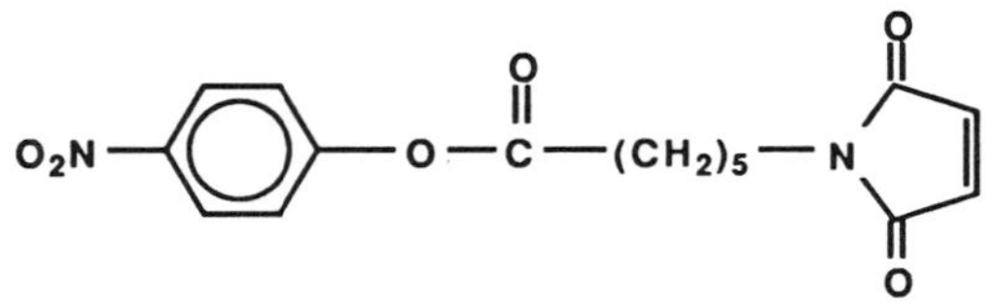

XXX. (2-Nitro-4-sulfonic acid-phenyl)-6-maleimidocaproate

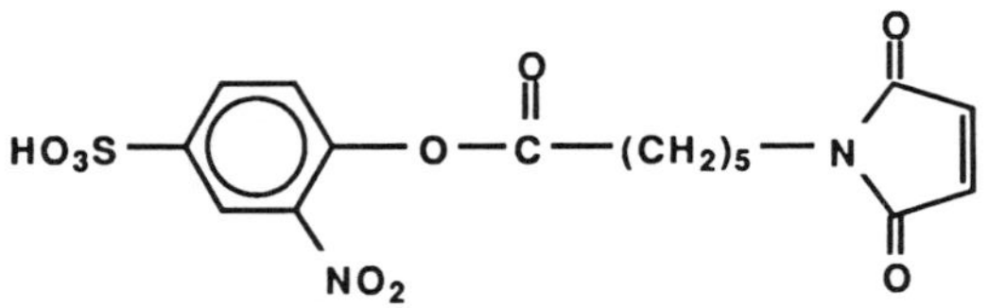

XXXI. p-Nitrophenyliodoacetate

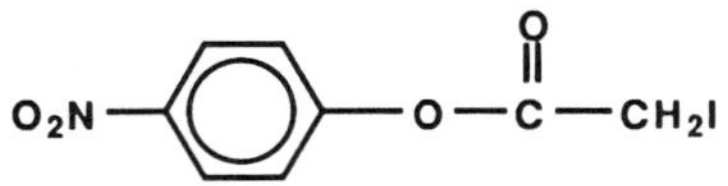

XXXII. p-Nitrophenylbromoacetate

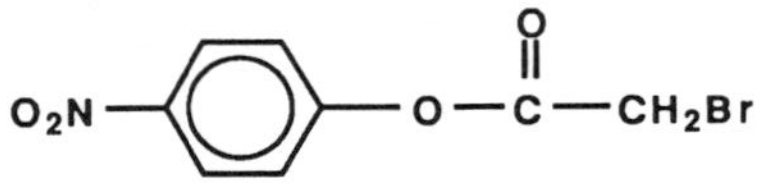

XXXIII. 2,4-Dinitrophenyl-p-(β-nitro-vinyl)benzoate

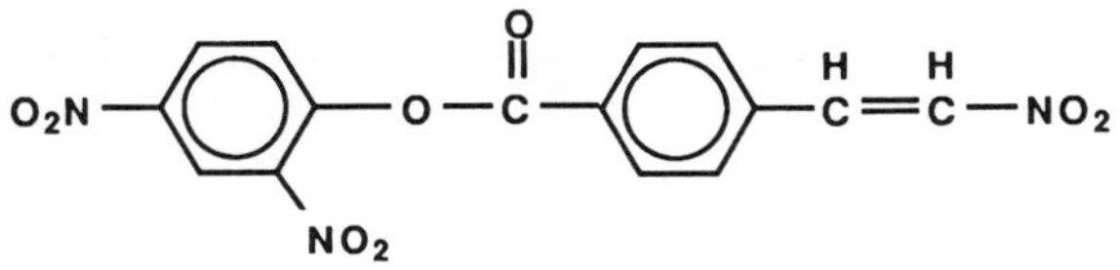

Table 1 (continued)

GROUP SELECTIVE HETEROBIFUNCTIONAL CROSS-LINKERS

Name (Abbreviation)	Structure	Cleavable/ condition	Reference
A. AMINO AND SULFHYDRYL GROUP DIRECTED BIFUNCTIONAL REAGENTS			
XXXIV. N-(3-Fluoro-4,6-dinitrophenyl)cystamine	(structure: fluoro-dinitrophenyl ring with NH—$(CH_2)_2$—S—S—$(CH_2)_2$—NH_2; O_2N and NO_2 substituents)	—	22
XXXV. Methyl 3-(4-pyridyldithio)propionimidate HCl	CH_3—O—$\overset{\overset{+}{H_2N}\ Cl^-}{\underset{\|}{C}}$—$CH_2$—$CH_2$—$S$—$S$—(pyridyl)	+/thiol	23
XXXVI. Ethyl iodoacetimidate HCl	CH_3CH_2—O—$\overset{\overset{+}{H_2N}\ Cl^-}{\underset{\|}{C}}$—$CH_2I$	—	24
XXXVII. Ethyl bromoacetimidate HCl	CH_3CH_2—O—$\overset{\overset{+}{H_2N}\ Cl^-}{\underset{\|}{C}}$—$CH_2Br$	—	25
XXXVIII. Ethyl chloroacetimidate HCl	CH_3CH_2—O—$\overset{\overset{+}{H_2N}\ Cl^-}{\underset{\|}{C}}$—$CH_2Cl$	—	24

26, 27

28

28

28

27, 29

—

—

—

—

—

XXXIX. N-(4-Azidocarbonyl-3-hydroxy-phenyl)maleimide ; 2-Hydroxy-4-(N-maleimido)benzoylazide (HMB)

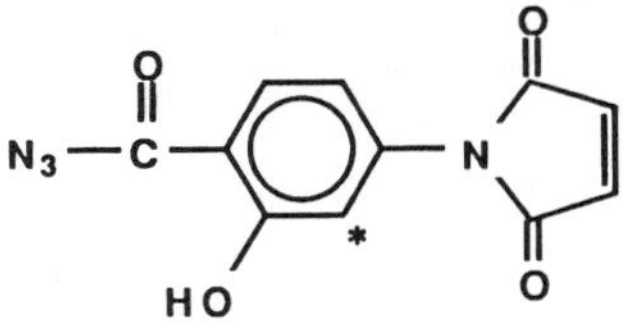

XL. 4-Maleimidobenzoylchloride

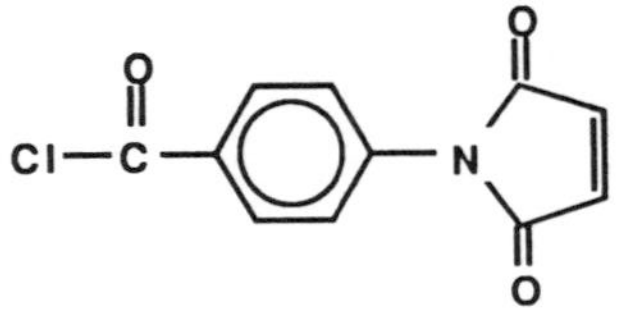

XLI. 2-Chloro-4-maleimidobenzoyl-chloride

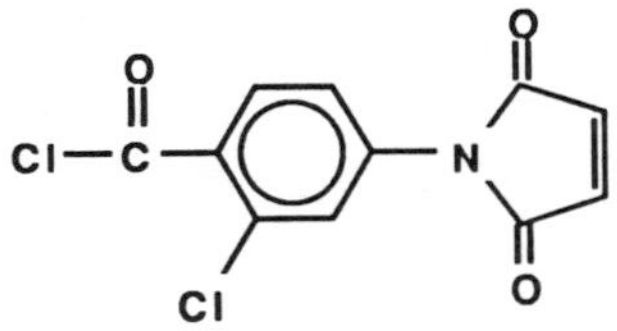

XLII. 2-Acetoxy-4-maleimidobenzoyl-chloride

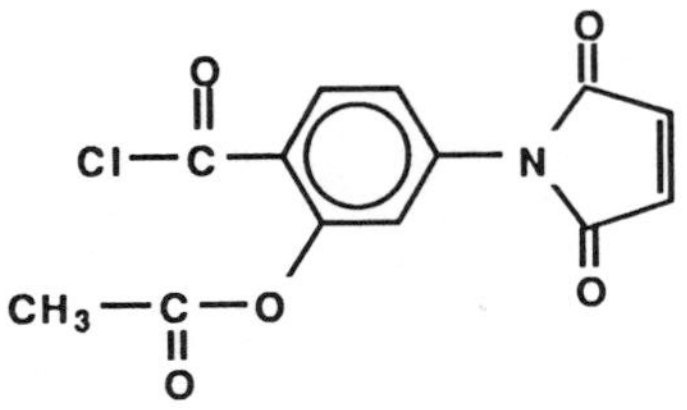

XLIII. 4-Chloroacetylphenylmaleimide

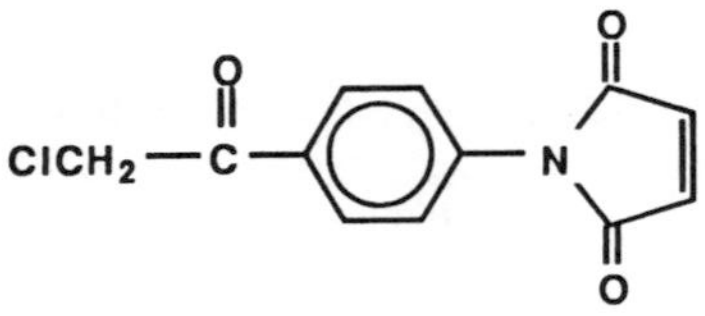

Table 1 (continued)

GROUP SELECTIVE HETEROBIFUNCTIONAL CROSS-LINKERS

Name (Abbreviation)	Structure	Cleavable/ condition	Reference

A. **AMINO AND SULFHYDRYL GROUP DIRECTED BIFUNCTIONAL REAGENTS**

XLIV. 2-Bromoethylmaleimide		—	30
XLV. N-[4-{(2,5-Dihydro-2,5-dioxo-3-furanyl)methyl}thiophenyl-2,5-di-hydro-2,5-dioxo-1H-pyrrole-1-hexanamide		+/H$^+$	31
XLVI. Epichlorohydrin		—	32
XLVII. 2-(p-Nitrophenyl)allyl-4-nitro-3-carboxyphenylsulfide		—	33

33

34

XLVIII. 2-(p-Nitrophenyl)allyltrimethyl-
ammonium iodide

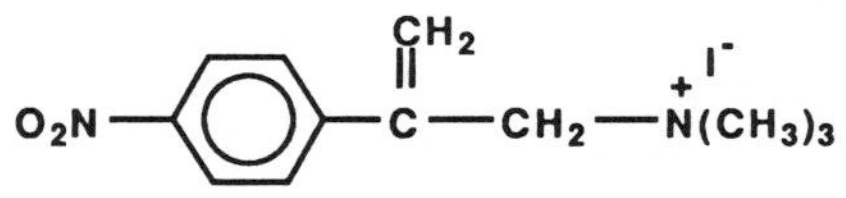

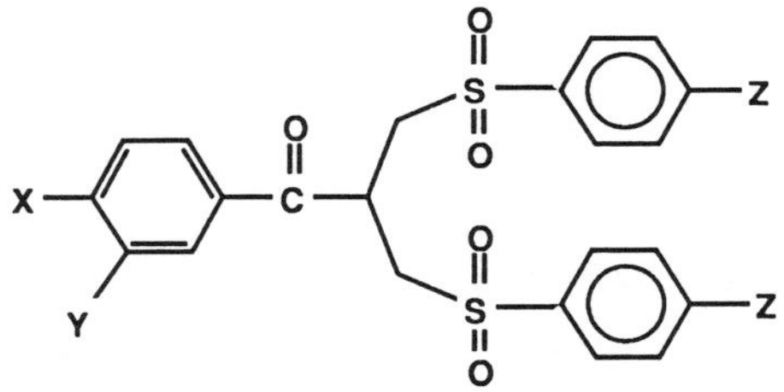

XLIX. X = H, Y = H, Z = Cl : α,α-Bis[[(p-chlorophenyl)sulfonyl]methyl]acetophenone

L. X = Cl, Y = H, Z = Cl : α,α-Bis[[(p-chlorophenyl)sulfonyl]methyl]-p-chloroacetophenone

LI. X = NO₂, Y = H, Z = Cl : α,α-Bis[[(p-chlorophenyl)sulfonyl]methyl]-4-nitroacetophenone

LII. X = NO₂, Y = H, Z = CH₃ : α,α-Bis[(p-tolysulfonyl)methyl]-4-nitroacetophenone

LIII. X = H, Y = NO₂, Z = Cl : α,α-Bis[[(p-chlorophenyl)sulfonyl]methyl]-m-nitroacetophenone

LIV. X = H, Y = NO₂, Z = CH₃ : α,α-Bis[(p-tolysulfonyl)methyl]-m-nitroacetophenone

Table 1 (continued)

GROUP SELECTIVE HETEROBIFUNCTIONAL CROSS-LINKERS

Name (Abbreviation)	Structure	Cleavable/ condition	Reference

A. AMINO AND SULFHYDRYL GROUP DIRECTED BIFUNCTIONAL REAGENTS

LV. X = COOH, Y = H, Z = CH₃ : 4-[2,2-Bis{(p-tolylsulfonyl)methyl}acetyl]benzoic acid

LVI. X = 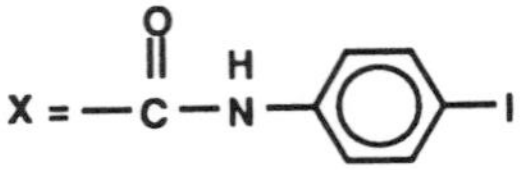 , Y = H, Z = CH₃ : N-[4[2,2-{(p-Tolylsulfonyl)methyl}acetyl]benzoyl]-4-iodoaniline

LVII. X = NH₂, Y = H, Z = CH₃ : α,α-Bis{(p-tolylsulfonyl)methyl}p-aminoacetophenone

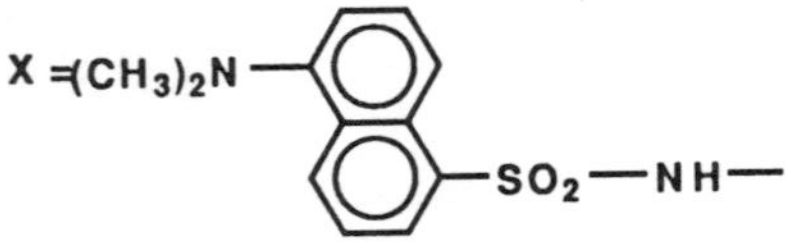

, Y = H, Z = CH₃ : N-[{5-(Dimethylamino)naphthyl}sulfonyl]-α,α-bis{(p-tolylsulfonyl)methyl}-p-aminoacetophenone

LVIII. X ={CH₃)₂N—

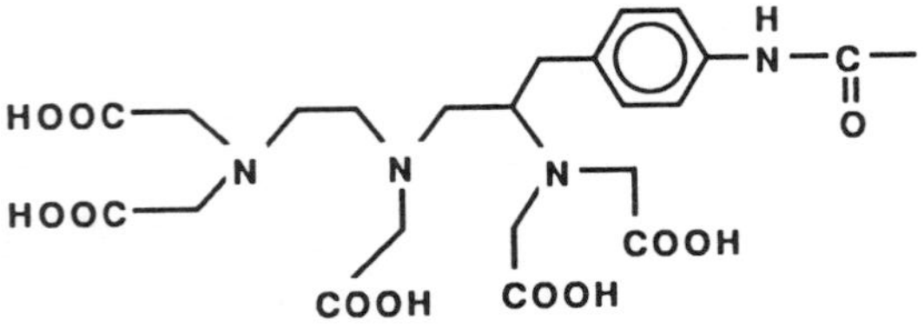

, Y=H, Z=CH₃ : N-[4-{2,2-Bis(p-tolylsulfonyl)methyl}-acetyl]benzoyl-1-(p-aminobenzyl)di-ethylenetriaminepentaacetic acid

LIX. X=

B. CARBOXYL AND EITHER SULFHYDRYL OR AMINO GROUP DIRECTED BIFUNCTIONAL REAGENTS

LX. Pyridyl-2,2'-dithiobenzyldiazoacetate (PDD)   +/thiol   43

LXI. 1-Diazoacetyl-1-bromo-2-phenylethane   –   44

LXII. p-Nitrophenyl diazoacetate   –   45

LXIII. p-Nitrophenyl diazopyruvate   –   46

C. CARBONYL AND SULFHYDRYL GROUP DIRECTED BIFUNCTIONAL REAGENTS

LXIV. 1-(Aminooxy)-4-[(3-nitro-2-pyridyl)dithio]butane   +/thiol   47

Table 1 (continued)

GROUP SELECTIVE HETEROBIFUNCTIONAL CROSS-LINKERS

Name (Abbreviation)	Structure	Cleavable/ condition	Reference

C. CARBONYL AND SULFHYDRYL GROUP DIRECTED BIFUNCTIONAL REAGENTS

LXV. 1-(Aminooxy)-4-{(3-nitro-2-pyridyl)dithio}but-2-ene

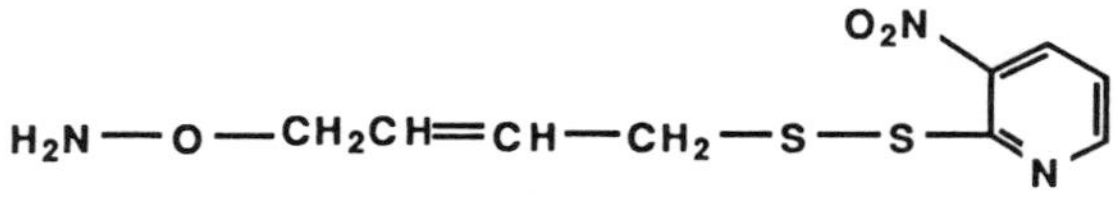

$H_2N-O-CH_2CH=CH-CH_2-S-S-$ (3-nitro-2-pyridyl)

+/thiol   47

D. MISCELLANEOUS HETEROBIFUNCTIONAL CROSS-LINKERS

LXVI. 2-Methyl-N'-benzenesulfonyl-N^4-bromoacetylquinonediimide

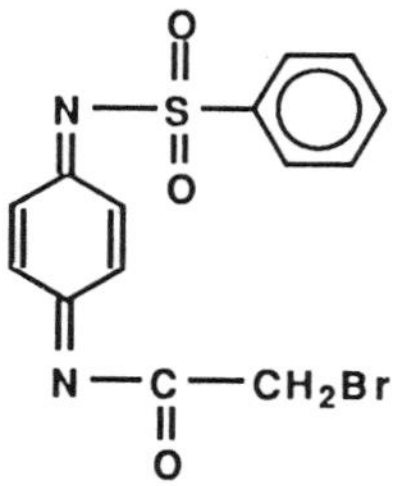

—   48

LXVII. N-Hydroxysuccinimidyl-p-formylbenzoate

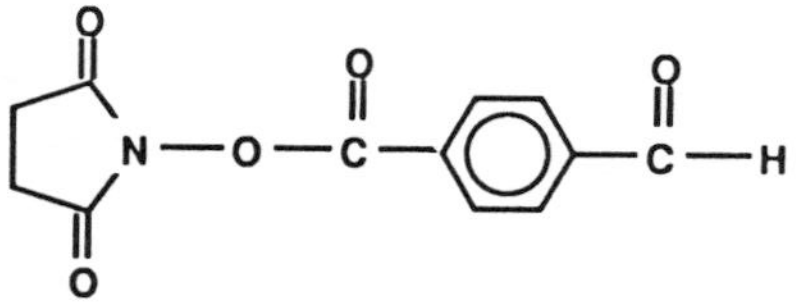

—   49

50-52

—

53

—

LXVIII. Methyl-4-(6-formyl-3-azido-phenoxy)butyrimidate HCl (FAPOB)

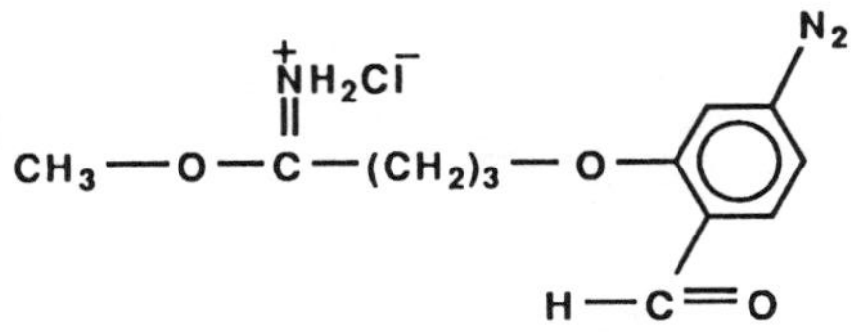

LXIX. Acrolein

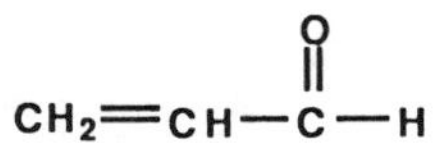

Table 2

PHOTOACTIVATABLE HETEROBIFUNCTIONAL CROSS-LINKING REAGENTS

Name (Abbreviation)	Structure	Cleavable/ condition	Reference

A. AMINO GROUP ANCHORED PHOTOSENSITIVE REAGENTS

Name (Abbreviation)	Structure	Cleavable/ condition	Reference
LXX. N-Succinimidyl-4-azidobenzoate (NHS-ABA, HSAB)		—	62-64
LXXI. N-Succinimidyl-4-azido-salicylate (NHS-ASA)		—	65
LXXII. N-Succinimidyl-N'-(4-azido-salicyl)-6-aminocaproate (NHS-ASC)		—	65
LXXIII. N-Succinimidyl-5-azido-2-nitro-benzoate (NHS-ANBA); N-5-Azido-2-nitrobenzoyloxy-succinimide (ANB-NOS)		—	66

67

67

65

74

68, 69

—

—

—

—

—

LXXIV. N-Succinimidyl-4-azidobenzo-
ylglycinate (NHS-ABG)

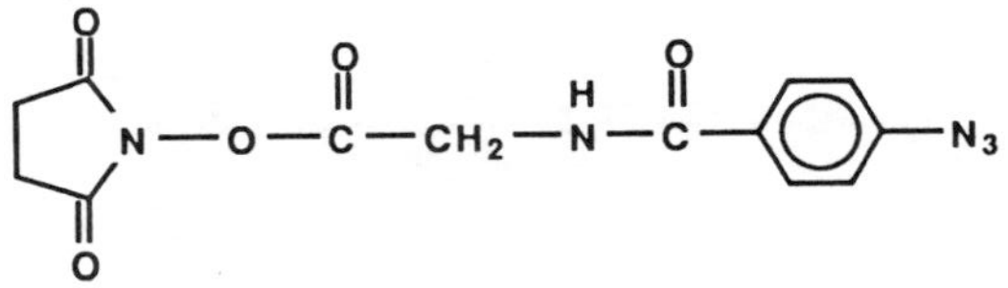

LXXV. N-Succinimidyl-4-azido-
benzoylglycylglycinate
(NHS-ABGG)

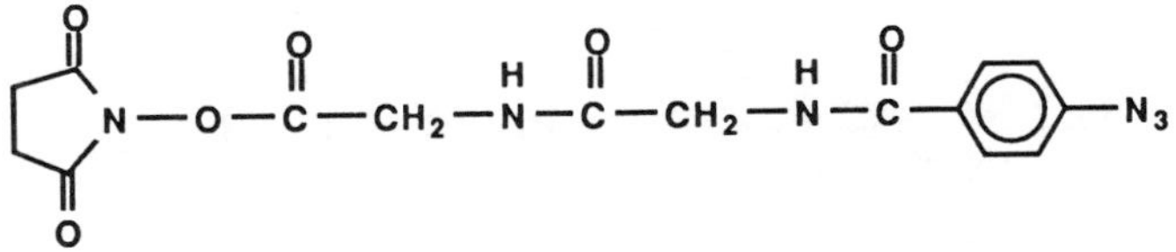

LXXVI. N-Succinimidyl-4-azido-
benzoylglycyltyrosinate
(NHS-ABGT)

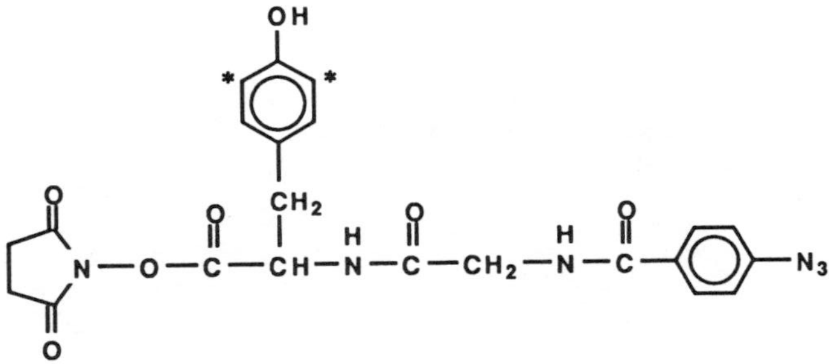

LXXVII. N-Succinimidyl-(4-azido-2-
nitrophenyl)glycinate

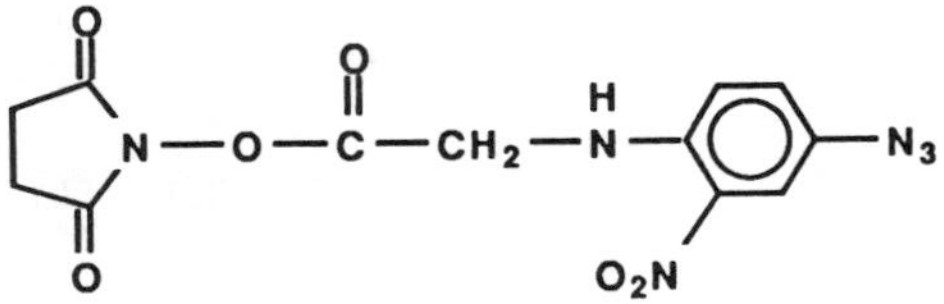

LXXVIII. N-Succinimidyl-(4-azido-2-
nitrphenyl)-γ-aminobutyrate
(NHS-ANAB)

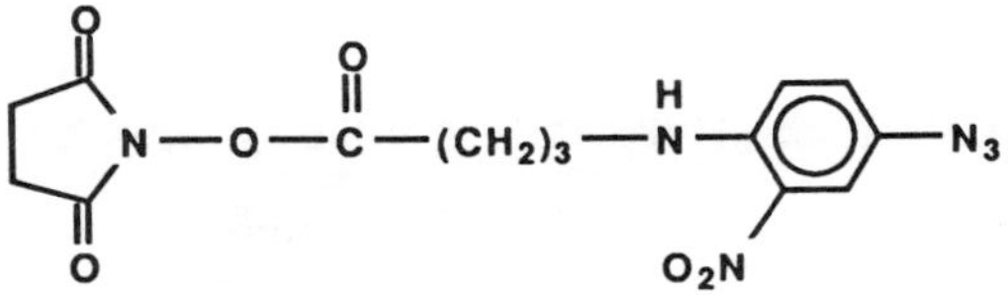

Table 2 (continued)

PHOTOACTIVATABLE HETEROBIFUNCTIONAL CROSS-LINKING REAGENTS

Name (Abbreviation)	Structure	Cleavable/condition	Reference

A. AMINO GROUP ANCHORED PHOTOSENSITIVE REAGENTS

Name (Abbreviation)	Structure	Cleavable/condition	Reference
LXXIX. N-Succinimidyl-6-(4'-azido-2'-nitrophenylamino)hexanoate (SANPAH) (Loman's reagent II)		—	70
LXXX. Sulfosuccinimidyl-6-(4'-azido-2'-nitrophenylamino)hexanoate (Sulfo-SANPAH)		—	71
LXXXI. N-Succinimidyl-N'-(4-azido-nitrophenyl)dodecanoate		—	72
LXXXII. N-Succinimidyl-2-{(4-azido-phenyl)dithio}acetate (NHS-APDA)		+/thiol	73

LXXXIII. Sulfosuccinimidyl-3-{(4-azidophenyl)dithio}propionate (Sulfo-SADP)

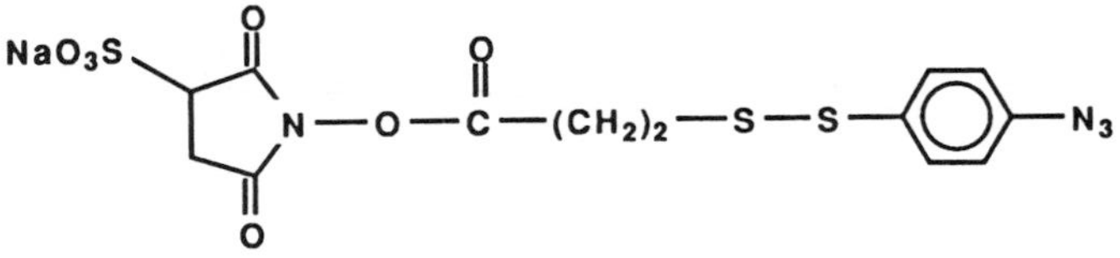

+/thiol   8, 73, 75

LXXXIV. N-Succinimidyl-3-{(4-azidophenyl)dithio}propionate (NHS-APDP); N-Succinimidyl-(4-azidophenyl)1,3'-dithiopropionate (SADP)

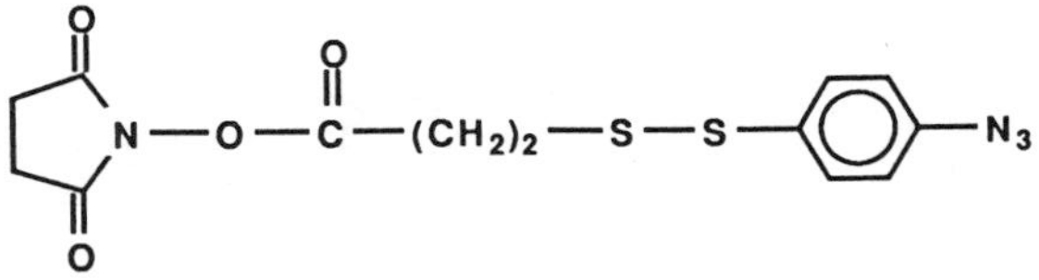

+/thiol   76

LXXXV. N-{4-(p-Azidophenylazo)-benzoyl}-3-aminopropyl-N'-oxysuccinimide ester

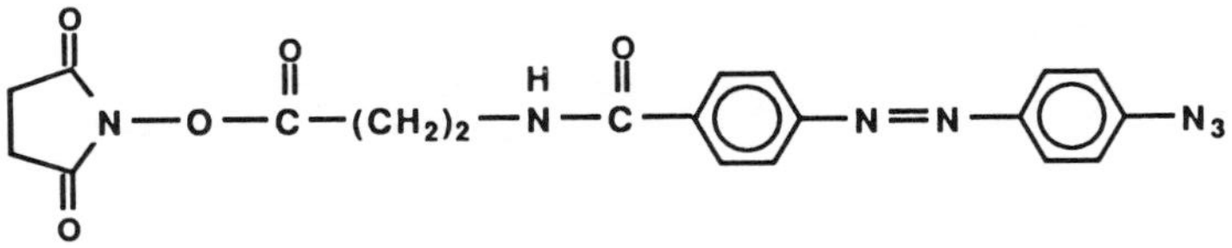

+/dithionite 77

LXXXVI. N-{4-(p-Azidp-o-iodo-phenylazo)benzoyl}-3-aminopropyl-N'-oxysuccinimide ester

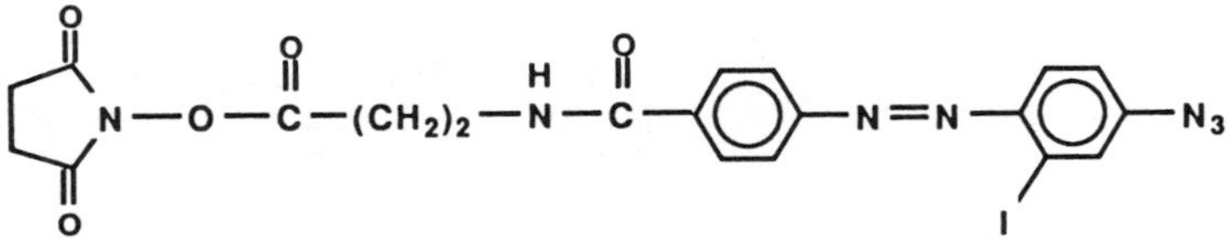

+/dithionite 78

LXXXVII. N-{4-p-Azidophenyl-azo)-benzoyl}-3-amino-hexyl-N'-oxysuccinimide ester

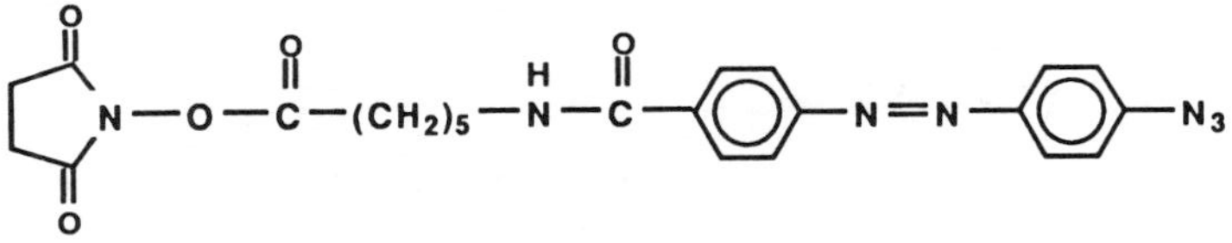

+/dithionite 77, 79

Table 2 (continued)

PHOTOACTIVATABLE HETEROBIFUNCTIONAL CROSS-LINKING REAGENTS

Name (Abbreviation)	Structure	Cleavable/condition	Reference
A. AMINO GROUP ANCHORED PHOTOSENSITIVE REAGENTS			
LXXXVIII. N-[(4-p-Azidophenyl-azo)benzoyl]-3-amino-undecyl-N'-oxysuccin-imide ester		+/dithionite	79
LXXXIX. 3-{[(2-Nitro-4-azidophenyl)-2-aminoethyldithio]-N-succinimidylpropionate (NAP-AEDSP); N-Succinimidyl-3-{(2-nitro-4-azidophenyl)-2-amino-ethyldithio}propionate (SNAP)		+/thiol	80
XC. Sulfosuccinimidyl-2-(p-azidosalicylamino)ethyl-1,3'-dithiopropionate (SASD)		+/thiol	8, 65 81, 82
XCI. Sulfosuccinimidyl-2-(m-azido-o-nitrobenzamido)-ethyl-1,3'-dithiopropionate (SAND)		+/thiol	66, 83

84

XCII. N-Succinimidyl-N-{N'-(4-azidobenzoyl)-tyrosyl}-β-alanine

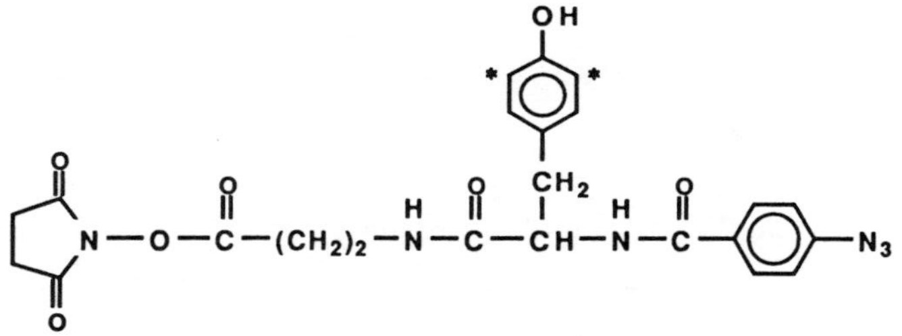

84

XCIII. N-Succinimidyl-N-{N'-(3-azido-benzoyl)tyrosyl}-β-alanine

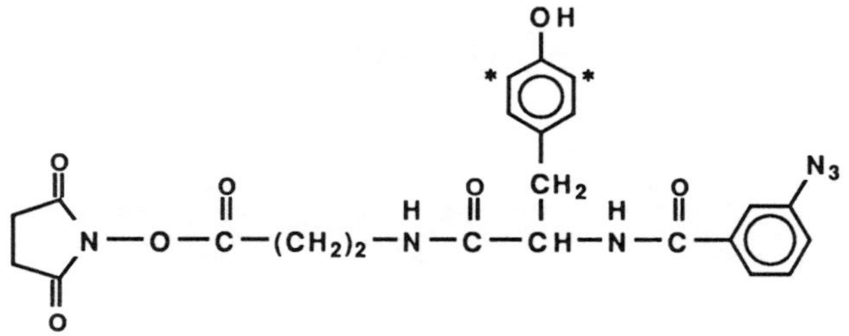

84

XCIV. N-Succinimidyl-N-{N'-(3-azido-5-nitrobenzoyl)tyrosyl}-β-alanine

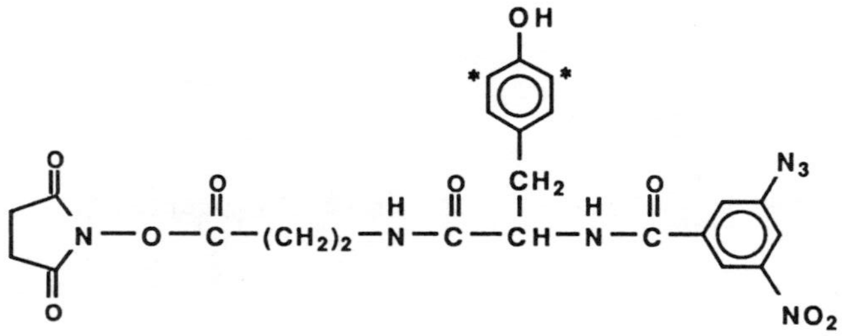

85

Salicylate azides

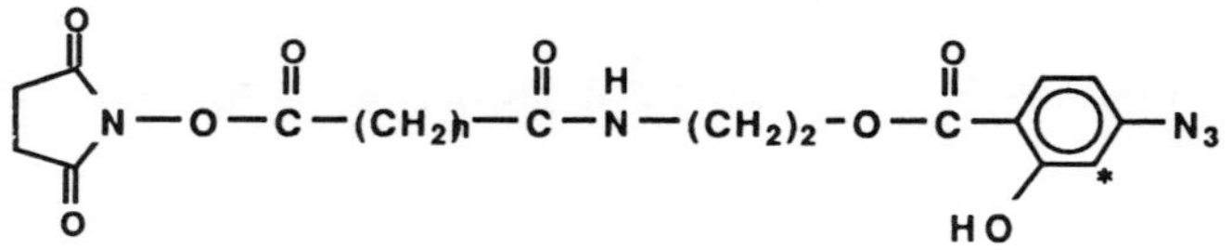

Table 2 (continued)

PHOTOACTIVATABLE HETEROBIFUNCTIONAL CROSS-LINKING REAGENTS

Name (Abbreviation)	Structure	Cleavable/condition	Reference
A. AMINO GROUP ANCHORED PHOTOSENSITIVE REAGENTS			
XCV. n=2 : N-succinimidyl N-[2-{(4-azidosalicyloyl)oxy}ethyl]succinimate			
XCVI. n=4 : N-succinimidyl N-[2-{(4-azidosalicyloyl)oxy}ethyl]adipamate			
XCVII. n=6 : N-succinimidyl N-[2-{(4-azidosalicyloyl)oxy}ethyl]suberamate			
XCVIII. Methyl-4-azidobenzimidate HCl (MABI)		—	71, 86, 87
XCIX. Methyl-3-(4-azidophenyl)acet-imidate HCl (MAPA)		—	2, 87
C. Ethyl N-(5-azido-2-nitrobenzoyl)amino-acetimidate HCl (ABNA)		—	66

		50-52
CI. Methyl 4-(6-formyl-3-azidophenoxy)-butyrimidate HCl (FAPOB)	—	50-52
CII. Methyl-{3-(4-azidophenyl)dithio}-propionimidate HCl (MADP)	+/thiol	73, 88, 89
CIII. Methyl-4-{(4-azidophenyl)dithio}-butyrimidate HCl (MADB)	+/thiol	90
CIV. Ethyl-(4-azidophenyl)-1,4-dithio-butyrimidate HCl (EADB)	+/thiol	66, 90
CV. 4-Azidoiodobenzene	—	91
CVI. 4-Fluoro-3-nitrophenylazide (FNA, FNPA)	—	68, 92, 93

Table 2 (continued)

PHOTOACTIVATABLE HETEROBIFUNCTIONAL CROSS-LINKING REAGENTS

Name (Abbreviation)	Structure	Cleavable/ condition	Reference
A. AMINO GROUP ANCHORED PHOTOSENSITIVE REAGENTS			
CVII. 2,4-Dinitro-5-fluorophenylazide (DNFA)		—	94
CVIII. p-Azidophenylisothiocyanate		—	95
CIX. 1-Azido-5-naphthaleneisothiocyanate; 5-Isothiolcyanato-1-naphthalene azide		—	95
CX. Benzophenone-4-isothiocyanate		—	96
CXI. 2-Diazo-3,3,3-trifluropropionyl chloride (DTPC)		—	55

CXII. p-Nitrophenyl-2-diazo-3,3,3-tri-
fluoropropionate (NDTFP)
— 56-59

CXIII. p-Nitrophenyldiazoacetate
— 45

CXIV. p-Nitrophenyl-3-diazopyruvate
— 46

B. SULFHYDRYL GROUP ANCHORED PHOTOACTIVATABLE REAGENTS

CXV. 4-Azidophenylmaleimide (APM)
— 26, 27, 29

CXVI. p-Azidophenacyl bromide (APB)
— 97-98

CXVII. 4-(Bromoaminoethyl)-3-nitro-
phenylazide (BANPA)
— 99

Table 2 (continued)

PHOTOACTIVATABLE HETEROBIFUNCTIONAL CROSS-LINKING REAGENTS

Name (Abbreviation)	Structure	Cleavable/condition	Reference
B. SULFHYDRYL GROUP ANCHORED PHOTACTIVATABLE REAGENTS			
CXVIII. 4-Azidophenylsulfenyl chloride		+/thiol	73, 128
CXIX. 2-Nitro-4-azidophenylsulfenyl chloride (NAPSCl)		+/thiol	100
CXX. N-(4-Azidophenylthio)phthalimide (APTP)		+/thiol	73, 90
CXXI. Di-N-(2-nitro-4-azidophenyl)-cystamine-S,S-dioxide (DNCO)		+/thiol	101

CXXII. 4,4'-Dithiobisphenylazide

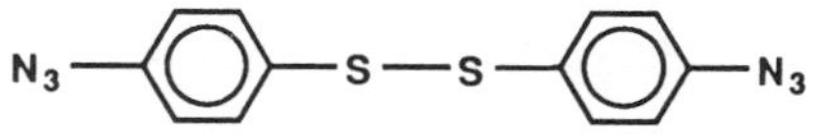

– 8, 74, 102

CXXIII. Pyridyl-2,2'-dithiobenzyl-diazoacetate (PDD)

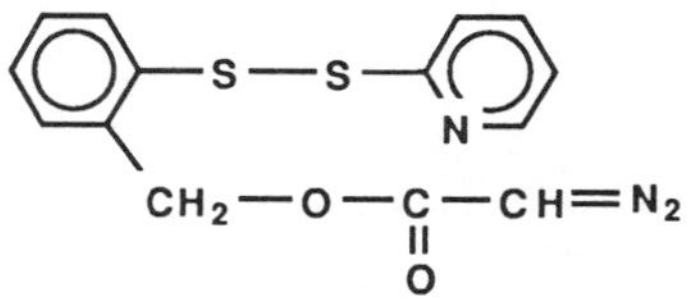

+/thiol 43

CXXIV. N-(4-Azidobenzoyl-2-glycyl)-S-(2-thiopyridyl)cysteine (AGTC)

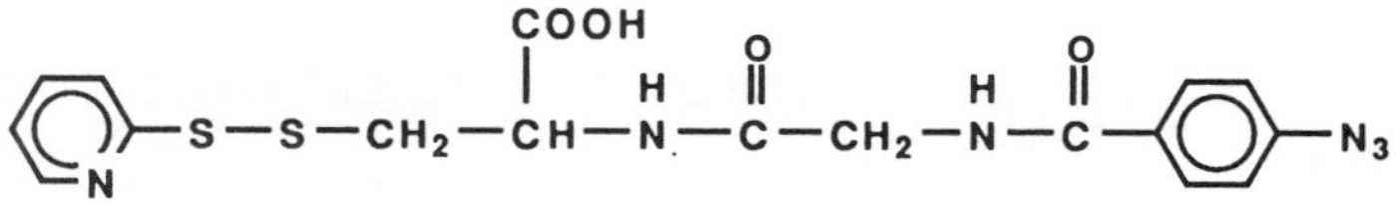

+/thiol 102–104

CXXV. N-(3-Iodo-4-azidophenyl-propionamide-S-(2-thio-pyridyl)cysteine

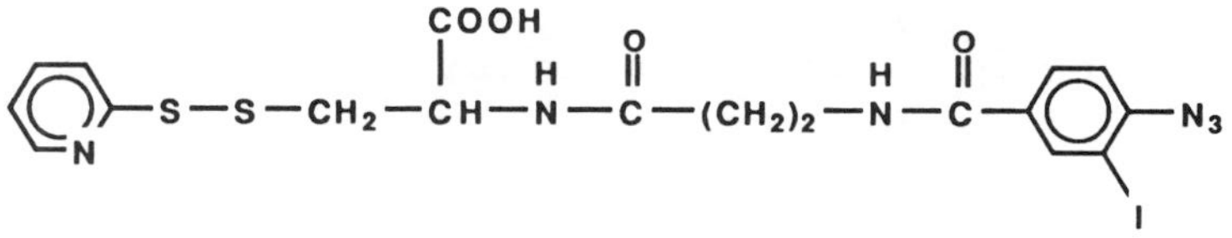

– 105

CXXVI. 3-(4-Azido-2-nitrobenzoyl-seleno)-propionic acid (ANBSP); 2-Carboethylseleno-4-azido-2-nitrobenzoate

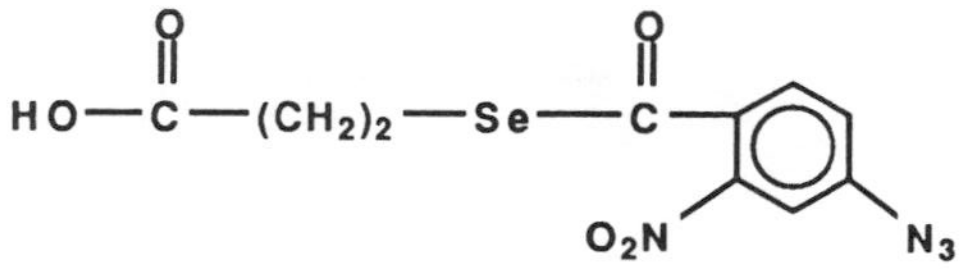

– 27, 106

CXXVII. Benzophenone-4-maleimide

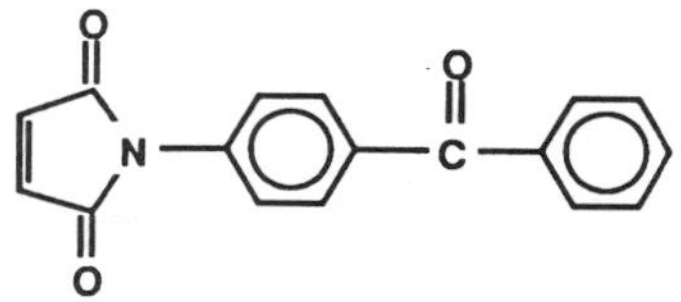

– 107–108

Table 2 (continued)

PHOTOACTIVATABLE HETEROBIFUNCTIONAL CROSS-LINKING REAGENTS

Name (Abbreviation)	Structure	Cleavable/ condition	Reference
B. SULFHYDRYL GROUP ANCHORED PHOTACTIVATABLE REAGENTS			
CXXVIII. Benzophenone-4-iodo-acetamide		—	109
CXXIX. N-(Maleimidomethyl)-2-(o-methoxy-p-nitrophenoxy)-carboxamidoethane		—	110
C. GUANIDINYL GROUP ANCHORED PHOTOACTIVATABLE REAGENT			
CXXX. 4-Azidophenylglyoxal (APG)		—	111, 112

D. CARBOXYL AND CARBOXAMIDE GROUPS ANCHORED PHOTOACTIVATABLE REAGENTS

Name	Structure	Condition	Ref.
CXXXI. N-(4-Azido-2-nitrophenyl)-ethylenediamine; N-(β-Aminoethyl)-4-azido-2-nitroaniline	H_2N—$(CH_2)_2$—NH— (aryl with N_3, O_2N)	—	98, 113
CXXXII. N-(5-Azido-2-nitrophenyl)-ethylenediamine	H_2N—$(CH_2)_2$—NH— (aryl with N_3, O_2N)	—	114, 115
CXXXIII. N-{β-(β'-Aminoethyl-dithioethyl)}-4-azido-2-nitroaniline	H_2N—CH_2CH_2—S—S—CH_2CH_2—NH— (aryl with N_3, O_2N)	+/thiol	113
CXXXIV. N-(4-Azido-2-nitrophenyl-β-aminoethyl)-N'-(β-aminoethyl)tartramide	H_2N—CH_2CH_2—NH—C(=O)—CH(OH)—CH(OH)—C(=O)—NH—CH_2CH_2—NH— (aryl with N_3, O_2N)	+/IO_4^-	113

Table 2 (continued)

PHOTOACTIVATABLE HETEROBIFUNCTIONAL CROSS-LINKING REAGENTS

Name (Abbreviation)	Structure	Cleavable/ condition	Reference

E. PHOTOAFFINITY REAGENTS

CXXXV. 3'-Arylazido-β-alanine-δ-azido-ATP; 3'-o-[3-[N-Azido-(2-nitrophenyl)amino)-propionyl]-8-azidoadenosine-5'-triphosphate

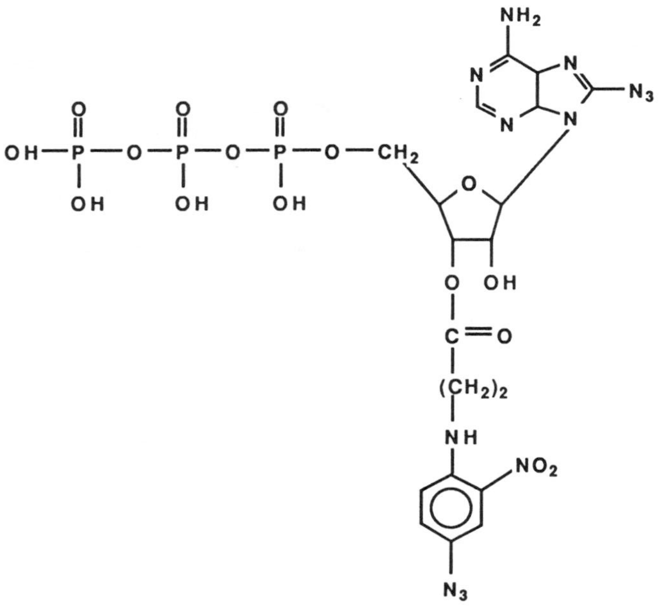

+/H⁺,OH⁻ 116

CXXXVI. 5'(p-Fluorosulfonylbenzoyl)-
8-azidoadenosine (FSBAzA)

+/H$^+$,OH$^-$ 117

* : Indicates position of iodination

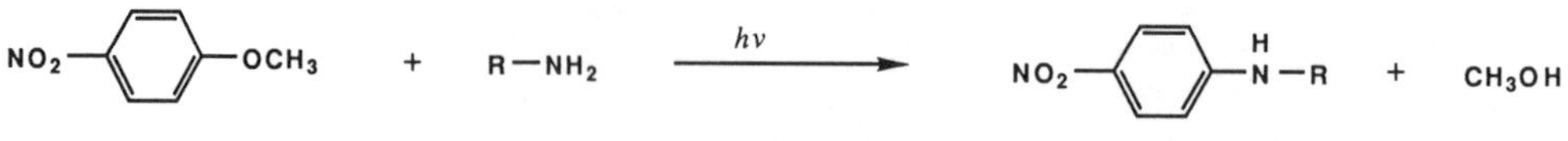

$$NO_2 - \langle \text{ring} \rangle - OCH_3 \quad + \quad R-NH_2 \quad \xrightarrow{h\nu} \quad NO_2 - \langle \text{ring} \rangle - \overset{H}{N} - R \quad + \quad CH_3OH$$

FIGURE 5.   Photochemical reaction of nitrophenyl ethers with amines.

Arylnitrenes have a half-life in the order of 10^{-2} to 10^{-4} seconds[96,119] and, therefore, the cross-linking reaction is expected to be terminated within a short period of time. Arylazides are susceptible to reduction to amino groups. They are not stable in the presence of thiols. The half-life of arylazides is 5 to 15 min in 10 mM dithiothreitol (pH 8.0) and over 24 h in 50 mM mercaptoethanol (pH 8.0).[107]

The benzophenone derivatives, compounds **CX**, **CXXVII** and **CXXVIII** constitute yet another class of photoaffinity compounds.[96,107-109,120,121] These compounds, as shown in Chapter 3, can form covalent adducts on irradiation with nearby amino acid residues leading to cross-linking. Unlike the azides, which are irreversibly photolyzed in most cases, the excited triplet state of benzophenones may be inert to water and may revert back to the starting material if no photoreaction takes place. Since benzophenones can be re-excited, their cross-linking efficiency can, in principle, reach 100%.[122] These compounds have been used to study ribosomal proteins,[109] troponin and tropomyosin,[120] thin filament proteins,[121] and virus induced proteins.[108]

Another class of photosensitive reagents are nitrophenyl ethers which have been proposed to be useful for cross-linking and labeling.[110] These compounds react quantitatively with amines at slightly alkaline conditions (pH 8) on irradiation with 366 nm light. The reaction involves the transfer of nitrophenyl group from the alcohol to the amine as shown in Figure 5 for methoxy-*p*-nitrobenzene. Like other photoreagents, these compounds are stable in the dark, but unlike other reagents, they are stable even upon irradiation in the absence of a nucleophile. Irradiation excites the compound to a triplet state with an extremely short lifetime of 10^{-7} to 10^{-9} s. The chemical reaction will have to occur during that time period, preventing the reagent from wandering around. Nonproductive deactivation regenerates the starting compound providing a relatively high yield as benzophenones.

With the photobifunctional cross-linker of this class, **CXXIX**, the 2-methoxy-4-nitro-phenyl ether is attached to the protein through the Michael addition reaction of a thiol group at the maleimide ring. Such reaction was demonstrated to occur at the γ-cysteine F9 of human fetal hemoglobin. On irradiation, the reagent yielded γ-γ-cross-linked hemoglobin.[110]

One of the most important applications of these photoactivatable bifunctional reagents is in the identification of receptors. The photosensitive agent is first anchored onto the protein in the dark according to the group specificity of the reagent as will be discussed below. The labeled polypeptide ligand is then allowed to bind to its specific receptors. On photolysis, cross-linking will occur with molecules directly interacting with or adjacent to the derivatized ligands. This technique has been used to identify binding sites for Con A,[2,87] epidermal growth factor,[88,89] insulin,[137] fibronectin,[138] bungarotoxin,[72] choriogonadotropin,[67] calmodulin,[139] interleukin-3,[82] glucagon,[140] nerve growth factor,[62,63] and parathyroid hormone,[141] just to mention a few.

A. AMINO GROUP ANCHORED PHOTOSENSITIVE REAGENTS

The majority of photoactivatable heterobifunctional cross-linkers (**LXX** to **CXIV**) are directed toward the amino group. They contain such classical amino group selective functionalities as *N*-hydroxysuccinimidyl ester (**LXX** to **XCVII**), imidoester (**XCVIII** to **CIV**), aryl halide (**CV** to **CVII**), isothiocyanate (**CVIII** to **CX**), acyl chloride (**CXI**), and p-nitrophenyl ester (**CXII** to **CXIV**).

A few of these compounds (**LXXI**, **LXXII**, **XC**, **XCII** to **XCVII**) contain the phenol ring and are directly iodinatable with reagents such as chloramine T. This provides a convenient way of introduction of the radioisotope [125]I. Some are cleavable. Compounds **LXXXII** to **LXXXIV, LXXXIX, XC,** and **CII** to **CIV** contain the disulfide bond and will be cleavable with excess mercaptans. The azo derivatives, **LXXXV** to **LXXXVIII,** are cleavable by dithionite. Of particular interest is FAPOB (**CI**) which has three reactive groups. The imidate ester and formyl moieties can react with amino groups, and as such may function as a homobifunctional reagent in the dark. On photolysis, the photosensitive phenylazide constitutes the third reactive group. Maasen et al.[50] have used the trifunctional reagent as an amino group directed cross-linker for linking the lysine residues of ribosomal proteins L7 and L12 using the imidate and aldehyde groups. The aldehyde group may also be used as a chromophor due to its contribution of absorbance at 325 nm. On reduction or Schiff base formation, the absorption band is diminished.

In addition to being used for the identification of cell surface receptors,[1,124] some of these compounds have been applied to study transferrin binding sites and ion channels.[125,126]

B. SULFHYDRYL GROUP ANCHORED PHOTOACTIVATABLE REAGENTS

Compounds that can undergo disulfide exchange (**CXXI** to **CXXIII**) with thiols are truly sulfhydryl specific. The cleavable amino group reagents with disulfide bonds mentioned above are therefore also thiol reagents. The cross-linked products formed are cleavable in the presence of an excess mercaptan. The seleno ester (**CXXVI**) also selectively reacts with a free sulfhydryl group.[119] Upon nucleophilic substitution by thiols, thiol esters are formed. The liberated selenol readily forms diselenides providing a very favorable equilibrium for the reaction. The protein thiol-ester can be cleaved with excess free thiols or amines, thus providing a means of identifying the labeled amino acids of proteins after photolysis. The labeled amino acids will be within a span of 7 Å from the cysteine residue anchor.

Reagents with maleimido groups (**CXV** and **CXXVII**), alkyl halides (**CXVI, CXVII** and **CXXVIII**) and thiol ethers (**CXVIII** to **CXX**) undergo nucleophilic reaction with the thiolate ion. However, reactions with other nucleophiles constitute the major side reaction, particularly at high pHs. For example, the phenylsulfenyl chloride derivative, **CXIX,** was synthesized for the modification of tryptophan residues.[100] Proteins lacking thiol groups may be first thiolated with 2-iminothiolane followed by reaction with the cross-linking reagent.[127]

In addition to being used for identifying receptors, these thiol reagents have been used to study troponin,[103,104] α-tropomyosin,[104] and cytochrome c.[128] The benzophenone, **CXXVII,** has been used to study conformational changes in myosin subfragment 1.[129]

C. GUANIDINYL GROUP ANCHORED PHOTOACTIVATABLE REAGENTS

p-Azidophenylglyoxal (**CXXX**) contains the glyoxal function which is directed toward the guanidinyl group of arginine. Ngo et al.[112] have used this reagent to inhibit LDH, lysozyme, alcohol-dehydrogenase as an arginine specific reagent. Politz et al.[130] have reacted the reagent with guanosine and cross-linked RNA to proteins in 30S ribosomal subunits after photolysis.

D. CARBOXYL AND CARBOXAMIDE GROUP ANCHORED
PHOTOACTIVATABLE REAGENTS

The photoactivatable reagents, **CXXXI** to **CXXXIV,** containing a free amino group can be coupled to proteins through the carboxyl groups or the γ-carboxamide moiety of glutamine. In the presence of a carbodiimide, condensation occurs between the carboxyl group of a protein and the amino group of the bifunctional reagent. After labeling, the proteins may be photolyzed to activate the arylazide.[115] In the presence of transglutaminase (see Chapter 6), the amines are introduced covalently at the γ-carboxamide group of peptide-bound glutamine residues.[131-133] These compounds have been incorporated into substance P, glucagon and casein in this manner.[113]

A.

B.

C. $(CH_3O)_2CH_2CH_2-\overset{H}{N}-\overset{O}{\overset{\|}{C}}-(CH_2)_4-\overset{O}{\overset{\|}{C}}-NHNH_2$

$\xrightarrow{\quad N_2O_4 \quad}$

$(CH_3O)_2CH_2CH_2-\overset{H}{N}-\overset{O}{\overset{\|}{C}}-(CH_2)_4-\overset{O}{\overset{\|}{C}}-N_3$

$\xrightarrow{\quad 50\% \text{ TFA} \quad}$

$(CH_3O)_2CH_2CH_2-\overset{H}{N}-\overset{O}{\overset{\|}{C}}-(CH_2)_4-\overset{O}{\overset{\|}{C}}-N_3$

$\boxed{R}-NH_2$ →

$O=\overset{\|}{C}-CH_2CH_2-\overset{H}{N}-\overset{O}{\overset{\|}{C}}-(CH_2)_4-\overset{O}{\overset{\|}{C}}-\overset{H}{N}-\boxed{R}$

$\boxed{P}-NH_2$ →

$\boxed{P}-N=CH-CH_2CH_2-\overset{H}{N}-\overset{O}{\overset{\|}{C}}-(CH_2)_4-\overset{O}{\overset{\|}{C}}-\overset{H}{N}-\boxed{R}$

$\xrightarrow{\quad \text{pyridine borane} \quad}$

$\boxed{P}-\overset{H}{N}-CH_2-CH_2CH_2-\overset{H}{N}-\overset{O}{\overset{\|}{C}}-(CH_2)_4-\overset{O}{\overset{\|}{C}}-\overset{H}{N}-\boxed{R}$

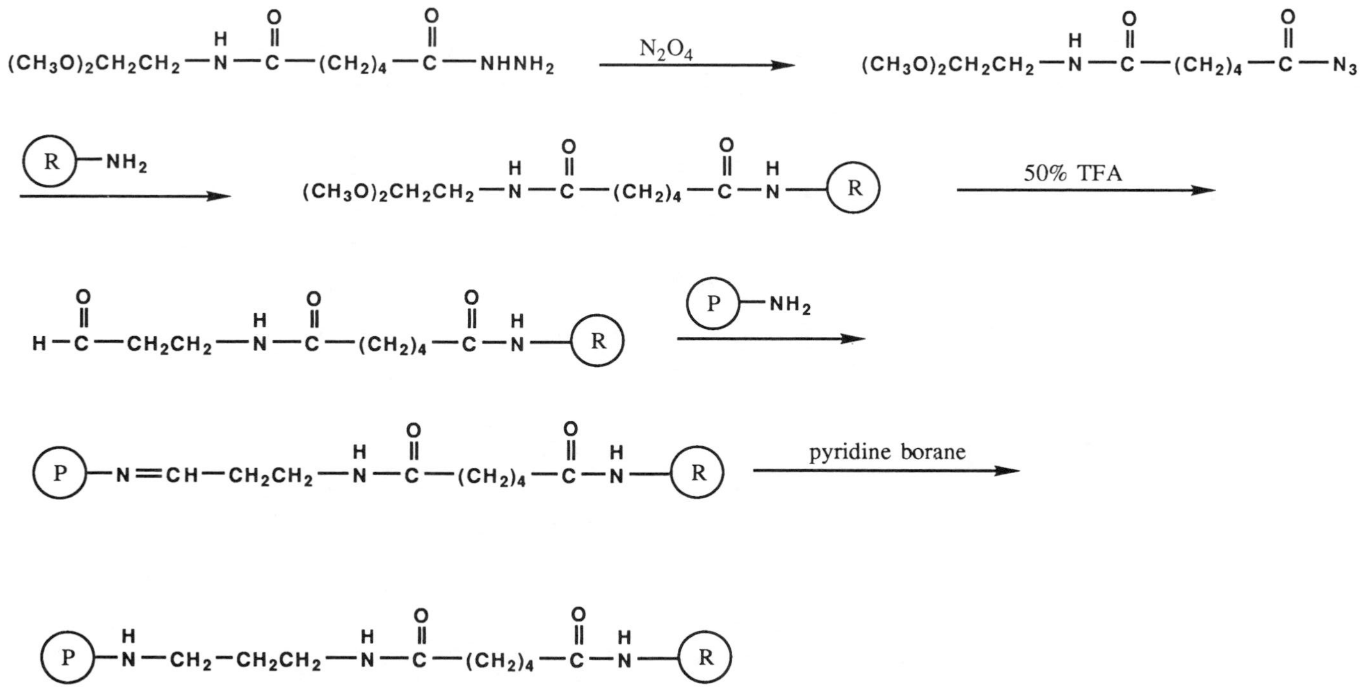

FIGURE 6. Cross-linking reactions of disguised reagents. (A) 3-amino-4-methoxyphenylvinyl sulfone; (B) ethyl *N*-(carbamoylcyanomethyl)acetimidate; and (C) *N*-(2,2-dimethoxyethyl)-5-(hydrazide carbonyl)pentanamide.

E. PHOTOAFFINITY REAGENTS

The photoactivatable affinity labels, diN$_3$ATP (**CXXXV**) and FSBAzA (**CXXXVI**), are ATP/adenosine analogs and would be expected to bind to ATP-binding proteins. The recognition of the adenosine binding site on proteins is the basis of affinity labeling. diN^{3}ATP has been shown to bind to the ATP binding site of bacterial F$_1$ATPases and actually serves as a substrate for the enzyme.[116] On photolysis, the 8-azidoadenosine moiety labels the adenine binding site which is located on the β-subunit. The other photoactivatable moiety, azidophenyl, interacts with the neighboring polypeptide, which is the α-subunit. Thus, diN$_3$ATP cross-links the α- and β-subunits of F$_1$ATPase on photoactivation.

FSBAzA, **CXXXVI**, has been shown to bind to the adenine nucleotide binding site of glutamate dehydrogenase.[117] The electrophilic fluorosulfonyl moiety is capable of reacting with amino acid side chain nucleophiles at the binding site. This nucleophilic substitution reaction takes place in the dark and anchors the photoaffinity label to the protein nucleotide binding site. On photolysis, the azido group is activated and reacts with the neighboring amino acids, making a cross-link between the nucleophile and the adenine binding residues.

IV. DISGUISED OR MASKED REAGENTS

Disguised or masked reagents are compounds that can be easily converted to heterobifunctional reagents in a simple reaction. These reagents are usually disguised as monofunctional agents. After reacting with a macromolecule, the label is converted to another reactive group. For example, 3-amino-4-methoxyphenyl vinyl sulfone has been used to couple enzymes catalase, trypsin, chymotrypsin, and ribonuclease to cellulose.[134] The first step involves a Michael addition reaction of a hydroxyl group to the vinyl sulfone. Treatment of the label with sodium nitrite in acid converts the amino group on the benzene ring to a diazonium compound which will react with tyrosine residues of the proteins to complete the cross-linking process (Figure 6A).

Similarly, ethyl *N*-(carbamoylcyanomethyl)acetimidate can be used as a disguised reagent. Under mild conditions, protein amino groups attack the compound which cyclizes to form an aminoimidazoyl derivative. The amino group can be diazotized and react with the tyrosyl residue of another protein as is shown in Figure 6B. By this method, antigens have been coupled to anti-human group O erythrocyte IgG.[135]

Another masked reagent is *N*-(2,2-dimethoxyethyl)-5-(hydrazidecarbonyl)pentanamide which has been used to couple glycopeptide to proteins.[136] The reagent is first converted to acylazide on treatment with dinitrogen tetraoxide. After reaction with an amino-containing compound, the blocked methyl acetal is removed with 50% trifluoroacetic acid to generate an aldehyde group. Cross-linking is achieved between the aldehyde and a protein amino group by reductive alkylation (Figure 6C).

REFERENCES

1. **Ji, T. H.,** The application of chemical crosslinking for studies on cell membranes and the identification of surface reporters, *Biochim. Biophys. Acta,* 559, 39, 1979.
2. **Ji, T. H.,** Crosslinking of lectins and receptors in membranes with hetero-bifunctional crosslinking reagents, in *Membranes and Neoplasia: New Approaches and Strategies,* Marchesi, V. T., Ed., Alan R. Liss, New York, 1976, 171.
3. **Carlsson, J., Drevin, H., and Axen, R.,** Protein thiolation and reversible protein-protein conjugation, *N*-succinimidyl 3-(2-pyridyldithio)propionate, a new heterobifunctional reagent, *Biochem. J.,* 173, 723, 1978.
4. **Kitagawa, T., Shimozono, T., Aikawa, T., Yoshida, T., and Nishimura, H.,** Preparation and characterization of hetero-bifunctional cross-linking reagents for protein modifications, *Chem. Pharm. Bull.,* 29, 1130, 1981.
5. **Keller, O. and Rudinger, J.,** Preparation and some properties of maleimido acids and malcoyl derivatives of peptides, *Helv. Chim. Acta,* 58, 531, 1975.
6. **Yoshitake, S., Yamada, Y., Ishikawa, E., and Masseyeff, R.,** Conjugation of glucose oxidase from *Aspergillus niger* and rabbit antibodies using N-hydroxysuccinimide ester of *N*-(4-carboxycyclohexylmethyl)maleimide, *Eur. J. Biochem.,* 101, 395, 1979.
7. **Staros, J. V. and Anjanejulu, P. S.,** Membrane-impermeable cross-linking reagents, *Methods Enzymol.,* 172, 609, 1989.
8. Pierce Chemical Company, BioResearch Products Technical Bulletin, Vol. V, Pierce Chemical Co., Rockford, IL, 1983.
9. **Kitagawa, T. and Aikawa, T.,** Enzyme coupled immunoassay of insulin using a novel coupling reagent, *J. Biochem.,* 79, 233, 1976.
10. **Srinivasachar, K. and Neville, D. M., Jr.,** New protein cross-linking reagents that are cleaved by mild acid, *Biochemistry,* 28, 2501, 1989.
11. **Weltman, J. K., Johnson, S. A., Langevin, J., and Riester, E. F.,** *N*-succinimidyl-(4-iodoacetyl)aminobenzoate: a new heterobifunctional crosslinker, *Biotechniques,* 1, 148, 1983.
12. **Rector, E. S., Schwenk, R. J., Tse, K. S., and Sehon, A. H.,** A method for the preparation of protein-protein conjugates of predetermined composition, *J. Immunol. Methods,* 24, 321, 1978.
13. **Cuatrecasas, P., Wilchek, M., and Anfinsen, C. B.,** Affinity labeling of the active site of staphylococcal nuclease. Reactions with bromoacetylated substrate analogs, *J. Biol. Chem.,* 244, 4316, 1969.
14. **Fasold, H., Baumert, H., and Fink, G.,** Comparison of hydrophobic and strongly hydrophilic cleavable crosslinking reagents in intermolecular bond formation in aggregates of proteins or protein-RNA, in *Protein Crosslinking: Biochemical and Molecular Aspects,* Plenum Press, New York, 1976, 207.
15. **Fink, G., Fasold, H., and Rammel, W.,** Reagents suitable for the crosslinking of nucleic acids to proteins, *Anal. Biochem.,* 108, 394, 1980.
16. **McKenzie, J. A., Raison, R. L., and Rivett, D. E.,** Development of a bifunctional crosslinking agent with potential for the preparation of immunotoxins, *J. Protein Chem.,* 7, 581, 1988.
17. **Kriwaczek, V. M., Bonnafous, J. C., Mueller, M., and Schwyzer, R.,** Tobacco mosaic virus as a carrier for small molecules. II. Cooperative affinity labeling of membrane vesicles with a TMV angiotensin conjugate, *Helv. Chim. Acta,* 61, 1241, 1978.
18. **Bjorn, M. J., Groetsema, G., and Scalapino, L.,** Antibody-Pseudomonas exotoxin A conjugates cytotoxic to human breast cancer cells *in vitro, Cancer Res.,* 46, 3262, 1986.
19. **Lorand, L., Brannen, W. T., and Rule, N. G.,** Thrombin-catalyzed hydrolysis of *p*-nitrophenyl esters, *Arch. Biochem. Biophys.,* 96, 147, 1962.
20. **Kriwaczek, V. M., Eberle, A. N., Mueller, M., and Schwyzer, R.,** Tobacco mosaic virus as a carrier for small molecules. I. The preparation and characterization of a TMV/α-melanotropin conjugate, *Helv. Chim. Acta,* 61, 1232, 1978.
21. **Fujii, N., Hayashi, Y., Katakura, S., Akaji, K., Yajima, H., Inouye, A., and Segawa, T.,** Studies on peptides. CXXVIII. Application of new heterobifunctional crosslinking reagents for the preparation of neurokinin (A and B)-BSA (bovine serum albumin) conjugates, *Int. J. Pept. Protein Res.,* 26, 121, 1985.
22. **Peters, K., and Richards, F. M.,** Chemical cross-linking: reagents and problems in studies of membrane structure, *Annu. Rev. Biochem.,* 46, 523, 1977.
23. **King, T. P., Li, Y., and Kochoumian, L.,** Preparation of protein conjugates via intermolecular disulfide bond formation, *Biochemistry,* 17, 1499, 1978.
24. **Olomucki, M. and Diopoh, J.,** New protein reagents. I. Ethyl chloroacetimidate. Its properties and its reaction with ribonuclease, *Biochim. Biophys. Acta,* 263, 312, 1972.
25. **Diopoh, J. and Olomucki, I. M.,** Ethyl bromoacetimidate, a NH$_2$-specific heterobifunctional reagent, *Hoppe-Seyler's Z. Physiol. Chem.,* 360, 1257, 1979.
26. **Trommer, W. E., Kolkenbrock, H., and Pfleiderer, G.,** Synthesis and properties of a new selective bifunctional cross-linking reagent, *Hoppe-Seyler's Z. Physiol. Chem.,* 356, 1455, 1975.

27. **Trommer, W. E., Friebel, K., Kiltz, H.-H., and Koldenbrock, H.-J.,** Synthesis and application of new bifunctional reagents, in *Protein Cross-Linking: Biochemical and Molecular Aspects,* Friedman, M., Ed., Plenum Press, New York, 1976, chap. 10.

28. **Hermentin, P., Doenges, R., Gronski, P., Bosslet, K., Kraemer, H. P., Hoffmann, D., Zilag, H., Steinstraesser, A., Schwarz, A., Kuhlmann, L., Lüben, G., Seiler, F. R.,** Attachman of rhodosaminyl anthracyclinone-type anthracylines to the hinge region of monoclonal antibodies, *Bioconjugate Chem.,* 1, 100, 1990.

29. **Trommer, W. E. and Hendrick, M.,** The formation of maleimides by a new mild cyclization procedure, *Synthesis,* 484, 1973.

30. **Simon, S. R. and Konigsberg, W. H.,** Chemical modification of hemoglobins: a study of conformation restraint by internal bridging, *Proc. Natl. Acad. Sci. U.S.A.,* 56, 749, 1966.

31. **Blätter, W. A., Enzi, B. S., Lambert, J. M., and Senter, P. D.,** New heterobifunctional protein cross-linking reagent that forms an acid-labile link, *Biochemistry,* 24, 1517, 1985.

32. **Bäumert, H. G. and Fasold, H.,** Cross-linking techniques, *Methods Enzymol.,* 172, 584, 1989.

33. **Mitra, S. and Lawton, R. G.,** Reagent for cross-linking of proteins by equilibrium transfer alkylation, *J. Am. Chem. Soc.,* 101, 3097, 1979.

34. **Liberatore, F. A., Comeau, R. D., Mckearin, J. M., Pearson, D. A., Belonga III, B. Q., Brocchini, S. J., Kath, J., Phillips, T., Oswell, K., and Lawton, R. G.,** Site-directed chemical modification and cross-linking of a monoclonal antibody using equilibrium transfer alkylating cross-link reagents, *Bioconjugate Chem.,* 1, 36, 1990.

35. **Hashida, S., Imagawa, M., Inoue, S., Ruan, K.-H., and Ishikawa, E.,** More useful maleimide compounds for the conjugation of Fab′ to horseradish peroxidase through thiol groups in the hinge, *J. Appl. Biochem.,* 6, 56, 1984.

36. **Yoshitake, S., Hamaguchi, Y., and Ishikawa, E.,** Efficient conjugation of rabbit Fab′ with β-D-galactosidase from *Escherichia coli, Scand. J. Immol.,* 10, 81, 1979.

37. **Haber, E., Quertermous, T., Matsueda, G. R., and Runge, M. S.,** Inovative approaches to plasminogen activator therapy, *Science,* 243, 51, 1989.

38. **Bode, C., Runge, M. S., Branscomb, E. E., Newell, J. B., Matsueda, G. R., and Harber, E.,** Antibody-directed fibrinogen: an antibody specific for both fibrin and tissue plasminogen activator, *J. Biol. Chem.,* 264, 944, 1989.

39. **Wawrzynczak, E. J. and Thorpe, P. E.,** Methods for preparing immunotoxins: effect of the linkage on activity and stability, in *Immunoconjugates, Antibody Conjugates in Radioimaging and Therapy of Cancer,* Vogel, C.-W., Ed., New York, Oxford University Press, 1987, 28.

40. **Marsh, J. W., Srinivasachar, K., and Neville, D. M., Jr.,** Antibody-toxin conjugation, in *Immunotoxins,* Frankel, A. E., Ed., Kluwer Academic Publishers, Boston, 1988, 213.

41. **Ishikawa, E., Imagawa, M., Hashida, S., Yoshitake, S., Hamaguchi, Y., and Ueno, T.,** Enzyme-labeling of antibodies and their fragments for enzyme immunoassay and immunohistochemical staining, *J. Immunoassay,* 4, 209, 1983.

42. **Ishikawa, E., Hashida, S., Kohno, T., and Tanaka, K.,** Methods for enzyme-labeling of antigens, antibodies and their fragments, in *Nonisotopic Immunoassays,* Ngo, T. T., Ed., Plenum Press, New York, 1988, 27.

43. **Henkin, J.,** Photolabeling reagent for thiol enzymes. Studies on rabbit muscle creatine kinase, *J. Biol. Chem.,* 252, 4293, 1977.

44. **Husain, S. S., Ferguson, J. B., and Fruton, J. S.,** Bifunctional inhibitors of pepsin, *Proc. Natl. Acad. Sci. U.S.A.,* 68, 2765, 1971.

45. **Shafer, J., Baronowsky, D., Laursen, R., Finn, F., and Westheimer, F. H.,** Products from the photolysis of diazoacetyl chymotrypsin, *J. Biol. Chem.,* 241, 421, 1966.

46. **Harrison, J. K., Lawton, R. G., and Gnegy, M. E.,** Development of a novel photoreactive calmodulin derivative. Cross-linking purified adenylate cyclase from bovine brain, *Biochemistry,* 28, 6023, 1989.

47. **Webb, R. R., II, and Kancko, E.,** Synthesis of 1-(aminooxy)-4-[(3-nitro-2-pyridyl)dithio]butane hydrochloride and 1-(aminooxy)-4-[(3-nitro-2-pyridyl)dithio]but-2-ene. Novel heterobifunctional cross-linking reagents, *Bioconjugate Chem.,* 1, 96, 1990.

48. **Liberatore, F. A., Comeau, R. D., and Lawton, R. G.,** Heterobifunctional cross-linking of a monoclonal antibody with 2-methyl-N^1-benzenesulfonyl-N^4-bromoacetylquinonediimide, *Biochem. Biophys. Res. Commun.,* 158, 640, 1989.

49. **Kraehenbuhl, J. P., Galardy, R. E., and Jamieson, J. D.,** Preparation and characterization of an immunoelectron microscope tracer consisting of a hemeoctapeptide coupled to F_{ab}, *J. Exp. Med.,* 139, 208, 1974.

50. **Maassen, J. A., Schop, E. N., and Moller, W.,** Structural analysis of ribosomal proteins L7/L12 by the heterobifunctional cross-linker: 4-(6-formyl-3-azidophenoxy)butyrimidate, *Biochemistry,* 20, 1020, 1981.

51. **Maassen, J. A.,** Cross-linking of ribosomal proteins by 4-(6-formyl-3-azido phenoxy)butyrimidate. A heterobifunctional cleavable cross-linker, *Biochemistry,* 18, 1288, 1979.

52. **Maassen, J. A. and Terhorst, C.,** Identification of a cell-surface protein involved in the binding site of sindbis virus on human lymphobastoic cell lines using a heterobifunctional cross-linker, *Eur. J. Biochem.,* 115, 153, 1981.

53. **Cater, C. W.,** The evaluation of aldehydes and other difunctional compounds as crosslinking agents for collagen, *J. Soc. Leather Trades Chem.,* 47, 259, 1963.

54. **Gilchrist, T. L. and Rees, C. W.,** *Carbenes, Nitrenes and Arynes (Studies in Modern Chemistry),* Nelson Publisher, London, 1969, 131.

55. **Chowdhry, V., Vaughan, R., and Westheimer, F. H.,** 2-Diazo-3,3,3-trifluoropropionyl chlorides: reagent for photoaffinity labeling, *Proc. Natl. Acad. Sci. U.S.A.,* 73, 1406, 1976.

56. **Pascual, A., Casanova, J. S., and Herbert, H.,** Photoaffinity labeling of thyroid hormone nuclear receptors in intact cells, *J. Biol. Chem.,* 257, 9640, 1982.

57. **Gupta, C. M., Radhadrishnan, R., Gerber, G. E., Olsen, W. L., Quay, S. C., and Khorana, H. G.,** Intermolecular cross-linking of fatty acylchains in phospholipids: use of photoactivable carbene precursors, *Proc. Natl. Acad. Sci. U.S.A.,* 76, 2595, 1979.

58. **Takagaki, Y., Gupta, C. M., and Khorana, H. G.,** Thiols and the diazo group in photoaffinity labels, *Biochem. Biophys. Res. Commun.,* 95, 589, 1980.

59. **Chowdhry, V. and Westheimer, F. H.,** Photoaffinity labeling of biological systems, *Annu. Rev. Biochem.,* 48, 293, 1979.

60. **Bayley, H. and Knowles, J. R.,** Photoaffinity labeling, *Methods Enzymol.,* 46, 69, 1977.

61. **Lutter, L. C., Ortanderl, F., and Fasold, H.,** The use of new series of cleavable protein-crosslinking on the *Escherichia coli* ribosomes, *FEBS Lett.,* 48, 288, 1974.

62. **Grob, P. M., Berlot, C. H., and Bothwill, M. A.,** Affinity labeling and partial purification of nerve growth factor receptors from rat pheochromocytoma and human melanoma cells, *Proc. Natl. Acad. Sci. U.S.A.,* 80, 6819, 1983.

63. **Puma, P., Buxser, S. E., Watson, L., Kellcher, D. J., and Johnson, G. L.,** Purification of the receptor for nerve growth factor from A875 melanoma cells by affinity chromatography, *J. Biol. Chem.,* 258, 3370, 1983.

64. **Borst, D. W. and Sayare, M.,** Photoactivated crosslinking of prolactin to hepatic membrane binding sites, *Biochem. Biophys. Res. Commun.,* 105, 194, 1982.

65. **Ji, T. H. and Ji, I.,** Macromolecular photoaffinity labeling with radioactive photoactivable heterobifunctional reagent, *Anal. Biochem.,* 121, 286, 1982.

66. **Lewis, R. V., Roberts, M. F., Dennis, E. A., and Allison, W. S.,** Photoactivated heterobifunctional cross-linking reagents which demonstrate the aggregation state of phospholipase A$_2$, *Biochemistry,* 16, 5650, 1977.

67. **Ji, I. and Ji, T. H.,** Both α and β subunits of human choriogonadotropin photoaffinity label the hormone receptor, *Proc. Natl. Acad. Sci. U.S.A.,* 78, 5465, 1981.

68. **Yaqub, M. and Guire, P.,** Covalent immobilization of L-asparaginase with photochemical reagent, *J. Biomed. Materials Res.,* 8, 291, 1974.

69. **Guire, P., Fliger, D., and Hodgson, J.,** Photochemical coupling of enzymes to mammalian cells, *Pharmacol. Res. Commun.,* 9, 131, 1977.

70. **Ballmer-Hofer, K., Schlup, V., Burn, P., and Burger, M. M.,** Isolation of *in situ* crosslinked ligand-receptor complexes using an anticrosslinker specific antibody, *Anal. Biochem.,* 126, 246, 1982.

71. **Schmidt, R. R. and Betz, H.,** Cross-linking of β-bungarotoxin to chick brain membranes. Identification of subunits of a putative valtage-gated K$^+$ channel, *Biochemistry,* 28, 8346, 1989.

72. **Witzemann, V., Muchmore, D., and Raftery, M. A.,** Affinity directed cross-linking of membrane-bound acetylcholine receptor polypeptides with photolabile α-bungarotoxin derivatives, *Biochemistry,* 18, 5511, 1979.

73. **Vanin, E. F. and Ji, T. H.,** Synthesis and application of cleavable photoactivable heterobifunctional reagents, *Biochemistry,* 20, 6754, 1981.

74. **Schwartz, I. and Offengand, J.,** *E. coli* tRNA^{Phe} modified at the 3-(3-amino-3-carboxypropyl)uridine with a photoaffinity label is fully functional for aminoacylation and for ribosomal interaction, *Biochim. Biophys. Acta,* 697, 330, 1982.

75. **Baenziger, J. U. and Fiete, D.,** Photoactivatable glycopeptide reagents for site-specific labeling of lectins, *J. Biol. Chem.,* 257, 4421, 1982.

76. **Galardy, R. E., Craig, L. C., Jamieson, J. D., and Printz, M. P.,** Photoaffinity labeling of peptide hormone binding sites, *J. Biol. Chem.,* 249, 3510, 1974.

77. **Jaffe, C. L., Lis, H., and Sharon, N.,** New cleavable photoreactive heterobifunctional cross-linking reagents for studying membrane organization, *Biochemistry,* 19, 4423, 1980.

78. **Denny, J. B. and Blobel, G.,** Iodine-125-labeled crosslinking reagent that is hydrophilic, photoactivatable and cleavable through an azo linkage, *Proc. Natl. Acad. Sci. U.S.A.,* 81, 5286, 1984.

79. **Schmitt, M., Painter, R. G., Jesaitis, A., Preissner, K., Sklar, L. A., and Cochrane, C. G.,** Human polymorphonuclear leukocytes, *J. Biol. Chem.,* 258, 649, 1983.

80. **Schwartz, M. A., Das, O. P., Hynes, R. O.,** A new radioactive cross-linking reagent for studying the interactions of proteins, *J. Biol. Chem.,* 257, 2343, 1982.

81. **Wollenweber, H. and Morrison, D. C.,** Synthesis and biochemical characterization of a photoactivatable, iodinatable, cleavable bacterial lipopolysaccharide derivative, *J. Biol. Chem.,* 260, 15068, 1985.

82. **Sorensen, P., Farber, N. M., and Krystal, G.,** Identification of the interleukin-3 receptor using an iodinatable, cleavable, photoreactive crosslinking agent, *J. Biol. Chem.,* 261, 9094, 1986.

83. **Hanson, C. V., Shen, C. J., and Hearst, J. E.,** Cross-linking of DNA *in situ* as a probe for chromatin structure, *Science,* 193, 62, 1976.

84. **Imai, N., Kometani, T., Crocker, P. J., Bowdan, J. B., Demir, A., Dwyer, L. D., Mann, D. M., Vanaman, T. C., and Watt, D. S.,** Photoaffinity heterobifunctional cross-linking reagents based on *N*-(azidobenzoyl)tyrosines, *Bioconjugate Chem.,* 1, 138, 1990.

85. **Imai, N., Dwyer, L. D., Komentani, T., Ji, T., Vanaman, T. C., and Watt, D. S.,** Photoaffinity heterobifunctional cross-linking reagents based on azide-substituted salicylates, *Bioconjugate Chem.,* 1, 144, 1990.

86. **Hinds, T. R. and Andreasen, T. J.,** Photochemical cross-linking of azidocalmodulin to the (Ca^{2+} + Mg^{2+})ATPase of the erythrocyte membrane, *J. Biol. Chem.,* 256, 7877, 1981.

87. **Ji, T. H.,** A novel approach to the identification of surface receptors, *J. Biol. Chem.,* 252, 1566, 1977.

88. **Das, M. and Fox, C. F.,** Molecular mechanism of mitogen action: Processing of receptor induced by epidermal growth factor, *Proc. Natl. Acad. Sci. U.S.A.,* 75, 2644, 1978.

89. **Das, M., Miyakawa, T., Fox, C. F., Pruss, R. M., Aharono, V. A., and Herschman, H. R.,** Specific radiolabeling of a cell surface receptor for epidermal growth factor, *Proc. Natl. Acad. Sci. U.S.A.,* 74, 2790, 1977.

90. **Moreland, R. B., Smith, P. K., Fujimoto, E. K., and Dockter, M. E.,** Synthesis and characterization of *N*-(4-azidophenylthio)-phthalimide, *Analyt. Biochem.,* 121, 321, 1982.

91. **Harris, R. and Findlay, J. B.,** Investigation of the organization of the major proteins in bovine myelin membranes. Use of chemical probes and bifunctional crosslinking reagents, *Biochim. Biophys. Acta,* 732, 75, 1983.

92. **Fleet, G. W. J., Knowles, J. R., and Porter, R. R.,** The antibody binding site. Labelling of a specific antibody against the photo-precursor of an aryl nitrene, *Biochem. J.,* 128, 499, 1972.

93. **Fleet, G. W. J., Porter, R. R., and Knowles, J. R.,** Affinity labelling of antibodies with aryl nitrene as reactive group, *Nature (London),* 224, 511, 1969.

94. **Wilson, D. F., Miyata, Y., Erecinska, M., and Vanderkooi, J. A.,** An aryl azide suitable for photoaffinity labeling of amine groups in proteins, *Arch. Biochim. Biophys.,* 171, 104, 1975.

95. **Sigrist, H., Allegrini, P. R., Kempf, C., Schnippering, C., and Zahler, P.,** 5-Isothiocyanato-1-naphthalene azide and *p*-azido-phenylisothiocyanate, *Eur. J. Biochem.,* 125, 197, 1982.

96. **DeGraff, B. A., Gillespie, D. W., and Sundberg, R. J.,** Phenyl nitrene. A flash photolytic investigation of the reaction with secondary amines, *J. Am. Chem. Soc.,* 96, 7491, 1974.

97. **Hixson, S. H. and Hixson, S. S.,** *p*-Azidophenylacyl bromide, a versatile photolabile bifunctional reagent: reaction with glyceraldehyde-3-phosphate dehydrogenase, *Biochemistry,* 14, 4251, 1975.

98. **Schwartz, I. and Offengand, J.,** Photo-affinity labeling of tRNA binding sites in macromolecules. I. Linking of the phenacyl-*p*-azide of 4-thiouridine in *Escherichia coli* valyl-tRNA to 16S RNA at the ribosomal P site, *Proc. Natl. Acad. Sci. U.S.A.,* 71, 3951, 1974.

99. **Rudnick, G., Kaback, H. R., and Weil, R.,** Photoinactivation of the β-galactoside transport system in *Escherichia coli* membrane vesicles with an impermeant azidophenylgalactoside, *J. Biol. Chem.,* 250, 6847, 1975.

100. **Demoliou, C. D. and Epand, R. M.,** Synthesis and characterization of a heterobifunctional photoaffinity reagent for modification of tryptophan residues and its application to the preparation of a photoreactive glucagon derivative, *Biochemistry,* 19, 4539, 1980.

101. **Huang, C.-K. and Richards, F. M.,** Reaction of a lipid-soluble, unsymmetrical, cleavable, cross-linking reagent with muscle aldolase and erythrocyte membrane proteins, *J. Biol. Chem.,* 252, 5514, 1977.

102. **Chong, P. C. S. and Hodges, R. S.,** A new heterobifunctional cross-linking reagent for the study of biological interaction between proteins. I. Design, synthesis and characterization, *J. Biol. Chem.,* 256, 5064, 1981.

103. **Chong, P. C. S. and Hodges, R. S.,** A new heterobifunctional cross-linking reagent for the study of biological interaction between proteins. II. Application to the troponin C-troponin I interaction, *J. Biol. Chem.,* 256, 5071, 1981.

104. **Chong, P. C. S. and Hodges, R. S.,** Photochemical cross-linking between rabbit skeleton troponin and alpha-tropomyosin, *J. Biol. Chem.,* 257, 9152, 1982.

105. **Ohanasekaran, N., Wessling-Resnick, M., Kelleher, D. J., Johnson, G. L., and Ruoho, A. E.,** Mapping of the carboxyl terminus within the tertiary structure of transducin's alpha subunit using the heterobifunctional cross-linking reagent, ^{125}I-*N*-(3-iodo-4-azidophenylpropionamido-*S*-(2-thiopyridyl) cysteine, *J. Biol. Chem.,* 263, 17942, 1988.

106. **Friebel, K., Huth, H., Jany, K. D., and Trummer, W. E.,** Semireversible cross-linking: synthesis and application of a novel heterobifunctional reagent, *Hoppe-Seyler's Z. Physiol. Chem.,* 362, 421, 1981.

107. **Staros, J. V., Bayley, H., Standring, D. N., and Knowles, J. R.,** Reduction of aryl azides by thiols: implication for the use of photoaffinity reagents, *Biochem. Biophys. Res. Commun.,* 80, 568, 1978.

108. **Harnish, D. G., Leung, W.-C., Rawls, W. E.,** Characterization of polypeptides immunoprecipitable from Pichinde virus-infected BHK-21 cells, *J. Virol.,* 38, 840, 1981.

109. **Odom, O. W., Deng, H.-Y., Subramanian, A. R., and Hardesty, B.,** Relaxation time, interthiol distance, and mechanism of action of ribosomal protein S1(1), *Arch. Biochem. Biophys.,* 230, 178, 1984.

110. **Jelenc, P. C., Cantor, C. R., and Simon, S. R.,** High yield photoreagents for protein crosslinking and affinity labeling, *Proc. Natl. Acad. Sci. U.S.A.,* 75, 3564, 1978.

111. **Vanin, E. F., Burkhard, S. J., and Kaiser, I. I.,** *p*-Azidophenylglyoxal. A heterobifunctional photosensitive reagent, *FEBS Lett.,* 124, 89, 1981.

112. **Ngo, T. T., Yam, C. F., Lenhoff, H. M., and Ivy, J.,** *p*-Azidophenylglyoxal: a heterobifunctional photoactivatable cross-linking reagent selective for arginine residue, *J. Biol. Chem.,* 256, 11313, 1981.

113. **Gorman, J. J. and Folk, J. E.,** Transglutaminase amine substrates for photochemical labeling and cleavable cross-linking of proteins, *J. Biol. Chem.,* 255, 1175, 1980.

114. **Drafler, F. L. and Marinetti, G. V.,** Synthesis of a photoaffinity probe for the β-adrenergic receptor, *Biochem. Biophys. Res. Commun.,* 79, 1, 1977.

115. **Das, M. and Fox, F.,** Chemical cross-linking in biology, *Annu. Rev. Biophys. Bioeng.,* 8, 165, 1979.

116. **Schäfer, H.-J.,** Divalent azido-ATP analog for photoaffinity cross-linking of F₁ subunits, *Methods Enzymol.,* 126, 649, 1986.

117. **Dombroski, K. E. and Colman, R. F.,** 5'-(*p*-Fluorosulfonylbenzoyl)-8-azidoadenosine. A new bifunctional affinity label for nucleotide binding sites in proteins, *Arch. Biochem. Biophys.,* 275, 302, 1989.

118. **Knowles, J. R.,** Photogenerated reagents for biological receptor site labeling, *Acc. Chem. Res.,* 5, 155, 1972.

119. **Reiser, A., Willets, F. W., Terry, G. C., Williams, V., and Morley, R.,** Photolysis of aromatic acides. IV. Lifetimes of aromatic nitrenes and absolute rates of some of their reactions, *Trans. Faraday Soc.,* 64, 3265, 1968.

120. **Tao, T., Scheiner, C. J., and Lamkin, M.,** Site-specific photo-cross-linking studies on interactions between troponin and tropomyosin and between subunits of troponin, *Biochemistry,* 25, 7633, 1986.

121. **Tao, T., Lamkin, M., and Scheiner, C.,** Studies on the proximity relationships between thin filament proteins using benzophenone-4-maleimide as a site-specific photoreactive cross-linker, *Biophys. J.,* 45, 261, 1984.

122. **Campbell, P. and Gioannini, T. L.,** The use of benzophenone as a photoaffinity label. Labeling in *p*-benzoylphenylacetyl chymotrypsin at unit efficiency, *Photochem. Photobiol.,* 29, 883, 1979.

123. **Del Rosario, R. B., Wahl, R. L., Brocchini, S. J., Lawton, R. G., and Smith, R. H.,** Sulfhydryl site-specific cross-linking and labeling of monoclonal antibodies by a fluorescent equilibrium transfer alkylation cross-link reagent, *Bioconjugate Chem.,* 1, 51, 1990.

124. **Shephard, E., De Beer, F. C., von Holt, E., and Hapgood, J. P.,** The use of sulfosuccinimidyl-2-(*p*-azidosalicylamido)-1,3'-dithiopropionate as a crosslinking reagent to identify cell surface receptors, *Anal. Biochem.,* 168, 306, 1988.

125. **Steiner, M.,** Identification of the binding site for transferrin in human reticulocytes, *Biochem. Biophys. Res. Commun.,* 94, 861, 1980.

126. **Schidt, R. R. and Betz, H.,** Cross-linking of beta-bungarotoxin to chick brain membrane. Identification of subunits of a putative voltage-gated K⁺ channel, *Biochemistry,* 28, 8346, 1989.

127. **Erecinska, M.,** A new photoaffinity labeled derivative of mitochondrial cytochrome-C, *Biochem. Biophys. Res. Commun.,* 76, 495, 1977.

128. **Kiehm, D. J., and Ji, T. H.,** Photochemical cross-linking of cell membranes, *J. Biol. Chem.,* 252, 8524, 1977.

129. **Rajasekharan, K. N., Mayadevi, M., and Burke, M.,** Studies of ligand-induced conformational perturbations in myosin subfragment 1. An examination of the environment about the SH2 and SH1 thiols using a photoprobe, *J. Biol. Chem.,* 264, 10810, 1989.

130. **Politz, S. M., Noller, H. F., and McWhirter, P. D.,** (4-azido-phenyl)glyoxal, a novel heterobifunctional reagent: RNA-protein cross-linking in *E. coli* ribosomes, *Biochemistry,* 20, 372, 1981.

131. **Folk, J. E. and Chung, S. I.,** Molecular and catalytic properties of transglutaminases, *Adv. Enzymol.,* 38, 109, 1973.

132. **Folk, J. E., and Finlayson, J. S.,** The ε-(γ-glutamyl)lysine crosslink and the catalytic role of transglutaminase, *Adv. Protein Chem.,* 31, 1, 1977.

133. **Folk, J. E.,** Transglutaminases, *Annu. Rev. Biochem.,* 49, 517, 1980.

134. **Kennedy, J. H., Kricka, L. J., and Wilding, P.,** Protein-protein coupling reactions and the application of protein conjugates, *Clin. Chim. Acta,* 70, 1, 1976.

135. **Wilson, D. V. and Devey, M.,** A new coupling procedure for red cell-linked antigen antiglobulin reaction, *Int. Arch. Allergy Appl. Immunol.,* 44, 77, 1973.

136. **Lee, R. T., Wong, T.-C., Lee, R., Yue, L., and Lee, Y. C.,** Efficient coupling of glycopeptides to proteins with a heterobifunctional reagent, *Biochemistry,* 28, 1856, 1989.

137. **Yip, C. C., Yeung, C. W. T., and Moule, M.,** Photoaffinity labeling of insulin receptor of rat adipocyte plasma membrane, *J. Biol. Chem.,* 253, 1743, 1978.

138. **Perkins, M. E., Ji, T. H., and Hynes, R. O.,** Cross-linking of fibronectin to sulfated proteoglycan at the cell surface, *Cell,* 16, 941, 1979.

139. **Andreasen, T. J., Keller, C. H., LaPorte, D. C., Edelman, A. M., and Storm, D. R.,** Preparation of azidocalmodulin: a photoaffinity labeling for calmodulin binding proteins, *Proc. Natl. Acad. Sci. U.S.A.,* 78, 2782, 1981.

140. **Johnson, G. L., MacAndrew, Jr., V. I., and Pilch, P. F.,** Identification of glucagon receptor in rat liver membranes by photoaffinity crosslinking, *Proc. Natl. Acad. Sci. U.S.A.,* 78, 875, 1981.

141. **Coltrera, M. D., Potts, J. T., and Rosenblatt, M.,** Identification of a renal receptor for parathyroid hormone by photoaffinity radiolabeling using a synthetic analogue, *J. Biol. Chem.,* 256, 10555, 1981.

Chapter 6

ZERO-LENGTH CROSS-LINKING REAGENTS

I. INTRODUCTION

Zero-length cross-linking reagents are a special class of compounds different from, but similar in function to, the homo- and heterobifunctional agents. These reagents induce direct joining of two intrinsic chemical groups of proteins without introduction of any extrinsic material. During the cross-linking reaction, atoms are eliminated from the reactants, thus shortening the distance between the two linked moieties. This is in opposition to other cross-linking reagents, homo- or heterobifunctional, where a spacer is always incorporated between the two cross-linked groups. Reagents that catalyze the formation of disulfide bonds, for example, are zero-length cross-linkers. Other reagents condense carboxyl and primary amino groups to form amide bonds, hydroxyl and carboxyl groups to form esters, thiol and carboxyl groups to form thioesters and so on. Many of these reagents simply act as activating agents converting one of the components, for example, carboxyl groups, into a reactive species. Examples of these reactions are multitudinous in organic synthesis where two molecules are condensed to form a new compound. Carboxylic acids, for instance, are activated to acyl chlorides for the synthesis of esters or amides. In the synthesis of polypeptides, the carboxyl group is activated by carbodiimide to a reactive intermediate which the amino group attacks. However, many of the activation reactions are too harsh to be used with native proteins. They either disrupt the three-dimensional structure of the protein or cause extensive modification of the amino acids. Only mild reagents that do not cause denaturation are useful for the cross-linking of proteins. This chapter will focus on these compounds. They are listed in Table 1. Other reagents that are not suitable for linking proteins but have been used to couple proteins to solid supports will be discussed in Chapter 12.

II. CARBOXYL GROUP ACTIVATING REAGENTS

There are many reagents that condense carboxyl and amino groups to form amide bonds. For the cross-linking of proteins, the most commonly used agents are carbodiimides,[1,2] Woodwards reagent K,[3,4] N-ethylbenzisoxazolium tetrafluoroborate,[5,6] ethylchloroformate,[7,8] diethylpyrocarbonate,[9] carbonyldiimidazole[10] and others. A feature common to all these reagents in the mechanism of action is the initial activation of the carboxyl group. The formation of an amide bond or an ester bond facilitated by these reagents proceeds in two steps. In the first step, the reagent forms a highly reactive adduct with the carboxyl group. During the subsequent reaction, nucleophilic attack at the activated species eliminates the activating moiety, resulting in the formation of a bond that does not involve the incorporation of the cross-linking agent. While the amino group is implied as the nucleophile in these reactions, it should be borne in mind that other amino acid nucleophilic side chains of serine, histidine, tyrosine, arginine, and cysteine residues may also be involved in these reactions. In addition to cross-linking of proteins, these compounds have also been used to immobilize proteins to solid supports (see Chapter 12).

A. CARBODIIMIDES

The basic reaction mechanism of carbodiimide-mediated modification of carboxyl group has been discussed in Chapter 2, Section IV.C.2. Figure 1 depicts the two-step reaction sequence of condensation between carboxyl and amino groups of proteins. During the first step, the carboxyl group is activated by the carbodiimide derivative to an *O*-acylisourea

Table 1

ZERO-LENGTH CROSS-LINKING REAGENTS

Name	Structure
A. Carboxyl group activating reagents	
I. Dihexylcarbodiimide (DCC)	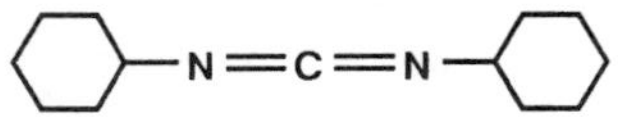
II. 1-Ethyl-3-(3-dimethylaminopropyl)-carbodiimide hydrochloride (EDC)	$CH_3CH_2-N{=}C{=}N-(CH_2)_3-\overset{H}{\underset{+}{N}}-(CH_3)_2Cl^-$
III. 1-Ethyl-3-(4-azonia-4,4-dimethylpentyl)carbodiimide iodide (EAC); 1-Ethyl-3-(3-dimethylaminopropyl) carbodiimide methiodide	$CH_3CH_2-N{=}C{=}N-(CH_2)_3-\overset{+}{N}(CH_3)_3I^-$
IV. 1-Cyclohexyl-3-(2-morpholinyl-(4)-ethyl)-carbodiimide metho-p-toluenesulfonate (CMC); N-Cyclohexyl-N'-[β-N-methylmorpholine) ethyl]-carbodiimide p-toluene sulfonate	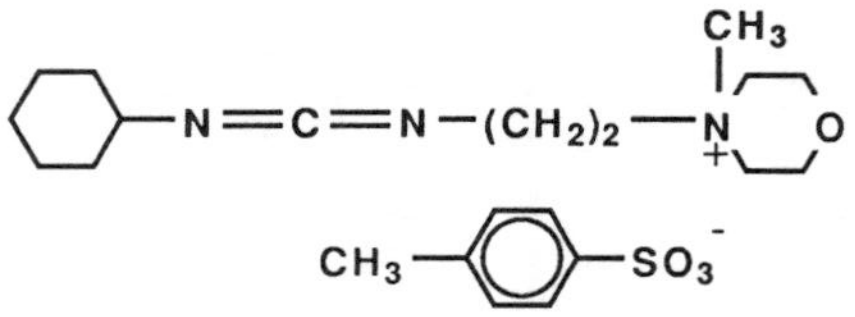
V. N-Benzyl-N'-3-dimethylaminopropyl-carbodiimide hydrochloride; 1-Benzyl-3-(3-dimethylsminopropyl) carbodiimide hydrochloride (BDC)	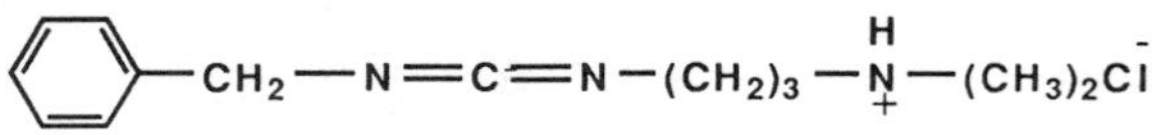

VI. N-Ethyl-3-phenylisoxazolium-3'-
sulfonate (Woodwards Reagent K)

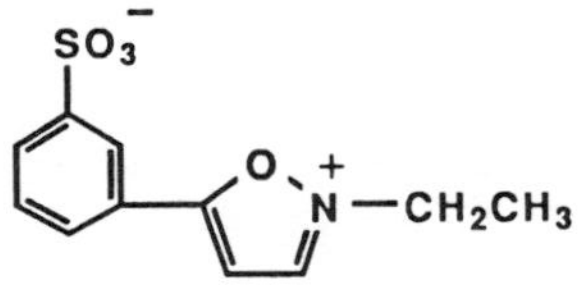

VII. N-Ethylbenzisoxazolium
tetrafluoroborate

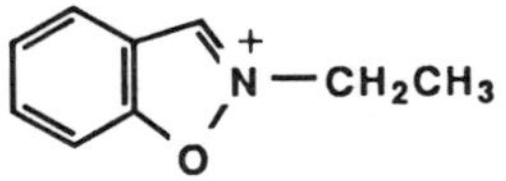

VIII. Ethylchloroformate

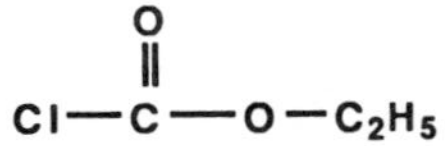

IX. p-Nitrophenylchloroformate

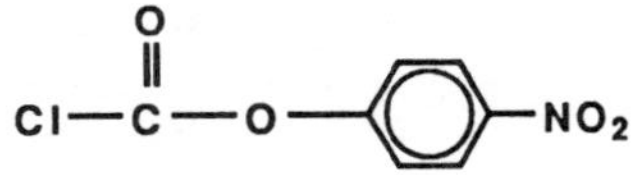

X. 1,1'-Carbonyldiimidazole

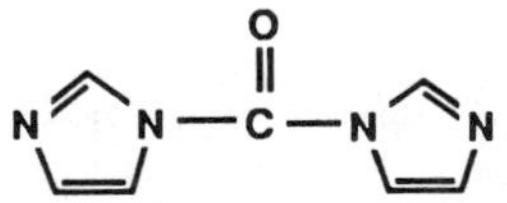

XI. N-(Ethoxycarbonyl)-2-ethoxy-
1,2-dihydroquinoline (EEDQ)

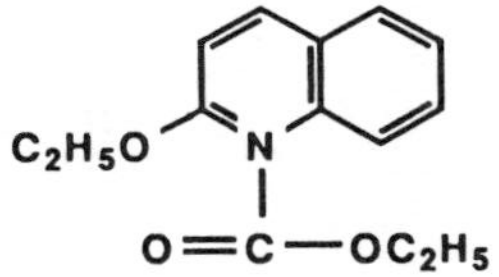

XII. N-(Isobutoxycarbonyl)-2-isobutoxy-
1,2-dihydroquinoline (IIDQ)

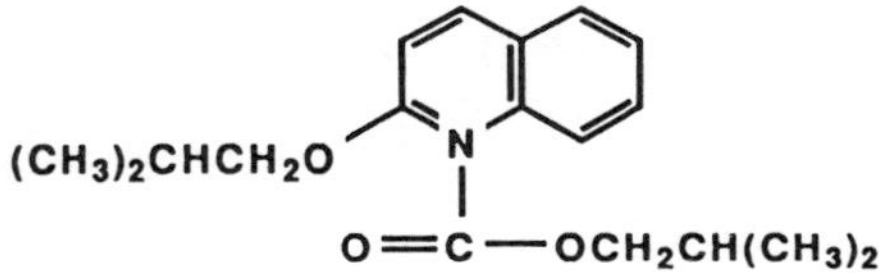

Table 1 (continued)

ZERO-LENGTH CROSS-LINKING REAGENTS

Name	Structure

B. Disulfide forming reagent

XIII. Cupric di(1,10-phenanthroline)
(CuP)

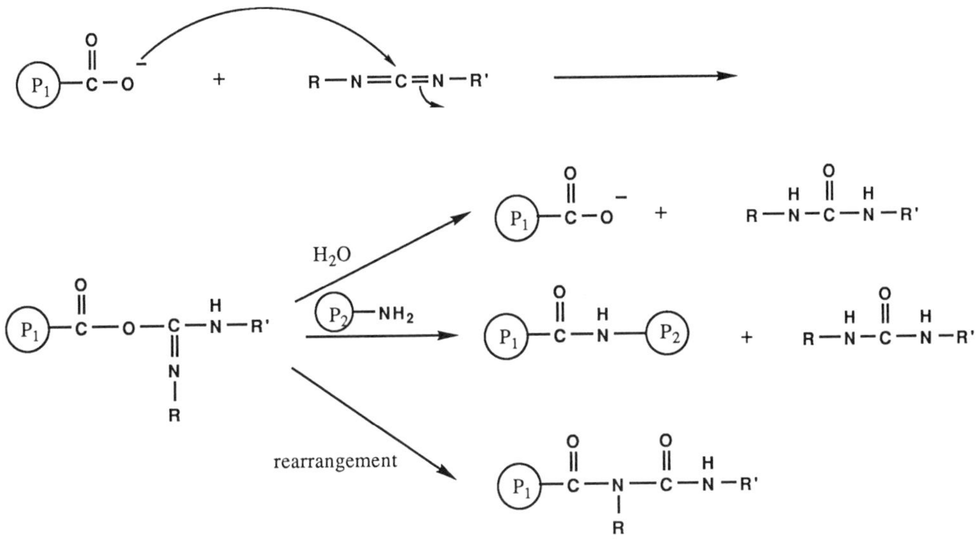

FIGURE 1. Cross-linking of proteins by carbodiimides. The O-acylisourea intermediate may hydrolyze to re-generate the original protein and urea derivative (top), react with an amino group to form a cross-linkage (middle), or rearrange to form a stable N-acylurea (bottom).

intermediate which can react further in a subsequent step in several ways. Nucleophilic attack of water will hydrolyze the intermediate to regenerate the free carboxyl group. Reaction with an amino group from a second protein will lead to a cross-link between the two proteins. Intramolecular cross-linking may also occur if the nucleophile is from the same protein. Without productive nucleophilic reaction, the O-acylisourea may undergo an intramolecular O- to N-acyl shift to form a more stable N-acylurea.

A large number of carbodiimides has been synthesized.[2,11] Those that are most commonly used in protein cross-linking because of their commercial availability are listed in Table 1 (Compounds I to V). These compounds have different stability in aqueous solutions. EDC (II), for instance, has a half-life of 37 h at pH 7 whereas EAC (III) is about tenfold less stable. Phosphate and phosphate-containing reagents as well as hydroxylamine and other amine derivatives increase the rate of loss of carbodiimides, and in some cases dramatically.[12] An extensive application of these carbodiimides can be found in the literature. For example, DCC (I) was used to cross-link calmodulin and myosin light chain kinase[13] as well as enzymes and antibodies.[14] The subunits of lutropin was cross-linked by EDC (II). This and other water-soluble carbodiimides have been used to cross-link heavy meromyosin to F-actin and between heavy meromyosin heads,[16,17] F-actin and myosin subfragment,[18] troponin C and troponin I,[19] F_1-ATPase,[20,21] ribosomal proteins and RNA,[22,23] voltage-gated K^+ channel subunits,[24] and cytochrome c and cytochrome b5.[25]

B. ISOXAZOLIUM DERIVATIVES

N-Ethyl-5-phenylisoxazolium-3'-sulfonate, compound VI (Table 1), was first synthesized by Woodward et al.,[3,4] and is thus known as Woodward's Reagent K. Under alkaline conditions at room temperature, the reagent is converted to a reactive ketoketenimine which reacts with the carboxylate anion to form an enol ester intermediate. This reactive intermediate then reacts with an amino group to form the cross-linked product as shown in Figure 2.

Woodward's Reagent K is not stable in aqueous medium as it is hydrolyzed rapidly above pH 3.0. The rate constant for the conversion of the isoxazolium salt to the ketoketenimine at pH 5.8 is 0.44 min^{-1} and the subsequent reaction step may occur at the same rate.[26,27] Although the reagent is used mostly as a carboxyl group modifying agent, it has

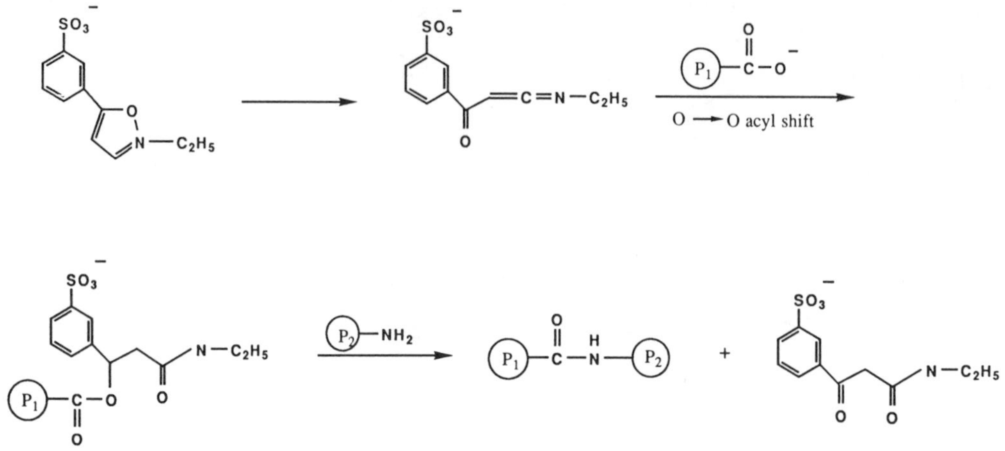

FIGURE 2.   Reaction mechanism of Woodward's reagent K.

been used to prepare polymeric α-chymotrypsin,[28] conjugate hemin and IgG,[29] label glutathione S-transferase with bilirubin[30] and so on.

A related compound, N-ethylbenzisoxazolium tetrafluoroborate (**VII**), has also been shown to condense carboxyl and amino groups to form an amide bond.[5] As in the case of Woodward's Reagent K, ethylbenzisozazolium decomposes to ethylbenzoketoketenimine which reacts with various nucleophiles including the carboxylate ion. The initial product of the reaction with the carboxyl group rearranges very rapidly to form an enol ester which can acylate an amino group to form an amide bond. Although the reagent was used in peptide synthesis,[4] its application in protein cross-linking is limited due to the availability of Woodward's Reagent K. This compound has been used to cross-link synthetic polypeptides,[21,22] angiotensin and polylysine.[6]

C. CHLOROFORMATES

Ethylchloroformate, **VIII**, and p-nitrophenylchloroformate, **IX**, which are referred to as homobifunctional cross-linking agents in Chapter 4, can function as a zero-length cross-linker. When the carboxyl group attacks the reagent first instead of the amino group, a mixed acid anhydride is formed. This active anhydride intermediate then transfers the carboxyl group to a nucleophile, such as an amino group to form an amide bond (Figure 3). Because of the reactivity of the anhydride, hydrolysis is the major competing reaction.

While these reagents have been used to prepare insoluble protein polymers[7,8] and the coupling of fluorescent or paramagnetic probes to gangliosides,[31,32] they have also been used for the immobilization of proteins.[32] In a similar token, isobutyl chloroformate has been used for the synthesis of affinity chromatography medium. Other chloroformates such as 2,4,5-trichlorophenyl chloroformates are also potential zero-length cross-linkers.[33]

D. CARBONYLDIIMIDAZOLE

Like the chloroformates, 1,1'-carbonyldiimidazole (**X**) functions both as a homobifunctional and a zero-length cross-linker. As a homobifunctional reagent, a carbonyl group is incorporated between two cross-linked amino groups. When the carboxyl group is first activated to form a mixed anhydride, subsequent reaction with an amino group will lead to an amide bond formation without the incorporation of any atoms from carbonyldiimidazole as shown in Figure 4.

Although the majority of applications of carbonyldiimidazole lies in the immobilization of proteins to solid matrices, few uses of the reagent as a zero-length cross-linker have appeared.[10,34]

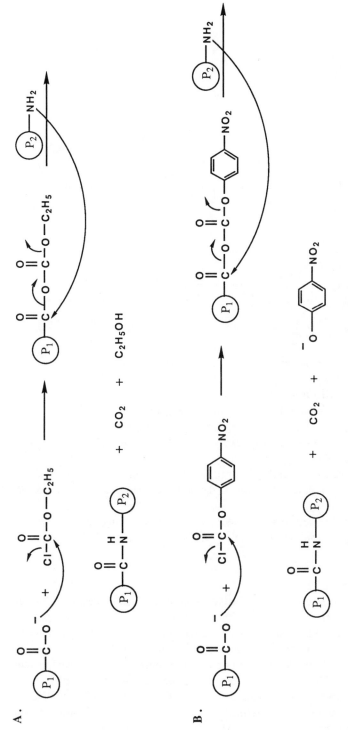

FIGURE 3. Cross-linking reaction with chloroformates: (A) ethylchloroformate; (B) *p*-nitrophenylchloroformate.

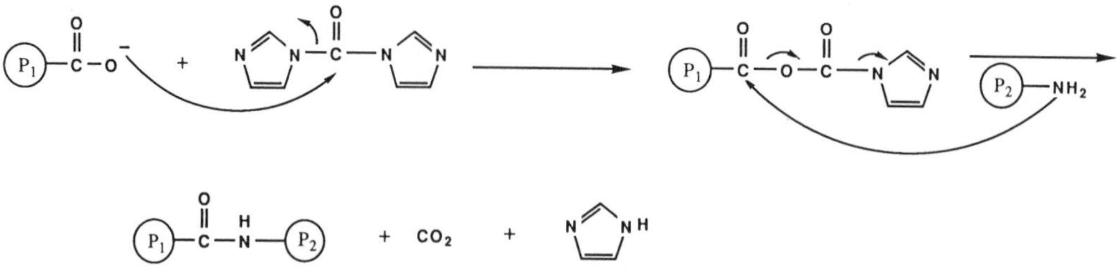

FIGURE 4.   Carbonyldiimidazole as a zero-length cross-linker.

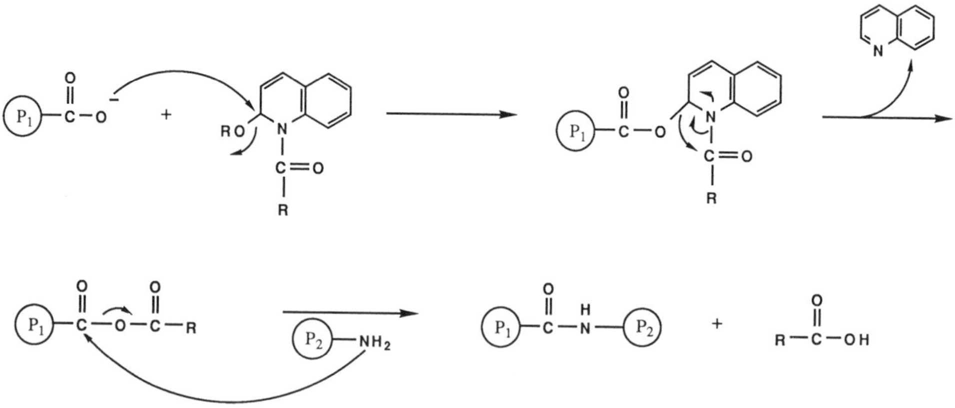

FIGURE 5.   Cross-linking reaction of *N*-carbalkoxydihydroquinoline.

E. *N*-CARBALKOXYDIHYDROQUINOLINES

N-(Ethoxycarbonyl)-2-ethoxy-1,2-dihydroquinoline, EEDQ (**XI**), was developed by Bel-leau et al.[35] as a carboxyl group reagent for the study of adrenergic receptors. This and another reagent, *N*-(isobutoxycarbonyl)-2-isobutoxy-1,2-dihydroquinoline, IIDQ (**XII**), also function as a zero-length cross-linker by activating the carboxyl group through a mixed anhydride as shown in Figure 5. The carboxyl group first forms an acyldihydroquinoline which rearranges to a mixed anhydride. Reaction of the activated anhydride with an amino group leads to the formation of an amide bond.

While EEDQ is commonly used to access the carboxyl group function at the active centers of enzymes, it has been used to cross-link F_1-actin and myosin S-1,[36] and to label F_1-ATPase with aniline.[37]

F. MISCELLANEOUS REAGENTS

Several other reagents that have been classified as homobifunctional reagent in Chapter 4 may also act as zero-length cross-linkers. The mechanism of action of these reagents is not known. Tetranitromethane, for example, has been reported in several occasions to produce polymeric aggregates of proteins after treatment.[38,39] It is possible that the reagent may function as a zero-length cross-linker through oxidation or nitration of some amino acid side chains. Potassium nitrosyldisulfonate may act the same way. Another reagent, die-thylpyrocarbonate, has also been found to induce amide bond formation and a putative *N*-carboxyanhydride has been proposed as an intermediate which might undergo transamidation with carboxyl groups to yield amide cross-links.[9] More likely, however, the carboxyl group may react with the reagent first to form a mixed anhydride which leads to an amide formation with an amino group, a reaction mechanism similar to the other carboxyl group activating reagents mentioned above.

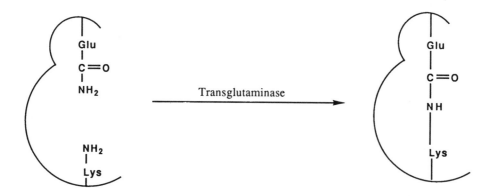

FIGURE 6.   Transglutaminase catalyzed reaction.

In nonaqueous solutions, many other reagents have been used to activate carboxyl groups. These reagents are used in most cases for peptide synthesis.[40] However, the reaction conditions are usually too vigorous for protein cross-linking. Some of the reagents are used for coupling of proteins to solid matrices which will be discussed in Chapter 12.

III. REAGENTS FOR CARBOHYDRATE ACTIVATION

For completeness, it may be mentioned that cross-linking by reductive alkylation between aldehydes derived from carbohydrates and amino groups is also a zero-length coupling process. The Schiff base formation does not incorporate any extrinsic atoms. Reagents that oxidize vicinal diols of carbohydrates to dialdehydes may be regarded as zero-length cross-linkers. The most commonly used mild oxidation agent is sodium periodate as discussed in Chapter 2. Reagents used for stabilization of the Schiff base are also discussed in that chapter. This procedure of reductive alkylation has been used to prepare immunoconjugates of horseradish peroxidase for immunoassays,[41] and ferritin-avidin conjugates for electron microscopic cytochemistry.[42]

VI. REAGENTS FOR DISULFIDE FORMATION

Free sulfhydryl groups may be cross-linked by oxidation to form disulfide bonds. Such linkage may occur intermolecularly or intramolecularly. Any oxidizing agent that facilitates the disulfide bond formation may be regarded as a zero-length cross-linker. This may be air, iodide or hydrogen peroxide. The reaction is also catalyzed by *bis*-1,10-phenanthroline complex of cupric ion, CuP, (**XIII**, Table 1).[43-46] With membrane proteins, micromolar concentrations of CuP is suffice to complete the reaction in few minutes.[44-47]

V. ENZYME AS ZERO-LENGTH CROSS-LINKERS

A. TRANSGLUTAMINASE

Transglutaminase as mentioned in Chapters 4 and 5 catalyzes the incorporation of various exogenous amines into proteins by the acyl-transfer reaction between γ-carboxamides of peptide-bound glutamines and exogenous amine substrates. When the amine is afforded as the ϵ-amino group of lysine residues of proteins, an amide bond cross-linkage is formed between the proteins as shown in Figure 6.[48] The enzyme has been isolated from liver, platelet, hair follicle, prostate, epidermis, umbilical vein endothelial cells and erythrocyte.[49-51] There is good evidence that the reaction proceeds through an acyl-enzyme intermediate in which an active site sulfhydryl group forms a thioester with the glutamyl group

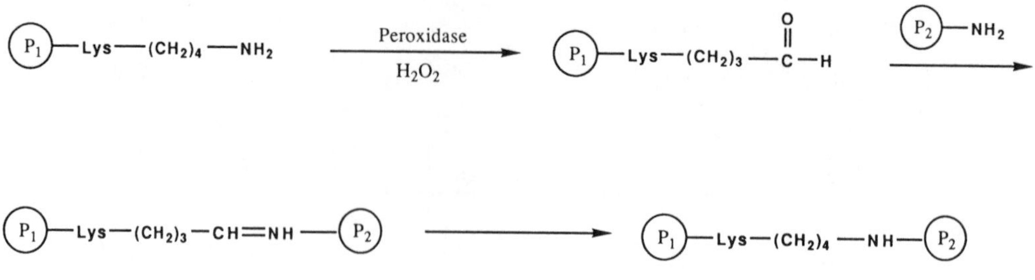

FIGURE 7. Putative mechanism of protein cross-linking by peroxidase.

of the substrate.[51] The active acyl intermediate then reacts with an adjacent amino group. Thus, the enzyme functions as a zero-length cross-linker similar to carbodiimide, Woodward's reagent K and ethyl chloroformate. The substrate specificity for the glutamine is high and only protein-bound glutamines are cross-linked such as those in fibrin,[52] fibrinogen,[50,53,54] fibrinogen and lipoprotein,[55] hair,[56] casein[57] soybean proteins,[58] and hemoglobin.[59] On the other hand, the substrate specificity for the amine donor is much less stringent and transglutaminase has been used for attachment of a number of primary amine-containing labels and probes into proteins as already discussed in Chapters 4 and 5. The potential use of transglutaminase as a zero-length protein cross-linker may be explored further.

B. PEROXIDASE

Stahmann et al.[60-62] have reported that horseradish peroxidase induced cross-linking of various proteins including cytochrome D, bovine serum albumin, catalase, ovalbumin, β-lactoglobulin and pepsin. In the presence hydrogen peroxide and a hydrogen donor such as benzidine, *p*-anisidine, *o*-phenylene diamine or pyrogallol, the peroxidase oxidatively deaminated some lysyl residues of the protein to form lysyl aldehyde. The aldehyde presumably forms a Schiff base with another amino group. Upon reduction, the lysinonorleucine formed a cross-link between the proteins as shown in Figure 7. Such lysinonorleucine has been isolated from cross-linked elastin.[63] The application of peroxidase in studies of cross-linking of proteins has not, however, been investigated.

C. XANTHINE OXIDASE

Another enzyme that has been found to cause protein-protein cross-linking is xanthine oxidase.[64] In the presence of xanthine and ferric ion, xanthine oxidase causes protein cross-linking of isolated erythrocyte membranes. Thiol-reducible bonds as well as nonreducible bonds are generated. The mechanism of the cross-linking is not known, although the involvement of hydrogen peroxide in the formation of disulfide bonds is speculated. Whether this enzyme system can be used for cross-linking of isolated proteins other than in erythrocyte membranes awaits to be studied.

REFERENCES

1. **Carraway, K. L. and Koshland, D. E., Jr.,** Carbodiimide modification of proteins, *Methods Enzymol.*, 25, 616, 1972.
2. **Kurzer, F. and Douraghi-Zadeh, K.,** Advances in the chemistry of carbodiimides, *Chem. Rev.*, 67, 107, 1967.
3. **Woodward, R. B., Olofson, R. A., and Mayer, H.,** A new synthesis of peptides, *J. Am. Chem. Soc.*, 83, 1007, 1961.

4. **Woodward, R. B. and Olofson, R. A.,** The reaction of isoxazolium salts with bases, *J. Am. Chem. Soc.,* 83, 1010, 1961.

5. **Kemp, D. S. and Woodward, R. B.,** *N*-Ethylbenzisoxazolium cation. I. Preparation and reactions with nucleophilic species, *Tetrahedron,* 21, 3019, 1965.

6. **Goodfriend, T., Fasman, G., Kemp, D., and Levine, L.,** Immunochemical studies of angiotensin, *Immunochemistry,* 3, 223, 1966.

7. **Patramani, I., Katsiri, K., Pistevou, E., Kalogerakos, T., Pawlatos, M., and Evangelopoulos, A. E.,** Glutamic-aspartic transaminase-antitransaminase interaction: a method for antienzyme purification, *Eur. J. Biochem.,* 11, 28, 1969.

8. **Avrameas, S. and Ternynck, T.,** Biological active water-insoluble protein polymers. I. Their use for isolation of antigens and antibodies, *J. Biol. Chem.,* 242, 1651, 1967.

9. **Wolf, B., Lesnaw, J. A., and Reichmann, M. E.,** Mechanism of the irreversible inactivation of bovine pancreatic ribonuclease by diethylpyrocarbonate. General reaction of diethylpyrocarbonate with proteins, *Eur. J. Biochem.,* 13, 519, 1970.

10. **Chang, S. I. and Hammes, G. G.,** Interaction of spin labeled nicotinamide adenine dinucleotide phosphate with chicken liver fatty acid synthase, *Biochemistry,* 25, 4661, 1986.

11. **Sheehan, J. C., Cruickshank, P. A., and Boshart, G. L.,** A convenient synthesis of water-soluble carbodiimides, *J. Organ. Chem.,* 26, 2525, 1961.

12. **Gilles, M. A., Hudson, A. Q., and Borders, Jr., C. L.,** Stability of water-soluble carbodiimides in aqueous solution, *Anal. Biochem.,* 184, 244, 1990.

13. **Zot, H. G. and Puett, D.,** An enzymatically active cross-linked complex of calmodulin and rabbit skeletal muscle myosin light chain kinase, *J. Biol. Chem.,* 264, 15552, 1989.

14. **Clyne, D. H., Norris, S. H., Modesto, R. R., Pesce, A. J., and Polak, V. E.,** Antibody enzyme conjugates. The preparation of intermolecular conjugates of horseradish peroxidase and antibody and their use in immunohistology of renal cortex, *J. Histochem. Cytochem.,* 21, 233, 1973.

15. **Parsons, T. F. and Pierce, J. G.,** Biologically active covalently cross-linked glycoprotein hormones and the effects of modification of the COOH-terminal region of their α subunits, *J. Biol. Chem.,* 254, 6010, 1979.

16. **Onishi, H., Maita, T., Matsuda, G., and Fujiwara, K.,** Evidence for the association between two myosin heads in rigor acto-smooth muscle heavy meromyosin, *Biochemistry,* 28, 1898, 1989.

17. **Onishi, H., Maita, T., Matsuda, G., and Fujiwara, K.,** Carbodiimide-catalyzed cross-linking sites in the heads of gizzard heavy meromyosin attached to F-actin, *Biochemistry,* 28, 1905, 1989.

18. **Takashi, R.,** A novel actin label. A fluorescent probe at glutamine-41 and its consequence, *Biochemistry,* 27, 939, 1988.

19. **Leszyk, J., Grabarek, Z., Gergely, J., and Collins, J. H.,** Characterization of zero-length cross-links between rabbit skeletal muscle troponin C and troponin I. Evidence for direct interaction between the inhibitory region of troponin I and the NH_2-terminal, regulatory domain of troponin C, *Biochemistry,* 29, 299, 1990.

20. **Bragg, P. D. and Hou, C.,** Chemical crosslinking of alpha subunits in the F_1 adenosine triphosphatase of *Escherichia coli, Arch. Biochem. Biophys.,* 244, 361, 1986.

21. **Satre, M., Lunardi, J., Dianoux, A.-C., Dupuis, A., Issartel, J. P., Klein, G., Pougeois, R., and Vignais, P. V.,** Modifiers of F_1-ATPases and associated peptides, *Methods Enzymol.,* 126, 712, 1986.

22. **Chiaruttini, C., Expert-Benzancon, A., Hayes, D., and Ehresmann, B.,** Protein-RNA crosslinking in *Escherichia coli* 30S ribosomal subunits. Identification of a 16S rRNA fragment cross-linked to protein S12 by the use of the chemical crosslinking reagent 1-ethyl-3-dimethylaminopropylcarbodiimide, *Nucleic Acid Res.,* 10, 7657, 1982.

23. **Chiarutini, C., Milet, M., Hayes, D. H., and Expert-Bezancon, A.,** Crosslinking of ribosomal proteins S4, S5, S7, S8, S11, S12 and S18 to domains 1 and 2 of 16SrRNA in the *Escherichia coli* 30S particle, *Biochemie,* 7, 839, 1989.

24. **Schidt, R. R. and Betz, H.,** Cross-linking of beta-bungarotoxin to chick brain membrane. Identification of subunits of a putative voltage-gated K^+ channel, *Biochemistry,* 28, 8346, 1989.

25. **Mauk, M. R. and Mauk, A. G.,** Crosslinking of cytochrome C and cytochrome b5 with a water-soluble carbodiimide. Reaction conditions, product analysis and critique of the technique, *Eur. J. Biochem.,* 186, 473, 1989.

26. **Dunn, B. M. and Affinsen, C. B.,** Kinetics of Woodward's Reagent K hydrolysis and reaction with staphylococcal nuclease, *J. Biol. Chem.,* 249, 3717, 1974.

27. **Kuimov, A. N. and Kochetov, G. A.,** Conversion of Woodward's Reagent K in an aqueous medium: mathematical analysis applied to enzyme modification, *Anal. Biochem.,* 172, 56, 1988.

28. **Patel, R. P. and Price, S.,** Derivatives of proteins. I. Polymerization of α-chymotrypsin by use of *N*-ethyl-5-phenylisoxazolium-3'-sulfonate, *Biopolymers,* 5, 583, 1967.

29. **Pikuleva, I. A. and Turko, I. V.,** A new method of preparing hemin conjugate with rabbit IgG, *Bioorganicheskaia Khimiia,* 15, 1480, 1989.

30. **Boyer, T. D.,** Covalent labeling of the nonsubstrate ligand-binding site of glutathione *S*-transferase with bilirubin-Woodward's reagent K, *J. Biol. Chem.,* 261, 5363, 1986.

31. **Acquotti, D., Sonnino, S., Masserini, M., Casella, L., Fronz, G., and Tettamanti, G.,** A new chemical procedure for the preparation of ganliosides carrying fluorescent or paramagnetic probes on the lipid moiety, *Chem. Phys. Lipids,* 40, 71, 1986.

32. **Aleix, J. A., Swaminathan, B., Minnich, S. A., and Wallshein, V. A.,** Enzyme immunoassays. Binding of salmonella antigens to activated microtiter plates, *J. Immunol.,* 6, 391, 1985.

33. **Veronese, F. M., Largajolli, R., Boccu, E., Benassi, C. A., Schiavon, O.,** Surface modification of proteins. Activation of monomethoxy-polyethylene glycols by phenylchloroformates and modification of ribonuclease and superoxide dismutase, *Appl. Biochem. Biotechnol.,* 11, 141, 1985.

34. **Barthing, G. J., Chattopadhyay, S. K., Barker, C. W., Farrester, L. J., and Brown, H. D.,** Preparation and properties of horseradish peroxidase cross-linked in nonaqueous media, *Int. J. Peptide Protein Res.,* 6, 287, 1974.

35. **Belleau, B., DiTullio, V., and Jodin, D.,** The mechanism of irreversible adrenergic blockade by *N*-carbethoxydihydroquinolines. Model studies with typical serine hydrolases, *Biochem. Pharm.,* 18, 1039, 1969.

36. **Bertrand, R., Chaussepied, P., Kassab, R., Boyer, M., Roustan, C., and Beuyamin, Y.,** Cross-linking of the skeletal myosin subfragment 1 heavy chain to the *N*-terminal actin segment of residues 40-113, *Biochemistry,* 27, 5728, 1988.

37. **Laikind, P. K., Hill, F. C., Allison, W. S.,** The use of ^{3}H-aniline to identify the essential carboxyl group in the bovine mitrochondrial F_1-ATPase that reacts with 1-(ethoxycarbonyl)-2-ethoxy-1,2-dihydroquinoline, *Arch. Biochem. Biophys.,* 240, 904, 1985.

38. **Doyle, R. J., Bello, J., and Rohott, O. A.,** Probable protein cross-linking with tetranitromethane, *Biochim. Biophys. Acta,* 160, 274, 1968.

39. **Boesel, R. W. and Carpenter, F. H.,** Crosslinking during nitration of bovine insulin with tetranitromethane, *Biochem. Biophys. Res. Commun.,* 38, 678, 1970.

40. **Bodanszky, M.,** In search of new methods in peptide synthesis. A review of the last three decades, *Int. J. Pept. Prot. Res.,* 25, 449, 1985.

41. **Wilson, M. B. and Nakane, P. K.,** Recent developments in the periodate method of conjugating horseradish peroxidase (HRPO) to antibodies, in *Immunofluorescence and Related Staining Techniques,* Knapp, W., Holuber, K., and Wilck, G., Eds., Elsevier/North Biomedical Press, Amsterdam, 1978, 215.

42. **Bayer, E. A., Skutelsky, E., Wynne, D., and Wilchek, M.,** Preparation of ferritin-avidin conjugates by reductive alkylation for use in electron microscopy cytochemistry, *J. Histochem.,* 24, 933, 1976.

43. **Peters, K. and Richards, F. M.,** Chemical cross-linking: Reagents and problems in studies of membrane structure, *Annu. Rev. Biochem.,* 46, 523, 1977.

44. **Ji, T. H.,** The application of chemical crosslinking for studies on cell membranes and the identification of surface reporters, *Biochim. Biophys. Acta,* 559, 39, 1979.

45. **Huang, C. K. and Richards, F. M.,** Reaction of a lipid-soluble, unsymmetrical, cleavable, cross-linking reagent with muscle aldolase and erythrocyte membrane proteins, *J. Biol. Chem.,* 252, 5514, 1977.

46. **Quinlan, R. A. and Franke, W. W.,** Heteropolymer filaments of vimentin and desmin in vascular smooth muscle tissue and cultured baby hamster kidney cells demonstrated by chemical crosslinking, *Proc. Natl. Acad. Sci. U.S.A.,* 79, 3452, 1982.

47. **Steck, T. L.,** Cross-linking the major proteins of the isolated erythrocyte membrane, *J. Mol. Biol.,* 66, 295, 1972.

48. **Folk, J. E. and Finlayson, J. S.,** The ε-(γ-glutamyl)lysine crosslink and the catalytic role of transglutaminase, *Adv. Prot. Chem.,* 31, 1, 1977.

49. **Folk, J. E. and Chung, S. I.,** Molecular and catalytic properties of transglutaminases, *Adv. Enzymol.,* 38, 109, 1973.

50. **Martinez, J., Rich, E., and Barsigian, C.,** Transglutaminase-mediated cross-linking of fibrinogen by human umbilical vein endothial cells, *J. Biol. Chem.,* 264, 20502, 1989.

51. **Folk, J. E., Cole, P. W., and Mullooly, J. P.,** Mechanism of action of guinea pig liver transglutaminase. IV. The trimethylacyl enzyme, *J. Biol. Chem.,* 242, 4329, 1967.

52. **McKee, P. A., Schwartz, M. L., Pizzo, S. V., and Hill, R. L.,** Cross-linking of fibrin by fibrin-stabilizing factor, *Ann. N.Y. Acad. Sci.,* 202, 127, 1972.

53. **Barsigian, C., Fellin, F. M., Jain, A., and Martinez, J.,** Dissociation of fibrinogen and fibronectin binding from transglutaminase-mediated cross-linking at the hepatocyte surface, *J. Biol. Chem.,* 263, 14015, 1988.

54. **Martinez, J., Rich, E., and Barsigian, C.,** Transglutaminase-mediated cross-linking of fibrinogen by human umbilical vein endothelial cells, *J. Biol. Chem.,* 264, 20502, 1989.

55. **Bowness, J. M., Tarr, A. H., and Wiebe, R. I.,** Transglutaminase-catalyzed cross-linking. A potential mechanism for the interaction of fibrinogen, low density lipoprotein and arterial type III procollagen, *Thromb. Res.,* 54, 357, 1989.

56. **Wajda, I. J., Hanbauer, I., Manigault, I., and Lajtha, A.,** Chromogranins as substrate for transglutaminase, *Biochem. Pharmacol.,* 20, 3197, 1971.
57. **Ikura, K., Kometani, T., Yoshikawa, M., Sasaki, R., and Chiba, H.,** Cross-linking of casein components by transglutaminase, *Agric. Biol. Chem.,* 44, 1567, 1980.
58. **Ikura, K., Kometani, T., Sasaki, R., Chiba, H.,** Cross-linking of soybean 7S and 11S proteins by transglutaminase, *Agric. Biol. Chem.,* 44, 2979, 1980.
59. **Uy, R. and Wold, F.,** Introduction of artificial crosslinks into proteins, in *Protein Crosslinking: Biochemical and Molecular Aspects,* Friedman, M., Ed., Plenum Press, N. Y., 1976, 170.
60. **Stahmann, M. A.,** Cross-linking of protein by peroxidase, in *Protein Cross-Linking: Nutritional and Medical Consequences,* Friedman, M., Ed., Plenum Press, New York, 1977, 285.
61. **Stahmann, M. A. and Spencer, A. K.,** Deamination of protein lysyl ϵ-amine groups by peroxidase *in vitro, Biopolymers,* 16, 1299, 1977.
62. **Stahmann, M. A., Spencer, A. K., and Honold, G. R.,** Cross-linking of proteins *in vitro* by peroxidase, *Biopolymers,* 16, 1307, 1977.
63. **Franzblau, C., Sinex, F. M., and Faris, B.,** Identification of a new cross-linking amino acid in elastin, *Biochem. Biophys. Res. Commun.,* 21, 575, 1965.
64. **Girotti, A. W., Thomas, J. P., and Jordan, J. E.,** Xanthine oxidase-catalyzed cross-linking of cell membrane proteins, *Arch. Biochem. Biophys.,* 251, 639, 1986.

Chapter 7

PROCEDURES, ANALYSIS, AND COMPLICATIONS

I. INTRODUCTION

With the diversity of cross-linking reagents, not only in their varying degree of selectivity but also in their different chemical reaction mechanisms, and the vast number of systems in which they have been employed, the procedures of conjugation and cross-linking are multifarious. A broad classification of the reaction types may be useful for the understanding of the application of these reagents. It is possible, however, to generalize the reaction procedures for common functional groups. An individual cross-linker may require specific reaction conditions. These conditions may be found in the literature. In the case of a new reagent or application, the conditions may be formulated after consultation of the general procedures. In some cases, different protocols may give rise to different products, or ratios of products. In other cases, complications may occur due to undesirable side reactions. It is advisable in some cases to vary the conditions in order to obtain the best results.

II. PROCEDURES FOR CROSS-LINKING

A. TYPES OF REACTIONS

1. One-Step Reactions

The simplest procedure in which a bifunctional reagent is added to a mixture of proteins to be coupled has been employed by many investigators. Modesto and Pesce[1] have demonstrated that the rate of addition of reagent influences the yield of conjugation. Slow addition of the reagent over a period of time, rather than addition of all of the reagent at one time, increases yields of coupled proteins. This procedure is most applicable to protein aggregates such as those in the membrane preparations and ribosomes. Since these proteins are situated next to each other, cross-linking will occur when the reagents react. For example, in the study of microsomal proteins, dithiobis(succinimidylpropionate) is added directly to a suspension of microsomes.[2] In the preparation of immunoconjugates where the proteins are free in solution, the procedure is less desirable. Both homopolymers or heteropolymers will be produced. In addition, each of these proteins to be coupled may have different reactivity towards the reagent resulting homopolymerization of one of the components. IgG, for example, has a higher reactivity with 4,4'-difluoro-3,3'-dinitrodiphenyl sulfone than horseradish peroxidase. The reagent selectively reacts with IgG giving rise to homopolymerization. The yield of conjugation between peroxidase and IgG is very low in the one-step procedure.[3] Under these circumstances, a two-step reaction procedure would be a better choice.

2. Two-Step Reactions

In this procedure one of the proteins to be conjugated is first reacted with the cross-linker. The unreacted reagent is then removed prior to addition of the second protein. This procedure takes advantage of the differential reactivities of functional groups in heterobifunctional reagents as well as the differential selectivity of homobifunctional reagents towards the two proteins to be coupled.

Practically all heterobifunctional cross-linkers, particularly the photosensitive reagents, are used according to a two-step process. For example, in the preparation of β-galactosidase-IgG conjugate with *m*-maleimidobenzoyl-*N*-hydroxysuccinimide (MBS), IgG, containing no free thiol group, is first labeled with the reagent through an amino group reaction with the

N-hydroxysuccinimide ester. After the removal of excess reagent, either by dialysis or gel filtration, β-galactosidase which contains free thiol groups is added to react with the maleimide moiety of labeled IgG, resulting in the desired immunoconjugate product.[3] Many other immunoconjugates are prepared in a similar two-step procedure as will be presented in Chapter 10.

A notable example of a homobifunctional reagent that shows differential reactivities usable in a two-step coupling reaction is toluene-2,4-diisocyanate. The para-isocyanate group is much more reactive than that at the ortho-position, owing to the steric hindrance of the latter by the methyl group.[4] As a first step in the reaction, the protein is mixed with the reagent at 0°C where modification of the protein will take place with the para-isocyanate group. After the reaction, a second protein to be cross-linked is added and the temperature raised to 37°C. At this stage, the ortho-isocyanate group will react to link the proteins together.

Similar differences in reactivity are seen in 1,5-difluoro-2,4-dinitrobenzene, bis(4-fluoro-3-nitro)sulfone and 2,4-dichloro-6-methoxy-*s*-triazine, probably due to electronic effects after nucleophilic replacement. For 2,4-dichloro-6-methoxy-*s*-triazine, coupling occurs at pH 7 with tyrosine residues, however the second chloro group will only react with a tyrosine residue of a second protein in alkaline conditions.[5] Similarly, in acidic conditions only one of the diazo groups of *bis*-diazotised *o*-dianisidine is reactive. For coupling of ferritin to rabbit gamma-globulin, the first step of the reaction is carried out at pH 5 and the second at pH 9.4.[6]

Differential reactivity of a homobifunctional reagent toward different proteins has also been used in two-step reactions. Glutaraldehyde, for example, reacts with γ-immunoglobulins much faster than horseradish peroxidase. Reaction of horseradish peroxidase with excess of glutaraldehyde constitutes the first step of the reaction. After removal of the excess reagent, the immunoglobulin is added to generate the enzyme-immunoglobulin conjugate. Self-coupling of horseradish peroxidase is minimal due to the unavailability of reactive groups.[7] Ferritin has also been coupled to γ-immunoglobulins under similar conditions.[8]

3. Three-Step Reactions

This procedure involves an extra step for the preparation of proteins to be coupled, for instance, the introduction of a thiol group as discussed in Chapter 2. Figure 1 shows an example of a three-step coupling process. In the preparation for coupling, IgG is first labeled with pyrrole-α-acyl azide. The second protein, albumin, is reacted with the cross-linker, *bis*-diazotised *p*-phenylenediamine in an acidic condition. After isolation, the pyrrole-modified IgG and the diazo-albumin are mixed to react at pH 6. Specific reaction occurs between the diazo group and the pyrrole ring as shown in Figure 1. The method reduces the side reactions by adjusting the coupling conditions. At acidic pHs, the first diazotized group is reactive whereas the second group is reactive at higher pH.[9]

Many immunoconjugates and immunotoxins are prepared by this three-step reaction process. Examples can be found in Chapter 10 and 11.

4. Multi-Step Reactions

Multi-step reactions involve several preparation procedures of proteins for the coupling process. Cross-linking effected by masked or disguised cross-linkers as eluded to in Chapter 5 proceeds through several steps. During the procedure, extra steps are necessary to generate the functional group needed for cross-linking process. The reactions are demonstrated in Chapter 5.

Various other multi-step cross-linking reaction schemes are possible. For example, some immunotoxins are prepared by using *N*-succinimidyl 3-(2-pyridyldithio)propionate (SPDP) to modified both proteins. One of these labeled proteins is then reduced to generate a free

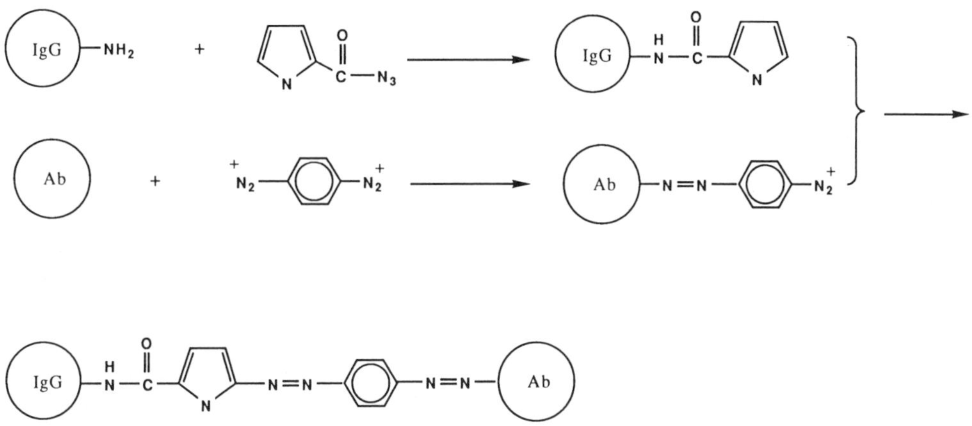

FIGURE 1.   Coupling of albumin and IgG by a three-step reaction procedure.

thiol group and purified. The two modified proteins are finally mixed to allow the cross-linking to proceed. These multi-step reactions are illustrated in Chapter 11.

B. GENERAL CONDITIONS FOR CROSS-LINKING

Cross-linking conditions are highly dependent on the type of reagent used and the particular system under investigation. Reaction time ranges from minutes to hours. The reagent concentration varies with its relative reactivity and the stability of each reagent. Some reagents are hydrolyzed readily and consequently may require an excess quantity. Cross-linking can generally be carried out in buffers such as phosphate-buffered saline or isotonic phosphate. Because reagent type dictates the reaction conditions, the following discussion represents generalized parameters for the most frequently used reagents only. For specific applications, the reader is referred to the specific literature.

1. Imidoesters

Imidates are generally readily soluble in aqueous solutions and are hydrolyzed rapidly with a pH-dependent half-life ranging from several minutes to half an hour. Below pH 8.5, the half-life for ethyl acetimidate is 2 to 5 min. The rate of hydrolysis increases substantially at higher pH values.[10,11] Up to a 100-fold excess of reagent, concentration range 0.1 to 10 mM, is required for complete reaction. To circumvent the degradation problem, incremental additions of reagents may be used. Imidoesters react over a wide pH range from 7 to 10[10-14] and temperature from 0 to 40°C.[10,11,15] Alkaline pH increases the rate of the reaction of imidates with amines to form amidines.[10,11,16,17] Reaction rate decreases several fold as the temperature drops from 39 to 25°C and again from 25 to near 0°C.[10] At or below zero, amidination occurs at considerably slower rates and requires longer reaction times of several hours to overnight.[15,18,19]

2. N-Hydroxysuccinimide Esters

The half-life of hydrolysis of N-hydroxysuccinimide (NHS) esters is approximately 10 min at pH 8.6[20] and 4 to 5 h at pH 7.[21] The optimal pH for the reaction is pH 9. At pH 7, the reactivity is only half of that at pH 9. Temperature has little effect on the reaction, allowing it to react efficiently at near freezing. NHS esters have been used in the concentration range 0.05 to 9 mM with reaction time from minutes to hours depending on the conditions used.[21-23]

3. Photosensitive Reagents

All the photosensitive reagents are generally sensitive to ultraviolet radiation. They must be handled in the dark or under dim or red light conditions. Under white fluorescent lamps, they have a half-life of about 6 h. Azides are also sensitive to sulfhydryl reducing agents such as dithiothreitol, 2-mercaptoethanol and glutathione, since they can be reduced to the corresponding amine.[24,25] This reduction is pH dependent. In 10 mM dithiothreitol, various azides are found to have a half-life of 5 to 15 min at pH 8. At pH 10, the rate is increased 12-fold.[25] With 50 mM glutathione or 2-mercaptoethanol at pH 8, the azides were reduced 60 to 70% and 10 to 20%, respectively, in 24 h. Therefore, if reducing conditions are required during cross-linking, the use of 2-mercaptoethanol is recommended.

Aryl azides are generally insoluble in aqueous buffers and may be solubilized with the aid of water-miscible organic solvents such as acetone, methanol, ethanol, dioxane, dimethylformamide, pyridine, acetonitrile and dimethyl sulfoxide.[22,26-30] The final concentration of organic solvent may be as high as 20%. Alternatively, a fine powder of the reagent may be added to the reaction mixture, although the rate of reaction may be reduced.

The photolysis of photosensitive cross-linkers and their subsequent chemical reactions are temperature independent. A common method to photoactivate azides is irradiation with a short-wavelength UV lamp, such as the mineralight UVS-11. The half-time of photolysis is usually on the order of 10 to 50 s with the sample positioned 1 cm from the lamp.[31] In general, complete photolysis is possible in less than 10 min.[32] An alternative method is flash photolysis using camera electronic flash units.[13,19] In a typical experiment, less than 10 flashes are sufficient to photolyze reagents associated with proteins.[33] With systems that are sensitive to radiation, longer wavelengths must be used. This can be easily achieved by using glass-filters,[34] which are particularly useful for activating aryl azides with nitro substituents.[32,34] In this case, irradiation must be lengthened to many minutes or even hours, depending on the molar absorptivity of the aryl azides at the longer wavelengths.[22,26,35,36]

4. Cleavage of Cross-Linked Complexes

Cleavage of the cross-linkers requires specific conditions depending on the type of bonding in the reagents. A general approach is given below for some of the common linkages.

a. Disulfide Linkages

Cleavage of disulfide bonds can be easily achieved by incubating with sulfhydryl compounds such as mercaptoethanol, dithiothreitol or dithioerythritol at concentrations of about 10 to 100 mM, between pH 7 to 9 at 25 to 37°C for 10 to 30 min. Occasionally, concentrations of reducing reagents up to 0.4 M may be used.[37] Common buffers such as Tris and phosphate, as well as detergents such as Triton X-100 and sodium dodecylsulfate do not interfere with the cleavage reaction. Disulfide bonds can also be conveniently cleaved during electrophoresis by addition of a reducing agent to the electrophoretic buffer.

b. Glycol Bonds

Glycol bridges can be cleaved by 15 mM sodium periodate, pH 7.5 for 4 to 5 h at 25°C.[38] Buffers such as triethanolamine and phosphate, as well as SDS do not interfere. But Tris cannot be used since it reacts with sodium periodate.

c. Azo Bonds

Azo linkages can be cleaved by reduction with 0.1 M sodium dithionite in 0.15 M NaCl, buffered at pH 8.0 with 0.1 M NaHCO$_3$ for 25 min.[39] Disulfide reducing agents do not interfere with this process.

d. Sulfone Linkages

Sulfone bonds can be cleaved by reduction in 100 mM sodium phosphate adjusted to pH 11.6 with Tris, 6 M urea, 0.1% SDS, and 2 mM dithiothreitol for 2 h at 37°C. The presence of dithiothreitol is not absolutely necessary.[40]

e. Ester Bonds

Theoretically esters and thioesters can be hydrolyzed under both acidic and alkaline conditions. They are most conveniently cleaved by 1 M hydroxylamine, in 50 mM Tris, pH 7.5 to 8.5, 25 mM CaCl$_2$ and 1 mM benzamidine for 3 to 6 h at 25 to 37°C.[41]

III. ANALYSIS OF CROSS-LINKED PRODUCTS

Almost in all cases, it is necessary to analyze the products after the cross-linking reaction to assess the extent of coupling. Isolation and purification of a particular conjugate of interest may also be one of the requirements of cross-linking. At the same time, characterization of the conjugate with regards to its composition, identity, catalytic activity, and immunological activity cannot be avoided in some studies. In general, the characterization of the product requires the same procedures as the characterization of proteins modified with monofunctional reagents. Excess reagent is first eliminated by dialysis or gel filtration. The cross-linked components are then isolated and characterized by biochemical and biophysical parameters. The following biochemical techniques that have been employed in various studies will be briefly described.

A. GEL-FILTRATION CHROMATOGRAPHY

Gel-filtration chromatography has been used both for purification and characterization of protein-conjugates. The removal of excess low-molecular weight bifunctional reagents from reaction mixture can be achieved by passage-down columns of Sephadex G-25, G-50, G-75 or Bio-Gel P-2 or P-10.[42-45] The characterization of the molecular weight distribution of proteins in the crude reaction mixtures have been achieved by Sephadex G-100, G-200, Sepharose 6-B, Bio-Gel P-100, P-150, P-300, A-1.5m, and A-5m.[1,46-50] The use of thin layer gel-filtration chromatography, using Sephadex G-200 superfine, for the rapid simultaneous assessment of protein-coupling reactions has also been described. With the advent of high pressure liquid chromatography (HPLC), protein components and conjugates have been isolated, purified, and characterized with this technique.

B. ELECTROPHORESIS

Various electrophoretic techniques such as paper,[51-53] continuous flow,[54] moving boundary,[4] agar gel[55,70] and SDS-polyacrylamide gel electrophoresis[56-58] have been used to assess the cross-linking reactions. New bands which appear after treatment with a cross-linking reagent do not necessarily represent cross-linked products since they could also result from intrachain cross-linking, monofunctional chemical modification or nondissociable aggregation. To resolve this problem, cleavable reagents are particularly helpful. Intrachain cross-linked molecules will revert to their original mobility upon cleavage. The same technique will also resolve the complication of overestimation of the cross-linked molecular masses.[80]

Components of cross-linked complexes can be identified after cleavage either on one-dimensional[59,60] or two-dimensional gels.[38,61] In the one-dimensional gel analysis, the band of a cross-linked complex is sliced out from unstained gels, treated with a cleaving reagent, placed on top of fresh gel, and electrophoresed. In this procedure, the cross-linked complexes are effectively isolated prior to cleavage and electrophoresis.

Immunoelectrophoresis is another technique that has been used to assess the immunological properties of conjugates.[46,55,56,62,63] For enzyme-antibody conjugates, the immunological and enzymatic activity has been demonstrated by staining immunoelectrophoretic plates for proteins and for enzyme activity.[55,64,65]

C. ULTRACENTRIFUGATION

Preparative separations of conjugated and nonconjugated proteins in crude reaction mixtures can be achieved by ultracentrifugation or density gradient ultracentrifugation.[43,63,66] Ultracentrifugation has also been used to determine the average molecular weight of conjugates,[67] and the enzyme-to-antibody ratios in various enzyme-antibody conjugates.[55]

D. IMMUNODIFFUSION

The immunological properties of cross-linked products have been assessed with Ouchterlony[46,54,70] and radial diffusion techniques.[56]

E. CHEMICAL ANALYSIS

Amino acid analysis has been used to determine the ratio of two proteins in a conjugate[68] and their site of attachment to each other.[69] For intramolecular cross-linking, the ratio of the number of amino acid residues modified to the number of reagent molecules incorporated will determine if the reagent reacted bifunctionally or monofunctionally. Determination of the location of covalent cross-links introduced into the protein is based on the general technique for the location of disulfide bonds. The conjugate is first digested by various chemical and enzymatic methods. After isolation of the cross-linked peptide or peptides, which is facilitated by radioactivity or chromaphoric marker of the reagent, amino acid analysis or sequencing can be performed to locate the modified residue. Like disulfide bonds, the use of cleavable cross-linkers will greatly facilitate the determination.

Special characteristics of conjugates have also been utilized for assessment of cross-linking reactions. For example, ferritin-protein conjugates have been characterized by measuring the iron content and the total protein concentration.[54,70]

F. ISOTOPE LABELING ANALYSIS

Radioisotope labeling in conjunction with assays of biological activity, offers an important means whereby the composition of a conjugate and the proportion of conjugated material which retains biological activity can be assessed. ^{125}I-labeled IgG and ^{125}I-labeled Fab have been used to determine the extent of conjugation with ferritin and horseradish peroxidase.[8,55]

III. REACTION COMPLICATIONS

A. GENERAL CONSIDERATIONS

Cross-linking of two different proteins with a bifunctional reagents can give rise to intra- and intermolecular products. Intermolecularly, a range of products including the desired 1:1-conjugate, and the undesired poly-conjugates and polymers of each of the reactant proteins can occur. Other possible products include aggregates of the various newly formed dimers, oligomers and polymers. These side reactions may lead to the loss of catalytic or immunologic activities of the original reactants. Intrachain cross-linking may give rise to a variety of complications, including change of structure which may lead to different mobility on polyacrylamide gel electrophoresis. To address some of these problems, cleavable reagents are particularly helpful. Upon cleavage, the molecule should return to its original molecular state, including electrophoretic mobility. In most of the applications of cross-linking reagents, a parallel experiment using the analogous monofunctional reagent should be carried out. These experiments will elucidate whether chemical modification itself is the cause of the problem, such as the loss of biological activity or immunogenicity. If such is the case, a different reagent may be chosen. Alternatively, protection of enzyme-active sites or immunological activities may be carried out with inhibitors,[71] substrates,[72] or antigens.[73-75]

While sulfhydryl cleavable reagents have many advantages, there are also several disadvantages. Major advantages are (1) rapid cleavage of the disulfide bond under mild conditions, (2) quantitative completion of the cleavage reaction, (3) ability to be cleaved both before and after electrophoresis, and (4) the specificity of the reduction reaction. The disadvantages to the use of these reagents are: (1) they are susceptible to disulfide exchange with the possibility of linking noninteracting molecules; (2) their use precludes the application of reducing agents for the isolation of cross-linked complexes; and (3) they cannot be used in a system which is sensitive to oxidation and would normally be kept under reducing conditions. Disulfide exchange usually involves the presence of free sulfhydryl groups which must be present in significant excess of disulfides.[76] This reaction can be decreased by lowering the pH of the reaction below the pK_a of sulfhydryl groups.[77-79]

For other cleavable reagents, there are also advantages and disadvantages. The major disadvantages with the use of glycol reagents include: (1) the reduced rate of cleavage relative to that obtained with disulfides; (2) the difficulty in obtaining complete cleavage; (3) the lack of specificity of the cleavage reaction, namely, the carbohydrate portions of glycoproteins can also be disrupted; and (4) the oxidative side reaction of the carbohydrates may lead to potential formation of Schiff bases with protein amino groups. A cross-linking reagent will have to be carefully chosen to suit the particular system under investigation to reduce the minimum complications.

B. IMMUNOGENICITY

In the application of cross-linking reagents to prepare antigen-carrier conjugates, the major concern is the effect of the reagent on the immunogenicity of the antigen. Peters et al.[81] have systematically investigated the problem with four cross-linking reagents. In a model system using angiotensin and tetanus toxoid, it is found that cross-linking does not affect the immunogenicity of both the peptide and the carrier in inducing antibody production. Antibodies are also induced against the cross-linking reagent. However, flexible nonaromatic linkers of succinimidyl 6-(N-maleimido)-n-hexanoate and succinimidyl 3-(2-pyridyldithio)propionate show the least immunogenicity. It would seem feasible to assume that cross-linking does not affect the antigenicity or immunogenicity of an antigen and that flexible cross-linkers are the best choices for this application.

C. STABILITY

The stability of immunoconjugates and immunotoxins is of paramount importance in the application of these cross-linked proteins. In general, immobilization of proteins tends to increase their stability towards both mechanical and thermal denaturation.[82] The same observation is also reported for cross-linked proteins.[83] In fact, higher activity of β-galactosidase was obtained when cross-linked with glutaraldehyde and dimethyladipimidate.

Some sense of the stability of a conjugate may be obtained from the Arrhenius equation (Equation 1):

$$\ln k = A - E_a/RT \tag{1}$$

When the natural logarithm of the rate constants, k, is plotted against the reciprocal of the absolute temperatures, T, as shown in Figure 2, the slope of the line gives the value of E_a/R where R is the known gas constant. The intercept is equal to A, the Arrhenius constant. The magnitude of E_a, the activation energy of the process, provides an information about the stability of the conjugate towards denaturation. The larger the value, the more stable

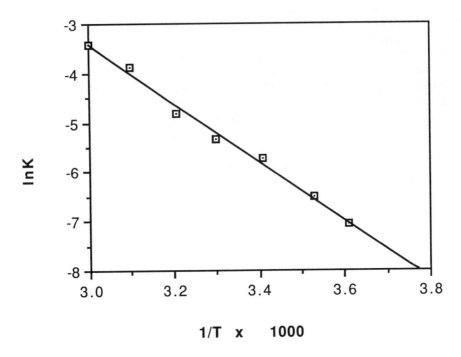

FIGURE 2.   An Arrhenius plot of inactivation of horseradish peroxidase.

TABLE 1
Heat of Inactivation of Horseradish
Peroxidase Conjugates

Conjugates	E_a(kcal/mol)
Horseradish peroxidase	35
Peroxidase-IgG	51
Peroxidase-Jacalin	43

the conjugate, since more energy is required for the inactivation. For horseradish peroxidase, peroxidase-IgG conjugate and peroxidase-jacalin conjugate, cross-linked with glutaralde-hyde, the values of E_a are shown in Table 1.[84] These values indicate that conjugated horseradish peroxidase is more stable than as a free entity.

The rate constants of inactivation of the conjugate at various temperatures (e.g., 20 to 70°C) required to make the Arrhenius plot are determined by measuring the rate of inactivation at that temperature. In most all the instances, the rate of denaturation is in accordance with that of a reaction of the first order (Equation 2):

$$\ln v = C - kt \tag{2}$$

Figure 3 shows a plot of the logarithm of the initial activity, v, of horseradish peroxidase incubated at the indicated temperatures vs. the time of inactivation, t. From such a plot, the rate constants of inactivation, k, at various temperatures are obtained from the slopes of the lines.

It is also possible to predict the half-life of a conjugate at a certain temperature from the Arrhenius equation. The rate constant of inactivation at that temperature of interest is

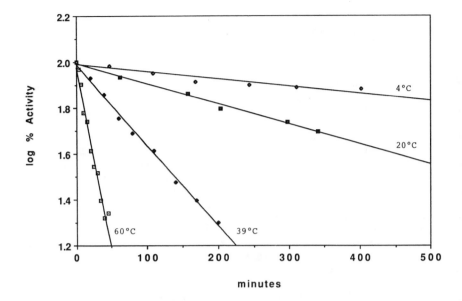

FIGURE 3. A first order plot for the inactivation of horseradish peroxidase. The log of % initial activity of the enzyme catalyzed reaction is plotted against the time of incubation at the indicated temperature (°C).

determined by extrapolation of the plot. The relationship of the half-life to the rate constant is shown in the following equation (Equation 3):

$$t_{1/2} = 0.693/k \tag{3}$$

For horseradish peroxidase at the conditions where the rate constants are determined, the half-life at 4°C is found to be 21 years. This mechanism may be useful for predicating the stability of an immunoconjugate or immunotoxin at a certain temperature.

REFERENCES

1. **Modesto, R. R. and Pesce, A. J.**, The reaction of 4,4'-difluoro-3,3'-dinitrodiphenyl sulfone with γ-globulin and horseradish peroxidase, *Biochim. Biophys. Acta*, 229, 384, 1971.
2. **Baskin, L. S. and Yang, C. S.**, Cross-linking studies of the protein topography of rat liver microsomes, *Biochim. Biophys. Acta*, 684, 263, 1982.
3. **O'Sullivan, M. J., Gnemmi, E., Morris, D., Chieregatti, G., Simmonds, A. D., Simmons, M., Bridges, J. W., and Marks, V.**, Comparison of two methods of preparing enzyme-antibody conjugates. Application of these conjugates for enzyme immunoassay, *Anal. Biochem.*, 100, 100, 1979.
4. **Schick, A. F. and Singer, S. J.**, On the formation of covalent linkages between two protein molecules, *J. Biol. Chem.*, 236, 2477, 1961.
5. **Agarwal, K. L., Grudzinski, S., Kenner, G. W., Rogers, N. H., Sheppard, R. C., and McGuigan, J. E.**, Immunochemical differentiation between gastrin and related peptide hormones through a novel conjugation of peptides to proteins, *Experientia*, 27, 514, 1971.
6. **Borek, F.**, A new two-stage method for cross-linking proteins, *Nature (London)*, 191, 1293, 1961.
7. **Nakane, P. K., Sri Ram, J., and Pierce, G. B.**, Enzyme-labeled antibodies: preparation and application for the localization of antigens, *J. Histochem. Cytochem.*, 14, 789, 1966.

8. **Otto, H., Takamiya, H., and Vogt, A.,** Two-stage method for crosslinking antibody globulin to ferritin by glutaraldehyde. Comparison between the one-stage and the two-stage method, *J. Immunol. Methods,* 3, 137, 1973.

9. **Howard, A. N. and Wild, F.,** A two-stage method of cross-linking proteins suitable for use in serological techniques, *Br. J. Exp. Pathol.,* 38, 640, 1957.

10. **Hunter, M. J. and Ludwig, M. L.,** The reaction of imidoesters with proteins and related small molecules, *J. Am. Chem. Soc.,* 84, 3491, 1962.

11. **Browne, D. T. and Kent, S. B. H.,** Formation of nonamidine products in the reaction of primary amines with imido esters, *Biochem. Biophys. Res. Commun.,* 67, 126, 1975.

12. **Liu, S. C., Fairbanks, G., and Palek, J.,** Spontaneous reversible protein cross-linking in the human erythrocyte membrane. Temperature and pH dependence, *Biochemistry,* 16, 4066, 1977.

13. **Kiehm, D. J. and Ji, T. H.,** Photochemical cross-linking of cell membranes. A test for natural and random collisional cross-links by millisecond cross-linking, *J. Biol. Chem.,* 252, 8524, 1977.

14. **Ji, T. H.,** The application of chemical crosslinking for studies of cell membrane and the identification of surface reporters, *Biochim. Biophys. Acta,* 559, 39, 1979.

15. **Haller, I. and Henning, U.,** Cell envelope and shape of *Escherichia coli* K12. Crosslinking with dimethyl imidoesters of the whole cell wall, *Proc. Natl. Acad. Sci. U.S.A.,* 71, 2018, 1974.

16. **Peters, K. and Richards, R. M.,** Chemical cross-linking. Reagents and problems in studies of membrane structure, *Annu. Rev. Biochem.,* 46, 523, 1977.

17. **Hand, E. S. and Jencks, W. P.,** Mechanism of the reaction of imido esters with amines, *J. Am. Chem. Soc.,* 84, 3505, 1962.

18. **Dutton, A., Adams, M., and Singer, S. J.,** Bifunctional imido-esters as cross-linking reagents, *Biochem. Biophys. Res. Commun.,* 23, 730, 1966.

19. **Carpenter, F. H. and Harrington, K. T.,** Intermolecular cross-linking of monomeric proteins and cross-linking of oligomeric proteins as a probe of quaternary structure, *J. Biol. Chem.,* 247, 5580, 1972.

20. **Cuatrecasas, P. and Parikh, I.,** Absorbents for affinity chromatography. Use of *N*-hydroxysuccinimide ester of agarose, *Biochemistry,* 11, 2291, 1972.

21. **Lomant, A. J. and Fairbanks, G.,** Chemical probes of extended biological structures. Synthesis and properties of the cleavable protein cross-linking reagent [³⁵]dithiobis(succinimidylpropionate), *J. Mol. Biol.,* 104, 243, 1976.

22. **Lewis, R. V., Roberts, M. F., Dennis, E. A., and Allison, W. S.,** Photoactivated heterobifunctional cross-linking reagents which demonstrate the aggregation state of phospholipase A₂, *Biochemistry,* 16, 5650, 1977.

23. **Smith, R. J., Capaldi, R. A., Muchmore, D., and Dahlquist, F.,** Cross-linking of ubiquinone cytochrome c reductase (complex III) with periodate-cleavable bifunctional reagents, *Biochemistry,* 17, 3719, 1978.

24. **Cartwright, I. L., Hutchinson, D. W., and Armstrong, V. W.,** The reaction between thiols and 8-azidoadenosine derivatives, *Nucleic Acids Res.,* 3, 2331, 1976.

25. **Staros, J. V., Bayley, H., Standring, D. N., and Knowles, J. R.,** Reduction of aryl azides by thiols: implications for the use of photoaffinity reagents, *Biochem. Biophys. Res. Commun.,* 80, 568, 1978.

26. **Fleet, G. W. J., Knowles, J. R., and Porter, R. R.,** The antibody binding site. Labelling of a specific antibody against the photo-precursor of an aryl nitrene, *Biochem. J.,* 128, 499, 1972.

27. **Hixson, S. H. and Hixson, S. S.,** *p*-Azidophenacyl bromide, a versatile photolabile bifunctional reagent. Reaction with glyceraldehyde-3-phosphate dehydrogenase, *Biochemistry,* 14, 4251, 1975.

28. **Henkin, J.,** Studies on rabbit muscle creatine kinase. Photolabeling reagent for thiol enzymes, *J. Biol. Chem.,* 252, 4293, 1977.

29. **Erecinska, M., Vanderkooi, J. M., and Wilson, D. F.,** Cytochrome c interactions with membranes. A photoaffinity labeled cytochrome c, *Arch. Biochem. Biophys.,* 171, 108, 1975.

30. **Mikkelsen, R. B. and Wallach, D. F. H.,** Photoactivated cross-linking of proteins within the erythrocyte membrane core, *J. Biol. Chem.,* 251, 7413, 1976.

31. **Ji, T. H.,** A novel approach to the identification of surface receptors. The use of photosensitive hetero-bifunctional cross-linking reagents, *J. Biol. Chem.,* 252, 1566, 1977.

32. **Miyakawa, T., Takemoto, L. J., and Fox, C. F.,** Membrane permeability of bifunctional, amino site-specific, cross-linking reagents, *J. Supramol. Struct.,* 8, 303, 1978.

33. **Middaugh, C. R. and Ji, T. H.,** A photochemical crosslinking study of the subunit structure of membrane-associated spectrin, *Eur. J. Biochem.,* 110, 587, 1980.

34. **Lee, T. K., Wong, L.-J. C. and Wong, S. S.,** Photoaffinity labeling of lactose synthase with a UDP-galactose analogue, *J. Biol. Chem.,* 258, 13166, 1983.

35. **Mas, M. T., Wang, J. K., and Hargrave, P. A.,** Topography of rhodopsin in rod out segment disk membranes. Photochemical labeling with *N*-(4-azido-2-nitrophenyl)2-aminoethanesulfonate, *Biochemistry,* 19, 684, 1980.

36. **Staros, J. V. and Richards, F. M.,** Photochemical labeling of the surface proteins of human erythrocytes, *Biochemistry,* 13, 2720, 1974.

37. **Ruoho, A., Bartlett, P. A., Dutton, A., and Singer, S. J.,** Disulfide-bridge bifunctional imidoester as a reversible cross-linking reagent, *Biochem. Biophys. Res. Commun.,* 63, 417, 1975.
38. **Lutter, L. C., Ortanderl, F., and Fasold, H.,** Use of a new series of cleavable protein-crosslinkers on the *Escherichia coli* ribosomes, *FEBS Lett.,* 48, 288, 1974.
39. **Jaffe, C. L., Lis, H., and Sharon, N.,** New cleavable photoreactive heterobifunctional cross-linking reagents for studying membrane organization, *Biochemistry,* 19, 4423, 1980.
40. **Zarling, D. A., Watson, D., Bach, F. H.,** Mapping of lymphocyte surface polypeptide antigens by chemical cross-linking with BSOCOES, *J. Immunol.,* 124, 913, 1980.
41. **Abdella, P. M., Smith, P. K., and Royer, G. P.,** A new cleavable reagent for cross-linking and reversible immobilization of proteins, *Biochem. Biophys. Res. Commun.,* 87, 734, 1979.
42. **Marfey, P. S., Uziel, M., and Little, J.,** Reaction of bovine pancreatic ribonuclease A with 1,5-difluoro-2,4-dinitrobenzene. II. Structure of an intramolecularly bridged derivative, *J. Biol. Chem.,* 240, 3270, 1965.
43. **Van Weemen, B. K. and Schuurs, A. H. W. M.,** Immunoassay using antigen-enzyme conjugates, *FEBS Lett.,* 15, 523, 1971.
44. **Marfey, P. S., Nowak, H., Uziel, M., and Yphantis, D. A.,** Reaction of bovine pancreatic ribonuclease A with 1,5-difluoro-2,4-dinitrobenzene. I. Preparation of monomeric intramolecularly bridged derivatives, *J. Biol. Chem.,* 240, 3264, 1965.
45. **Marfey, P. S. and King, M. V.,** Chemical modification of ribonuclease A crystals. I. Reaction with 1,5-difluoro-2,4-dinitrobenzene, *Biochim. Biophys. Acta,* 105, 178, 1965.
46. **Habeeb, A. F. S. A. and Hiramoto, R.,** Reaction of proteins with glutaraldehyde, *Arch. Biochem. Biophys.,* 126, 16, 1968.
47. **Miedema, K., Boelhouwer, J., and Otten, J. W.,** Determinations of proteins and hormones in serum by an immunoassay using antigen-enzyme conjugates, *Clin. Chim. Acta,* 40, 187, 1972.
48. **Engvall, E., Jonsson, K., and Perlmann, P.,** Enzyme-linked immunosorbent assay. II. Quantitative assay of protein antigen, immunoglobulin G, by means of enzyme-labelled antigen and antibody-coated tubes, *Biochim. Biophys. Acta,* 251, 427, 1971.
49. **Olsen, B. R., Berg, R. A., Kishida, Y., and Prockop, D. J.,** Collagen synthesis, localization of prolyl hydroxylase in tendon cells detected with ferritin-labeled antibodies, *Science,* 182, 825, 1973.
50. **Nakane, P. K. and Pierce, G. B., Jr.,** Enzyme-labeled antibodies for the light and electron microscopic localization of tissue antigens, *J. Cell Biol.,* 33, 307, 1967.
51. **Smith, C. W. and Metzger, J. F.,** Studies of ferritin conjugates used in immune electron microscopy, *Biochim. Biophys. Acta,* 47, 587, 1961.
52. **Sri Ram, J.,** Protein-protein conjugates and some novel polyamine acid derivatives employing a bifunctional reagent, *Biochim. Biophys. Acta,* 78, 228, 1963.
53. **Tawde, S. S. and Sri Ram, J.,** Conjugation of antibody to ferritin by means of p,p'-difluoro-m,m'-dinitrodiphenylsulfone, *Arch. Biochem. Biophys.,* 97, 429, 1962.
54. **Borek, F. and Silverstein, A. M.,** Characterization and purification of ferritin-antibody globulin conjugates, *J. Immunol.,* 87, 555, 1961.
55. **Avrameas, S.,** Coupling of enzymes to proteins with glutaraldehyde. Use of the conjugates for the detection of antigens and antibodies, *Immunochemistry,* 6, 43, 1969.
56. **Kraehenbuhl, J. P., Oe Grandi, P. B., and Campiche, M. A.,** Ultrastructural localization of intracellular antigen using enzyme-labeled antibody fragments, *J. Cell Biol.,* 50, 432, 1971.
57. **Avrameas, S. and Ternynck, T.,** Peroxidase-labeled antibody and Fab conjugates with enhanced intra-cellular penetration, *Immunochemistry,* 8, 1175, 1971.
68. **Davies, G. E. and Stark, G. R.,** Use of dimethyl suberimidate, a cross-linking reagent, in studying the subunit structure of oligomeric proteins, *Proc. Natl. Acad. Sci. U.S.A.,* 66, 651, 1970.
59. **Steck, T. L.,** Cross-linking the major proteins of the isolated erythrocyte membranes, *J. Mol. Biol.,* 66, 295, 1972.
60. **Ji, T. H. and Ji, I.,** Crosslinking of glycoproteins in human erythrocyte ghosts, *J. Mol. Biol.,* 86, 124, 1974.
61. **Wang, K. and Richards, F. M.,** An approach to nearest neighbor analysis of membrane proteins. Application to the human erythrocyte membrane of a method employing cleavable cross-linking, *J. Biol. Chem.,* 249, 8005, 1974.
62. **Hsu, K. C.,** Ferritin-labeled antigens and antibodies, in *Methods in Immunology and Immunochemistry,* Williams, C. A. and Chase, M. W., Eds., Vol. 1, Academic Press, New York, 1967, 397.
63. **Clyne, D. H., Norris, S. H., Modesto, R. R., Pesce, A. J., and Pollak, V. E.,** Antibody enzyme conjugates. Preparation of intermolecular conjugates of horseradish peroxidase and antibody and their use in immunohistology of renal cortex, *J. Histochem. Cytochem.,* 21, 233, 1973.
64. **Kraehenbuhl, J. P., Galardy, R. E., and Jamieson, J. D.,** Preparation and characterization of an immunoelectron microscope tracer consisting of a heme-octapeptide coupled to Fab, *J. Exp. Med.,* 139, 208, 1974.

65. **Mendell, J. R., Whitaker, J. N., and Engel, W. K.,** Skeletal muscle binding site of antistriated muscle antibody in myasthenia gravis. Electron microscopic immunohistochemical study using peroxidase conjugated antibody fragments, *J. Immunol.,* 111, 847, 1973.

66. **Rifkind, R. A., Hsu, K. C., and Morgan, C.,** Immunochemical staining for electron microscopy, *J. Histochem. Cytochem.,* 12, 131, 1964.

67. **Anderer, F. A. and Schlumberger, H. D.,** Antigenic properties of protein cross-linked by multidiazonium compounds, *Immunochemistry,* 6, 1, 1969.

68. **Goodfriend, T. L., Levine, L., and Fasman, G. D.,** Antibodies to bradykinin and angiotensin. A use of carbodiimide in immunology, *Science,* 144, 1344, 1964.

69. **Tawde, S. S., Sri Ram, J., and Iyengar, M. R.,** Physicochemical and immunochemical studies on the reaction of bovine serum albumin with *p,p'*-difluoro-*m,m'*-dinitrophenyl sulfone, *Arch. Biochem. Biophys.,* 100, 270, 1963.

70. **Sri Ram, J., Tawde, S. S., Pierce, Jr., G. B., and Midgley, Jr., A. R.,** Preparation of antibody-ferritin conjugates for immuno electron microscopy, *J. Cell Biol.,* 17, 673, 1963.

71. **Jansen, E. F., Tomimatsu, Y., and Olson, A. C.,** Cross-linking of α-chymotrypsin and other proteins by reaction with glutaraldehyde, *Arch. Biochem. Biophys.,* 144, 394, 1971.

72. **Nicolson, G. L. and Singer, S. J.,** The distribution and asymmetry of mammalian cell surface saccharides utilizing ferritin-conjugated plant agglutinins as specific saccharide stains, *J. Cell Biol.,* 60, 236, 1974.

73. **Donati, E. J., Petrali, J. P., and Sternberger, L. A.,** Formation of vaccinia antigen studied by immunouranium and immunodiazothioether osmium tetroxide techniques, *Exp. Mol. Pathol.,* 3 (Suppl.), 59, 1966.

74. **Mannik, M. and Downey, W.,** Conjugation of horseradish peroxidase to Fab fragments, *J. Immunol. Methods,* 3, 233, 1973.

75. **Kraehenbuhl, J. P. and Jamieson, J. D.,** Solid-phase conjugation of ferratin to Fab-fragments of immunoglobulin G for use in antigen localization on thin sections, *Proc. Natl. Acad. Sci. U.S.A.,* 69, 1771, 1972.

76. **Liu, T. Y.,** The role of sulfur in proteins, in *The Proteins,* Vol. 3, 3rd ed., Neurath, H. and Hill, R. L., Eds., Academic Press, New York, 1977, 239.

77. **Fava, A., Iliceto, A., and Camera, E.,** Kinetics of the thioldisulfide exchange, *J. Am. Chem. Soc.,* 79, 833, 1957.

78. **Eldjarn, L. and Pihl, A.,** Equilibrium constants and oxidation-reduction potentials of some thiol-disulfide systems, *J. Am. Chem. Soc.,* 79, 4589, 1957.

79. **Barker, R.,** Equilibrium constants and oxidation-reduction potentials of some thiol-disulfide systems, in *Organic Chemistry of Biological Compounds,* Prentice-Hall, Englewood Cliffs, NJ, 1971, 352.

80. **Richard, C., Han, K. K., Yang, H. L., Zhu, D. X., Balduyck, M., and Mizon, J.,** Evidence for the overestimation of molecular masses of proteins after chemical modification and chemical crosslinks on sodium dodecyl sulfate/polyacrylamide gel electrophoresis (SDS-PAGE), *Biomed. Chromatogr.,* 3, 131, 1989.

81. **Peters, J. M., Hazendonk, T. G., Beuvery, E. C., and Tesser, G. I.,** Comparison of four bifunctional reagents for coupling peptides to proteins and the effect of the three moieties on the immunogenicity of the conjugates, *J. Immunol. Methods,* 120, 133, 1989.

82. **Demers, A. G. and Wong, S. S.,** Increased stability of galactosyltransferase on immobilization, *J. Appl. Biochem.,* 7, 122, 1985.

83. **Khare, S. K. and Gupta, M. N.,** A crosslinked preparation of *E. coli* β-D-galactosidase, *Appl. Biochem. Biotechnol.,* 16, 1, 1987.

84. **Wong, S. S., Losiewicz, M., and Wong, L.-J. C.,** submitted, 1991.

Chapter 8

APPLICATION OF CHEMICAL CROSS-LINKING TO SOLUBLE PROTEINS

I. INTRODUCTION

Elucidation of the location of intra- and interchain chemical cross-links in proteins has helped unravel the folding of the polypeptide chains. The most common naturally occurring inter- and intramolecular linkage is the disulfide bond. Other examples include transglutaminase-catalyzed formation of amide bond between γ-carboximide groups of glutamines and the ε-amino groups of lysines in fibrin,[1] and the extreme complex cross-links in collagen.[2,3] Determination of the locations of these cross-linkages has contributed to the understanding of the tertiary and quaternary structures of proteins. It follows, therefore, that the introduction of stable covalent linkages between amino acid residues in the native states of proteins should provide additional means for the study of inter-residue distances, the relationship between various protein domains, the conformational states of a protein, and the interaction of polypeptide chains in solutions.

The application of bifunctional reagents to proteins will result in intramolecular as well as intermolecular cross-linking (see Chapter 1). Each mode of reaction provides different ways of gathering information, and the reaction condition can be adjusted to favor the yield of one product over the other. Intramolecular cross-linking can be enhanced by low protein concentrations (<0.1 mg/ml) and high protein to reagent ratios. The opposite will increase the yield of intermolecular linkages. From the product of intramolecular cross-linking, determination of the location of each cross-link will give specific information on its three-dimensional folding of the polypeptide. Studies of the physical and biological properties of protein derivatives with different extents of cross-linking should also reveal the effect of covalent bonds in stabilizing the tertiary structure. In order to draw meaningful conclusions from cross-linking studies, the effect of such cross-linking with respect to the catalytic and biological function must be assessed. The protein should be studied with the corresponding monofunctional analogs. That is, if the cross-linking reagent is x-R-R-x or x-R-R-y, the cross-linked product should be compared with the reaction products of x-R and/or x-Y. For example, the effect of methyl acetimidate is compared with dimethylsuberimidate.[4]

Intermolecular cross-linking provides two different types of products. Cross-linking between identical proteins yields homopolymers. Intermolecular linking between different proteins yield heteropolymers. These derivatives provide excellent means for the study of protein-protein interactions in multisubunit protein systems and could provide practical use of stable active insoluble proteins.[5]

A large volume of work on the application of cross-linking reagents to the study of soluble proteins has been published in recent years. This chapter will selectively present some aspects of the information obtainable from such studies. The application of cross-linking to membrane proteins will be considered in the next chapter.

II. INTER-RESIDUE DISTANCE DETERMINATIONS

A. MOLECULAR DISTANCES OF CROSS-LINKING REAGENTS

For the measurement of distances between reactive amino acid residues, cross-linking reagents of various chain lengths containing different numbers of carbon atoms have been synthesized. These homologous series include the imidoester,[6] N-maleimido,[7] azido acyl,[8] N-succinimidyl,[9] and thiosulfonate[10] functional groups. As an estimate of the maximum span

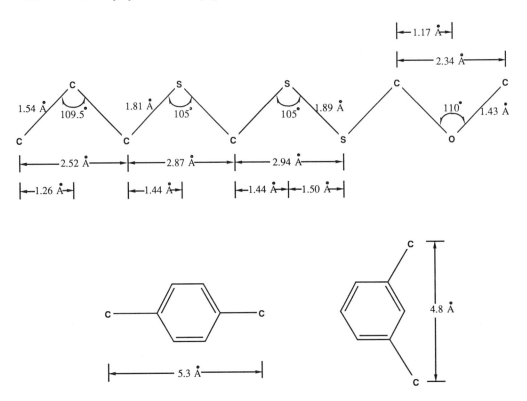

FIGURE 1.   Various projected distances between different atoms.

of these molecules, the following calculations may be noted. Because of the hybridization of the carbon atom, the length of the carbon chain is not the sum of the individual bond distances. Taking the C-C bond length to be 1.54 Å and a bond angle of 109.5°, one can calculate that the distance between alternate carbons is about 2.52 Å. This translates into 1.26 Å for the "projected" C-C bond (Figure 1). For the thioether bond, this is 1.44 Å for the C-S projected distance. Since the C-O bond is shorter (1.43 Å), the projected C-O distance is only 1.17 Å. On the other hand, the S-S bond length is the longest, being 1.89 Å, and the projected distance is estimated to be 1.50 Å as shown in Figure 1. It should be noted that these values are calculated from methylene carbons. Any substituents on the carbon that affect its bond angle or bond length will affect the value of the projected distance. For aromatic compounds, the distance between two groups bonded to the ortho positions is about 4.8 Å, that between the para-positions is about 5.3 Å. From these values, the distance span of any cross-linked species can be estimated. For example, *bis*[2-(succimidyloxycarbonyloxy)ethyl]sulfone (see Chapter 4 for structure **XLII**) can be estimated to have a maximum span of 10.4 Å between the reactive carbons. The distance between two cross-linked groups will be about 13 Å, since a bond is formed between the reactive carbon of the reagent and a reactive group in a protein. Thus, the maximum distance between two coupled groups is that of the span of the reagent plus two bond lengths.

As the C-C single bond is freely rotatable, the molecule can assume various conformations. Many of these conformational states will have spans much less than the estimated maximum distance. It is therefore possible for a cross-linker to join two groups within the calculated distance but not beyond.

B. MEASUREMENT OF INTER-RESIDUE DISTANCES

The strategic approach for measuring the distance between reactive amino acid residues in a protein is to cross-link the groups with a homologous series of cross-linkers of different lengths. Only the cross-linker with molecular span equal or greater than the inter-residue distance will react. Not only must the flexibility of the cross-linking reagents be considered, but the ability of the reactive amino acid residues to move around must also be taken into account. For example, in cross-linking the two sulfhydryl groups in myosin subfragment 1, Wells et al.[11,12] have used rigid *bis*-maleimide cross-linkers that span from 5 to 14 Å. All of the reagents reacted with the protein, indicating the possibility of the thiol groups to assume different positions.

Many studies which focus on measuring the interresidue distances have been published. The following are some examples. Husain and Lowe[13] have found that 1,3-dibromoacetone cross-linked the active-site cysteine and histidine in ficin and stem-bromelain. From the molecular structure of the reagent, the authors concluded that the active-site residues were about 5 Å apart from each other.

In a study of tryptophan synthase, Heilmann and Holzner[7] isolated and modified one of the nonidentical subunits with *N*-succinimidyl-3-(2-pyridyldithio)propionate, SPDP, to generate a sulfhydryl group. After the reaction, the subunits were mixed to form the tetrameric complex. A series of homologous *bis*-maleimide cross-linkers were used to cross-link the subunits. Only *bis*-*N*-maleimido-1,6-hexane and *bis*-*N*-maleimido-1,8-octane were found to be effective. These compounds have a maximum span of 13.9 Å and 16.4 Å, respectively. Together with the incorporated portion of SPDP, the distance between the reactive groups in the nonidentical subunits was estimated to be at least 10-20 Å.

A series of *bis*-imidates was used to cross-link phycobiliprotein complexes by Rümbeli et al.[14] The highest yield was obtained with dimethyl pimelimidate which has a maximum cross-linking distance of 10 Å. From the X-ray crystallographic data of phycocyanin, the authors were able to identify the cross-linked amino acid residues.

III. NEAREST NEIGHBOR ANALYSIS

The principle behind the nearest neighbor analysis lies in the fact that proteins are cross-linked only when they are within the reaction distance of the cross-linking reagents. In a protein complex where the subunits associate to form a distinct organization, only those subunits situated next to each other will be linked together. By analyzing which proteins are coupled, a topographical model of their location in the complex may be deduced. This powerful premise of probing the protein-protein interaction sites has been applied to investigate the arrangement of the subunits in a protein complex. The following examples will serve to illustrate the usefulness of this method.

A. QUATERNARY STRUCTURES OF MULTI-SUBUNIT PROTEINS

Perhaps the best model to demonstrate the use of cross-linking reagents in elucidation of the quaternary structure of a protein complex is F_1-adenosine triphosphatase (F_1-ATPase). F_1-ATPase is part of the energy-transduction system of mitochondria, chloroplasts, chromatophores, and bacteria, and can be released from membranes as a Ca^{2+}, Mg^{2+} dependent ATPase. The solubilized F_1-ATPase consists of five different subunits, α, β, γ, δ and ϵ. The stoichiometry of the subunits is $\alpha_3\beta_3\gamma\delta\epsilon$. Chemical cross-linking has been used to study the arrangement of these subunits. Using dithiobis(succinimidyl propionate), methyl-4-mercaptobutyrimidate, dimethyl-3,3'-dithiobispropionimidate, disuccinimidyl tartarate and cupric 1,10-phenanthrolinate, Bragg and Hou[15] have obtained the following cross-linked subunit dimers of F_1-ATPase from *Escherichia coli*: $\alpha\alpha$, $\beta\beta$, $\alpha\beta$, $\alpha\delta$, $\beta\gamma$, $\beta\delta$, $\beta\epsilon$ and $\gamma\epsilon$. More recently, the same investigators have used 2,2'- and 3,3'-dithiobis(succinimidyl propionate), 3,3'-dithiobis(sulfosuccinimidyl propionate), disuccinimidyl tartarate, dimethyl adipimidate, 1-ethyl-3-[3-(dimethylamino)propyl]carbodiimide (EDC) and 1,2:3,4-diepoxybutane to study

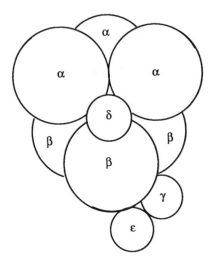

FIGURE 2.   Possible structure of F_1-ATPase.

the subunits and found cross-linked dimers of αα, αβ, βγ, αδ, βε and γε. The presence of trimer ααδ was also noted.[16] These findings were consistent with a model where the three α and the three β subunits stack in two planes as a triangular antiprism as shown in Figure 2. A similar conclusion was reached by Joshi and Wang[17] who first used 1-fluoro-2,4-dinitro-benzene to modify the F_1-ATPase from bovine heart mitochondria. After reduction with sodium hydrosulfite, the modified complex was cross-linked with EDC. A cross-linked ββ dimer was detected. (See Figure 2.)

Chemical cross-linking has also been used to study the structure of F_1-ATPase from another source, *Micrococcus lysodeikticus*. Muñoz et al.[18] have used dimethyl suberimidate and dithiobis(succinimidyl propionate) to explore the nearest neighbor relationship of the subunits and found that γ subunit cross-linked with itself as well as with other subunits except β. The α subunit also cross-linked with itself and with other subunits. They also detected cross-linked $δ_2$. From these data, the authors proposed a subunit stoichiometry of $α_3β_3γ_2δ_2$. A spatial organization of three sets of α subunit stacked on another three sets of β subunit was derived. The other subunits were arranged around of this core.

Photoaffinity cross-linking has also been employed to investigate the relative location of the active site of F_1-ATPase from *Micrococcus luteus*. The bifunctional photosensitive substrate analog, 3′-arylazido-β-alanine-8-azido ATP (diN$_3$ATP) (see Chapter 5 for structure), was found to bind to the active site,[19,20] which was located in the β subunit. On photoactivation, diN$_3$ATP cross-linked the β and the α subunits. It was concluded that the catalytic nucleotide binding sites were located on the β subunits very close to the α subunits.

B. PROTEIN ORGANIZATION IN COMPLEX CELLULAR SYSTEMS

1. Ribosomal Protein Arrangements

Proteins within the large and small ribosomal subunits from *Escherichia coli* have been studied for many years with chemical cross-linking reagents to determine their arrangement in these cellular systems.[21-25] In most reports, 2-iminothiolane was used as the cross-linking reagent. Other reagents such as *p*-phenylenebis(maleimide), *bis*-imidoesters, both cleavable and noncleavable, tartryl-containing acyl azides and tetranitromethane have also been used. For the 30S subunit, Lambert et al.[25] have compiled all cross-linked protein pairs. A total of 33 pairs were documented. These cross-linking data were compared to the three-dimensional arrangement of the ribosomal proteins derived from immunoelectron microscopy and neutron scattering.[26] Although there were cross-links that were incompatible with the derived model, many of the chemical data were in agreement as illustrated in Figure 3. In this figure,

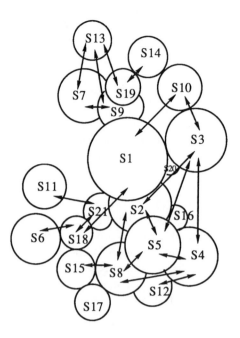

FIGURE 3. Cross-linking of the proteins in 30S
ribosomal subunit. The cross-linked protein-pairs are
indicated by two-headed arrows. The arrangement
of the proteins is derived from neutron map.

the arrangement of the proteins in 30S ribosomal subunit based on the neutron map is
presented. The cross-linked protein pairs that are compatible with this model are indicated
by the double-headed arrows.

For the 50S ribosomal subunit, similar results have been derived. Traut et al.[24] have
collected all the cross-linking data revealing 81 different cross-linked protein pairs. Cross-
links between 50S and 30S proteins have also been obtained. Comparison of these data with
the protein topography obtained from immunoelectron microscopy showed many incompatible distances.[27] More recently, Stöffler et al.[28,29] have used immunoblotting technique
to analyze protein-protein cross-links using dimethylsuberimidate and 2-iminothiolane. A
total of 14 cross-links have been identified. These data further enable the location of the
proteins in the 50S subunit to be determined.

Not only have the proteins in the 30S and 50S ribosomal subunits been cross-linked,
RNA-protein coupling has also been obtained. In the 30S ribosomal subunit, proteins S4,
S5, S7, S8, S11, S12 and S18 are efficiently cross-linked to domains 1 and 2 of 16S RNA
using 1-ethyl-3(3-dimethylaminopropyl) carbodiimide.[30] These results have provided information on the three dimensional relationship between the RNA and the protein arrangements
in the ribosomal subunit.[31]

In addition to ribosomes of *Escherichia coli,* 40S and 60S ribosomal subunits isolated
from rat liver have also been studied with cross-linking agents, dimethyl suberimidate and
dimethyl 3,3'-dithiobispropionimidate.[32,33] Seven protein pairs were obtained from the 40S
subunit and 14 pairs were obtained from the 60S subunit.

2. Contractile Protein System

Another complex cellular organization in which the application of cross-linking reagents
has provided valuable information is the muscle contractile system. The contractile system
is made up of myofibrils which is composed of thin and thick filaments. Thin filaments are

made up primarily of three types of proteins: actin, tropomyosin and troponin in a ratio of 7:1:1. *In vitro,* actin molecules exist in two states: globular G-actin and fibrous F-actin. Troponin consists of a complex of three separate proteins: troponin T, troponin C, and troponin I. The basic structure of the thin filament consists of two strands of F-actin polymers intertwined in the conformation of a double helix, two double stranded helixes of tropomyosin and a troponin complex. Thick filaments consist primarily of myosin which has two globular heads, possessing the ATPase activity. Treatment of myosin with trypsin gives two components: light meromyosin which contains the tail part of the native molecule, and heavy meromyosin, the globular end of the molecule. Interaction of these various protein components under different conditions has been extensively investigated using cross-linking reagents.

Leszyk et al.[34] have used EDC to cross-link troponin C and troponin I of the thin filament. Troponin C was first activated with EDC and *N*-hydroxysuccinimide, and then mixed with troponin I. Analysis of the cross-linked peptides revealed that the regulatory Ca^{2+}-binding site II in the *N*-terminal domain of troponin C is closely associated with the inhibitory region of troponin I. The interaction between the inhibitory segment of troponin I and an α-helical segment of troponin C adjacent to Ca^{2+}-binding site III has also been documented by 4-maleimidobenzophenone.[35]

The structural organization of the thick filament myosin has been investigated by several groups. Chantler and Bower[36] have used 4,4'-dimaleimidylstilbene-2,2'-disulfonate to study the relative locations and surface topologies of the myosin light chains on the myosin molecule. Mercenaria myosin and scallop pure hybrid myosin possessing Mercenaria light chains were reacted with the cross-linker. Regulatory light chain homodimers were obtained irrespective of the presence or absence of calcium and/or ATP. Analysis revealed that the translationally equivalent residues (Lys-50) were cross-linked, indicating that these sites on the two heads of myosin could come within 18 Å of each other, the span of the reacted cross-linker.

A photo-activatable reagent, benzophenone-4-iodoacetamide, was used to study the spatial relationship of various domains in myosin subfragment 1.[37] One of the two reactive thiols, SH-1, in the myosin subfragment 1 was modified with the reagent. Photolysis revealed a cross-link between the *N*-terminal 25 kDa fragment and the C-terminal 20 kDa fragment. In the presence of Mg^{2+}-ATP, an additional cross-link between the C-terminal 20 kDa fragment and the middle 50 kDa fragment was obtained. These data suggest a conformational change induced by nucleotide binding. Conformational change induced by Ca^{2+} on troponin C was also detected using the cross-linking agent, 1,3-difluoro-4,6-dinitrobenzene.[38] Troponin C was first modified at the Cys-98 residue. In the presence of Ca^{2+}, the second reactive group of the reagent reacted with Lys-90. In the absence of Ca^{2+}, two additional cross-links were obtained, reaction with Tyr-109 and Lys-136 has also taken place. Thus Ca^{2+} binding causes a change in the distance between the residues.

Studies on the interaction between the thin and thick filaments have also involved cross-linking reagents. Arata[39] cross-linked myosin subfragment 1 to F-actin in the presence and absence of nucleotides using EDC. Fluorescence energy transfer analysis revealed that in the presence of a nucleotide, there was an increase in distance between SH1 thiol of subfragment 1 and Cys-374 of actin. The same zero-length cross-linker was used to join F-actin and heavy meromyosin.[40] Two cross-linked products were obtained. One was formed between F-actin and the C-terminal 68 kDa fragment of heavy meromyosin. The other was a cross-linked dimer of the N-terminal 24 kDa fragment of heavy meromyosin. These data were interpreted to mean that the two heads of heavy meromyosin were in contact with each other.

C. PROTEIN-PROTEIN INTERACTIONS

Detection of protein-protein interactions in solution may be illustrated by the lactose synthase and the calmodulin regulatory systems. Lactose synthase is composed of two subunits, galactosyltransferase and α-lactalbumin. The substrate specificity of galactosyl-transferase is regulated by association and dissociation of α-lactalbumin.[41] Lactose is synthesized when the two proteins associate. The interaction between galactosyltransferase and α-lactalbumin in solution was detected by cross-linking agent, dimethyl pimelimidate.[42] Cross-linked complex was obtained in the presence of substrates or inhibitors of galacto-syltransferase. Thus N-acetylglucosamine, or a combination of N-acetylglucosamine, Mn^{2+}, and UDP, or Mn^{2+} and UDP-galactose facilitated the formation of the cross-linked species. The isolation of a cross-linked complex provided a direct evidence for the association of these two proteins.

Calmodulin is another regulatory protein which mediates a wide variety of calcium-dependent cellular processes.[43] A series of enzymes are regulated by calmodulin through a calcium-dependent interaction with the catalytic domains of the enzymes. Photolabile derivatives of calmodulin have been synthesized and used to label calmodulin-binding proteins. For example, Harrison et al.[44] have prepared a photoreactive derivative of calmodulin using the heterobifunctional cross-linking agent p-nitrophenyl 3-diazopyruvate. The interaction of calmodulin with adenylate kinase was demonstrated by the formation of a cross-link product on photolysis. The process was Ca^{2+}-dependent. Cross-linking occurred only in the presence of Ca^{2+}.

Other calmodulin-enzyme complexes have also been detected by cross-linking. N-Suc-cinimidyl 3-(2-pyridylthio)propionate was used to cross-link calmodulin and cyclic nucleo-tide phosphodiesterase in the presence of Ca^{2+}. Calmodulin was first modified with the reagent through an amino group. The 3-(2-pyridyldithio)propionyl-substituted calmodulin was then mixed with cyclic nucleotide phosphodiesterase. In the presence of Ca^{2+} an enzymatically active cross-linked complex was obtained.[45] Zot and Puett[46] have used N,N'-dicyclohexyl carbodiimide to study the interaction between bovine testes calmodulin and rabbit skeletal muscle myosin light chain kinase. A cross-linked complex between the proteins was isolated in the presence of Ca^{2+}. This complex was active without exogeneous Ca^{2+} or calmodulin.

IV. CONFORMATIONAL ANALYSIS

A. DETECTION OF CONFORMATIONAL CHANGES

Conformational changes of proteins on binding of ligand or effectors can be detected by cross-linking reagents. Already mentioned above are the structural changes of myosin subfragment 1 on binding Mg^{2+}-ATP[37] and of troponin C induced by Ca^{2+}. In these cases, different cross-linked products were obtained in the presence and absence of the ligand. Additional examples are provided by the insulin receptor[47] and the glycogen phosphorylase system.[48,49]

The conformational states of the insulin receptor were demonstrated by cross-linking with 4,4'-dithiostilbene-2,2'-disulfonate. The insulin receptor is a disulfide-linked hetero-tetramer with two α subunits and two β subunits. Insulin binds to the α subunits which induces autophosphorylation by the β subunits where the tyrosine kinase activity is located. Binding of insulin to the α-subunit of purified insulin receptor also increased the rate of cross-linking, especially the formation of β-β dimers.[47] These results support a conformational change following insulin binding, leading to the activation of autophosphorylation.

Structural changes of the highly regulated glycogen phosphorylase system were revealed by cross-linking with a homologous series of bis-imidoesters of different lengths. In the presence of various activators, inhibitors and substrates, glycogen phosphorylase b exhibited different cross-linked patterns.[48] The effect of these ligands on the conformation of the enzyme was documented. Similar results were obtained with phosphorylase b and a.[49] These data reflect structural changes on binding of effectors and were used to construct a model of allostericity for the enzyme system.

B. CONFORMATION LOCK

While conformational changes of proteins lead to different cross-linking patterns, cross-linking will in turn prevent structural changes from taking place. Thus, a specific conformation of a protein may be locked in by intramolecular cross-linking. Cross-linked hemoglobin provides the most obvious example. Native hemoglobin is in an equilibrium state between high ligand affinity (R) and low ligand affinity (T) forms. Cross-linking of human deoxy-hemoglobin (I-state) between $\alpha_1 Lys^{99}$ and $\alpha_2 Lys^{99}$ residues by *bis*(3,5-dibromosali-cyl)fumarate lowered its oxygen affinity.[50] Cooperativity was slightly reduced and all heterotropic effects were diminished. These results were attributable to the locking of the hemoglobin molecule in the T-state conformation; although an argument that the reduced oxygen affinity arose from smaller binding constants for both T- and R-states was presented.[50]

When hemoglobin S was cross-linked with dimethyl adipimidate, there was an increase in oxygen affinity in either the presence or absence of 2,3-diphosphoglycerate, slight decrease in Bohr effect and cooperativity, and a small but significant destabilization of the confor-mation of deoxyhemoglobin.[51] More importantly, the modification of both α and β^s subunits increased the solubility of cross-linked hemoglobin S. This anti-sickling effect possibly was a consequence of locking the molecule in a specific conformation.

Another example of conformation lock by cross-linking reagents is presented by cross-linking α_2-macroglobulin with *cis*-dichlorodiammineplatinum(II) (*cis*-DDP) which caused extensive intersubunit cross-links.[52] Treatment of native α_2-macroglobulin with protease results in cleavage of the molecule and subsequently in a conformational change in the inhibitor. Treatment of cross-linked α_2-macroglobulin with trypsin leads to complete subunit cleavage, however, no conformational change, receptor recognition site exposure, or the appearance of thiol groups are detectable. These results demonstrate that cross-linking of α_2-macroglobulin by *cis*-DDP locks the molecule in the native or slow conformation.

V. STRUCTURAL STABILITY

Cross-linking has been investigated as a means for stabilizing protein structures.[53-56] Glyceraldehyde-3-phosphate dehydrogenase and asparaginase were cross-linked with dicar-boxylic acids in the presence of water soluble carbodiimide as well as a series of homologous *bis*-imidoesters. The cross-linked enzymes were shown to have increased thermal stability as well as increased stability against other denaturing factors such as urea, guanidinium chloride, detergents, and different solvents.

The stability of lactate dehydrogenase cross-linked with glutaraldehyde was also reported to be enhanced. Cross-linking was shown not to affect the pH-dependent inactivation, but the effect on thermal inactivation and guanidinium denaturation was greatly reduced.[57] Himmel et al. have also demonstrated that intramolecular cross-linking of amyloglucosidase with various bifunctional reagents more than doubled the half-life of the enzyme at 65°C.[58]

It seems in general that intramolecular cross-linking confers increased thermal and chemical stability to the structure of proteins.

REFERENCES

1. **Furlan, M.**, Structure of fibrinogen and fibrin, in *Fibrinogen, Fibrin Stabilisation and Fibrinolysis: Clinical, Biochemical and Laboratory Aspects,* Francis, J. L., Ed., Ellis Horwood, Chichester, England, 1988, 16.
2. **Eyre, D. R., Paz, M. A., and Gallop, P. M.,** Cross-linking in collagen and elastin, *Annu. Rev. Biochem.,* 53, 717, 1984.
3. **Mimni, M. E., Ed.,** *Collagen,* Vol. I, CRC Press, Boca Raton, FL, 1988.
4. **Monneron, A. and d'Alayer, J.,** Effects of imido-esters on membrane-bound adenylate cyclase, *FEBS Lett.,* 122, 241, 1980.
5. **Kennedy, J. F. and Scabral, J. M. S.,** Immobilized enzymes, in *Solid Phase Biochemistry: Analytical and Synthetic Aspects,* Scouten, W. H., Ed., Wiley-Interscience, New York, 1983, 253.
6. **Ji, T. H.,** Cross-linking of glycolipids in erythrocyte ghost membrane, *J. Biol. Chem.,* 249, 7841, 1974.
7. **Heilmann, H.-D. and Holzner, M.,** The spatial organization of the active sites of the bifunctional oligomeric enzyme tryptophan synthase: cross-linking by a novel method, *Biochem. Biophys. Res. Commun.,* 99, 1146, 1981.
8. **Lutter, L. C., Ordanderl, F., and Fasold, H.,** The use of a new series of cleavable protein-crosslinkers on the *Escherichia coli* ribosome, *FEBS Lett.,* 48, 288, 1974.
9. **Hill, M., Bechet, J.-J., and d'Albis, A.,** Disuccinimidyl esters as bifunctional crosslinking reagents for proteins, *FEBS Lett.,* 102, 282, 1979.
10. **Bloxham, D. P. and Sharma, R. P.,** The development of S,S'-polymethylene bis(methanesulfonates) as reversible cross-linking reagents for thiol groups and their use to form stable catalytically active cross-linked dimers within glyceraldehyde 3-phosphate dehydrogenase, *Biochem. J.,* 181, 355, 1979.
11. **Wells, J. A., Knoeber, C., Sheldon, M. C., Werber, M. W., and Yount, R. G.,** Cross-linking of myosin subfragment 1. Nucleotide-enhanced modification by a variety of bifunctional reagents, *J. Biol. Chem.,* 255, 11135, 1980.
12. **Wells, J. A. and Yount, R. G.,** Chemical modification of myosin by active-site trapping of metal-nucleotides with thiol crosslinking reagents, *Methods Enzymol.,* 85, 93, 1982.
13. **Husain, S. S. and Lowe, G.,** Evidence for histidine in the active sites of ficin and stem-bromelain, *Biochem. J.,* 110, 53, 1968.
14. **Rümbeli, R., Wirth, M., and Zuber, H.,** Crosslinking of phycobiliproteins from cyanobacterium Mastigocladus laminosus with bis-imidates: localization of an intersubunit and an intrasubunit crosslink in C-phycocyanin, *Biol. Chem. Hoppe-Seyler,* 368, 1179, 1987.
15. **Bragg, P. D. and Hou, C.,** A cross-linking study of the Ca^{2+}, Mg^{2+}-activated adenosine triphosphatase of *Escherichia coli, Eur. J. Biochem.,* 106, 495, 1980.
16. **Bragg, P. D. and Hou, C.,** Chemical crosslinking of α-subunits in the F_1-adenosine triphosphate of *Escherichia coli, Arch. Biochem. Biophys.,* 244, 361, 1986.
17. **Joshi, V. K. and Wang, J. H.,** Cross-linking study of the quaternary fine structure of mitochondrial F_1-ATPase, *J. Biol. Chem.,* 262, 15721, 1987.
18. **Muñoz, E., Palacios, P., Marquet, A., Andreu, J. M.,** Structure of F_1-ATPase (BF_1 factor) from Micrococcus lysodeikticus. 4. Cross-linking study with diimido esters, *Mol. Cell. Biochem.,* 33, 3, 1980.
19. **Schäfer, H.-J., Scheurich, P., Rathgeber, G., and Dose, K.,** Fluorescent photoaffinity labeling of F_1 ATPase from Micrococcus luteus with 8-azido-1, $^1/_2$ N-^{6}-etheno-adenosine 5'-triphosphate, *Anal. Biochem.,* 104, 106, 1980.
20. **Schäfer, H.-J. and Dose, K.,** Photoaffinity cross-linking of the coupling factor 1 from *Micrococcus luteus* by 3'-arylazido-8-azido-ATP, *J. Biol. Chem.,* 259, 15301, 1984.
21. **Bickle, T., Hershey, J. W. B., and Traut, R. R.,** Spatial arrangement of ribosomal proteins: reaction of the *Escherichia coli* 30S subunit with *bis*-imidoesters, *Proc. Natl. Acad. Sci. U.S.A.,* 69, 1327, 1972.
22. **Lutter, L. C., Bode, U., Kurland, C. G., and Stöffler, G.,** Ribosomal protein neighborhoods. III. Cooperativity of ribosome assembly, *Mol. Gen. Genet.,* 129, 167, 1974.
23. **Traut, R. R., Lambert, J. M., Boileau, G., and Kenny, J. W.,** Protein topography of *Escherichia coli* ribosomal subunits as inferred from protein crosslinking in *Ribosomes: Structure, Function and Genetics,* Chambliss, G., Craven, G. R., Davies, J., Davis, K., Kahan, L., and Nomura, M., Eds., University Park Press, Baltimore, 1980, 89.
24. **Traut, R. R., Tewari, D. S., Sommer, A., Gavino, G. R., Olson, H. M., and Glitz, D. G.,** Protein topography of ribosomal functional domains: effect of monoclonal antibodies to different epitopes in *Escherichia coli* protein L7/L12 on ribosome function and structure, in *Structure, Function and Genetics of Ribosomes,* Hardesty, B. and Kramer, G., Eds., Springer-Verlag, New York, 1986, 286.
25. **Lambert, J. M., Boileau, G., Cover, J. A., and Traut, R. R.,** Cross-links between ribosomal proteins of 30S subunits in 70S tight couples and in 30S subunits, *Biochemistry,* 22, 3923, 1983.
26. **Capel, M. S., Kjeldgaard, M., Engelman, D. M., and Moore, P. B.,** Positions of S2, S13, S16, S17, S19 and S21 in the 30S ribosomal subunit of *Escherichia coli, J. Mol. Biol.,* 200, 65, 1988.

27. **Stöffler, G. and Stöffler-Meilicke, M.,** Immuno electron microscopy on Escherichia coli ribosomes, in *Structure, Function and Genetics of Ribosomes,* Hadesty, B. and Dramer, G., Eds., Springer-Verlag, New York, 1986, 28.

28. **Redl, B., Walleczek, J., Stöffler-Meilicke, M., Stöffler, G.,** Immunoblotting analysis of protein-protein crosslinks within the 50S ribosomal subunit of *Escherichia coli.* A study using dimethylsuberimidate as crosslinking reagent, *Eur. J. Biochem.,* 181, 351, 1989.

29. **Walleczek, J., Redl, B., Stöffler-Meilicke, M., and Stöffler, G.,** Protein-protein cross-linking of the 50S ribosomal subunit of Escherichia coli using 2-iminothiolane. Identification of cross-links by immunoblotting techniques, *J. Biol. Chem.,* 264, 4231, 1989.

30. **Chiaruttini, C., Milet, M., Hayes, D. H., Expert-Bezancon, A.,** Crosslinking of ribosomal proteins S4, S5, S7, S8, S11 and S18 to domains 1 and 2 of 16S rRNA in the *Escherichia coli* 30S particle, *Biochemie,* 71, 839, 1989.

31. **Brimacomke, R., Atmadja, J., Stiege, W., and Schuler, D.,** A detailed model of the three-dimensional structure of *Escherichia coli* 16S ribosomal RNA in situ in the 30S subunit, *J. Mol. Biol.,* 199, 115, 1988.

32. **Terao, K., Uchiumi, T., Kobayashi, Y., and Ogata, K.,** Identification of neighbouring protein pairs in the rat liver 40S-ribosomal subunits cross-linked with dimethyl suberimidate, *Biochim. Biophys. Acta,* 621, 72, 1980.

33. **Uchiumi, T., Terao, K., and Ogata, K.,** Identification of neighboring protein pairs in rat liver 60S ribosomal subunits cross-linked with dimethyl suberimidate or dimethyl 3,3'-dithiobispropionimidate, *J. Biochem.,* 88, 1033, 1980.

34. **Leszyk, J., Grabarek, Z., Gergely, J., and Collins, J. H.,** Characterization of zero-length cross-link between rabbit skeletal muscle troponin C and tropnin I: evidence for direct interaction between the inhibitory region of troponin I and the NH₂-terminal, regulatory domain of troponin C, *Biochemistry,* 29, 299, 1990.

35. **Leszyk, J., Collins, J. H., Leavis, P. C., and Tao, T.,** Cross-linking of rabbit skeletal muscle troponin subunits: labeling of cysteine-98 of troponin C with 4-maleimidobenzophenone and analysis of products formed in the binary complex with troponin T and the ternary complex with troponin I and T, *Biochemistry,* 27, 6983, 1988.

36. **Chantler, P. D. and Bower, S. M.,** Cross-linking between translationally equivalent sites on two heads of myosin. Relationship to energy transfer results between the same pair of sites, *J. Biol. Chem.,* 263, 938, 1988.

37. **Lu, R. C., Moo, L., and Wong, A. G.,** Both the 25 kDa and 50 kDa domains in myosin subfragment 1 are close to the reactive thiol, *Proc. Natl. Acad. Sci. U.S.A.,* 83, 6392, 1986.

38. **Kareva, V. V., Dobrovol'sky, A. B., Baratova, L. A., Friedrich, P., and Gusev, N. B.,** Ca²⁺-induced structural change in the Ca²⁺/Mg²⁺ domain of troponin C detected by crosslinking, *Biochim. Biophys. Acta,* 869, 322, 1986.

39. **Arata, T.,** Structure of the actin-myosin complex produced by crosslinking in the presence of ATP, *J. Mol. Biol.,* 191, 107, 1986.

40. **Onishi, H., Maita, T., Matsuda, G., and Fujiwara, K.,** Evidence for the association between two myosin heads in rigor acto-smooth muscle heavy meromyosin, *Biochemistry,* 28, 1898, 1989.

41. **Lambright, D. G., Lee, T. K., and Wong, S. S.,** Association-dissociation modulation of enzyme activity: the case of lactose synthase, *Biochemistry,* 24, 910, 1985.

42. **Brew, K., Shaper, J. H., Olsen, K. W., Trayer, I. P., and Hill, R. L.,** Cross-linking of the components of lactose synthetase with dimethylpimelimidate, *J. Biol. Chem.,* 250, 1434, 1975.

43. **Stoclet, J.-C., Gérard, D., Kilhoffer, M.-C., Lugnier, C., Miller, R., and Schaeffer, P.,** Calmodulin and its role in intracellular calcium regulation. *Prog. Neurobiol.,* 29, 321, 1987.

44. **Harrison, J. K., Lawton, R. G., and Gnegy, M. E.,** Development of a novel photoreactive calmodulin derivative: cross-linking of purified adenylate cyclase from bovine brain, *Biochemistry,* 28, 6023, 1989.

45. **Kincaid, R. L.,** Preparation of an enzymatically active cross-linked complex between brain cyclic nucleotide phosphodiesterase and 3-(2-pyridyldithio)propionyl-substituted calmodulin, *Biochemistry,* 23, 1143, 1984.

46. **Zot, H. G. and Puett, D.,** An enzymatically active cross-linked complex of calmodulin and rabbit skeletal muscle myosin light chain kinase, *J. Biol. Chem.,* 264, 15552, 1989.

47. **Schenker, E. and Kohanski, R. A.,** Conformational states of the insulin receptor, *Biochem. Biophys. Res. Commun.,* 157, 140, 1988.

48. **Hajdu, J., Dombrádi, V., Bot, G., and Friedrich, P.,** Structural changes in glycogen phosphorylase as revealed by cross-linking with bifunctional diimidates: phosphorylase b, *Biochemistry,* 18, 4037, 1979.

49. **Bombrádi, V., Hajdu, J., Bot, G., and Friedrich, P.,** Structural changes in glycogen phosphorylase as revealed by cross-linking with bifunctional diimidates. Phospho-dephospho hybrid and phosphorylase a, *Biochemistry,* 19, 2295, 1980.

50. **Vandegriff, K. D., Medina, F., Marini, M. A., and Winslow, R. M.,** Equilibrium oxygen binding to human hemoglobin cross-linked between the α chains by bis(3,5-dibromosalicyl)fumarate, *J. Biol. Chem.,* 264, 17824, 1989.

51. **Pennathur-Das, R., Heath, R. H., Mentzer, W. C., and Lubin, B. H.,** Modification of hemoglobin S with dimethyl adipimidate. Contribution of individual reacted subunits to changes in properties, *Biochim. Biophys. Acta,* 704, 389, 1982.

52. **Roche, P. A., Jensen, P. E. H., and Pizzo, S. V.,** Intersubunit cross-linking by cis-dichlorodiammine-platinum(II) stabilizes an α_2-macroglobulin ''nascent'' state: evidence that thiol ester bond cleavage correlates with receptor recognition site exposure, *Biochemistry,* 27, 759, 1988.

53. **Torchilin, V. P. and Martinek, K.,** Enzyme stabilization without carriers, *Enzyme Microb. Technol.,* 1, 74, 1979.

54. **Torchilin, V. P., Trubetskoy, V. S., Omel'yanenko, V. G., and Martinek, K.,** Stabilization of subunit enzymes by intersubunit cross-linking with bifunctional reagents: studies with glyceraldehyde-3-dehydrogenase, *J. Mol. Catal.,* 19, 291, 1983.

55. **Trubetskoy, V. S. and Torchilin, V. P.,** Artificial and natural thermostabilization of subunit enzymes. Do they have similar mechanism? *Int. J. Biochem.,* 17, 661, 1985.

56. **Torchilin, V. P. and Trubetskoy, V. S.,** Stabilization of subunit enzymes by intramolecular crosslinking with bifunctional reagents, *Ann. N.Y. Acad. Sci.,* 434, 27, 1984.

57. **Gottschalk, N. and Jaenicke, R.,** Chemically crosslinked lactate dehydrogenase: stability and reconstitution after glutaraldehyde fixation, *Biotechnol. Appl. Biochem.,* 9, 389, 1987.

58. **Tatsumoto, K., Oh, K. K., Baker, J. O., and Himmel, M. E.,** Enhanced stability of glucoamylase through chemical cross-linking, *Appl. Biochem. Biotechnol.,* 20, 293, 1989.

Chapter 9

CROSS-LINKING OF MEMBRANE PROTEINS

I. INTRODUCTION

Biological membranes are an assembly of proteins, carbohydrates, and lipids. The lipids form a bilayer providing a milieu where proteins and carbohydrates, which are usually components of glycoproteins or glycolipids, interact with one another both within and without the inner and outer surfaces. The spatial organization and the interactions of these molecules have been investigated by electron microscopy,[1] immuno-electronmicroscopy,[2] fluorescence spectroscopy,[3] electron spin resonance,[4] magnetic resonance,[5] optical and Raman spectroscopy.[6] In complement with these techniques, the application of cross-linking reagents has added a third dimension in predicting subunit structures of molecules and molecular associations in biological membranes. The use of cross-linking reagents to identify membrane receptors for macromolecular ligands has become an established technique.

The first application of chemical cross-linking to membrane was carried out by Berg et al.[7] in 1965 who used difluorodinitrobenzene to cross-link erythrocytes. However, it was not until early 1970s when sodium dodecyl sulfate-polyacrylamide gel electrophoresis (SDS-PAGE) was applied for the systemic analysis of cross-linked products,[8] that its potential was recognized. The introduction of cleavable reagents containing disulfide, glycol, and other bonds further made the analysis of components of cross-linked products simple and easy.[9] The advent of cross-linking of membrane components has been the subject of reviews during the past two decades.[10-16]

In this chapter, the general application of cross-linking reagents to biological membranes will be discussed. Because of the diversity and extensive application of these reagents, only a general summary of recent examples will be presented to demonstrate the usefulness of this method. Many of the earlier applications have already been reviewed.[10-16]

II. GENERAL METHODOLOGY OF MEMBRANE CROSS-LINKING

A. CHOICE OF CROSS-LINKING REAGENTS

The large variety of cross-linking reagents listed in Chapters 4 to 6 provides a wide spectrum of choices. Successful application of a reagent depends on the objective and the system to be investigated. Many factors must be considered. For example, random collision-dependent cross-linking as a result of membrane fluidity has been the major objection to the use of cross-linking in membranes.[10,17] To minimize the problem, photosensitive reagents may be used. Hydrophobicity may play a very important part. Hydrophobic and hydrophilic reagents have the potential to probe different regions of the membrane as well as different sites of cross-linked components. For instance, different cross-linking patterns are obtained when reagents of varying hydrophobicity are used.[18] This is observed for complement components[19,20] and acetylcholine receptor.[21,22] It has been demonstrated that different cross-linking molecules have different preference in orientation and location in a membrane.[12] Spin labeled bis(N-hydroxysuccinimide) esters of chain lengths 15 to 29 Å tend to orient in a location at the interface between membrane and water, whereas spin labeled dicarboxylic acids orient extended across a phospholipid bilayer.[23] To study membrane proteins that interact in the cytoplasmic or extracytoplasmic domains of the lipid bilayer, membrane-impermeant cross-linking reagents have been developed.[16] These compounds bear hydrophilic or charged groups which make them lipid insoluble. Group specificity also provides another dimension of cross-linking. Sulfhydryl reagents, for example, favor cross-linking of integral membrane proteins of erythrocytes, while imidates prefer peripheral membrane proteins. Last, but not least, the size of the reagent provides a different extent of cross-linking. Cross-linkers shorter than 5 Å usually yield few intermolecular cross-links, whereas extensive cross-linking is frequently achieved with reagents greater than 11 Å in length. Thus, a homologous series may be used to map the interacting neighbors.[11]

B. REACTION PROCEDURES

Membrane cross-linking is similar to soluble protein cross-linking in general. It can usually be achieved by simply mixing a cross-linking reagent with a membrane preparation. However, there is no general rule for the cross-linking procedure because of the diversity of membrane systems and the many different variety of reagents. A preliminary experiment may result in too much or too little cross-linking. In essence, one topographically correct cross-link is better than a multiplicity of cross-linked species. The optimum condition for a particular situation is empirical and often constitutes the major experimental effort in a membrane cross-linking study.

III. POSSIBLE ARTIFACTS OF MEMBRANE CROSS-LINKING

Possible cross-linking artifacts of membrane components may arise from the perturbation of the system as a result of introduction of cross-linking agents. Very often the reagents are used at millimolar or higher concentrations.[24] In addition, organic solvents are frequently required for the dissolution of water-insoluble compounds in the biological aqueous medium. It is therefore important to assure that the membrane biological system is not affected by the presence of these solvents. For the identification of receptors, the introduction of multivalent ligands, such as lectins and antibodies, may induce membrane structural alterations or aggregations of membrane proteins.[25,26] This is though to be a consequence of the rapid lateral diffusion of membrane molecules.[3,27-29] This type of movement of proteins in the bilayer results in frequent collisions with one another.[30] These molecules may be cross-linked during their collisional processes. If this occurs, it poses a serious problem of distinguishing cross-linked products of natural, stable complexes from those of collisional products. Fortunately, Ji and Middaugh have shown that such collisional cross-linking occurs only at extremely high protein concentrations and for long cross-linking periods.[31] Furthermore, the use of photosensitive heterobifunctional reagents and the conduction of these experiments at low temperatures considerably reduce the risk of this complication.[17,32] Theoretical treatment of random collisional cross-linking is also possible, at least in the case of rhodopsin.[33] By comparing the experimental results with theoretical predictions of cross-linked patterns for monomeric, dimeric, and tetrameric proteins, it is concluded that rhodopsin exists as a monomer in the dark-adapted state.

Other artifactual cross-linking may occur between soluble cytoplasmic proteins and membrane components. For example, hemoglobin in intact erythrocytes can be cross-linked to spectrin at elevated temperatures during extended incubation periods.[9] Another possibility is the cross-linking of proteins in two adjacent cells. The frequency of this artifact would be expected to increase with cell population. However, the significance of such occurrence is unknown.

IV. IDENTIFICATION OF MEMBRANE RECEPTORS

A. GENERAL APPROACHES

Identification of membrane surface receptors has been achieved by the method of macromolecular affinity labeling introduced by Ji.[34,35] Basically, a macromolecular ligand is cross-linked to its membrane receptor by means of heterobifunctional photosensitive agents. The reagent is first bound to the ligand through conventional functional group-selective reactions such as that between an amino group and an imidoester. After purification, the labeled ligand is then incubated with membrane preparations containing its receptors. Cross-linking is achieved by photoactivation of the photosensitive group. The cross-linked products are isolated and the receptors are characterized and identified.

In this approach, heterobifunctional reagents with an imidate or N-hydroxysuccinimide ester at one end and an arylazide at the other are best choices. They have been successfully employed to investigate various receptors. Many of these have been reviewed.[10-16] Some of the recent applications will be mentioned below. Conventional bifunctional reagents have also been used to identify membrane receptors, although they indiscriminately cross-link other membrane constituents.

Complications in receptor cross-linking may arise during the introduction of the reagents to molecular ligands. The ligands may be inactivated during labeling either by denaturation or by reacting with the essential group. They may also undergo modification-induced conformational changes. All of these will either reduce or destroy their ability to bind to their receptors. The functional integrity of the ligand should be carefully monitored before cross-linking. Reagents attached to a locus near the binding sites will have the best chance to yield successful cross-linking. Reagents attached to hydrophobic centers are generally ineffective. Thus, sulfhydryl reagents, which are generally more hydrophobic than imidoester, should be critically evaluated.

Identification of receptors in isolated ligand-receptor complexes can be achieved by SDS-PAGE. To facilitate the process, radioactive labeling has been used. Radioactivity can be introduced into either the ligand or membrane surface proteins or both by iodination with ^{125}I. Alternatively, radioactive cross-linking reagents may be used. Another technique for the identification of cell-surface receptors is the use of antibodies directed against the cross-linker.[36]

B. SPECIFIC EXAMPLES

1. Use of Photosensitive Heterobifunctional Cross-Linkers

Shephard et al.[37] have used sulfosuccinimidyl-2-(p-azidosalicylamido)-1,3'-dithiopropionate (SASD, see Chapter 5 for structure **XC**) to identify phytohemagglutinin cell surface receptors. The reagent can be iodinated with ^{125}I using iodogen to facilitate the identification of cross-linked products. In this study lectin phytohemagglutinin was first labeled with the radio-iodinated SASD through a displacement reaction of the sulfosuccinimidyl group. Blood mononuclear cells were then incubated with the labeled phytohemagglutinin at 4°C for 30 min, after which the unbound ligand was removed by washing the cells. The ligand-bound cells were irradiated at ice temperature to activate the photosensitive group. Membrane proteins were solubilized with Triton X-100 and separated on SDS-PAGE under reducing conditions. Several major bands arising specifically from photocross-linking were identified. This procedure of using cleavable iodinated heterobifunctional cross-linking reagent, under optimized conditions may be employed as a general approach to demonstrate the presence of receptors on the plasma membrane of intact cells.

Another photosensitive heterobifunctional cross-linking reagent, N-succinimidyl-6-(4'-azido-2'-nitrophenylamino)hexanoate (SANPAH, see Chapter 5 for structure **LXXIX**), was used for the identification of D$_1$-dopamine receptor.[38] In this case the D$_1$-antagonist SCH-38548, (R,S)-5-(3'-aminophenyl)-8-chloro-2,3,4,5-tetrahydro-3-methyl-[1H]-3-benzazepin-7-ol, was radio-iodinated by chloramine T. The labeled SCH-38548 bound to the D$_1$-dopamine receptor of the rat striatum membrane with high affinity. After incubation of the membrane with radio-iodinated SCH-38548, the membrane was washed to remove unbound radioactivity. The cross-linker SANDPAH was then added to react with SCH-38548 and the mixture was photolyzed to complete the cross-linking reaction. Membrane proteins were then solubilized with sodium dodecyl sulfate (SDS) and analyzed by SDS-PAGE for radioactive fragments. Covalent incorporation of SCH-38548 into a protein of 72 kDa was determined.

SANPAH was also used for the identification of A_2 adenosine receptor.[39] A ^{125}I-labeled adenosine analog, 2-[4-(2-{2-[(4-aminophenyl)methylcarbonylamino]ethylaminocarbonyl}ethyl)-phenyl]-ethylamine-5'-N-ethylcarboxamidoadenosine (PAPA-APEC), was found to bind to the A_2 adenosine receptor in bovine striated membrane with high affinity and selectivity. After incubation with striatal membranes, unbound radio-iodinated PAPA-APEC was removed by washing the membrane preparation. SANPAH was then added to the membrane/ligand suspension and incubated for 5 min. At the end of this period, the suspension was irradiated. Membrane proteins were again solubilized with SDS and analyzed on SDS-PAGE. A single specifically radiolabeled protein with an apparent molecular mass of 45 kDa was identified as the A_2 adenosine receptor.

Vasopressin V_2 receptors from porcine kidney membrane was identified with N-succinimidyl 4-azidobenzoate (HSAB, see Chapter 5 for structure **LXX**).[40] The receptor-containing membrane was solubilized from the membrane with egg lysolecithin and incubated with ^{125}I-vasopressin. After binding, the ^{125}I-vasopressin-receptor complex was isolated by gel filtration and mixed with HSAB. After 10 min at room temperature, the mixture was irradiated to effect photocross-linking. The reaction product was lyophilized and resuspended in SDS-containing buffer and analyzed on SDS-PAGE. A 62 kDa band specifically labeled with vasopressin was identified.

HSAB was also employed to analyze the receptor for murine-interferon-γ.[41] Thymoma cell line, EL-4 cells were mixed with radio-iodinated interferon-γ (^{125}I-IFN-γ) for 16 h at 4°C. After incubation cells were washed to remove free ^{125}I-IFN-γ. Different concentrations of HSAB were added and incubated in the dark for 1 h, after which the cells were irradiated at 366 nm for 10 min. The cell membranes were then solubilized and analyzed on SDS-PAGE. At low HSAB concentrations (<50 μM), the results showed only a single band of about 95 kDa. At high concentrations of HSAB (500 μM), two bands of 95 and 110 kDa were detected. These results were interpreted to show that a receptor of about 80 kDa existed and that ^{125}I-IFN-γ could bind as a monomer (16 kDa) or as a dimer (32 kDa) to the receptor.

2. Use of Other Bifunctional Cross-Linkers

While various bifunctional reagents have different applications for the study of membrane proteins,[9-16] disuccinimidyl suberate (DSS, see Chapter 4 for structure **XXXI**) has acquired an acclaimed popularity for the investigation of membrane receptors in recent years. Murthy et al.[42] used DSS to characterize murine interleukin 3 (mIL-3) receptor. For studies with intact cells, mIL-3-dependent cell line B6SUtA$_1$ were first incubated for 2 h at 4°C with ^{125}I-mIL-3. The cells were then washed and DSS was added and mixed for 1 h. After that time period, the membrane proteins were solubilized with SDS-containing buffer and analyzed by SDS-PAGE. The results showed two radiolabeled species with molecular weights of 140 (p140) and 70 (p70) kDa after correction for mIL-3. When the cross-linked cells were further incubated at 37°C, the intensity of p140 decreased relative to p70. The authors speculated that the receptor was a 140 kDa glycoprotein which was converted to a 70 kDa surface protein on mIL-3 binding.

DSS was also used for characterization of atrial natriuretic peptide (ANP) receptors in bovine ventricular sarcolemma.[43] Membranes isolated from bovine heart were incubated for 1 h with radio-iodinated ANP with or without an ANP analog des[QSGLG]ANP which bound with high affinity to C but not B receptors. The membranes were pelleted and washed and then resuspended with DSS for 40 min on ice. The cross-linking reaction was terminated by addition of 1 M ammonium acetate and pelleting. Membrane proteins were solubilized by SDS and analyzed on SDS-PAGE. Two radiolabeled bands of 120 kDa and 65 kDa were obtained in the absence of 2-mercaptoethanol. Both bands were lost if the samples were reduced with 2-mercaptoethanol before electrophoresis. These bands also disappeared in the presence of des[QSGLS]ANP. It is speculated that there might be a third receptor subtype present in the cell membrane.

Similar procedure using DSS have been applied for the study of substance P receptor in guinea pig lung tissues,[44] ovarian and testicular LH/hCG receptors,[45,46] insulin and insulin-like growth factor I receptors in human adrenal gland[47] and platelets,[48] thyrotropin receptors[49] endothelin receptor[50] and many other membrane receptors.[51]

Other nonphotosensitive cross-linker that have been used in conjunction with DSS include the cleavable DSS analog, dithiobis(succinimidyl propionate) (DSP) and ethylenegly-cobis(succinimidyl succinate) (EGS, see Chapter 4 for structures **XLIII** and **XL**). DSP was used by Zhang and Menon to characterize rat Leydig cells gonadotropin receptor.[46] The procedure for cross-linking of human gonadotropin (hCG) to its cell surface used was identical to that for DSS. In fact, DSP was used as a confirmatory counterpart of DSS to ensure that cross-linked products effected by DSS contained hCG. This was achieved by treatment of the sample with 2-mercaptoethanol before electrophoresis. No cross-linked band was observed under this condition.

EGS was used with DSP for the identification of mutogenic fibrinogen receptor (MFR).[52] Radio-iodinated (^{125}I) hemopoietic cell lines Raji and JM were incubated with fibrinogen for 30 min at 20°C before addition of EGS. Cross-linking was allowed for only one minute. Cell proteins were extracted with Triton X-100. Fibrinogen-MFR conjugates were immunoprecipitated by anti-fibrinogen polyclonal antibodies. The immunoprecipitate was dissociated and the cross-linked conjugate analyzed on SDS-PAGE. A membrane protein of 92 kDa was identified as the MFR. When parallel experiments were carried out with DSP, similar results were obtained.

V. NEAREST NEIGHBOR ANALYSIS OF MEMBRANE COMPONENTS

The approach to determine proteins that lie in close proximity to each other or associate in the membrane is similar to that in solution. The following examples will illustrate the general principle of application of cross-linking reagents in this area.

A. STRUCTURAL ANALYSIS

Structural organization of membrane protein complexes has been investigated by cross-linking reagents since the 1970s. One of the most outstanding applications is the structural study of virus proteins. The approach flourished during the late 1970s and the early 1980s when numerous publication on virus proteins structures appeared. These include avian sarcoma and leukemia virus,[53] murine and feline leukemia virus,[54] mature murine mammary virus,[55] Rous sarcoma virus,[56] Mengo virus,[57] paramyxoviruses,[58] Sindis virus,[59] and coliphage M13.[60]

As an example, a recent study on herpes simplex virus will be described.[61] Herpes simplex virus type 1 (HSV-1) contains several glycoproteins in its virion envelope. Chemical cross-linking of these glycoproteins was reported earlier.[62] The structural organization of one of these, glycoprotein C (gC) was studied with EGS and dimethyl 3,3'-dithiobispropionimidate (DTBP, see Chapter 4 for structure **XIX**) to analyze the association with its nearest neighbor. Human embryonic lung cells were radiolabeled biosynthetically with [^{3}H]arginine and [^{3}H]proline. The labeled cells were infected with HSV-1 strain KOS and treated with various concentrations of the homobifunctional cross-linkers EGS. A parallel experiment with DTBP was carried out. After reaction, the cells were lyzed and gC was immunoprecipitated with monoclonal antibodies specific for gC. The gC-containing immunoprecipitates were analyzed by SDS-PAGE. A band corresponding to gC was detected together with higher molecular weight complexes. Two dimensional SDS-PAGE analyses of the products cross-linked with DTBP under reducing and nonreducing conditions revealed that gC was the only component of these high molecular weight complexes. The results indicated that gC molecules may be localized in the infected cell membrane as dimers.

Structural studies on the env gene product of HIV was investigated by Schawaller et al.[63] EGS was used to cross-link membrane associated glycoproteins. After SDS-PAGE analysis, it was found that gp160 was a tetramer of identical subunits and that its proteolytic product gp41 was also tetrameric.

In addition to viruses, membrane proteins in other microorganisms have been investigated with cross-linkers. The glycoproteins of *Plasmodium falciparum* merozoites were studied in relationship to the soluble proteins in the parasite.[64] Merozoites were isolated from *P. falciparum* cultures labeled with [³H]mannose and [³⁵S]methionine and incubated with DSP at a final concentration of 0.2 mM for 7 min at room temperature. After solubilization, the cross-linked product was immunoprecipitated with monoclonal antibodies directed against the major merozoite surface glycoprotein, Pf200. The immunoprecipitates were analyzed by SDS-PAGE with and without dithiothreitol. Immunoblotting analysis of the electrophoresed product revealed that Pf200 was cross-linked with S-antigen, one of the three soluble *P. falciparum* antigens.

Many other membrane complexes in different cells from various origins have been investigated. The structure of phospholamban in cardiac sarcoplasmic reticulum (SR) membranes was studied using *p*-azidophenylacyl bromide (APB) and ethyl-(4-azidophenyl)1,4-dithiobutyrimidate (EADB, see Chapter 5 for structures).[65] Phospholamban is an intrinsic membrane protein in cardiac SR and is believed to be involved in the regulation of calcium pump. For cross-linking studies, isolated cardiac SR membranes were incubated with EADB at a final concentration of 4 mg/ml at room temperature in the dark. After 3 h, the sample was irradiated for 20 min on a Spectroline transilluminator. Cross-linking with APB was similarly carried out. The cross-linked membrane proteins were dissolved in SDS-containing buffer and resolved on 10% SDS-PAGE. The bands were analyzed by Western blot method. The results showed that phospholamban might be present as monomers and dimers in the native SR, and might interact with Ca^{2+}-ATPase. Such direct interaction between phospholamban and ATPase was further documented by James et al.[66] using a cleavable photosensitive heterobifunctional cross-linker, *N*-[4-(*p*-azido-m-iodophenylazo)benzoyl]-3-aminopropyl *N'*-oxysulfosuccinimide ester.

Human placental ectoemzyme 5'-nucleotidase was cross-linked with EGS and SANPAH and analyzed by Western blot.[67] Cross-linking of the purified enzyme indicated a dimeric structure. However, cross-linking of the plasma membrane-bound enzyme revealed a monomer coupled to an unknown protein of about 30 kDa.

Mitochondrial membrane bound rat brain hexokinase is another enzyme that has been studied for its structure.[68] The enzyme labeled with sulfosuccinimidyl-2-(*m*-azido-*o*-nitrobenzamido) ethyl-1,3'-dithiopropionate (SAND, see Chapter 5 for structure **XCI**) through an amino group was mixed with a mitochondrial preparation. After incubation for 20 min at 0°C, the mitochondrial suspensions were isolated and irradiated. The cross-linked product was analyzed by SDS-PAGE. Immunoblotting of the electrophoresed products revealed that hexokinase were bound to the mitochondrial membrane as a tetrameric enzyme.

In addition to direct analysis of cross-linked products, another approach to study the structure of membrane proteins involves the investigation of the kinetics of cross-linking reactions. The time-dependent cross-linking patterns detected on SDS-PAGE can be achieved by either changing the duration of photolysis, addition of a reactive substrate or termination of the reaction by denaturation. Information about the structures of membrane proteins is derived from the various possible cross-linked complexes as a function of time. Such analysis has been applied to erythrocyte spectrin.[69,70] It was found that dimers accumulated before trimers and tetramers, and as tetramers increased, dimers diminished.

B. MEMBRANE PROTEIN-PROTEIN INTERACTIONS

Interaction of membrane proteins can be shown by chemical cross-linking. Proteins and lipoproteins that are involved in the interactions between the inner membrane, outer membrane and murein layers of *Escherichia* cells were studied by Leduc et al.[71] The cells were mixed with DSP at a reagent to protein ratio of 2:1. After 30 min at room temperature, the cell envelopes obtained after lysis was mixed in 8% SDS at 100°C to isolate murein in the presence of 2-mercaptoethanol. The solubilized preparation was analyzed on SDS-PAGE. Nine murein-associated peptides were identified—five of which were lipoproteins. The results suggested that membrane lipoproteins may play a significant role in the structural integration of the murein.

The interaction of ras oncogene products (p21^{ras}) with other proteins was studied by de Gunzburg et al.[72] The p21^{ras} proteins are expressed in most, if not all, tissues as membrane-associated proteins that bind guanine nucleotides and possess an intrinsic GTPase activity. They are thought to be involved in the transduction of incoming growth signals to several messenger-producing enzymes within the cell. To identify proteins that might interact with p21^{ras}, de Gunzburg et al. used DSP to cross-link Rat-1 fibroblasts and derived cell lines over-expressing p21^{ras}. Cell proteins were labeled by culturing in [^{35}S]methionine-containing medium. The cells were harvested and incubated with 1 mM DSP for 30 min on ice. After washing, the proteins were solubilized with Triton X-100 and SDS. p21^{ras}-containing protein complexes were immunoprecipitated with polyclonal anti-p21^{ras} antibodies and analyzed on 5-15% SDS-PAGE. A protein of 60 kDa was found associated with p21^{ras}. The same protein was noted to be associated with p21^{ras} in numerous mammalian cell lines using the *in vivo* cross-linking technique.

Chemical cross-linking has also been used to investigate the interaction between soluble and membrane-bound proteins.[73,74] The interaction of creatine kinase and hexokinase with the mitochondrial membrane was studied by Font et al.[73] Cross-linking experiments were carried out for 5 min at 30°C on isolated mitochondria with DMS, EGS, Cu^{2+}-phenanthroline, SANPAH or methyl 4-azidobenzoimidate (MABI, see Chapter 5 for structure **XCVIII**). For the latter two photosensitive labels, the mitochondrial suspensions were further photoactivated for 5 min. After cross-linking, the enzyme activity in the soluble fraction was measured and analyzed by SDS-PAGE. The results showed that both the activity and concentration of hexokinase and mitochondrial creatine kinase in the soluble fraction were decreased, indicating that the soluble proteins were associated with membrane components.

The interaction of calcium-activated neutral proteinase (CANP) with erythrocyte membrane components was demonstrated by cross-linking with 4,4′-dithiobisphenylazide (DTBPA, see Chapter 5 for structure **CXXII**).[74] Erythrocytes were reacted with DTBPA in the dark for 30 min on ice and then irradiated with 365 nm VU light. After irradiation, the cells were lyzed and membrane proteins were solubilized with 1% Triton X-100. CANP-linked proteins were selectively isolated by immunoprecipitation with polyclonal anti-CANP antibodies and analyzed by SDS-PAGE. In the calcium-free incubation medium, CANP was cross-linked to cytosolic soluble proteins, such as hemoglobin. With calcium ion in the incubation medium, CANP was found cross-linked to membrane skeletal proteins such as spectrin, band 3, 4.1, 4.2 and 6 proteins. These data demonstrated the effect of calcium on the dynamic interaction of CANP and the membrane proteins.

VI. MEMBRANE PROTEIN CONFORMATIONAL ANALYSIS

Like soluble proteins, conformational changes of membrane proteins can be detected by chemical cross-linking techniques. Numerous studies on structural changes of various membrane components have been reported. For example, band 3 protein of erythrocyte membrane was found to undergo structural changes on reacting with pyridoxal 5'-phosphate (PLP).[75] Cross-linking of intact erythrocyte with *bis*(sulfosuccinimidyl)suberate (BS^3, see Chapter 4 for structure **XXXII**) revealed a mixture of band 3 protein dimers and tetramers. Treatment of the red blood cells with PLP prior to conjugation provided exclusive tetrameric band 3 protein. The results indicated that PLP induced a conformational change in the band 3 protein.

In another study, the band 3 protein anion antiport was found similarly affected by the transport inhibitor, 4,4'-dinitrostilbene-2,2'-disulfonate.[76] In the presence of the inhibitor, only cross-linked tetrameric product was obtained. A model was proposed wherein band 3 transport-site ligands allosterically modulated the global conformation of a tetrameric porter between two quatenary structures.

VII. STABILIZATION OF MEMBRANE PROTEINS

The use of cross-linking reagents to stabilize membrane proteins is examplified by the superoxide-generating respiratory burst oxidase system. This NADPH-oxidase of neutrophil plasma membrane is highly unstable, thus hampering a detailed study of its components. However, Tamura et al.[77] have used EDC to enhance its stability. Cross-linking increased its half-life from 2 min to over 20 min at 37°C. Its stability towards high salt and detergent was also increased. Cross-linking also stabilized the V_{max} and K_m of the enzyme, which were noted to increase on storage in its native state. The increased stability paves the way to further study the enzyme system.

REFERENCES

1. **Tanaka, K., Misushima, A., Kashima, Y., Nakadera, T., and Fukudome, H.,** Ultra-high resolution scanning electron microscopy of biological materials, *Prog. Clin. Biol. Res., 295,* 21, 1989.
2. **Lococq, J. M., Pryde, J. G., Berger, E. G., and Warren, G.,** A mitotic form of the golgi apparatus identified using immunoelectronmicroscopy, *Prog. Clin. Biol. Res., 270,* 43, 1988.
3. **Devenport, L., Wang, J. Z., and Knutson, J. R.,** Studies of lipid fluctuations using polarized fluorescence spectroscopy, *Prog. Clin. Biol. Res., 292,* 97, 1989.
4. **Butterfield, D. A.,** ESR studies of transmembrane signaling process: modulation of skeletal protein-protein interactions and their influence on the physical state of cell-surface carbohydrates in human erythrocyte membranes, *Prog. Clin. Biol. Res., 292,* 53, 1989.
5. **Smith, I. C. P., Baenziger, A. M., and Jarrell, H. C.,** Deuterium NMR as a monitor of organization and dynamic at the surface of membranes: the glycolipids, *Prog. Clin. Biol. Res., 292,* 13, 1989.
6. **Van de Ven, M., Kattenberg, M., Van Ginkel, G., and Levine, Y. K.,** Study of the orientational ordering of carotenoids in lipid bilayers by resonance-raman spectroscopy, *Biophys. J., 45,* 1203, 1984.
7. **Berg, H. C., Diamond, J. M., and Marfey, P. S.,** Erythrocyte membrane: chemical modification, *Science, 150,* 64, 1965.
8. **Steck, T. L.,** Cross-linking the major proteins of the isolated erythrocyte membrane, *J. Mol. Biol., 66,* 295, 1972.
9. **Wang, K. and Richards, F. M.,** Approach to nearest neighbor analysis membrane proteins. Application to the human erythrocyte membrane of a method employing cleavable crosslinkages, *J. Biol. Chem., 249,* 8005, 1974.
10. **Ji, T. H.,** The application of chemical crosslinking for studies on cell membranes and the identification of surface reporters, *Biochim. Biophys. Acta, 559,* 39, 1979.

11. **Middaugh, C. R., Vanin, E. F., and Ji, T. H.,** Chemical cross-linking of cell membranes, *Mol. Cell. Biochem.,* 50, 115, 1983.
12. **Gaffney, B. J.,** Chemical and biochemical crosslinking of membrane components, *Biochim. Biophys. Acta,* 822, 289, 1985.
13. **Peters, K. and Richards, F. M.,** Chemical cross-linking: Reagents and problems in studies of membrane structure, *Annu. Rev. Biochem.,* 46, 523, 1977.
14. **Das, M. and Fox, C. F.,** Chemical cross-linking in biology, *Annu. Rev. Biophys. Bioeng.,* 8, 165, 1979.
15. **Shaw, A. B. and Marinetti, G. V.,** Cross-linking of erythrocyte membrane proteins and phospholipids by chemical probes, *Membrane Biochem.,* 3, 1, 1980.
16. **Staros, J. V.,** Membrane-impermeant cross-linking reagents: probes of the structure and dynamics of membrane proteins, *Acc. Chem. Res.,* 21, 435, 1988.
17. **Kiehm, D. J. and Ji, T. H.,** Photochemical cross-linking of cell membranes. A test for natural and random collisional cross-linking by millisecond crosslinking, *J. Biol. Chem.,* 252, 8524, 1977.
18. **Schramm, H. J. and Duffler, T.,** Synthesis and application of cleavable and hydrophilic cross-linking reagents, in *Protein Cross-Linking. Biochemical and Molecular Aspects,* Part A, Friedman, M., Ed., Plenum Publishing, New York, 1977, 197.
19. **Hu, V. W., Esser, A. F., Podack, E. R., and Wisnieski, B. J.,** The membrane attack mechanism of complement: photolabeling reveals insertion of terminal proteins into target membrane, *J. Immunol.,* 127, 380, 1981.
20. **Ishida, B., Wisnieski, B. J., Lavine, C. H., and Esser, A. F.,** Photolabeling of a hydrophobic domain of the ninth component of human complement, *J. Biol. Chem.,* 257, 10551, 1982.
21. **Gonzalez-Ros, J. M., Farach, M. C., and Martinez-Carrion, M.,** Ligand-induced effects at regions of acetylcholine receptor accessible to membrane lipids, *Biochemistry,* 22, 3807, 1983.
22. **Middlemas, D. S. and Raftery, M. A.,** Exposure of acetylcholine receptor to the lipid bilayer, *Biochem. Biophys. Res. Commun.,* 115, 1075, 1983.
23. **Gaffney, B. J., Willingham, G. L., and Schepp, R. S.,** Synthesis and membrane interactions of spin-labeled bifunctional reagents, *Biochemistry,* 22, 881, 1983.
24. **Whiteley, N. M. and Berg, H. C.,** Amidination of the outer and inner surfaces of the human erythrocyte membrane, *J. Mol. Biol.,* 87, 541, 1974.
25. **Unanue, E. R. and Karnofsky, M. J.,** Redistribution and fate of Ig complexes on surface of B lymphocytes. Functional implications and mechanisms, *Transplant. Rev.,* 14, 184, 1973.
26. **Nicolson, G. L.,** The interactions of lectins with animal cell surfaces, *Int. Rev. Cytol.,* 39, 89, 1974.
27. **Singer, S. J. and Nicolson, G. L.,** The fluid mosaic model of the structure of cell membranes. Cell membranes are viewed as two-dimensional solutions of oriented globular proteins and lipids, *Science,* 175, 720, 1972.
28. **Edidin, M.,** Rotational and translational diffusion of membranes, *Annu. Rev. Biophys. Bioeng.,* 3, 179, 1974.
29. **Schlessinger, J., Koppel, D. E., Axelrod, D., Jacobson, K., Webb, W. W., and Elson, E. L.,** Lateral transport on cell membranes. Mobility of concanavalin A receptors on myoblasts, *Proc. Natl. Acad. Sci. U.S.A.,* 73, 2409, 1976.
30. **Adams, G. and Delbruck, M.,** Reduction of dimensionality in biological diffusion processes, in *Structural Chemistry and Molecular Biology,* Rich, A. and Davidson, N., Eds., W. H. Freeman, San Francisco, 1968, 198.
31. **Ji, T. H. and Middaugh, C. R.,** Does random collisional cross-linking occur? *Biochim. Biophys. Acta,* 603, 371, 1980.
32. **Edidin, M. and Fambrough, D.,** Fluidity of the surface of cultured muscle fibers. Rapid lateral diffusion of marked surface antigens, *J. Cell. Biol.,* 57, 27, 1973.
33. **Downer, N. W.,** Cross-linking of dark-adapted frog photoreceptor disk membranes. Evidence for monomeric rhodopsin, *Biophys. J.,* 47, 285, 1985.
34. **Ji, T. H.,** Crosslinking of lectins and receptors in membranes with heterobifunctional crosslinking reagents, in *Membranes and Neoplasia: New Approaches and Strategies,* Marchesi, V. T., Ed., Alan R. Liss, New York, 1976, 171.
35. **Ji, T. H.,** A novel approach to the identification of surface receptors. The use of photosensitive heterobifunctional cross-linking reagent, *J. Biol. Chem.,* 252, 1566, 1977.
36. **Ballmer-Hofer, K., Schlup, V., Burn, P., and Burger, M. M.,** Isolation of *in situ* crosslinked ligand-receptor complexes using an anticrosslinker specific antibody, *Anal. Biochem.,* 126, 246, 1982.
37. **Shephard, E. G., de Beer, Fl. C., von Holt, C., and Hopgood, J. P.,** The use of sulfosuccinimidyl-2-(p-azidosalicylamido)-1,3'-dithiopropionate as a crosslinking reagent to identify cell surface receptors, *Anal. Biochem.,* 168, 306, 1988.
38. **Amlaiky, N., Berger, J. G., Chang, W., McQuade, R. J., Caron, M. G.,** Identification of the binding subunit of the D_1-dopamine receptor by photoaffinity crosslinking, *Mol. Pharmacol.,* 31, 129, 1987.

39. **Barrington, W. W., Jacobson, K. A., Hutchison, A. J., Williams, M., and Stiles, G. L.,** Identification of the A$_2$ adenosine receptor binding subunit by photoaffinity crosslinking, *Proc. Natl. Acad. Sci. U.S.A.,* 86, 6572, 1989.

40. **Aiyar, N., Valinski, W., Nambi, P., Minnich, M., Stassen, F. L., and Crooke, S. T.,** Solubilization of a guanine nucleotide-sensitive form of vasopressin V$_2$ receptors from porcine kidney, *Arch. Biochem. Biophys.,* 268, 698, 1989.

41. **Cofano, F., Landolfo, S., Appella, E., and Ullrich, S. J.,** Analysis of murine interferon-γ binding to its receptor on intact cells and solubilized membranes. Identification of an 80 kDa receptor, *FEBS Lett.,* 242, 233, 1989.

42. **Murthy, S. C., Mui, A. L., and Krystal, G.,** Characterization of the interleukin 3 receptor, *Exp. Hematol.,* 18, 11, 1990.

43. **McCartney, S., Aiton, J. F., and Cramb, G.,** Characterization of atrial natriuretic peptide receptors in bovine ventricular sarcolemma, *Biochem. Biophys. Res. Commun.,* 167, 1361, 1990.

44. **Coats, S. R. and Gerard, N. P.,** Characterization of the substance P receptor in guinea pig lung tissues, *Am. J. Respir. Cell. Mol. Biol.,* 1, 269, 1989.

45. **Dufau, M. L., Minegishi, T., Buczko, E. S., Delgado, C. J., and Zhang, R.,** Characterization and structure of ovarian and testicular LH/hCG receptors, *J. Steroid Biochem.,* 33, 715, 1989.

46. **Zhang, Q.-Y. and Menon, K. M. J.,** Characterization of rat Leydig cell gonadotropin receptor structure by affinity cross-linking, *J. Biol. Chem.,* 263, 1002, 1986.

47. **Pillion, D. J., Arnold, P., Yang, M., Stockhard, C. R., and Grizzle, W. E.,** Receptors for insulin and insulin-like growth factor-I in the human adrenal gland, *Biochem. Biophys. Res. Commun.,* 165, 204, 1989.

48. **Hartmann, K., Baier, T. G., Loibl, R., Schmitt, A., and Schonberg, D.,** Demonstration of type I insulin-like growth factor receptors on human platelets, *J. Recept. Res.,* 9, 181, 1989.

49. **McQuade, R., Thomas, C. G., Jr., and Nayfeh, S. N.,** Further studies on the covalent crosslinking of tyrotropin to its receptor. Evidence that both the alpha and beta subunits of thyrotropin are crosslinked to the receptor, *Arch. Biochem. Biophys.,* 252, 409, 1987.

50. **Schvartz, I., Ittoop, O., and Hazum, E.,** Identification of endothelin receptors by chemical cross-linking, *Endocrinology,* 126, 1829, 1990.

51. **Gronemeyer, H., Govindan, M. V.,** Affinity labelling of steroid hormone receptors, *Mol. Cell. Endocrinol.,* 46, 1, 1986.

52. **Levesque, J.-P., Hatzfeld, J., and Hatzfeld, A.,** A mitogenic fibrinogen receptor that differs from glycoprotein IIb-IIIa. Identification by affinity chromatography and by covalent cross-linking, *J. Biol. Chem.,* 265, 328, 1990.

53. **Pepinsky, R. B., Cappiello, D., Wilkowski, C., and Vogt, V. M.,** Chemical crosslinking of proteins in avian sarcoma and leukemia viruses, *Virology,* 102, 205, 1980.

54. **Pinter, A. and Fleissner, E.,** Structural studies of retroviruses: characterization of oligomeric complexes of murine and feline leukemia virus envelope and core components formed upon cross-linking, *J. Virol.,* 30, 157, 1979.

55. **Racevskis, J. and Sarkar, N. H.,** Murine mammary tumor virus structural protein interactions: formation of oligomeric complexes with cleavable cross-linking agents, *J. Virol.,* 35, 937, 1980.

56. **Gebhardt, A., Bosch, J. V., Ziemiecki, A., and Friis, R.,** Rous sarcoma virus p19 and gp35 can be chemically crosslinked to high molecular weight complexes. An insight into virus assembly, *J. Mol. Biol.,* 174, 297, 1984.

57. **Hordern, J. S., Leonard, J. D., and Scraba, D. G.,** Structure of the Mengo virion. VI. Spatial relationships of the capsid polypeptides as determined by chemical cross-linking analyses, *Virology,* 97, 131, 1979.

58. **Markwell, M. A. K. and Fox, C. F.,** Protein-protein interactions within paramyxoviruses identified by native disulfide bonding or reversible chemical cross-linking, *J. Virol.,* 33, 152, 1980.

59. **Maassen, J. A. and Terhorst, C.,** Identification of a cell-surface protein involved in the building site of Sindis virus on human lymphoblastoic cell lines using a heterobifunctional cross-linker, *Eur. J. Biochem.,* 115, 153, 1981.

60. **Hu, V. W. and Wisnieski, B. J.,** Photoreactive labeling of M13 coat protein in model membranes by use of a glycolipid probe, *Proc. Natl. Acad. Sci. U.S.A.,* 76, 5460, 1979.

61. **Kikuchi, G. E., Glorioso, J. C., and Nairn, R.,** Cross-linking studies show that herpes simplex virus type 1 glycoprotein C molecules are clustered in the membrane of infected cells, *J. Gen. Virol.,* 71, 455, 1990.

62. **Zhu, Q. and Clourtney, R. J.,** Chemical crosslinking of glycoproteins on the envelope of herpes simplex virus, *Virology,* 167, 377, 1988.

63. **Schawaller, M., Smith, G. E., Skehel, J. J., and Wiley, D. C.,** Studies with crosslinking reagents on the oligomeric structure of the env glycoprotein of HIV, *Virology,* 172, 367, 1989.

64. **Perkins, M. E. and Rocco, L. J.,** Chemical crosslinking of *Plasmodium falciparum* glycoprotein, Pf200 (190-205 kDa), to the S-antigen at the merozoite surface, *Exp. Parasitol.,* 70, 207, 1990.

65. **Young, E. F., McKee, M. J., Ferguson, D. G., and Kranias, E. G.,** Structural characterization of phospholamban in cardiac sarcoplasmic reticulum membranes by cross-linking, *Memb. Biochem.,* 8, 95, 1989.

66. **James, P., Inui, M., Tada, M., Chiesi, M., and Carofoli, E.,** Nature and site of phospholamban regulation of the Ca^{2+} pump of sarcoplasmic reticulum, *Nature (London),* 342, 90, 1989.

67. **Buschette-Brambrink, S. and Gutensohn, W.,** Human placental ecto-5'-nucleotidase: isoforms and chemical crosslinking products of the membrane-bound and isolated enzyme, *Biol. Chem. Hoppe-Seyler,* 370, 67, 1989.

68. **Xie, G. and Wilson, J. E.,** Tetrameric structure of mitochondrially bound rat brain hexokinase: a crosslinking study, *Arch. Biochem. Biophys.,* 276, 285, 1990.

69. **Middaugh, C. R. and Ji, T. H.,** A photochemical crosslinking study of the subunit structure of membrane-associated spectrin, *Eur. J. Biochem.,* 110, 587, 1980.

70. **Ji, T. H., Kiehm, D. J., and Middaugh, C. R.,** Presence of spectrin tetramer on the erythrocyte membrane, *J. Biol. Chem.,* 255, 2990, 1980.

71. **Leduc, M., Joseleau-Petit, D., and Rothfield, L. I.,** Interactions of membrane lipoproteins with the murein sacculus of *Escherichia coli* as shown by chemical crosslinking studies of intact cells, *FEMS Microbiol. Lett.,* 60, 11, 1989.

72. **de Gunzburg, J., Riehl, R., and Weinberg, R. A.,** Identification of a protein associated with p21[ras] by chemical crosslinking, *Proc. Natl. Acad. Sci. U.S.A.,* 86, 4007, 1989.

73. **Font, B., Eichenberger, D., Goldschmidt, D., and Vial, C.,** Interaction of creatine kinase and hexokinase with the mitochondrial membranes, and self-association of creatine kinase: cross-linking studies, *Mol. Cell Biochem.,* 78, 131, 1987.

74. **Sakai, K., Hayashi, M., Kawashima, S., and Akanuma, H.,** Calcium-induced localization of calcium-activated neutral proteinase on plasma membranes, *Biochim. Biophys. Acta,* 985, 51, 1989.

75. **Salhany, J. M. and Sloan, R. L.,** Partial covalent labeling with pyridoxal 5'-phosphate induces *bis*(sulfosuccinimidyl)suberate crosslinking of band 3 protein tetramers in intact human red blood cells, *Biochem. Biophys. Res. Commun.,* 156, 1215, 1988.

76. **Salhany, J. M. and Sloan, R. L.,** Direct evidence for modulation of porter quaternary structure by transport site ligands, *Biochem. Biophys. Res. Commun.,* 159, 1337, 1989.

77. **Tamura, M., Tamura, T., Burnham, D. N., Uhlinger, D. J., and Lambeth, J. D.,** Stabilization of the superoxide-generating respiratory burst oxidase of human neutrophil plasma membrane by crosslinking with 1-ethyl-3-(3-dimethylaminopropyl) carbodiimide, *Arch. Biochem. Biophys.,* 275, 23, 1989.

Chapter 10

PREPARATION OF ENZYME IMMUNOCONJUGATES AND OTHER IMMUNOASSAY COMPONENTS

I. INTRODUCTION

Since the introduction of radioimmunoassay (RIA) by Yalow and Berson in 1959,[1] immunological methods have prevailed for the quantitation and detection of a wide variety of compounds. However, because of health hazards, difficulty in handling and disposal, as well as short shelf lives of radioisotopes, other labels have been developed to replace RIA. These include enzymes and fluorescent, chromophoric, and spin probes. The extremely fast advancement in this area is witnessed by the publication of several monographs on this subject.[2-7]

That enzyme activity could be used to determine the concentration of analytes was determined as early as 1968 by Miles and Hales.[8] Subsequently, enzyme-labeled antigens and antibodies were used in immunoassays.[9-11] Their use has improved reagent shelf life and avoided radiation hazards.

There are many types of enzyme immunoassays (EIAs) which have been treated in various textbooks and monographs;[2-7,12,13] broadly speaking, they can be divided into homogeneous (or separation free) and heterogeneous (or separation required) assays. The heterogeneous EIA is also known as enzyme-linked immunosorbent assay (ELISA). It encompasses competitive binding assays and immunoenzymometric or sandwich assays. The classical competitive EIA is analogous to traditional RIA. Either enzyme-antigen or enzyme-antibody conjugates are employed to measure the concentration of analytes, as shown in Figure 1. When enzyme-labeled antigen is used, it competes with the analyte for the antibody binding site (Figure 1A). Similarly, when enzyme-labeled antibody is used, the analyte competes for the labeled antibody (Figure 1B). The labeled antigen-antibody complex is separated and assayed for the enzyme activity which is inversely proportional to the analyte concentration.

In the sandwich or immunoenzymometric assays, enzyme-coupled antibodies are used (Figure 2). For detection of antigens, the analyte is sandwiched between two antibodies (Figure 2A). Similarly, for the quantitation of an antibody, the analyte is sandwiched between an antigen and a second antibody (Figure 2B). In either case, the enzyme activity of the isolated sandwich complex is proportional to the concentration of the analyte detected.

Several variations of the immunometric assays have been devised. These include the use of enzyme-coupled species-specific antibodies,[14] enzyme-labeled protein A,[15] and affinity column mediated immunometric assay.[16]

Homogeneous EIAs do not require separation of labeled antigen-antibody complexes. They depend on a change in enzyme-specific activity when enzyme-antigen conjugates bind to the antibody. There are many different versions of homogeneous EIA. The best known example is the enzyme multiplied immunoassay technique (EMIT), as illustrated in Figure 3. A decrease in enzyme activity is seen when an antibody is bound to the enzyme-antigen conjugate. In the presence of the antigen analyte, such complex formation is prevented. Thus, the amount of enzyme activity is proportional to the concentrate of analyte.

Not only does EIA require an enzyme-conjugated antigen and antibody, these conjugates are used in a variety of histochemical and cytochemical studies.[17] For example, immunostaining using horseradish peroxidase-labeled antibody (immunoperoxidase) has become a prevalent technique. Different methods, including direct and indirect immunoperoxidase

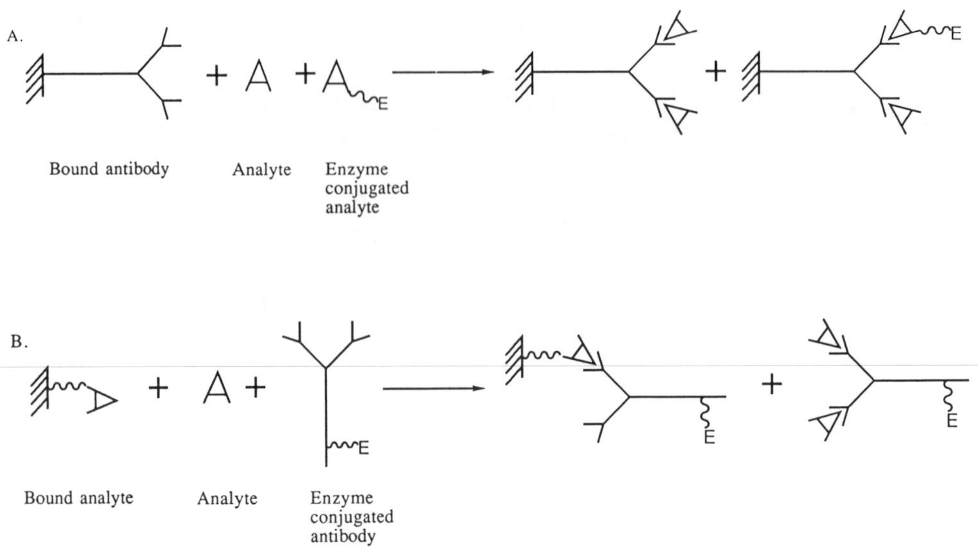

FIGURE 1.   Competitive EIA. (A) Enzyme-antigen conjugate is used to compete for the antibody; (B) enzyme-antibody conjugate is used. After separation, the amount of enzyme-labeled antigen-antibody complex is inversely proportional to the concentration of analyte present in the assay.

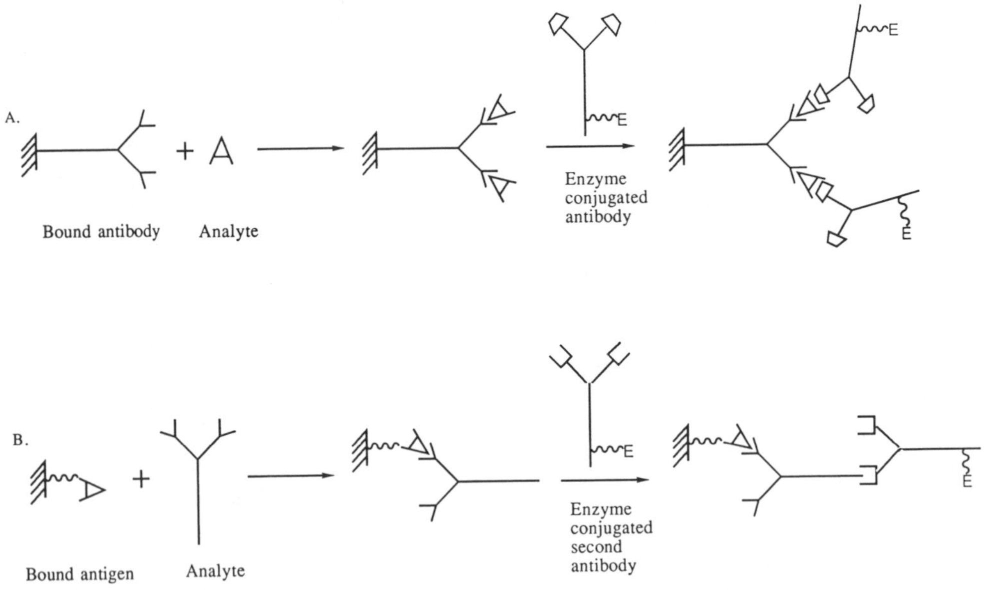

FIGURE 2.   Sandwich or immunoenzymometric assay. (A) Detection for antigen. Antigen analyte is sandwiched between two antibodies, one of which is coupled to enzyme. (B) Assay for antibody. Antibody analyte is sandwiched between immobilized antigen and enzyme-antibody conjugate. The sandwich is isolated and bound enzyme activity is measured.

procedures, two-stage protein A-peroxidase, and antigen-labeled methods have been devised, as shown in Figure 4. The immunoperoxidase methods are also applicable to electron microscopy.

It is obvious that enzyme-antigen and enzyme-antibody conjugates are of paramount importance in immunoassays. The ability to produce active and stable conjugates is critical

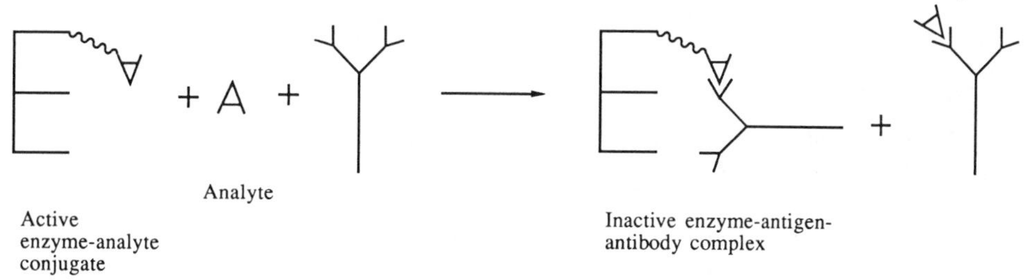

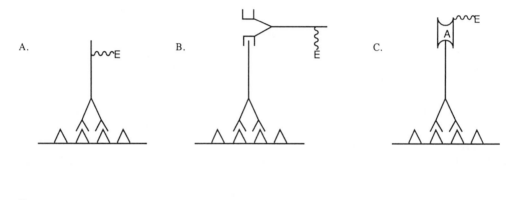

Analyte

Active
enzyme-analyte
conjugate

Inactive enzyme-antigen-
antibody complex

FIGURE 3. Enzyme multiplied immunoassay technique (EMIT). The enzyme activity of enzyme-antigen conjugate is inhibited when antibody is bound to the antigen. Competitive binding of analyte to the antibody protects the inactivation of enzyme.

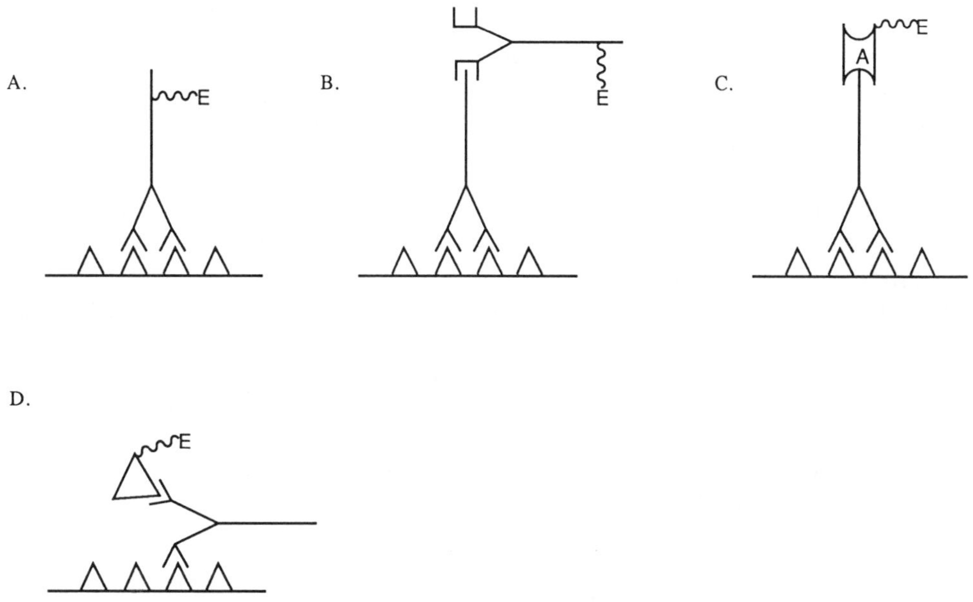

FIGURE 4. Immunoperoxidase staining methods. (A) Direct staining procedure. Peroxidase-antibody conjugate is used. (B) Indirect method. A peroxidase-second antibody conjugate is used. (C) Two-stage protein A-peroxidase method. Peroxidase-protein A conjugate is used. (D) Labeled antigen method. Peroxidase-antigen conjugate is used.

to such analytical techniques. This chapter is devoted to presenting the various versatile methods for the cross-linking of enzymes to antibodies and antigens.

In addition, avidin-biotin interaction has been used to amplify EIA.[18,19] As discussed in Chapter 4, the system is based on the principle that avidin possesses four binding sites and can act as a bridge between two different biotinylated proteins. Similarly, lectins which possess two or more active sites have been used to amplify immunoassays, as alluded to in Chapter 4.[20]

II. COMPONENTS OF ENZYME IMMUNOASSAYS

A. CHOICE OF ENZYMES

Theoretically, any enzyme can be used as a label in EIA. However, certain properties are more desirable than others. These include:

1. High substrate turnover rate, i.e., high specific activity and low Km
2. High stability, i.e., long shelf life
3. An easy, cheap, nontoxic and sensitive assay procedure
4. Reactive groups for coupling to other molecules
5. Cheap and easy purification
6. Enzyme activity not found in test fluids
7. Stable enzyme-labeled conjugates

Over 25 different enzymes have been used as labels in EIA. Some of the characteristics of these enzymes are listed in Table 1. It should be noted that the information given in the table is dependent on the source of the enzyme. The kinetic parameters may vary for enzymes isolated from different species. Readers who are interested should consult the literature for the enzymology of these proteins. The use of these enzymes in EIA is given in References 21 through 47. By far, the most widely employed enzymes are horseradish peroxidase (HRP), alkaline phosphatase (ALP), glucose oxidase (GO), glucose-6-phosphate dehydrogenase (G6PDH), and β-galactosidase (GS). HRP is cheap and readily available in pure form. There are many substrates available. In addition, it has 10 to 15% of carbohydrate which can be used for conjugation. GS has 20 free thiol groups that provide a useful, functional group for coupling reactions. It is gaining popularity because of the availability of fluorogenic substrates. G6PDH has been used in the EMIT type assays. Other enzymes have prevailed because of the reasons cited above. Since it is impossible to provide all the coupling procedures for all the enzymes, only the most commonly used enzymes are described to illustrate the versatility of the conjugation methods.

B. ANTIBODIES AND THEIR FRAGMENTS

The basic unit of an immunoglobulin (Ig) molecule consists of two identical light chains and two identical heavy chains. These chains are held together by noncovalent interactions as well as disulfide bonds. The light chain contains one variable domain and one constant domain, whereas the heavy chain contains a variable domain and three separate constant domains. There are five major classes of Igs: IgA, IgG, IgD, IgM, and IgE; each has different molecular complexities, as shown in Table 2.

Among these Igs, IgG is the most abundant, constituting 8 to 16 mg/ml in human serum. It is therefore the most widely used antibody in immunoassays. IgG is easily purified from serum or ascites by fractionation with sodium sulfate or ammonium sulfate (35 to 45%) followed by DEAE ion exchange chromatography[48,49] or affinity chromatography on a column of protein A,[48,50] protein G,[51] or recombinant protein A/G-Sepharose.[48]

IgG can be further fractionated into its subclasses. For example, mouse IgG can be fractionated into IgG_1, IgG_{2a}, IgG_{2b}, and IgG_3 by various affinity chromatography.[48-52] Unfortunately, it is beyond the scope of this book to cover the procedures for the purification of immunoglobulins. The reader is referred to literature for information.[53]

Fragmentation of IgG into antigen-binding (Fab) and effector-activating (Fc) fragments can be achieved by enzymatic cleavage of the hinge region between constant domains one and two of the heavy chain. Treatment of IgG with papain will produce two Fab and one Fc fragments,[54-57] whereas treatment with pepsin generates only $F(ab')_2$ fragment, as shown in Figure 5.[57-59] $F(ab')_2$ can be further reduced to Fab' in the presence of 2-mercaptoethy-

TABLE 1
Enzymes Used in EIA

Enzyme	Specific activity (U/mg)	K_m (mM)	Mol. wt (kDa)	pH	Ref.
Acetylcholine esterase (EC 3.1.1.7)	11,000	0.09 (acetylcholine)	54	7—8	21
Adenosine deaminase (EC 3.5.4.4)	437	0.04 (adenosine)	110	7—9	22
Alkaline phosphatase (EC 3.1.3.1)	350	0.2[a]	100	8—10	23
α-Amylase (EC 3.2.1.1)	2,240	1g/ml (starch)	97	5—9	24
β-Amylase (EC 3.2.1.2)	1,640	0.07 (amylose)	152	4—6	25
Catalase (EC 1.11.1.6)	9,000,000	1,100 (H_2O_2)	250	6—8	26
Carbonic anhydrase (EC 4.2.1.1)	30,000	2.8 (CO_2)	30	7—8	27
β-Galactosidase (EC 3.2.1.23)	340	1[b]	540	6—8	28
β-Glucosidase (EC 3.2.1.21)	38	0.08[c]	300	6—7	29
Glucose oxidase (EC 1.1.3.4)	80	33 (glucose)	186	4—7	30
Glucose-6-phosphate dehydrogenase (EC 1.1.1.49)	700	0.02[d]	128	7—8	31
Glucoamylase (EC 3.2.1.3)	37	0.03 (amylose)	48	4—6	32
Hexokinase (EC 2.7.1.1)	800	0.1 (glucose)	99	6—8	33
Horseradish peroxidase (EC 1.11.1.7)	4,500	0.2 (H_2O_2)	40	5—7	34
Inorganic phosphatase (EC 3.6.1.1)	1,433	0.05 (pyrophosphate)	63	6—8	35
Invertase (EC 3.2.1.26)	3,000	9.1 (glucose)	270	4—7	36
Δ5,3-Ketosteroid isomerase (EC 5.3.3.1)	88,000	0.3[e]	40	6—8	37
Lactoperoxidase (EC 1.11.1.7)	122	1(H_2O_2)	82	5—8	38
Lactate dehydrogenase (EC 1.1.1.27)	11,500	0.8 (pyruvate)	140	6—8	39
Luciferase (EC 1.13.12.7)	28,000	0.2(ATP) 0.02(luciferin)	100	6—8	40
Lysozyme (EC 3.2.1.17)	2×10^{-5}	4[f]	14.4	4—6	41
Malate dehydrogenase (EC 1.1.1.37)	1,000	0.3(malate) 0.1(NAD^+)	70	8—10	42
Penicillinase (EC 3.5.2.6)	5,400	0.05 (benzylpenicillin)	23	7	44
Phospholipase C (EC 3.1.4.3)	15	0.1 (phosphoinositol)	85	7—8	45
Pyruvate kinase (EC 2.7.1.40)	445	0.07(PEP)	237	6—8	46
Ribonuclease A (EC 3.1.27.5)	12,000	1.4(ApC)	14	6—8	31
Urease (EC 3.5.1.5)	2,130	11(urea)	483	6—8	47

[a] For p-nitrophenylphosphate.
[b] For o-nitrophenyl-β-D-galactopyranoside.
[c] For p-nitrophenyl-β-D-glucopyranoside.
[d] For glucose-6-phosphate.
[e] For Δ5-androstene-3,17-dione.
[f] For p-nitrophenyl-β-D-chitobioside.

lamine.[57,58] Enzyme-Fab' conjugates are more useful than the IgG-enzyme conjugate in both immunohistochemical staining of tissue sections and enzyme immunoassays.[60-62] They also give a lower nonspecific binding and a higher sensitivity in solid phase EIA[3] and more readily penetrate into tissue sections and provide a lower background staining.

Since Fab and F(ab')$_2$ fragments retain their antigen-binding capability, the use of these fragments reduces nonspecific binding due to the removal of Fc. In addition, the generation of free sulfhydryl group in Fab' fragment facilitates some of the coupling reactions.

TABLE 2
Immunoglobulin Classes

Class	IgA	IgG	IgD	IgM	IgE
Heavy chain	α	γ	δ	μ	ϵ
Subclasses	1,2	1,2,3 or 4[a]	None	1,2	None
Light chain	κ,λ	κ,λ	κ,λ	κ,λ	κ,λ
Formula	$\alpha_2\kappa_2$	$\gamma_2\kappa_2$	$\delta_2\kappa_2$	$\mu_{10}\kappa_{10}$	$\epsilon_2\kappa_2$
	$\alpha_2\lambda_2$	$\gamma_2\lambda_2$	$\delta_2\lambda_2$	$\mu_{10}\lambda_{10}$	$\epsilon_2\lambda_2$
	monomer or dimer	monomer	monomer	pentamer	monomer
J-Chain	On dimer	None	None	On pentamer	None
Mol. wt.	400,000	150,000	180,000	900,000	190,000
Human serum conc	1.4—3.5 mg/ml	8—16 mg/ml	0—0.14 mg/ml	0.5—2 mg/ml	<0.3 μg/ml

[a]   Human IgG has 4 subclasses, whereas mouse has IgG_1, IgG_{2a}, IgG_{2b} and IgG_3.

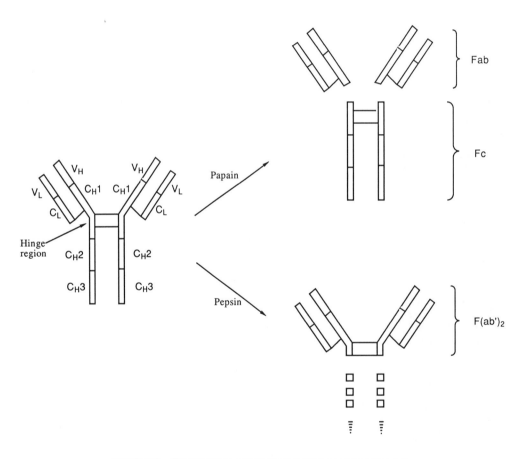

FIGURE 5.   Fragmentation of immunoglobulin by papain and pepsin.

III. CONJUGATION OF ENZYMES TO ANTIBODIES AND THEIR FRAGMENTS

A. INTRODUCTION OF THIOL GROUPS INTO IMMUNOGLOBULINS

Although both enzymes and antibodies contain various functional groups, the most reactive functionality is that of the thiol. The generation of free sulfhydryl groups has been achieved by two different methods. The first involves reductive cleavage of the native cystine residues in the protein with reagents such as dithiothreitol (DTT), and the second involves chemical introduction of thiol groups (thiolation). Reductive cleavage of disulfides can be employed to functionalize antibodies with no free thiol groups. This approach is feasible because reduction conditions can be kept mild enough not to significantly alter the functional features of Igs.[63] The concentration of reducing agents needed varies with the antibody, and the optimum levels have to be individually determined. Thiol groups in IgG can be generated by reduction of disulfide bonds in the hinge region. This is easily achieved by incubating the sample at 37°C for 2 h with 10 mM 2-mercaptoethylamine in 0.1 M sodium phosphate buffer, pH 6.0, containing 5 mM EDTA.[65] One can also reduce F(ab')$_2$ fragments to introduce a free thiol group into Fab'. Reduced Fab' from rabbit will contain one thiol group per molecule. Fab' from other animal sources may contain one to three sulfhydryl groups. The content of thiols can be measured by adding 0.2 mM 4,4'-dithiodipyridine to the sample. After incubation at room temperature for 20 min, the concentration of thiol may be obtained from the absorbance at 324 nm using a molar absorptivity of 19,800 mol^{-1}·cm^{-1}.[64]

Alternatively, thiol groups can be introduced into IgG and F(ab')$_2$ by mercaptosuccinylation. Succinylation is first carried out by incubating IgG or F(ab')$_2$ in 0.1 M sodium phosphate buffer, pH 6.5, with 100-fold molar excess of S-acetylmercaptosuccinic anhydride (previously dissolved in N,N-dimethylformamide) at room temperature for 30 min. Hydroxylamine is then added to attain a final concentration of 0.2 M. Free thiol will be obtained on incubation at 30°C for 5 minutes (see reaction in Chapter 2). The content of thiol groups can be measured as described above. Thiol groups are fairly stable in the presence of EDTA.[66] Other chemical methods of introduction of thiol groups using N-succinimidyl 3-(2-pyridyldithio)propionate (SPSP), 2-iminothiolane, and other compounds have been discussed in Chapter 2. Examples illustrating the involvement of these reactions in the preparation of immunoconjugates will be shown below.

Although various reagents have been used to cross-link the functional groups present in proteins, the most favorable cross-linkers are those directed toward the thiol and amino groups. Theoretically, any heterobifunctional reagents in this category can be used. However, there are certain preferences toward different enzymes and Ig fragments. Detailed experimental methods for the preparation of some of these conjugates are described below.

B. PREPARATION OF HORSERADISH PEROXIDASE IMMUNOCONJUGATES
1. Conjugation with Amino and Thiol Directed Cross-Linkers

Among the heterobifunctional reagents selective toward amino and thiol groups, N-succinimidyl 4-(N-maleimidomethyl)cyclohexane-1-carboxylate (SMCC), and N-succinimidyl 6-maleimidohexanoate (SMH) have been most favorably used to cross-link the amino group of peroxidase with the thiol group of Fab', reduced IgG, or thiolated F(ab')$_2$.[60,67-69] The strategy of the coupling reaction is shown in Figure 6.

In short, HRP (6 mg/ml) in 0.1 M sodium phosphate buffer, pH 7.0, is first reacted with 0.3 to 0.7 mg of cross-linker dissolved in N,N-dimethylformamide (total organic solvent 10%). The reaction is incubated for 0.5 to 1.0 h at 30°C. Labeled HRP can be isolated by gel filtration on Sephadex G-25 (equilibrated with 0.1 M sodium phosphate, pH 6.0), by gravity or centrifugation.[70] The number of maleimide groups introduced into the enzyme can be estimated by first reacting with excess 2-mercaptoethylamine followed by back titration with 4,4'-dithiodipyridine, as discussed above.[64]

Coupling of labeled HRP with Fab', reduced IgG, or thiolated IgG and F(ab')$_2$ can be achieved by incubating the components (1:1 mol/mol; 0.01 to 0.15 mM) in 0.1 M sodium phosphate buffer, pH 6.0, containing 2.5 mM EDTA, at 4°C for 20 h or at 30°C for 1 h. The conjugated product may be separated on LKB Ultrogel AcA 44.

Since the incorporation of cross-linker into HRP is limited and the number of thiols in immunoglobulins is small, this method affords 1:1 conjugates.

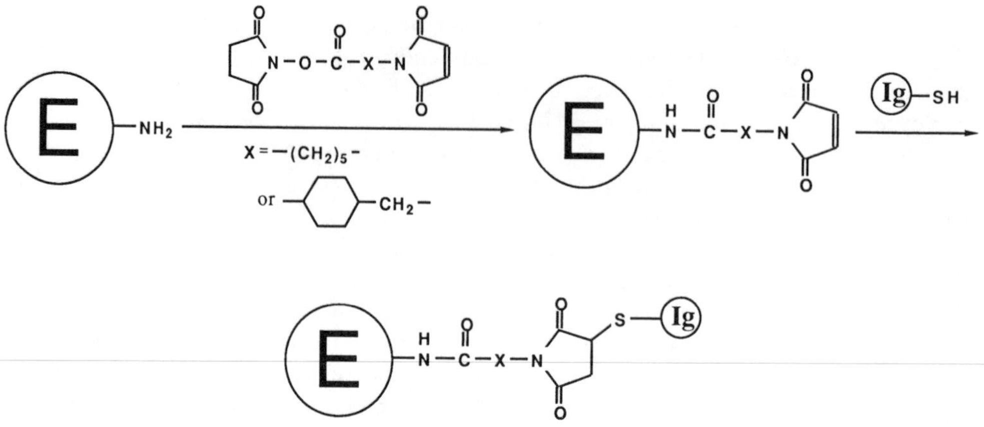

FIGURE 6.   Coupling of horseradish peroxidase (E) to thiol containing immunoglobulin (Ig) with *N*-succinimidyl 4-(*N*-maleimidomethyl) cyclohexane-1-carboxylate (SMCC) or *N*-succinimidyl 6-maleimidylhexanoate (SMH). Glucose-6-phosphate dehydrogenase and alkaline phosphatase may be similarly conjugated.

2. Conjugation Through Disulfide Formation

HRP and IgG or its fragments can be conjugated through a disulfide bond.[60] The reaction of SPDP with HRP will activate the amino groups through labeling with pyridyl disulfide which can react further with free thiol groups of Fab′ or thiolated IgG and F(ab′)$_2$ (Figure 7A). Alternatively, HRP can be thiolated with *S*-acetylmercaptosuccinic anhydride followed by hydroxylamine to generate a free thiol which can react with dithiopyridine introduced into Fab′ or thiolated IgG (Figure 7B).

Introduction of pyridyl disulfide groups into HRP can be achieved by incubating HRP (about 6 mg/ml) with 50-fold molar excess of SPDP (previously dissolved in ethanol) in 0.1 *M* sodium phosphate buffer, pH 7.5, for 30 minutes at 25°C. After the reaction, pyridyl disulfide labeled HRP can be isolated by gel filtration through a Sephadex G-25 column. Coupling of the labeled enzyme to Fab′ is carried out by mixing equimolar concentration of the components in 0.1 *M* sodium phosphate buffer, pH 6, containing 5 m*M* EDTA for 2.5 h at 30°C. The conjugate can be purified by gel filtration.

Thiolation of HRP can be carried out by reacting HRP (8 mg/ml) in 0.1 *M* sodium phosphate buffer, pH 7.5, with 200-fold excess *S*-acetylmercaptosuccinic anhydride (previously dissolved in *N,N*-dimethylformamide). After 30 min at 30°C, hydroxylamine (1:5 v/v) is added to hydrolyze the thioester. Thiolated HRP can be purified by gel filtration in the presence of 5 m*M* EDTA. To prepare for conjugation, Fab′ is activated with 10-fold excess of 4,4′-dithiodipyridine for 15 min and isolated by gel filtration. The two components are cross-linked by mixing similar to the procedure described above. Before purification by gel filtration, *N*-ethylmaleimide is added to terminate the reaction.

3. Conjugation with Glutaraldehyde

Glutaraldehyde readily reacts with amino groups of enzymes, antigens, and antibodies under mild conditions to form stable conjugates. However, the reaction of glutaraldehyde with HRP is much slower than antibodies. Thus a two-step procedure is adapted.[71] First, HRP (50 mg/ml) is modified with 0.2% glutaraldehyde in 0.1 *M*, pH 6.8, at ambient temperature for 18 h. Excess reagent and peroxidase polymers are then removed by gel filtration. The isolated glutaraldehyde-activated HRP (about 10 mg/ml) is then mixed with IgG, F(ab′)$_2$ or Fab′ (pretreated with *N*-ethylmaleimide to block the free thiol group) in a 1:0.2 molar ratio in 0.1 *M* sodium carbonate buffer, pH 9.5, containing 0.15 *M* NaCl. After 24 h at 4°C, L-lysine (15 m*M*) is added to terminate the reaction and the mixture is incubated

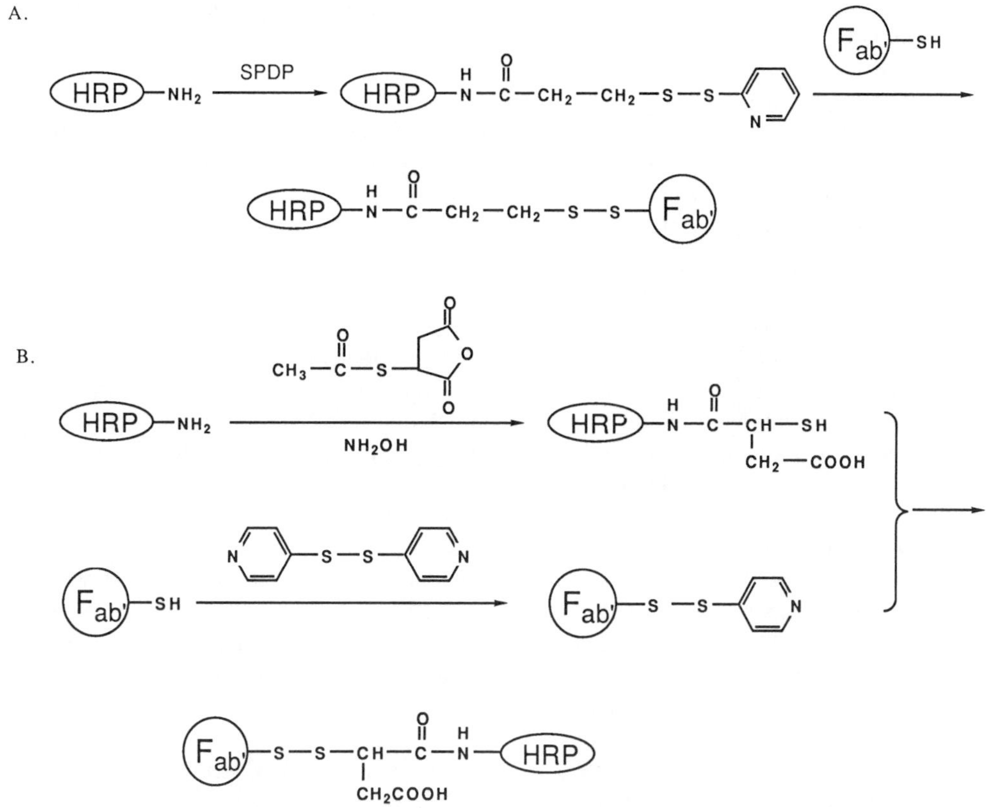

A.

B.

FIGURE 7. Coupling of horseradish peroxidase (HRP) and Fab' through a disulfide bond. (A) HRP is activated by SPDP and then reacts with Fab'; (B) Fab' is activated with dithiodipyridine and then reacts with thiolated HRP.

for another 2 h. Peroxidase-antibody conjugate can be isolated on Ultrogel AcA 44. The conjugate is composed mostly of 1:1 component.[71] However, only a small proportion of HRP is recovered in the conjugate. Some dimerization of glutaraldehyde-activated HRP has been reported.[72,73]

Direct one-step coupling of HRP and antibodies with glutaraldehyde has also been described.[74-76] However, the products are generally heterogeneous, high molecular-weight aggregates with less than 10% recovery of enzyme activity.[75] The two-step procedure is definitely preferred.

4. Conjugation Using Periodate Oxidation

Since HRP is a glycoprotein, oxidation of its carbohydrate moiety with sodium periodate will result in the formation of aldehyde groups which will form Schiff bases with amino groups, as discussed in Chapter 2. The enzyme has few amino groups, but self-coupling can be prevented. When IgG is added, stable conjugates can be obtained after reduction with sodium borohydride or other reducing agents (see Chapter 2).[77]

Basically, HRP (4 gm/ml) is oxidized with 16 mM NaIO$_4$ at room temperature for 10 to 20 min. At the end of that period, ethylene glycol is added to stop the reaction. Excess reagents are removed by either gel filtration with 1 mM sodium acetate buffer, pH 4.4, or dialysis against the same buffer overnight at 4°C. Oxidized peroxidase is then incubated with IgG, F(ab')$_2$, or Fab' (pretreated with N-ethylmaleimide) in a 1:1 molar ratio in 0.1 M sodium carbonate buffer, pH 9.5, containing 0.15 M NaCl, at 25°C for 2 h. The pH is adjusted, if necessary, with 0.2 M sodium bicarbonate. The Schiff base formed is reduced by adding freshly prepared sodium borohydride (final concentration about 0.2 mg/ml) and incubated at 4°C for 2 h. Peroxidase-Ig conjugate can be isolated by gel filtration on a column of Ultrogel AcA 44.

Tijssen and Kurstak[78] designed a simplified version of the periodate method. In this procedure, HRP (1 mg/ml) is oxidized with 4 to 8 mM sodium periodate in 0.1 M sodium carbonate in a sealed tube. After 2 h at ambient temperature, excess IgG (1:7 mg/mg) is added followed by dry Sephadex (one sixth combined weight of peroxidase and IgG) to destroy unreacted periodate. This mixture is incubated for 3 h at ambient temperature. Finally, the proteins are eluted from the Sephadex and reduced with sodium borohydride in two additions at 30 min intervals (total 0.5 mg/ml). Peroxidase-IgG conjugate is purified by first precipitation with 50% ammonium sulfate and then on Concanavalin A affinity column from which peroxidase-IgG can be eluted with 0.05 M α-methyl-D-mannopyranoside.

To prevent self-coupling, the reactive amino groups of HRP can be first irreversibly blocked by alkylation with dinitrofluorobenzene (DNFB).[79,80] After activation by periodate the amino group blocked HRP is then coupled to IgG, F(ab')$_2$, or Fab', as discussed above. With this procedure, a maximum of 5 to 6 mol of HRP could be bound per mole of IgG.[79,80]

5. Conjugation with Miscellaneous Cross-Linkers

Many other cross-linking reagents have been used to couple HRP to antibodies. However, most of these either give low yield or cause considerable loss of activity and are of only historical interest. For example, the use of *p,p'*-difluoro-*m,m'*-dinitrophenyl sulfone gives less than 1% yield.[81] Other reagents include cyanuric chloride,[82] toluene diisocyanate,[83,84] water soluble carbodiimides,[84,85] and *p*-benzoquinone.[86]

C. PREPARATION OF ALKALINE PHOSPHATASE IMMUNOCONJUGATES
1. Conjugation with Amino and Thiol Directed Reagents

Like horseradish peroxidase, alkaline phosphatase can be coupled to antibodies by SMCC or SMH, according to the scheme shown in Figure 6. ALP from calf intestine (4 mg/ml) is first labeled with the cross-linker (1:25 mol/mol) in 50 mM sodium borate buffer, pH 7.6, containing 1 mM MgCl$_2$ and 0.1 mM ZnCl$_2$, at 30°C for 30 min. The labeled enzyme is then isolated by gel filtration.[70] About 4 to 6 cross-linkers can be introduced per mol of enzyme, but up to 40 to 50% of activity may be lost.

Fivefold excess of Fab' (or reduced IgG, mercaptosucciniminidyl IgG, or F (ab')$_2$) is then incubated with labeled alkaline phosphatase (2 mg/ml) at 4°C for 20 h. The reaction is terminated by incubating with 0.2 mM 2-mercaptoethylamine for 20 min at room temperature. ALP immunoconjugates are isolated by gel filtration on Ultrogel AcA 34. A significant proportion of Ig remains unconjugated. To correct this situation, excess labeled alkaline phosphatase may be used. Complete conversion will avoid the difficulty in separating free IgG from the conjugate.

Recently, Jeanson et al.[87] have used SPDP to conjugate ALP and IgG. The reaction scheme is shown in Figure 8. The principle involves reaction of both ALP and IgG separately with excess SPDP for 30 min at room temperature. Excess reagent and free products are removed by dialysis. DTT (25 mM) is added to the labeled ALP to expose the sulfhydryl group. After 20 min, excess DTT is removed by gel filtration on Sephadex G-25. Modified IgG and ALP are mixed and incubated for 20 h. The conjugates can be isolated by gel filtration. The authors claimed that this method yielded better performing conjugates.

2. Conjugation with Glutaraldehyde

Both one-step and two-step procedures of coupling ALP to antibodies have been used. In the one-step method, the enzyme and IgG (2:1 wt/wt) or Fab' (4:1 wt/wt) are mixed with

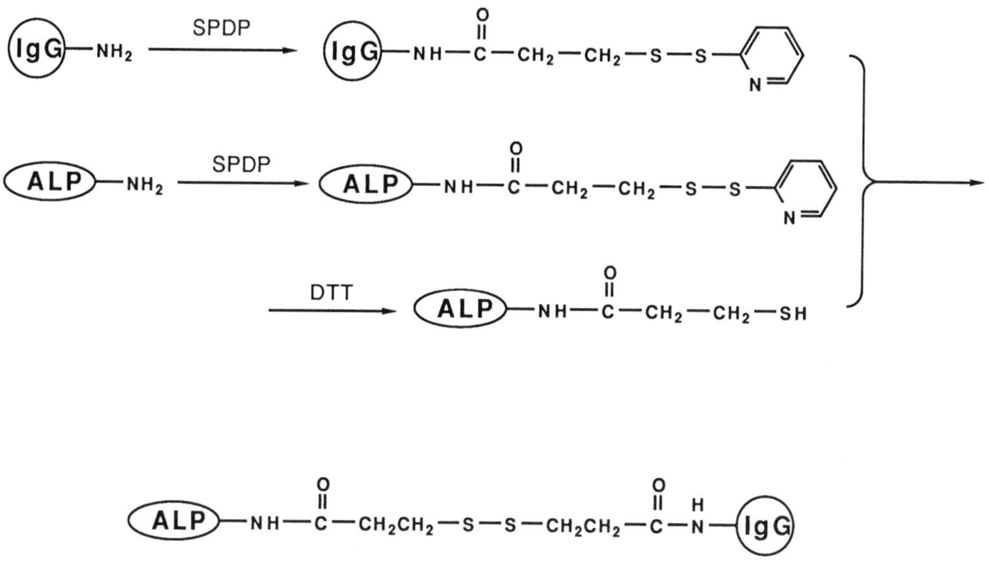

FIGURE 8. Conjugation of alkaline phosphatase (ALP) and IgG with SPDP.

0.2% (v/v) glutaraldehyde in 0.1 M phosphate buffer, pH 6.8, for 2 h at ambient temperature.[76,87,88] L-lysine (0.1 M) is used to block the reactive sites.[74,76] However, the conjugates are heterogeneous aggregates and the recovery of enzyme activity is low.[88] For the two-step procedure, ALP (2.5 mg/ml) in 50 mM potassium phosphate buffer, pH 7.2, is first treated with 0.2% glutaraldehyde for 50 min at 24°C and then mixed with IgG (1 mg/ml). The mixture is incubated for 75 min at 24°C before dialysis.

3. Conjugation with Periodate Oxidation

The procedure for coupling ALP to IgG with sodium periodate is similar to that for HRP.[87] ALP is first treated with DNFB (0.1%) for 2 h at 23°C to block its amino groups. Sodium periodate (20 mM) is then added. After 6 h, ethylene glycol is added to terminate the reaction. Free reagents are removed by dialysis. The oxidized ALP is then mixed with IgG and incubated for 24 h at 4°C. Monoethanolamine is then added and the conjugate dialyzed to get rid of free agents.

D. PREPARATION OF β-D-GALACTOSIDASE IMMUNOCONJUGATES

1. Conjugation with Amino and Thiol Directed Reagents

In addition to SMCC and SMH, other amino and thiol group directed reagents such as maleimidobenzoyl-N-hydroxysuccinimide ester (MBS) have been used to cross-link GS to IgG. These reagents are of particular value when one of the proteins involved has no free thiol groups, e.g., IgG. Although maleimide groups attached to benzene rings are labile at neutral pH,[68] MBS has been successfully applied to GS at pH 6. The principle of the reaction is shown in Figure 9. Since GS contains free thiol groups, IgG is first modified with MBS. IgG (1 mg/ml) in 0.1 M phosphate buffer, pH 6.0, containing 0.05 M sodium chloride, is first reacted with 20 mg/ml MBS (dissolved in dioxin, final organic solvent 10%) at 25°C for 1 h. Labeled IgG is removed from unreacted reagent by gel filtration on Sephadex G-25, equilibrated with 0.1 M phosphate buffer, pH 6.0, containing 20 mM magnesium chloride and 0.05 mM sodium chloride, and incubated with equal weight of GS. After 1 h at 30°C, the reaction is stopped by adding a reducing agent such as 10 mM 2-mercaptoethanol.[89] The conjugate may be purified by gel filtration on Ultrogel AcA 34.

GS-antibody conjugates are also prepared using N-succinimidyl 4-maleimidobutyrate (GMBS).[90] The reaction is similar to MBS except that GMBS is used.

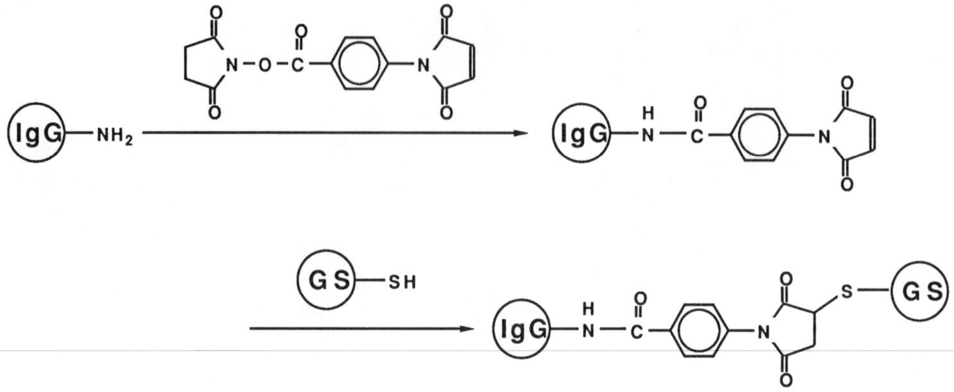

FIGURE 9.   Conjugation of β-D-galactosidase (GS) and IgG with maleimidobenzoyl-*N*-hydroxysuccinimide ester (MBS).

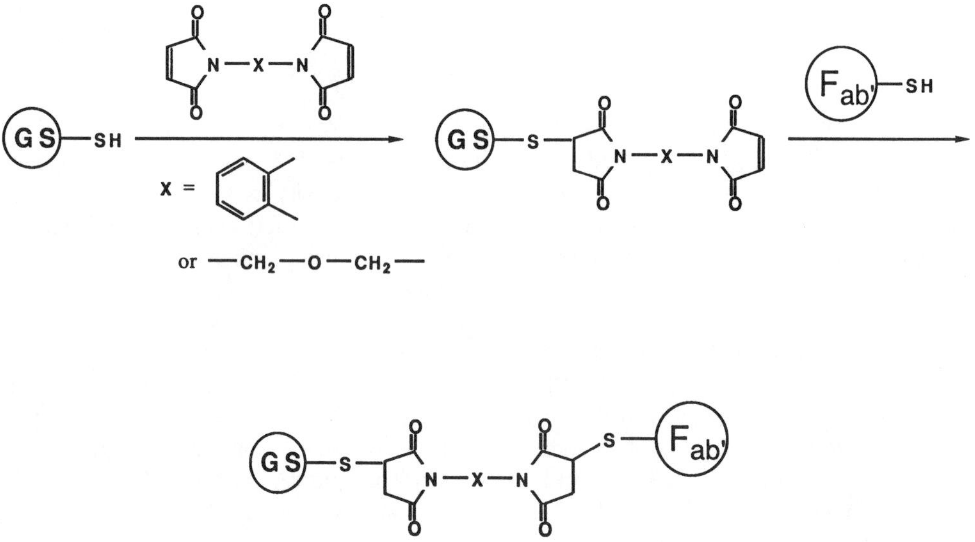

FIGURE 10.   Conjugation of β-D-galactosidase (GS) and Fab′ with dimaleimide, *N*,*N*′-*o*-phenylenedimaleimide or *N*,*N*′-oxydimethylene dimaleimide.

2. Conjugation with Thiol Group Directed Dimaleimides

Coupling the thiol groups of GS and Fab′, reduced or thiolated IgG or F(ab′)$_2$ can be achieved with dimaleimide bifunctional reagents such as *N*,*N*′-*o*-phenylenedimaleimide or *N*,*N*′-oxydimethylenedimaleimide.[91,92] Either the enzyme or the immunoglobulin can be labeled first according to the scheme depicted in Figure 10. For example, the enzyme (5 mg/ml) is reacted with 0.5 mg/ml cross-linker (previously dissolved in *N*,*N*′-dimethylfor-mamide) in 0.1 *M* sodium phosphate buffer, pH 6.0. After incubation at 30°C for 20 min, the labeled enzyme is isolated by gel filtration. In general, about 13 to 16 maleimide groups may be introduced per GS molecule with less than 10% loss of enzyme activity.

The labeled enzyme (1 mg/ml) is incubated with Fab′, reduced or thiolated IgG, or F(ab′)$_2$ (about 1:4 mol/mol) in 0.1 *M* sodium phosphate buffer, pH 6.0, containing 2.5 m*M* EDTA, at 4°C for about 20 h. GS-antibody conjugates may be isolated by gel filtration on Sepharose 6B. The use of excess Ig to completely convert GS to the conjugate is important, since the enzyme with a molecular weight of 540,000 cannot be easily separated from the conjugate by gel filtration.

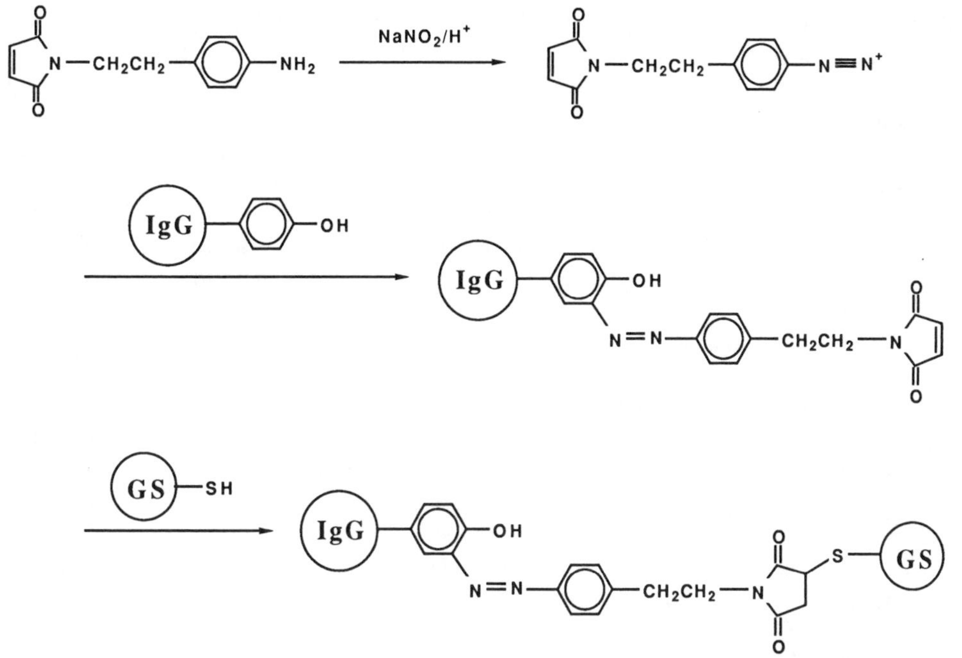

FIGURE 11. Conjugation of IgG and β-D-galactosidase with N-[β-(4-diazophenyl)ethyl]maleimide (DPEM).

3. Conjugation with Phenolate and Thiol Group Directed Reagent

Fujiwara et al.[90] used N-[β-(4-diazophenyl)ethyl]maleimide (DPEM) to conjugate IgG and GS. The diazo functional group will react selectively with tyrosine and histidine residues and the maleimide group with thiol groups, as shown in Figure 11. DPEM is derived from N-[β-(4-aminophenyl)ethyl]maleimide with sodium nitrite and acetic acid. The diazotized compound reacts with IgG through the diazo functional group at pH 9 for 20 min at 20°C. The labeled antibody is isolated by gel filtration and then mixed with GS. Conjugation is complete after 40 min at 30°C or can be terminated with mercaptoethanol.

4. Conjugation with Glutaraldehyde

The one-step procedure of glutaraldehyde coupling has been applied to conjugate GS to antibodies.[94] However, due to the heterogeneous cross-linking nature of the procedure, the use of the method is limited.

E. PREPARATION OF GLUCOSE-6-PHOSPHATE DEHYDROGENASE IMMUNOCONJUGATES

Similar to horseradish peroxidase and alkaline phosphatase, SMCC and SMH have been used to couple G6PDH to Ig, according to the same scheme shown in Figure 6. The enzyme (1 mg/ml) is reacted with the cross-linker (1:2000 mol/mol) in 0.2 M, sodium phosphate buffer, pH 7.0, containing 4.8 mM glucose-6-phosphate and 2 mM NAD. After 30 min at 30°C, the labeled enzyme is isolated by gel filtration. An average of about five cross-linkers are introduced per mol of enzyme, but the enzyme may lose 50 to 70% of its activity.[3]

The labeled enzyme (0.3 to 0.5 mg/ml) is then incubated with Fab' (1:10) in 0.1 M sodium phosphate buffer, pH 6.0, containing 5 mM EDTA at 4°C for 20 h. The reaction is terminated by adding 1.5 mM 2-mercaptoethylamine, and the conjugate can be isolated on Ultrogel AcA 34.

F. PREPARATION OF GLUCOSE OXIDASE IMMUNOCONJUGATES

1. Coupling with *N*-Ethoxycarbonyl-2-Ethoxy-1,2-Dihydroquinoline (EEDQ)

GO can be coupled to IgG through the activation of its carboxyl groups with EEDQ, according to Gueson.[95] The amino groups of GO are first biotinylated (or blocked with DNFB) by incubating with D-biotinyl-*N*-hydroxysuccinimide ester (dissolved in *N,N'*-di-methylformamide) (1:400 mol/mol, organic solvent 10%) in 0.1 M NaHCO$_3$ containing 0.15 M NaCl, for 1 h at ambient temperature. After dialysis against 0.1 M KH$_2$PO$_4$ containing 0.15 M NaCl, 10 mg EEDQ in 0.05 ml dimethylformamide is added and incubated at ambient temperature for 1 h. The labeled enzyme is separated by gel filtration on Sephadex G-25 and mixed with 1 mg IgG in potassium phosphate saline buffer, pH 8.0, at 4°C for 24 h. The immunoconjugate is then dialyzed or purified.

2. Coupling with Amino and Thiol Directed Reagents

GO has been coupled to Fab', reduced IgG, or thiolated antibody by SMCC, as described for peroxidase. The enzyme is first treated with SMCC to introduce maleimide groups which react with the thiol group of Ig. Because of the number of reactive amino groups and the high molecular weight of GO (153,000 Da), excess antibody is used to convert all maleimide labeled GO to the conjugate. Conjugated and unconjugated GO are difficult to separate on gel filtration.[60] In place of SMCC, MBS, SMH, and SPDP may be used to prepare GO-Ig.

3. Coupling with Other Cross-Linkers

GO immunoconjugates have also been prepared by the one-step procedure with glutaraldehyde.[74,76] However, the recovery of the enzyme is low with extensive polymerization of IgG. Cyanuric chloride has also been used to couple IgG to GO, but the yield of the conjugate is low and loss of enzyme activity is often noted.[96] Periodate oxidation of the polysaccharide moiety of GO to aldehydes for cross-linking with amino groups of Ig is also applicable.[97]

G. PREPARATION OF OTHER ENZYME IMMUNOCONJUGATES

Various methods have been employed to prepare other immunoconjugates. Glutaraldehyde is probably the most widely used reagent.[74] Acetylcholinesterase,[21] inorganic pyrophosphatase,[35] lactoperoxidase,[38,98] lactate dehydrogenase,[39] glucoamylase,[32] ribonuclease,[99] and urease[47] immunoconjugates have been prepared with this reagent. Periodate oxidation is applicable to cross-link glucoamylase and IgG. Succinate *bis* (*N*-hydroxysuccinimide ester) was used to couple microperoxidase to IgG. Water soluble carbodiimide, 1-cyclohexyl-3-(2-morpholinyl-4-ethyl) carbodiimide, has been used to link phospholipase C and IgG.[45] Phospholipase C immunoconjugate has also been prepared with SMCC.[100] The same reagent was also used to conjugate Fab' and α-amylase.[24] It should be realized that a diversity of reagents can be used to prepare the immunoconjugates.

IV. COUPLING PROTEINS TO ANTIBODIES AND THEIR FRAGMENTS

Many proteins, in addition to enzymes, have been conjugated to Ig for various forms of immunoassays. Bovine serum albumin (BSA) is conjugated to IgG in a one-step reaction by mixing IgG (2 mg) and BSA (5 mg) with glutaraldehyde (0.1%) in 0.1 M phosphate buffer, pH 6.8, for 3 h at room temperature. Glycine (0.1 M) is added to terminate the reaction.[95]

Agglutinins of *Canavalia ensiformis* (concanavalin A), *Triticum valgare*, *Lens culinaris*, and *Ricinus cummunis* (4 mg/ml) are coupled to antibodies (2 mg/ml) with 0.02% glutaraldehyde in 0.1 M phosphate buffer, pH 6.8, containing 0.1 M lectin-specific saccharide, e.g., methyl-α-D-mannoside, galactose, and *N*-acetyl glucosamine.[95]

Phycobiliproteins from a variety of algae and cryptomonads have been coupled to antibodies by SPDP,[101,102] SMCC[103] and succinimidyl-4-(p-maleimidophenyl)butyrate (SMPB).[103] Free thiol groups may be introduced into both IgG and phycobiliproteins with iminothiolane,[101] S-acetylmercaptosuccinic anhydride,[104] SPDP,[102] or by mild reduction of endogeneous disulfides.[105] These proteins may be cross-linked through a disulfide bond similar to the scheme shown in Figure 7.

Ferritin-antibody conjugates have been prepared by various cross-linkers. These include p,p'-difluoro-m,m'-dinitrophenyl sulfone,[106] toluene diisocyanate,[107] and bis-diazotized benzidine.[108] Urokinase-type plasminogen activator has been cross-linked to F(ab')$_2$ with SPDP.[109]

V. COUPLING ENZYMES TO PROTEINS AND ANTIGENS

A. CONJUGATION OF ENZYMES AND PROTEINS

As mentioned in Chapter 4, the avidin-biotin system provides another dimension of EIA. Labeling of enzymes or other proteins with biotin is easily achieved by incubation with biotinyl-N-hydroxysuccinimide ester which is readily available commercially. The conjugation of avidin with enzymes can be attained by different cross-linkers. Thiolation of avidin can be carried out with 20-fold excess of S-acetylmercaptosuccinic anhydride dissolved in N,N-dimethylformamide (final organic solvent less than 10%) in 0.2 M sodium phosphate buffer, pH 7.0. After 30 min at 30°C, free thiol groups are generated by adjusting to 0.2 M hydroxylamine-HCl and 5 mM EDTA at pH 7.0. An average of 1.6 thiol groups are introduced per molecule of avidin.

Coupling the thiolated avidin to horseradish peroxidase can be carried out by any amino and thiol group directed cross-linkers, as described above. Avidin has also been coupled to enzymes using the one-step glutaraldehyde procedure.

Other proteins that have been coupled to enzymes include protein A and ferritin. Both have been used as a powerful tool in cytochemistry and immunochemistry. Coupling of alkaline phosphatase to protein A can be achieved by either the one- or two-step glutaraldehyde procedure.[110] In the one-step procedure, alkaline phosphatase (5 mg/ml) and protein A (5 mg/ml) in phosphate-buffered saline, pH 7.2, is added to glutaraldehyde in a final concentration of 0.2 to 0.5%. After incubation at room temperature for 2 to 3 h, the reaction is stopped by dialysis. In the two-step procedure, alkaline phosphatase (5 mg/ml) in phosphate buffered saline is first treated with 1% glutaraldehyde for 16 h. After removal of excess glutaraldehyde, protein A (5 mg/ml) is added and incubated for 16 h. The conjugate is fractionated by gel filtration on Ultragel.

A conjugate of HRP with ferritin is obtainable by periodate oxidation of the carbohydrate moiety of the enzyme, as described earlier.[111]

B. CONJUGATION OF ENZYMES AND ANTIGENS

Enzyme-antigen conjugates are used in EIA, particularly in the homogeneous immunoassays. Coupling of small organic antigens to enzymes is generally achieved by introducing active functional groups into the antigen through organic synthesis. Functional groups such as acyl chloride, reactive esters, and anhydrides are frequently employed. Occasionally, bifunctional cross-linkers may be used. Bifunctional reagents are particularly useful for antigens of larger molecular weight where organic synthetic modification is difficult. These antigens can be coupled to enzymes in the same way as IgG. For example, SMCC is used to couple vitamin B$_{12}$ to GS enzyme-donor fragment.[112] The maleimide group is introduced into B$_{12}$ by first reacting with SMCC. Conjugation is complete with the reaction between the thiol groups of the enzyme and the maleimide. The same reagent is used to couple acetylcholinesterase to substance P and atriopeptide. DPEM is used to conjugate neurotensin to GS. Neurotensin is first modified with DPEM and then coupled to GS.[113] GS has also

TABLE 3
Characteristics of Various Coupling Methods

Method	Conjugate	Polymerization	Molar ratio
NH$_2$/SH[a]	HRP-Fab'	No	1:1
NH$_2$/SH[a]	GS-Fab'	No	1—3:1
NH$_2$/SH[a]	ALP-Fab'	No	1—3:1
Glutaraldehyde	HRP-Fab'	No	1:1
	GS-Fab'	Yes	Polymer
	ALP-Fab'	Yes	Polymer
Periodate	HRP-Fab'	Mix	Polymer and monomer
NH$_2$/SH[a]	HRP-IgG	No	1—3:1
SPDP	GS-IgG	No	1—3:1
SPDP	ALP-IgG	No	1:1—3

[a] Conjugation by amino and thiol directed reagents.

Adapted from Ishikawa, E., Hashida, S., Kohno, T., Ranaka, K., in *Nonisotopic Immunoassay,* Ngo, T. T., Ed., Plenum Press, New York, 1988, 27.

been conjugated to insulin by the dimaleimide cross-linkers.[93] Sulfhydryl groups can be introduced into insulin by mercaptosuccinylation, as described earlier. Other reagents, such as difluoronitrobenzene, have been used to couple succininyl-cAMP to acetylcholinesterase, dicyclohexylcarbodiimide for conjugation of succinyl-GMP and acetylcholinesterase.[114] Gentamicin is coupled to catalase by water-soluble carbodiimide, 1-ethyl-3-(3-dimethylaminopropyl) carbodiimide.[26] Many other antigens have been coupled to enzymes and, in fact, proteins using various cross-linkers.

VI. CHARACTERIZATION OF CONJUGATION METHODS

Since functional groups are present in proteins, whether it be an enzyme or Ig, it is evident that specific coupling producing an exclusive homogeneous enzyme-protein conjugate is almost impossible. In fact, all conjugation reactions produce a mixture of heterogeneous immunoconjugates, such as enzyme-enzyme, enzyme-Ig and Ig-Ig conjugates, in addition to multiconjugates. Theoretically, the two-step procedure (see Chapter 7) provides better control of the reaction. Homogeneous conjugates or conjugates of limited heterogeneity may be expected.

Because of the limited number of amino groups (two to three) on HRP available for reaction, HRP immunoconjugates prepared by amino and thiol specific reagents largely contain a one-to-one ratio.[60,61] Similar conjugates are obtained using glutaraldehyde. However, the periodate oxidation method yielded polymers as well. Blockage of amino groups with DNFB before coupling greatly improves the result. Immunoconjugates of other enzymes are generally heterogeneous. GO, for example, has many reactive amino groups. The resulting conjugates consist of molecules with various numbers of Ig. Table 3 shows some of the complications. Although the enzymes such as GS and ALP are not polymerized with various cross-linkers, they are polymerized using glutaraldehyde.[60] When this occurs the yield is always low. Monomeric conjugates are best obtained with other methods.[115,116]

Various immunoconjugates prepared by different methods have been compared.[60,75,117,118] However, because of the diversity of the conditions used, it is difficult to access the validity of these comparisons. For example, Beyzavi et al.[117] found the periodate method to be most effective for preparing HRP-IgG conjugate, particularly at low pH, and the glutaraldehyde method to be more effective than periodate for conjugating ALP to antibodies. Ishikawa et al.,[118] on the other hand, reported polymerization of HRP with the periodate method. The results of conjugation are procedure dependent. The choice of a suitable reaction condition is important. In general, the methods described here provide relatively good results.

The stability of immunoconjugates has been studied.[118,119] Enzyme immunoconjugates have been reported to remain active after 3 to 4 years when stored at 4°C in the presence of 1 mg/ml bovine serum albumin.[118] Montoya and Castell[119] estimated the half-life of the whole HRP-IgG activity to be 9 years at 4°C. As revealed in Chapter 7, cross-linking generally increases the life expectancy of enzymes.

REFERENCES

1. **Yalow, R. S. and Berson, S. A.,** Assay of plasma insulin in human subjects by immunological methods, *Nature,* 184, 1648, 1959.
2. **Kricka, L. J.,** *Ligand-Binder Assays. Labels and Analytical Strategies,* Marcel Dekker, New York, 1985.
3. **Ngo, T. T., Ed.,** *Nonisotopic Immunoassay,* Plenum Press, New York, 1988.
4. **Maggio, E. T., Ed.,** *Enzyme-Immunoassay,* CRC Press, Boca Raton, FL, 1980.
5. **Kurstak, E.,** *Enzyme Immunodiagnosis,* Academic Press, New York, 1986.
6. **Ngo, T. T. and Lenhoff, H. M., Eds.,** *Enzyme-Mediated Immunoassay,* Plenum Press, New York, 1985.
7. **Engvall, E. and Pesce, A. J., Eds.,** *Quantitative Enzyme Immunoassays,* Blackwell Scientific, London, Great Britain, 1978.
8. **Miles, L. E. M. and Hales, C. N.,** Labelled antibodies and immunological assay systems, *Nature,* 219, 186, 1968.
9. **Engvall, E. and Perlmann, P.,** Enzyme-linked immunosorbent assay (ELISA). Quantitative assay for immunoglobulin G, *Immunochemistry,* 8, 871, 1971.
10. **Engvall, E. and Perlmann, P.,** Enzyme-linked immunosorbent assay (ELISA). III. Quantitation of specific antibodies by enzyme-labeled anti-immunoglobulin in antigen-coated tubes, *J. Immunol.,* 109, 129, 1972.
11. **Van Weemen, B. K. and Schuurs, A. H. W. M.,** Immunoassay using antigen enzyme conjugates, *FEBS Lett.,* 15, 232, 1971.
12. **Kaplan, L. A. and Pesce, A. J.,** *Clinical Chemistry: Theory, Analysis, and Correlation,* C. V. Mosby Company, St. Louis, MO, 1989, 191.
13. **Tietz, N. W., Ed.,** *Textbook of Clinical Chemistry,* W. B. Saunders, Philadelphia, 1986, 227.
14. **Voller, A., Bartlett, A., and Bidwell, D. E.,** Enzyme immunoassay with special reference to ELISA techniques, *J. Clin. Pathol.,* 31, 507, 1978.
15. **Schuurs, A. H. W. and Van Weemen, B. K.,** Enzyme-immunoassay: a powerful analytical tool, *J. Immunoassay,* 1, 229, 1980.
16. **Freytag, J. W.,** Affinity column mediated immunoenzymometric assays, in *Enzyme-Mediated Immunoassay,* Ngo, T. T. and Lenhoff, H. M., Eds., Plenum Press, New York, 1985, 277.
17. **Falini, B. and Taylor, C. R.,** New developments in immunoperoxidase techniques and their application, *Arch. Pathol. Lab. Med.,* 107, 105, 1983.
18. **Berman, J. W. and Bash, R. S.,** Amplification of the biotinavidin immunofluorescence technique, *J. Immunol. Methods,* 36, 335, 1980.
19. **Kendall, C., Ionescu-Matin, I., and Dreesman, G. R.,** Utilization of the biotin/avidin system to amplify the sensitivity of the enzyme-linked immunosorbent assay (ELISA), *J. Immunol. Methods,* 56, 329, 1983.
20. **Guesdon, J.-L. and Avrameas, S.,** Lectin-immunotests: quantitation and titration of antigens and antibodies using lectin-antibody conjugates, *J. Immunol. Methods,* 39, 1, 1980.
21. **Van der Waart, M. and Schuurs, A. H. W. M.,** Towards the development of a radioenzyme-immunoassay (REIA), *J. Anal. Chem.,* 279, 142, 1976.
22. **Gebauer, C. R. and Rechnitz, G. A.,** Immunoassay studies using adenosine deaminase enzyme with potentiometric rate measurement, *Anal. Lett.,* 14, 97, 1981.
23. **Mottolese, M., Lacona, A., and Natali, P. G.,** Use of protein A bearing *Staphylococcus aureus* in enzyme immunoassay (EIA), *Immunol. Commun.,* 9, 379, 1980.
24. **Shimura, T., Nakamura, T., Kawakami, A., Haga, M., and Kato, Y.,** A new type of enzyme immunosensor using antigen-bound membrane and multivalent antibody (Fab'-α-amylase conjugate), *Chem. Pharm. Bull. (Tokyo),* 34, 5020, 1986.
25. **Oellerich, M.,** Enzyme immunoassays in clinical chemistry: present status and trends, *J. Clin. Chem. Clin Biochem.,* 18, 197, 1980.

26. **Mattiasson, B., Svensson, K., Borrebaeck, C., Jonsson, S., and Kronvall, G.,** Non-equilibrium enzyme immunoassay of gentamicin, *Clin Chem.,* 24, 1770, 1978.

27. **O'Sullivan, M. J., Bridges, J. W., and Marks, V.,** Enzyme immunoassay: a review, *Ann. Clin. Biochem.,* 16, 221, 1979.

28. **Ishikawa, E. and Kato, K.,** Ultrasensitive enzyme immunoassay, *Scand. J. Immunol.,* 8 (Suppl. 7), 43, 1978.

29. **Scharpe, S. L., Cooreman, W. M., Bloome, W. J., and Laekeman, G. M.,** Quantitative enzyme immunoassay: current status, *Clin. Chem.,* 22, 733, 1976.

30. **Maiolini, B., Ferrua, B., and Masseyeff, R.,** Enzyme immunoassay of human alpha-fetoprotein, *J. Immunol. Methods,* 6, 355, 1975.

31. **Ullman, E. F. and Maggio, E. T.,** Principles of homogeneous enzyme-immunoassay, in *Enzyme Immunoassay,* Maggio, E. T., Ed., CRC Press, Boca Raton, FL, 1980, 105.

32. **Ishikawa, E.,** Enzyme immunoassay of insulin by fluorimetry of the insulin-glucoamylase complex, *J. Biochem.,* 73, 1319, 1973.

33. **Litman, D. J., Hanlon, T. M., and Ullman, E. F.,** Enzyme channelling immunoassay: a new homogeneous enzyme immunoassay technique, *Anal. Biochem.,* 106, 223, 1980.

34. **Van Weeman, B. K., and Schuurs, A. H. W. M.,** Immunoassay using antigen-enzyme conjugates, *FEBS Lett.,* 15, 232, 1971.

35. **Baykov, A. A., Kasho, V. N., and Avaeva, S. M.,** Inorganic pyrophosphatase as a label in heterogeneous enzyme immunoassay, *Anal. Biochem.,* 171, 271, 1988.

36. **Pain, D. and Surolia, A.,** Protein A-enzyme monoconjugate as a versatile tool for enzyme immunoassays, *FEBS Lett.,* 107, 73, 1979.

37. **Terouanne, B., Nicolas, J. C., Descomps, B., and Crastes de Paulet, A.,** Coupling of 5,3-ketosteroid isomerase to human placental lactogen with intermolecular disulfide bond formation: use of this conjugate for a sensitive enzyme immunoassay, *J. Immunol. Methods,* 35, 267, 1980.

38. **Pesce, A. J., Modesto, R. R., Ford, D. J., Sethi, K., Clyne, D. H., and Pollak, V. E.,** Preparation and analysis of peroxidase antibody and alkaline phosphatase antibody conjugate, in *Immunoenzymatic Techniques,* Feldmann, G., Druet, P., Bignon, J., and Avrameas, S., Eds., North-Holland, Amsterdam, 1976, 7.

39. **Casu, A. and Avrameas, S.,** Conjugation of lactic-dehydrogenase with protein by use of glutaraldehyde. Detection of antigens in the immunocompetent cells by means of the conjugates, *Ital. J. Biochem.,* 13, 166, 1969.

40. **Wannlund, J. and DeLuca, M.A.,** Bioluminescent immunoassays. Use of luciferase-antigen conjugates for determination of methotrexate and DNP, in *Bioluminescence and Chemiluminescence,* DeLuca, M. A. and McElroy, W. D., Eds., Academic Press, New York, 1981, 693.

41. **Rubenstein, K. E., Schneider, R. S., and Ullman, E. F.,** ''Homogeneous'' enzyme immunoassay. A new immunochemical technique., *Biochem. Biophys. Res. Commun.,* 47, 846, 1972.

42. **Ullman, E. F., Blakemore, J., Leute, R. K., Eimstad, W., and Jaklitsch, A.,** Homogeneous enzyme immunoassay for thyrosine, *Clin. Chem.,* 21, 1011 (Abstr.), 1975.

43. **Stott, R. A., Thorpe, G. H. G., Kricka, L. J., and Whitehead, T. P.,** Preparation of microperoxidase and its use as a catalytic label in luminescent immunoassays, in *Analytical Application of Bioluminescence and Chemiluminescence,* Kricka, L. J., Stanley, P. E., Thorpe, G. H. G., and Whitehead, T. P., Eds., Academic Press, London, 1984, 233.

44. **Kuanbaev, D. N., Chimirov, O. B., Temiralieva, G. A., Lukhnova, L., and Tursunov, A. N.,** An immunoglobulin-penicillinase conjugate for detecting the antigen of the causative agent of tularemia, *Lab. Delo,* 1, 44, 1990.

45. **Wei, R. and Riebe, S.,** Preparation of a phospholipase C-antihuman IgG conjugate and inhibition of its enzymatic activity by human IgG, *Clin. Chem.,* 23, 1386, 1977.

46. **Reichard, D. W. and Miller, R. J., Jr.,** Bioluminescent immunoassay: a new enzyme-linked analytical method for the quantitation of antigens, in *Bioluminescence and Chemiluminescence,* DeLuca, M. A. and McElroy, W. D., Eds., Academic Press, New York, 1981, 667.

47. **Chandler, H. M., Cox, J. C., Healey, K., MacGregor, A., Premier, R. R., and Hurrell, J. G. R.,** An investigation of the use of urease-antibody conjugates in enzyme immunoassays, *J. Immunol. Methods,* 53, 187, 1982.

48. **Goding, J. W.,** Methods used in antibody purification, in *Monoclonal Antibodies: Principles and Practice,* 2nd ed., Academic Press, New York, 1986, 108.

49. **Menozzi, F. D., Vanderpoorten, P., Dejaiffe, C., and Miller, A. O. A.,** One-step purification of mouse monoclonal antibodies by mass ion exchange chromatography on zetaprep, *J. Immunol. Methods,* 99, 229, 1987.

50. **Lindmark, R., Thoren-Tolling, K., and Sjoquist, J.,** Binding of immunoglobulins to protein A and immunoglobulin levels in mammalian sera, *J. Immunol. Methods,* 62, 1, 1983.

51. **Bjorck, L. and Kronvall, G.,** Purification and some properties of streptococcal protein G. A novel IgG-binding reagent, *J. Immunol.,* 133, 969, 1984.

52. **Eliasson, M., Eliasson, M., Andersson, R., Olsson, A., Wigzell, H., and Uhlen, M.,** Differential IgG-binding characteristics of staphylococcal protein A, streptococcal protein G, and A chimeric protein AG, *J. Immunol.,* 142, 575, 1989.

53. **Goldberg, B., Nochumson, S., and Sloshberg, S.,** A chromatographic method for purification of immunoglobulins, *Am. Biotech. Lab.,* 7, 27, 1989.

54. **Coulter, A. and Harris, R.,** Simplified preparation of rabbit Fab fragments, *J. Immunol. Methods,* 59, 199, 1983.

55. **Michaelsen, T. E.,** Methods for high yield of Fab and Fc fragments of hybridoma immunoglobulin G-1, immunoglobulin G-2a, immunoglobulin G-2b and immunoglobulin G-3 antibodies, *Scand. J. Immunol.,* 24, 471, 1986.

56. **Newkirk, M. M., Edmundson, A., Wistar, R., Jr., Klapper, K. G., and Capra, J. D.,** A new protocol to digest human IgM with papin that results in homogeneous Fab preparations that can be routinely crystallized, *Hybridoma,* 6, 453, 1987.

57. **Rousseaux, J., Rousseaux-Prévost, R., and Bazin, H.,** Optimal conditions for the preparation of Fab and F(ab')$_2$ fragments from monoclonal immunoglobulin G of different rat immunoglobulin G subclasses, *J. Immunol. Methods,* 64, 141, 1983.

58. **Parham, P.,** On the fragmentation of monoclonal IgG$_1$, IgG$_{2a}$, and IgG$_{2b}$ from BALB/c mice, *J. Immunol.,* 131, 2895, 1983.

59. **Lamoyi, E.,** Preparation of F(ab')$_2$ fragments from mouse IgG of various subclasses, *Methods Enzymol.,* 121, 652, 1986.

60. **Ishikawa, E., Imagawa, M., Hashida, S., Yoshitake, S., Hamaguchi, Y., and Ueno, T.,** Enzyme-labeling of antibodies and their fragments for enzyme immunoassay and immunohistochemical staining, *J. Immunoassay,* 4, 209, 1983.

61. **Ishikawa, E., Yoshitake, S., Imagawa, M., and Sumiyoshi, A.,** Preparation of monomeric Fab'-horseradish peroxidase conjugate using thiol groups in the hinge and its evaluation in enzyme immunoassay and immunohistochemical staining, *Ann. N.Y. Acad. Sci.,* 420, 74, 1983.

62. **Imagawa, M., Hashida, S., Ishikawa, E., and Sumiyoshi, A.,** Evaluation of Fab'-horseradish peroxidase conjugates prepared using pyridyl disulfide compounds, *J. Appl. Biochem.,* 4, 400, 1982.

63. **Kranz, D. M. and Voss, E. W., Jr.,** Restricted reassociation of heavy and light chains from hapten-specific monoclonal antibodies, *Proc. Natl. Acad. Sci. U.S.A.,* 78, 5807, 1981.

64. **Grassetti, D. R., and Murray, Jr., J. F.,** Determination of sulfhydryl groups with 2,2'- or 4,4'-dithiopyridine, *Arch. Biochem. Biophys.,* 103, 1132, 1967.

65. **Palmer, J. L. and Nisonoff, A.,** Dissociation of rabbit γ-globulin into half molecules after reduction of one labile disulfide bond, *Biochemistry,* 3, 863, 1964.

66. **Yoshitake, S., Hamaguchi, Y., and Ishikawa, E.,** Efficient conjugation of rabbit Fab' with β-D-galactosidase from *Escherichia coli, Scand. J. Immunol.,* 10, 81, 1979.

67. **Imagawa, M., Yoshitake, S., Hamaguchi, Y., Ishikawa, E., Niitsu, Y., Urushizaki, I., Kanazawa, R., Tachibana, S., Nakazawa, N., and Ogawa, H.,** Characteristics and evaluation of antibody-horseradish peroxidase conjugates prepared by using a maleimide compound, glutaraldehyde, and periodate, *J. Appl. Biochem.,* 4, 41, 1982.

68. **Hasida, S., Imagawa, M., Inoue, S., Ruan, K-H., and Ishikawa, E.,** More useful maleimide compounds for the conjugate of Fab' to horseradish peroxidase through thiol groups in the hinge, *J. Appl. Biochem.,* 6, 56, 1984.

69. **Weiss, E. and Van Regenmortel, M. H.,** Use of rabbit Fab'-peroxidase conjugates prepared by the maleimide method for detecting plant viruses by ELISA, *J. Virol. Methods,* 24, 11, 1989.

70. **Penefsky, H. S.,** A centrifuged-column procedure for the measurement of ligand binding by beef heart F1, *Methods Enzymol.,* 56, 527, 1979.

71. **Avrameas, S. and Ternynck, T.,** Peroxidase labelled antibody and Fab conjugates with enhanced intracellular penetration, *Immunochemistry,* 8, 1175, 1971.

72. **Mannik, M. and Downey, W.,** Studies on the conjugation of horseradish peroxidase to Fab fragments, *J. Immunol. Methods,* 3, 233, 1973.

73. **Boorsma, D. M. and Streefkerk, J. G.,** Peroxidase-conjugate chromatography. Isolation of conjugates prepared with glutaraldehyde or periodate using polyacrylamide-agarose gel, *J. Histochem. Cytochem.,* 24, 481, 1976.

74. **Avrameas, S., Ternynck, T., and Guesdon, J.-L.,** Coupling of enzyme to antibodies and antigens, in *Quantitative Enzyme Immunoassay,* Engvall, E. and Pesce, A. J., Eds., Blackwell Scientific, Oxford, 1978, 7.

75. **Boorsma, D. M. and Kalsbeek, G. L.,** A comparative study of horseradish peroxidase conjugates prepared with a one-step and a two-step method, *J. Histochem. Cytochem.,* 23, 200, 1975.

76. **Avrameas, S.,** Coupling of enzyme to proteins with glutaraldehyde. Use of the conjugates for the detection of antigens and antibodies, *Immunochemistry*, 6, 43, 1969.

77. **Wilson, M. B. and Nakane, P. K.,** Recent developments in the periodate method of conjugating horseradish peroxidase (HRPO) to antibodies, in *Immunofluorescence and Related Staining Techniques*, Knapp, W., Holubar, K., and Wick, G., Eds., Elsevier/North-Holland Biomedical Press, Amsterdam, 1978, 215.

78. **Tijssen, P. and Kurstak, E.,** High efficient and simple methods for the preparation of peroxidase and active peroxidase. Antibody conjugates for enzyme immunoassays, *Anal. Biochem.*, 136, 451, 1984.

79. **Nakane, P. K. and Kawaoi, A.,** Peroxidase-labeled antibody, a new method of conjugation, *J. Histochem. Cytochem.*, 22, 1084, 1974.

80. **Nakane, P. K.,** Recent progress in the peroxidase-labeled antibody method, *Ann. N.Y. Acad. Sci.*, 254, 203, 1975.

81. **Modesto, R. R. and Pesce, A. J.,** The reaction of 4,4'-difluoro 3,3'-dinitrophenyl sulfone with γ-globulin and horseradish peroxidase, *Biochim. Biophys. Acta*, 229, 384, 1971.

82. **Avrameas, S. and Lespinats, G.,** Enzymes couplées aux protéines; leur utilisation pour la détection des antigenes et des anticorps, *C. R. Acad. Sci. Paris*, 265, 1149, 1967.

83. **Modesto, R. R. and Pesce, A. J.,** Use of toluene diisocyanate for the preparation of a peroxidase-labeled antibody conjugate. Quantitation of the amount of diisocyanate bound, *Biochim. Biophys. Acta*, 295, 283, 1973.

84. **Clyne, D., Norris, S., Modesto, R. R., and Pesce, A. J.,** Antibody enzyme conjugates. The preparation of intermolecular conjugates of horseradish peroxidase and antibody and their use in immunohistology of renal cortex, *J. Histochem. Cytochem.*, 21, 233, 1973.

85. **Nakane, P. K., Sri Ram, J., and Pierce, G. B.,** Enzyme-labeled antibodies for light and electron microscopic localization of antigens, *J. Histochem. Cytochem.*, 14, 789, 1966.83.

86. **Ternynck, T. and Avrameas, S.,** Conjugation of *p*-benzoquinone treated enzymes with antibodies and Fab fragments, *Immunochemistry*, 14, 767, 1977.

87. **Jeanson, A., Cloes, J. M., Bouchet, M., and Rentier, B.,** Preparation of reproducible alkaline phosphatase-antibody conjugates for enzyme immunoassay using a heterobifunctional linking agent, *Anal. Biochem.*, 172, 392, 1988.

88. **Ford, D. J., Radin, R., and Pesce, A. J.,** Characterization of glutaraldehyde coupled alkaline phosphatase-antibody and lactoperoxidase-antibody conjugates, *Immunochemistry*, 15, 237, 1978.

89. **Kurstak, E.,** *Enzyme Immunodiagnosis*, Academic Press, Orlando, FL, 1986, 20.

90. **Fujiwara, K., Saita, T., and Kitagawa, T.,** The use of *N*-[β-(4-diazophenyl)ethyl]maleimide as a coupling agent in the preparation of enzyme-antibody conjugate, *J. Immunol. Methods*, 110, 47, 1988.

91. **Kato, K., Hamaguchi, Y., Fukui, H., and Ishikawa, E.,** Conjugation of rabbit anti-(human immunoglobulin G) antibody with β-D-galactosidase from *Escherichia coli* and its use for human immunoglobulin G assay, *Eur. J. Biochem.*, 62, 295, 1976.

92. **Kato, K., Fukui, H., Hamaguchi, Y., and Ishikawa, E.,** Enzyme-linked immunoassay: conjugation of the Fab' fragment of rabbit IgG with β-D-galactosidase from *E. coli* and its use for immunoassay, *J. Immunol.*, 116, 1554, 1976.

93. **Kato, K., Hamaguchi, Y., Fukui, M., and Ishikawa, E.,** Enzyme-linked immunoassay. I. Novel method for synthesis of the insulin-β-D-galactosidase conjugate and its applicability for insulin assay, *J. Biochem.*, 78, 235, 1975.

94. **Cameron, D. J. and Erlanger, B. F.,** An enzyme-linked procedure for the detection and estimation of surface receptors on cells, *J. Immunol.*, 116, 1313, 1976.

95. **Guesdon, J.-L.,** Amplification systems for enzyme immunoassay, in *Nonisotopic Immunoassay*, Ngo, T. T., Ed., Plenum Press, New York, 1988, 85.

96. **Engvall, E. and Perlmann, P.,** Enzyme-linked immunosorbent assay (ELISA). Quantitative assay of immunoglobulin G, *Immunochemistry*, 8, 871, 1971.

97. **Muzykantov, V. R., Sakharov, D. V., Sinitsyn, V. V., Domogatsky, S. P., Goncharov, N. V., and Danilov, S. M.,** Specific killing of human endothelial cells by antibody-conjugated glucose oxidase, *Anal. Biochem.*, 169, 383, 1988.

98. **Pene, J., Rousseau, V., Stanislawski, M.,** *In vitro* cytolysis of myeloma tumor cells with glucose oxidase and lactoperoxidase antibody conjugates, *Biochem. Int.*, 13, 233, 1986.

99. **Herrmann, J. E. and Morse, S. A.,** Conjugation of enzymes to anti-poliovirus globulin: effect of enzyme molecular weight on virus neuralization capacity, *Immunochemistry*, 11, 79, 1974.

100. **Lal, R. B., Brown, E. M., Seligmann, B. E., Edison, L. J., and Chused, T. M.,** Selective elimination of lymphocyte subpopulations by monoclonal antibody-enzyme conjugates, *J. Immunol. Methods*, 79, 307, 1985.

101. **Oi, V. T., Glazer, A. N., and Stryer, L.,** Fluorescent phycobiliprotein conjugates for analyses of cell and molecules, *J. Cell Biol.*, 93, 981, 1982.

102. **Kronick, M. N. and Grossman, P. D.,** Immunoassay techniques with fluorescent phycobiliprotein conjugates, *Clin. Chem.*, 29, 1582, 1983.

103. **Hardy, R. R.,** Purification and coupling of fluorescent proteins for use in flow cytometry, in *Handbook of Experimental Immunology,* 4th ed., Vol. 1, Weir, D. M., Herzenberg, L. A., Blackwell, C. C., and Herzenberg, L. A., Eds., Blackwell Scientific, Edinburgh, 1986, chap. 13.

104. **Houghton, R.,** A dichromatic polymerization-induced separation immunoassay for the simultaneous measurement of human serum IgG and IgM, presented at Conf. on Phycobiliproteins in Biology and Medicine, Seattle, September 9 to 10, 1985.

105. **Parks, D. R. and Herzenberg, L. A.,** Fluorescence activated cell sorting: theory, experimental optimization, and application in lymphoid cell biology, *Methods Enzymol.,* 108, 197, 1984.

106. **Tawde, S. S. and Sri Ram, J.,** Conjugation of antibody ferritin by means of p,p'-difluoro-*m,m'*-dinitrodiphenyl sulphone, *Arch. Biochem. Biophys.,* 97, 429, 1962.

107. **Singer, S. J. and Schick, A.,** The properties of specific stains for electron microscopy prepared by the conjugation of antibody molecules with ferritin, *J. Biophys. Biochem. Cytol.,* 9, 519, 1961.

108. **Gregory, D. W. and Williams, M. A.,** The preparation of ferritin-labeled antibodies and other protein-protein conjugates with *bis*-diazotized benzidine, *Biochim. Biophys. Acta,* 133, 319, 1967.

109. **Lijnen, H. R., Dewerchin, M., De Cock, F., and Collen, D.,** Effect of fibrin-targeting on clot lysis with urokinase-type plasminogen activator, *Thromb. Res.,* 57, 333, 1990.

110. **Engvall, E.,** Preparation of enzyme-labelled staphylococcal protein A and its use for detection of antibodies, in *Quantitative Enzyme Immunoassay,* Engvall, E. and Pesce, A. J., Eds., Blackwell Scientific, Oxford, 1978, 25.

111. **Denisov, V. N. and Metelitsa, D. I.,** Catalytic and immunochemical properties of ferritin conjugates with horseradish peroxidase, *Biokhimiia,* (Engl. Trans.) 52, 1070, 1987.

112. **Khanna, P. L., Dworschack, R. T., Manning, W. B., and Harris, J. D.,** A new homogeneous enzyme immunoassay using recombinant enzyme fragments, *Clin. Chim. Acta,* 185, 231, 1989.

113. **Fujiwara, K. and Saita, T.,** The use of *N*-[β-(4-diazophenyl)ethyl]maleimide as a heterobifunctional agent in developing enzyme immunoassay for neurotensin, *Anal. Biochem.,* 161, 157, 1987.

114. **Pradelles, P., Grassi, J., Chabardes, D., and Guiso, N.,** Enzyme immunoassays of adenosine cyclic 3',5'-monophosphate and guanosine cyclic 3',5'-monophosphate using acetylcholinesterase, *Anal. Chem.,* 61, 447, 1989.

115. **Imagawa, M., Hashida, S., Ishikawa, E., and Freytag, J. W.,** Preparation of a monomeric 2,4-ditrophenyl Fab'-β-D-galactosidase conjugate for immunoenzymometric assay, *J. Biochem.,* 96, 1727, 1984.

116. **Ioue, S., Hashida, S., Tanaka, K., Imagawa, M., and Ishikawa, E.,** Preparation of monomeric affinity-purified Fab'-β-D-galactosidase conjugate for immunoenzymometric assay, *Anal. Lett.,* 18(B11), 1331, 1985.

117. **Beyzavi, K., Hampton, S., Kwasowski, P., Fickling, S., Marks, V., and Clift, R.,** Comparison of horseradish peroxidase and alkaline phosphatase-labelled antibodies in enzyme immunoassays, *Ann. Clin. Biochem.,* 24, 145, 1987.

118. **Ishikawa, E., Hashida, S., Kohno, T., and Ranaka, K.,** Methods for enzyme-labeling of antigens, antibodies and their fragments, in *Nonisotopic Immunoassay,* Ngo, T. T., Ed., Plenum Press, New York, 1988, 27.

119. **Montoya, A. and Castell, J. V.,** Long-term storage of peroxidase-labelled immunoglobulins for use in enzyme immunoassay, *J. Immunol. Methods,* 99, 13, 1987.

Chapter 11

PREPARATION OF IMMUNOTOXINS AND OTHER DRUG CONJUGATES FOR TARGETING THERAPUTICS

I. INTRODUCTION

Drug conjugates developed as a result of systemic pharmacotherapy are target-specific cytotoxic agents. The concept involves coupling a therapeutic agent to a carrier molecule with specificity for a defined target cell population. Antibodies with high affinity for antigens are a natural choice as targeting moieties.[1,2] With the availability of high affinity monoclonal antibodies and their fragments, the prospects of antibody-targeting therapeutics have become promising. Toxic substances that have been conjugated to monoclonal antibodies include protein toxins, low-molecular-weight drugs, biological response modifiers, and radio-nuclides.[3] Antibody-toxin conjugates are frequently termed immunotoxins, whereas immunoconjugates consisting of antibodies and low-molecular-weight drugs such as methothrexate and adriamycin are called chemoimmunoconjugates. Immunomodulators contain biological response modifiers that are known to have regulatory functions such as lymphokines, growth factors, and complement-activating cobra venom factor (CVF). Radioimmunoconjugates consist of radioactive isotopes which may be used as therapeutics to kill cells by their radiation or used for imaging.

In addition to antibodies, other molecules that have specific receptors or binding sites on target cells have also been used as targeting agents. These include transferrin, α_2-macroglobulin, epidermal growth factor, and hormones. When hormones are used as the targeting agent, the term hormonotoxin is frequently used. Toxins have also been conjugated to antigens to selectively kill antigen-responsive B cells.[4,5]

The coupling of targeting agents with toxic moieties is most commonly performed with heterobifunctional cross-linking reagents. Earlier attempts to conjugate toxins and antibodies by homobifunctional reagents generated nonspecific cross-linking products. More specific and efficient cross-linking techniques have been developed.[6] This chapter will review the procedures that have been used to prepare immunotoxins and other cytotoxic drug conjugates. Theoretically, a cross-linking agent used to couple one toxin is applicable to others. In fact, the reader will find this is the case for most of the toxins to be described.

II. SELECTION OF CROSS-LINKING REAGENTS AND TOXINS

Like other proteins, the amino side chains of lysines, the carboxyl groups of aspartic and glutamic acids, the thiol group of cysteine, and the carbohydrate moiety of antibodies, toxins, and other drugs have been used to prepare drug conjugates. However, the side chains of hisitidines, methionine, arginine, serine, threonine, tryptophan, and tyrosine have not been fully utilized. In earlier studies, immunotoxins were prepared using homobifunctional reagents such as glutaraldehyde, diethylmalonimidate hydrochloride, and difluorodinitro-phenylsulfone that cross-linked the free amino groups found in abundance in both protein species. As has been discussed in earlier chapters, such cross-linking reagents produced a mixture of heterogeneous conjugate with extensive polymerization in both antibody and toxin. In addition, intramolecular cross-linking between toxin subunits generally destroys its cytotoxicity. The development of two-step coupling procedures using heterobifunctional reagents such as m-maleimido-benzoyl-N-hydroxysuccinimide ester (MBS) and N-succinim-idyl 3-(2-pyridyldithio)propionate (SPDP) has resulted in more efficient use of reactants, increased yields, and more desirable structural features in synthesized antibody-toxin conjugates. These reagents generally first involve generating a free thiol group on one protein followed by reacting selectively with a functionality on a second protein. The art of this type of conjugation will be illustrated below in detail.

The most commonly used toxins in the preparation of immunotoxins are diphtheria toxin (DT), ricin, abrin, modeccin, pokeweed antiviral protein (PAP), CVF, Pseudomonas exo-toxin (PE), gelonin, and other ribosome-inactivating proteins (RIPs). These proteins have been purified and characterized.[7-12] Some of these toxins such as ricin, abrin, modeccin, viscumin, and volkensin have two dissimilar A and B polypeptides attached through a disulfide bond and are referred to as true toxins. Only the A chain contains the enzymatic activity that is cytotoxic and the B chain is for receptor binding which is usually a carbo-hydrate. Other toxins such as PAP and gelonin contain a single polypeptide that have similar or identical enzymatic activity to the A chains of the true toxins. These hemitoxins can generate a true toxin when covalently bound to the B chain of ricin.[11] Diphtheria toxin, on the other hand, contains a single polypeptide, but can be cleaved with trypsin to generate fragment A on reduction. There are therefore two ways of preparing immunotoxins. The first is to link the whole intact toxin to the antibody or its fragment. The second is to couple the antibody to the isolated A chain or the single-polypeptide toxin. The method dictates the choice of cross-linker to be used. Since the A and B subunits are separated during the action of cell killing, intramolecular cross-linking of true toxins must be avoided. Similarly, the linkage of an antibody-A chain conjugate must be reducible as in an intact toxin to be cytotoxic. Another constraint on the choice of coupling agent is the retention of cytotoxicity of the toxin and the antigen-binding capacity of the antibody. Thus, each conjugate should be tested for both cytotoxicity and antigen binding activity. Limited modification of lysine residues in abrin, for example, reduces its galactose-binding affinity but the toxicity of the A chain is not affected.[6,13] On the other hand, limited modification of lysine residues of RIP from *Momordica charantia* reduces its ability to inhibit protein synthesis.[14] Furthermore, the cross-linking should be stable under *in vivo* conditions. The biological stability of immunotoxin *in vivo* might be the major limitation of its efficacy. Most of the active conjugates have been prepared with a disulfide link between the effector and the antibody. The disulfide bond may be disrupted by disulfide exchange with glutathione or other serum-borne or cellular factors. Generally SPDP-linked conjugates showed good stability in blood.[15]

Not only must the drug have a functional group available for conjugation or be capable of modification to contain such a group without affecting its chemotherapeutic activity, the drug must not have to be activated *in vivo* by some metabolic or catabolic step. These considerations have greatly reduced the number of drugs of choice.

III. CONJUGATION THROUGH DISULFIDE BOND

Because the linkage between the toxin and antibody is of critical importance in deter-mining the cytotoxicity and that a reducible bond between the antibody and the A chain is required for toxicity,[16,17] the most common amino acid side chains of toxins and immuno-globulins that have been used to link the two molecules together are the amino and sulfhydryl groups. The thiol is used to form reducible disulfide bonds. Amino groups are reactive, abundant, and in most cases expendable. That is, a limited number of amino groups can be modified without diminishing the biological activity of the protein. In cases where a free thiol is not available, such as in native antibodies, free sulfhydryl groups can be generated by reductive cleavage of native cystine residues with thiol reagents (e.g., dithiothreitol [DTT]) or by thiolation of the amino group using SPDP, 2-iminothiolane or methyl-3-mercaptopropionimidate as described in the last chapter. The conversion of an amino group into a sulfhydryl group has been discussed in Chapter 2. The following sections illustrate various possible ways for the preparation of drug conjugates.

A. COUPLING WITH SPDP

1. Preparation of Immunotoxins

SPDP is probably the most widely used cross-linker for preparing antibody-toxin conjugates. Not only has it been used to cross-link between amino and thiol groups, it has also been used to introduce a free thiol into immunoglobulins or other molecules that do not have a free sulfhydryl group. The reaction scheme is shown in Figure 1. First, the amino groups of both an immunoglobulin and a toxin are reacted with the reagent to form a carboxamide bond. One of the introduced 2-pyridyldisulfide group, for example, that on the antibody, is then reduced by DTT to generate a free thiol. The introduction of sulfhydryl groups into immunoglobulins is more preferable than the toxin in order to avoid the exposure of the latter to reducing conditions which could result in the separation of its A and B subunits. On mixing the modified proteins, the reduced thiol then nucleophilically displaces pyridine-2-thione from the 3-(2-pyridyldithio)propionyl group incorporated into toxins.[8] The resulting conjugate contains a disulfide bond derived from two molecules of SPDP. The yield of the 1:1 antibody:toxin conjugate is usually about 20 to 40%. A disadvantage of SPDP conjugation is the reported instability of the disulfide bond, which is prone to *in vivo* cleavage and exchange reactions.[17]

This method has been used to prepare immunotoxins of ricin,[18,19] abrin,[6,20] gelonin,[21-23] PAP,[24] saporin,[25] human lymphoblastoid interferon-α[26,27] and CVF.[28] The same approach has also been used to prepare F(ab')₂ conjugates with diphtheria toxin,[29] abrin,[30] PAP,[31] CVF[32] and saporin.[25] Complement component C_{3b} was conjugated to monoclonal antibodies in the same way.[33]

Alternatively, a free sulfhydryl group may be introduced into the toxin after reaction with SPDP followed by reduction. For the conjugation of monoclonal antibody to neocarzinostatin, a membrane-reactive anticancer protein, the chromophore free apoprotein was reacted with SPDP and reduced with DTT to generate free thiol groups. The thiol was coupled to SPDP activated antibody. After conjugation, the chromophore was reassociated with the apoprotein to generate active immunotoxin.[34] In a different version, Lambert et al.[35] prepared antibody-toxin conjugates by introducing free sulfhydryl groups into the toxin using 2-iminothiolane. The antibody is modified with 2-pyridyl disulfide using SPDP. Conjugation is achieved in a thiol-disulfide exchange reaction between the 2-pyridyl disulfide derivative of the antibody and the modified toxin. A similar procedure was used to prepare Pseudomonas toxin A conjugate with breast tumor-selective antibodies,[36] antibody-gelonin, and antibody-PAP conjugates.[37-39]

When a free thiol is present in the toxin, such as ricin A, there is no need to introduce a free thiol. SPDP is simply used to couple the components. The antibody is first activated with SPDP. After isolation, the 3-(2-pyridyldithio)propionyl-derived antibody is mixed with the toxin reduced with dithiothreitol to afford an immunotoxin. This approach has been used to link antibodies to A chains of abrin,[40-42] ricin,[43,44] recombinant ricin,[45] and diphtheria toxin[6] as well as the preparation of antibody-ricin B,[46,47] and ricin A-F(ab')₂ fragment conjugates.[16,48] Usually a yield of up to 50% of the 1:1 conjugates is attainable.

2. Preparation of Other Toxin Conjugates

In addition to immunoglobulins, other targeting molecules have also been conjugated to various toxins using SPDP. Very specific and potent conjugated cytotoxins have been formed from ligands such as transferrin, α₂-macroglobulin, epidermal growth factor (EGF), and other molecules for receptor-mediated endocytosis.[49-51] DT was conjugated to transferrin via a disulfide linkage following derivatization of both components using SPDP and forming thiolated toxin upon reduction according to Figure 1.[52] The same approach was used to prepare bovine luteinizing hormone-gelonin hormonotoxin where the hormone was activated

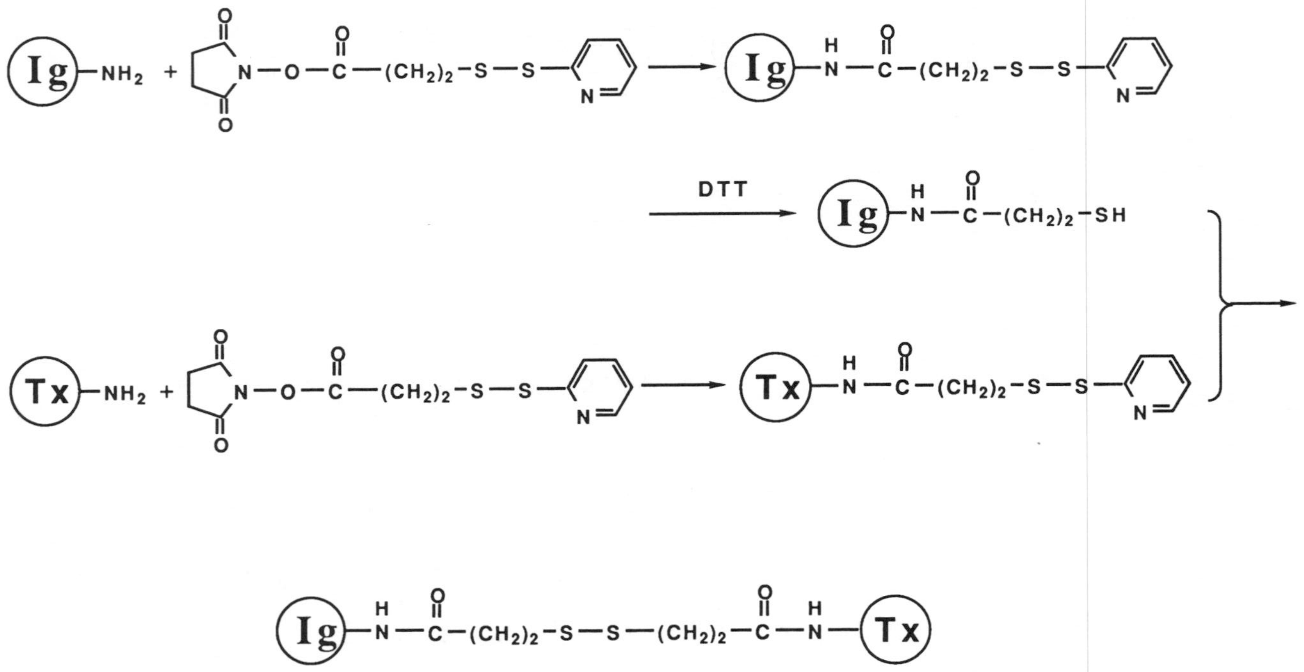

FIGURE 1.   Conjugation of holotoxins to antibodies with SPDP.

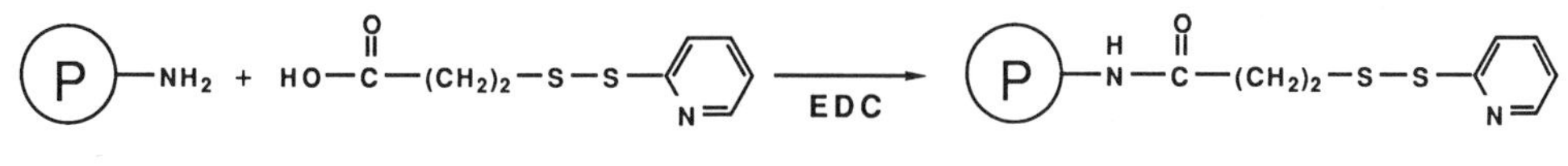

FIGURE 2.   Introduction of 2-pyridyldisulfide into proteins using 3-(2-pyridyldithio)propionate and carbodiimide.

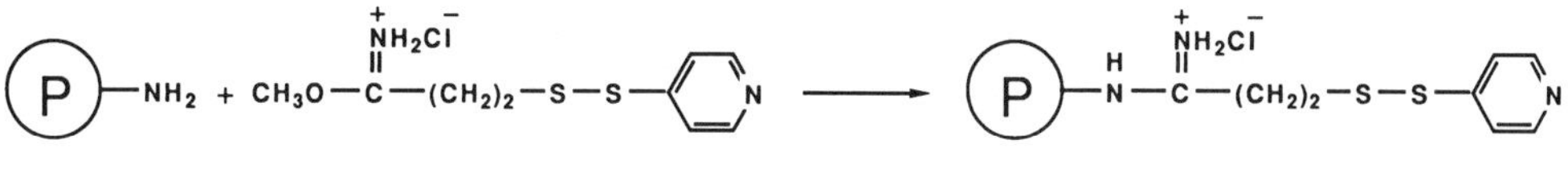

FIGURE 3.   Introduction of 4-pyridyldisulfide by methyl 3-(4-pyridyldithio)mercaptopropionimidate.

with SPDP and gelonin was thiolated.[53,54] In a similar synthesis, SPDP derivatized transferrin was reduced and disulfide linked to SPDP modified whole ricin.[55,56] Transferrin was also disulfide-linked to a monoclonal antibody directed against ricin A chain producing a hybrid that could simultaneously attach to transferrin receptors and reversibly bind the ricin A chain via its antibody component.[55] Similarly, ricin A chain was disulfide-linked in a 1:1 ratio with human α_2-macroglobulin that had been derivatized using SPDP.[57]

By the same procedure, the A chains of ricin and DT have been disulfide-coupled to pyridyldithiopropionate-derivatized fetuin and asialofetuin.[58,59] Asialoorosomucoid has been likewise disulfide-linked to diphtheria A chain.[60]

Stirpe et al.[61] linked the lectin concanavalin A to gelonin by a disulfide bond to form a moderately cytotoxic conjugate to HeLa Cells. A free thiol group was introduced into gelonin by reacting with SPDP followed by reduction, while concanavalin A was activated with pyridyldisulfide.

SPDP was also used to introduce a 2-pyridyldisulfide group to the amino terminus of EGF (a compact, disulfide-bonded, 6 K polypeptide with no lysine residues). This activated substituent underwent thiol-disulfide exchange with the free sulfhydryl of ricin A chain or diphtheria fragment A to form hormonotoxin.[62,63] SPDP reagent has been used to modify a single amino group of intact whole human chorionic gonadotropin (hCG), allowing for full retention of its receptor-binding activity. This derivatized hormone was then reacted with either diphtheria fragment A, ricin A-chain, or gelonin to produce the corresponding disulfide conjugates with much greater potency.[64] SPDP-activated monoclonal antibodies have been used to couple to methotrexate modified with cysteine or reduced cystamine.[65]

B. COUPLING WITH OTHER DISULFIDE GENERATING AGENTS
1. Use of Pyridyldisulfide

In addition to SPDP, introduction of pyridyl disulfide groups into proteins can also be achieved by using 3-(2-pyridyldithio)propionate and water soluble carbodiimide such as 1-ethyl-3-(3-dimethylaminopropyl) carbodiimide (EDC) according to the reaction shown in Figure 2.[66,67] Three to four dithiopyridyl substituents were linked to amino groups on transferrin in this way. Reacting this activated protein with the free sulfhydryl of isolated ricin A chain resulted in the formation of disulfide-linked human transferrin-ricin A-chain conjugate. Fractionation permitted isolation of an 11 K cross-linked species with one A-chain and one transferrin molecule.[68,69]

A direct method of incorporation of 4-pyridyl disulfide can be achieved using methyl 3-(4-pyridyldithio)mercaptopropionimidate.[70] Amidation occurs at the free amino groups of proteins as shown in Figure 3.

Another cross-linker containing the pyridyldisulfide that can be bonded to amino groups of antibodies is 4-succinimidyloxycarbonyl-α-methyl-α-(2-pyridyldithio)toluene (SMPT).

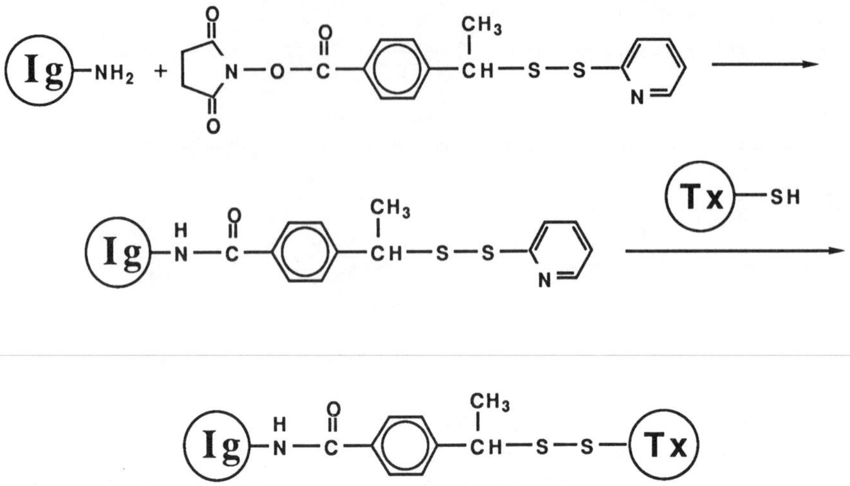

FIGURE 4.   Preparation of immunotoxins with SMPT.

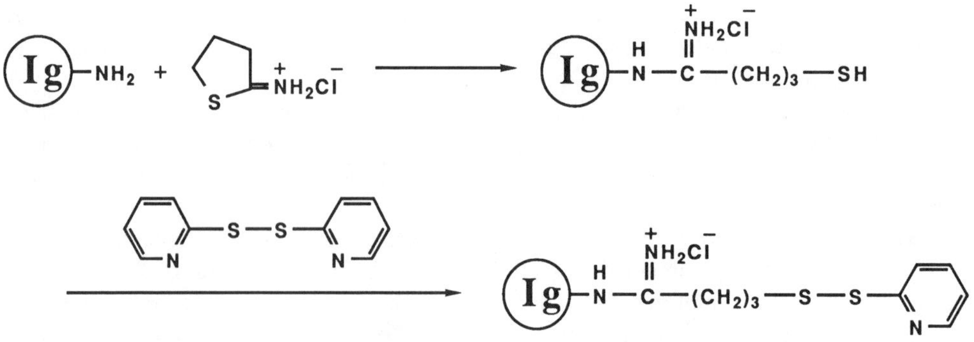

FIGURE 5.   Introduction of pyridyldisulfide with 2,2'-dipyridyldisulfide.

After purification the SMPT-derivatized antibody is reacted with deglycosylated ricin A-chain or abrin A-chain as illustrated in Figure 4. The immunotoxins contain shielded disulfide bonds that have improved stability *in vivo*.[71,72]

Free sulfhydryl groups may be directly activated to the pyridyldisulfide level by 2,2'-dipyridyl disulfide. The reaction involves a disulfide interchange, displacing 2-thiopyridine as shown in Figure 5. Free sulfhydryl groups are introduced into the proteins by thiolation as described above.

2. Use of Cystamine

Cystamine incorporated into proteins provides another means of conjugation through disulfide interchange with a free thiol. The labeling of proteins with cystamine is generally carried out by water-soluble carbodiimide (e.g., EDC) catalyzed amidation reaction of a carboxyl group as shown in Figure 6. Several conjugates are prepared this way.

As an alternative approach to conjugate EGF to DT, the carboxyl groups of EGF is modified with cystamine. This cystaminyl derivative was mixed with reduced diphtheria fragment A to produce a disulfide-coupled EGF-diphtheria A conjugate.[73] Similarly, insulin-diphtheria toxin A fragment conjugation was achieved by first derivatizing the carboxyl groups of porcine insulin with cystamine using EDC. An excess of cystaminyl-insulin was mixed with reduced fragment A to allow disulfide interchange, and the product was then separated from reactants by gel filtration in 7 *M* urea.[74]

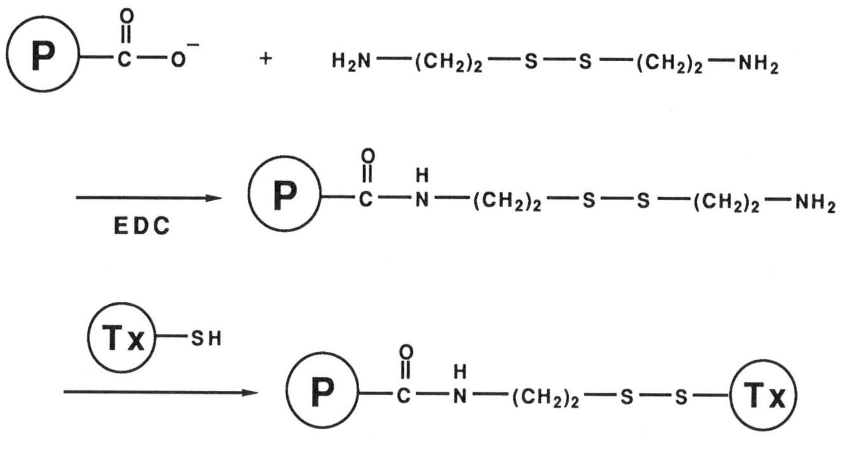

FIGURE 6. Modification of proteins with cystamine and cross-linking through disulfide interchange.

By the same token, Oeltmann and Forbes[75] linked anti-Thy-1.1 antibody to DT fragment A through a disulfide bond established by reacting the cystaminyl-antibody with the thiol of diphtheria toxin A.

In a slightly modified version, two diphtheria toxin related polypeptides, CRM26 and CRM45, were disulfide-linked to thyrotropin-releasing hormone, TRH. The histidyl imidazole of TRH (L-pyroglutamyl-L-histidyl-L-proline amide) was first modified with iodoacetylcystamine which was derived from iodoacetic acid and cystamine using EDC condensation as shown in Figure 7. The acetylcystaminylated TRH was then mixed with reduced CRM peptides to afford the conjugate.[76]

3. Use of *S*-Sulfonate

Another method of forming asymmetric disulfides is the reaction through *S*-sulfonate. Sulfite is easily replaced by an attacking free thiol. *S*-sulfonate can be introduced into proteins directly or by reacting a free thiol with Na_2SO_3-$Na_2S_4O_6$. Masuho et al.[77] introduced *S*-sulfonate into DT fragment A and cross-linked it to the monovalent Fab' fragment of a polyclonal antibody directed against L1210 cells. The reaction is shown in Figure 8.

DT fragment A was coupled to human placental lactogen using *S*-sulfonate introduced by methyl-5-bromovalerimidate as shown in Figure 9.[78] Methyl-5-bromovalerimidate was synthesized from 5-bromovaleryl nitrile by the Pinner method. After reaction with the toxin, the alkyl bromide was replaced by *S*-sulfonate. The 5-*S*-sulfomercaptovaleramidinated toxin was reacted with human placental lactogen to produce the desired hybrid. Ricin A-chain was also disulfide cross-linked to the β subunit of hCG utilizing the same methyl-5-bromovalerimidate method.[79]

4. Use of Ellman's Reagent

5,5'-Dithiobis(2-nitrobenzoic acid) (DTNB), Ellman's reagent, can be used to conjugate two free thiols by a double disulfide interchange reaction.[80] One thiol is first activated with DTNB to form mixed disulfide to which the second thiol reacts. The reaction is illustrated in Figure 10. Since one of the components, e.g., IgG, is generally devoid of a free sulfhydryl group, the protein is thiolated. For example, PE and antibodies were thiolated with 2-iminothiolane. Two moles of thiol groups can be introduced per mole of thiolated toxin which were then activated with DTNB. Conjugates were obtained on mixing the activated toxin with the thiolated antibody.[81-83] PE is similarly linked to epidermal growth factor instead of antibody.[84] Ricin A-chain immunotoxins was also prepared by thiolating the

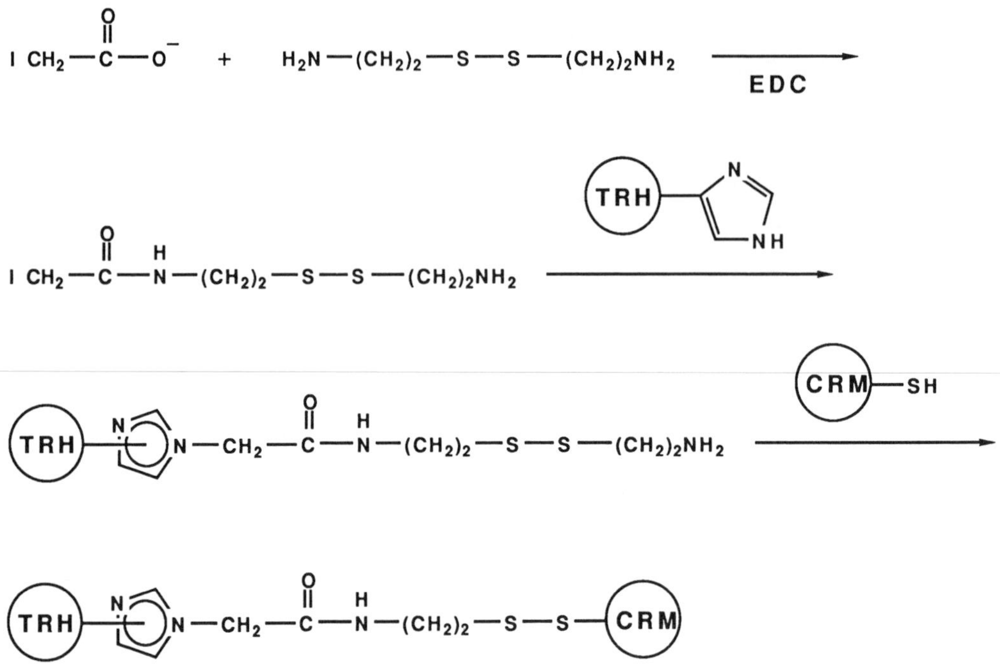

FIGURE 7. Reaction of TRH with iodoacetylcystamine and subsequent conjugation with CRM polypeptides.

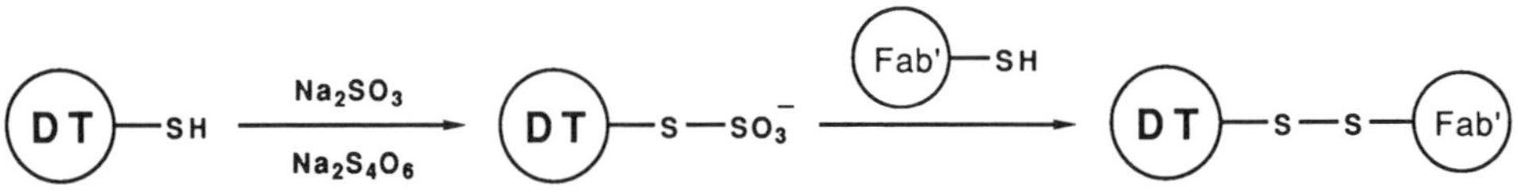

FIGURE 8. Coupling diphtheria toxin A to Fab' by means of S-sulfonate.

antibody with iminothiolane followed by reacting with DTNB.[81] Methyl 3-mercaptopropionimidate was used to thiolate rabbit antimouse IgG antibodies before reaction with DTNB for conjugation with ricin A-chain.[85]

Direct coupling with Ellman's reagent is possible if the protein contains free thiol groups. For example, reduced Fab' fragment can be coupled to ricin A chain by disulfide interchange formed through Ellman's reagent.[44,86] Reduced Fab' fragment is reacted with excess Ellman's reagent. After dialysis to remove excess reagent, the Fab' fragment containing the activated disulfide group is incubated with ricin A-chain. The progress of the reaction can be monitored by the production of 3-carboxylato-4-nitrothiophenolate anion released.

5. Use of Hindered Disulfide Bond

A new reagent, S-4-succinimidyl oxycarbonyl-α-methyl benzyl thiosulfate (SMBT), introduced by Wawrzynczak and Thorpe[71,87] provides a direct incorporation of S-sulfonate to amino groups of antibodies. The final conjugated disulfide bond has a much slower rate of hydrolysis with $t_{1/2} \approx 24$ h.[71,87] Because of the steric hindrance, direct displacement of the sulfite by the sulfhydryl group of toxin A-chain to form disulfide linked conjugates is relatively slow. To increase the efficiency, the S-sulfonate of the SMBT-labeled protein is first reduced by DTT and then activated by Ellman's reagent to form a activated mixed disulfide. The activated disulfide is then reacted with free thiols of toxin A chain as shown in Figure 11. Anti-Thy-1.1-abrin A-chain conjugate has been prepared by this method.[71,87]

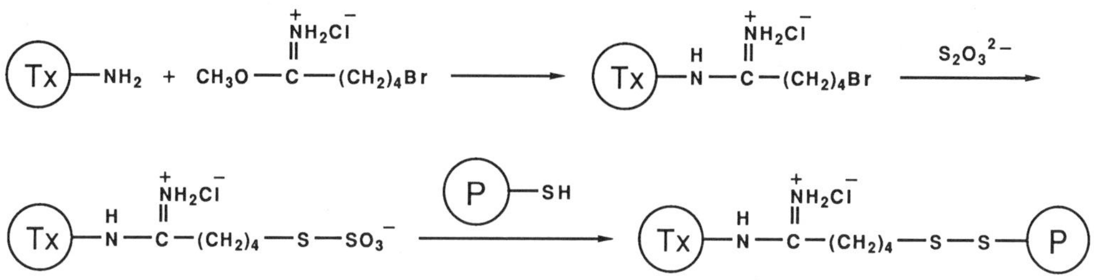

FIGURE 9. Cross-linking of toxin and protein with *S*-sulfonate introduced by methyl-5-bromovalerimidate.

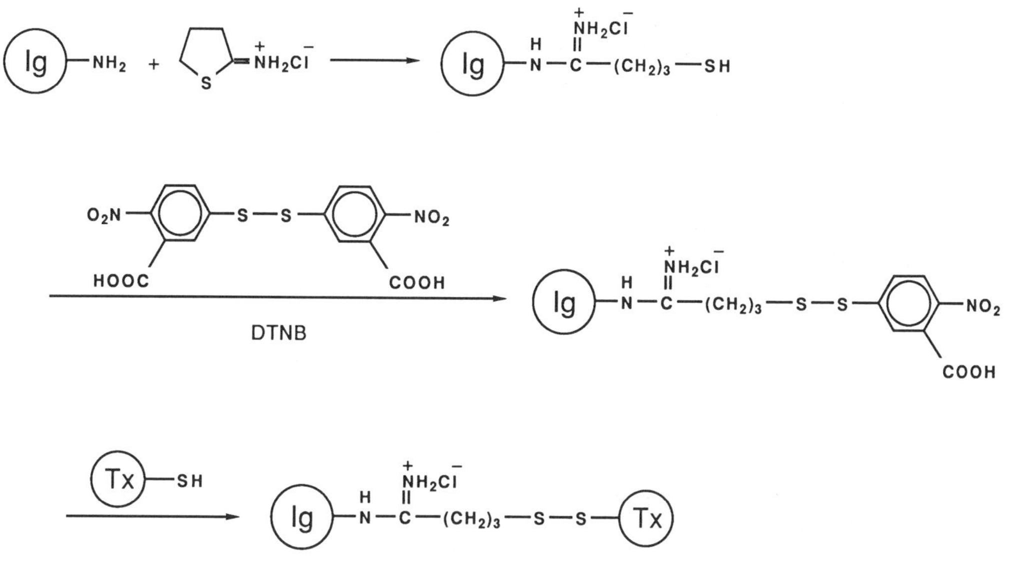

FIGURE 10. Coupling ricin A to antibody through disulfide interchange introduced by 2-iminothiolane and Ellman's reagent.

IV. CONJUGATION THROUGH THIOETHER LINKAGE

Covalent cross-linking of toxins and proteins through a thioether bond can be achieved by reacting free thiols with maleimides or alkyl halides introduced into proteins. The thioether linkage is stable under reducing conditions. Because it is probable that the A chains have to be released from the antibody in order to diffuse into the cells to inactivate them, the conjugates made without disulfide linkages are usually inactive unless the whole toxin molecules are used. In this case, the cytotoxic A-chain can still be released *in vivo*.

A. USE OF IODOACETYL COMPOUNDS

Iodoacetyl groups can be introduced into the toxin by using *N*-hydroxysuccinimidyl iodoacetate or *N*-succinimidyl(4-iodoacetyl)aminobenzoate (see Chapter 5 for structures). The incorporated alkyl iodide can react with thiols of SPDP-thiolated immunoglobulin to produce immunotoxins according to the reaction shown in Figure 12.

This procedure has been used to couple CVF[32] and ricin,[88] to several antibodies. The yield of 1:1 antibody:toxin conjugate are usually 20 to 40%.[88] The linkage of ricin to antibodies by this method produces impairment of the galactose binding sites of the B-chain of the toxin in the majority of conjugate molecules formed, reducing the tendency of the conjugates to bind to and kill cells nonspecifically.[88]

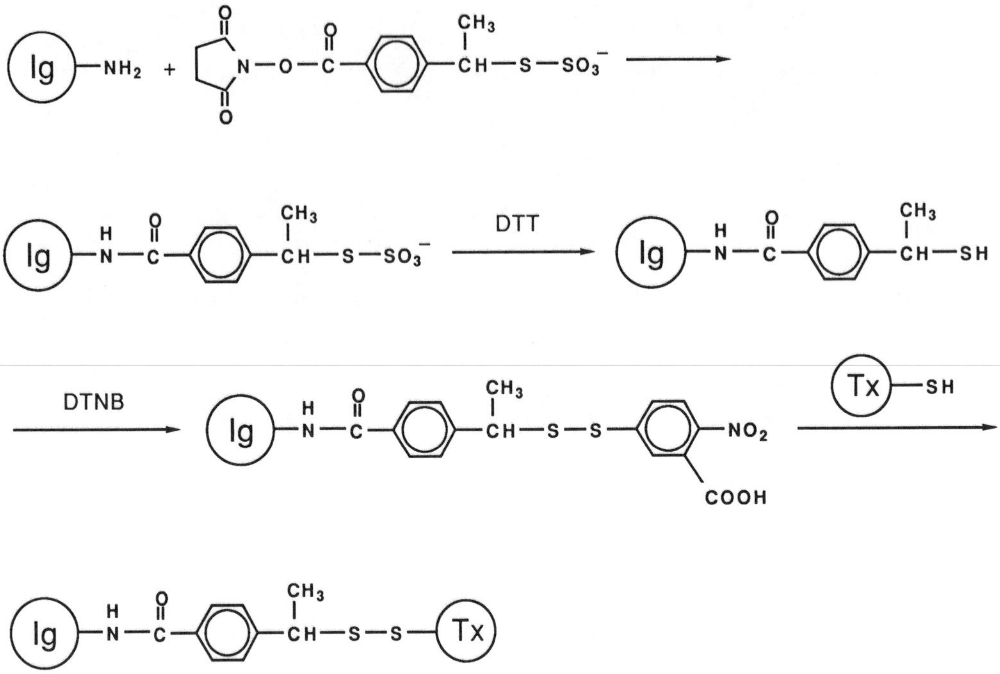

FIGURE 11.   Coupling of antibody and toxin by SMBT.

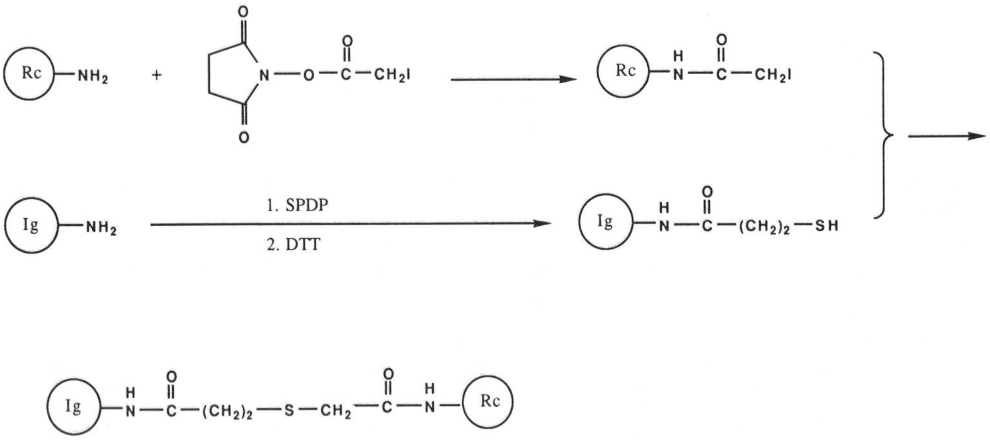

FIGURE 12.   Coupling of ricin to antibody through iodoacetyl group.

B. USE OF AMINO AND THIOL DIRECTED CROSS-LINKERS

Introduction of maleimido groups into proteins without any free thiols can be achieved with maleimido-containing amino and thiol-directed cross-linkers. Direct conjugation of antibodies and toxins using MBS has been used to prepare ricin immunotoxins.[79-92] The two-step reaction is generally applied. MBS is first reacted with the amino groups of ricin. After removal of excess reagent, the derivatized ricin is mixed with reduced or thiolated antibody for the sulfhydryls on the antibody to react with the maleimido group on ricin to form a thioether bond (Figure 13). Purification of the conjugate is simplified by the free galactose-binding site on the B chain of intact ricin.[92] Asialoorosomucoid-diphtheria A conjugate and CVF-Ig immunotoxin have also been prepared by this method.[32,60] Tetanus toxoid thiolated with SPDP and reduced with DTT has been coupled to ricin derivatized with MBS.[93]

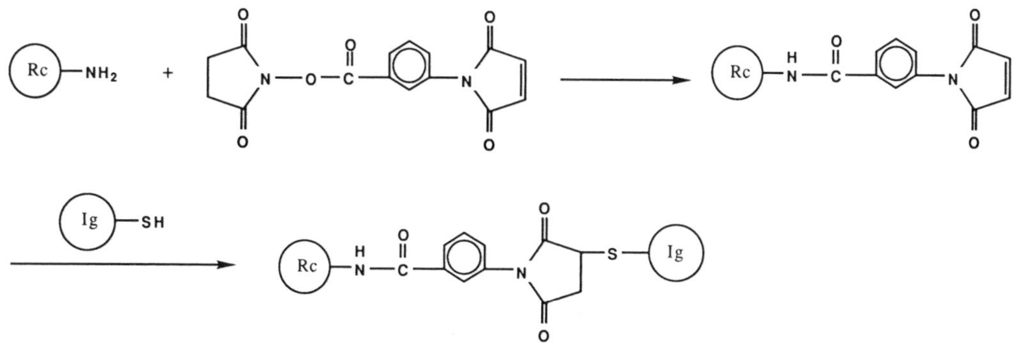

FIGURE 13. Conjugation of ricin and antibody through a thioether bond by MBS.

Other analogs of MBS have also been used. Sulfo-MBS (*m*-maleimidobenzoyl sulfos-uccinimide ester) reportedly increased the yield of antibody T101-ricin conjugate two- to fourfold as compared to MBS. This is probably contributable to the increased water solubility of the reagent. Another analog, sulfosuccinimidyl-4-(*p*-maleimidophenyl) butyrate (sulfo-SMPB, see Chapter 5 for structure), provides an extended chain length and also gives four-to sevenfold greater yield than with MBS. Sulfo-SMPB has been used to prepare thioether conjugates of PE-immunotoxin.[82]

Other amino and thiol directed heterobifunctional reagents that have been used for preparation of immunotoxins include succinimidyl 4-(*N*-maleimidomethyl)cyclohexane-1-carboxylate (SMCC), *N*-succinimidyl 4-maleimidobutyrate (GMBS, γ-*N*-maleimidobutyryl-*N*-hydroxysuccinimide ester), succinimidyl-4-(*p*-maleimidophenyl)butyrate (SMPB) and its sulfo-analog (sulfo-SMPB) and (2-nitro-4-sulfonic acid-phenyl)-6-maleimidocaproate (MAHNSA, maleimido-6-aminocaproyl 4-hydroxy-3-nitrobenzene sulfonic acid ester) (see Chapter 5 for structures and reaction mechanisms). Gelonin and PAP were thiolated with iminothiolane and conjugated to antibodies reacted previously with SMCC.[37] For the prep-aration of CVF immunotoxins, antibodies were first thiolated with iminothiolane before conjugation with SMCC, SMB, or MAHNSA.[32] MAHNSA was used to synthesize thioether-linked conjugate of monoclonal antibody and PE. The antibody was first reacted with the reagent. After purification, the derivatized antibody was mixed with iminothiolane thiolated PE.[94]

Myers et al.[95] compared the use of various aromatic and aliphatic maleimide cross-linkers on the preparation of anti-CD5-ricin immunotoxins. The results show that cross-linkers with aromatic moiety such as sulfo-SMPB gave higher yield. However, aliphatic GMBS cross-linker yielded the most toxic immunotoxin in cell-free translation assays.

V. CONJUGATION WITH ACTIVATED CHLORAMBUCIL

N-succinimidyl chlorambucil (4-[*bis*(2-chloroethyl)amino]-benzene butanoic acid *N*-hy-droxysuccinimide ester), contains functional groups that can cross-link amino groups of proteins. The compound is synthesized from chlorambucil and *N*-hydroxysuccinimide with dicyclohexyl carbodiimide. The activated ester first reacts with amino groups in immuno-globulins at 4°C at which the *N*-chloroethylamine groups are relative inert. After isolation, the substituted immunoglobulin is mixed at pH 9 with excess toxin at ambient temperature to promote the reaction between amino side chains of the toxin and the *bis*-2-chloroethylamine group to form a piperazine ring as shown in Figure 14.[6,96] This method has been used to prepare conjugates of diphtheria toxin,[6] abrin,[96,97] ricin[19] and RIP gelonin[6] to various anti-bodies. Six to eight molecules of chlorambucil can be introduced per molecule of immu-

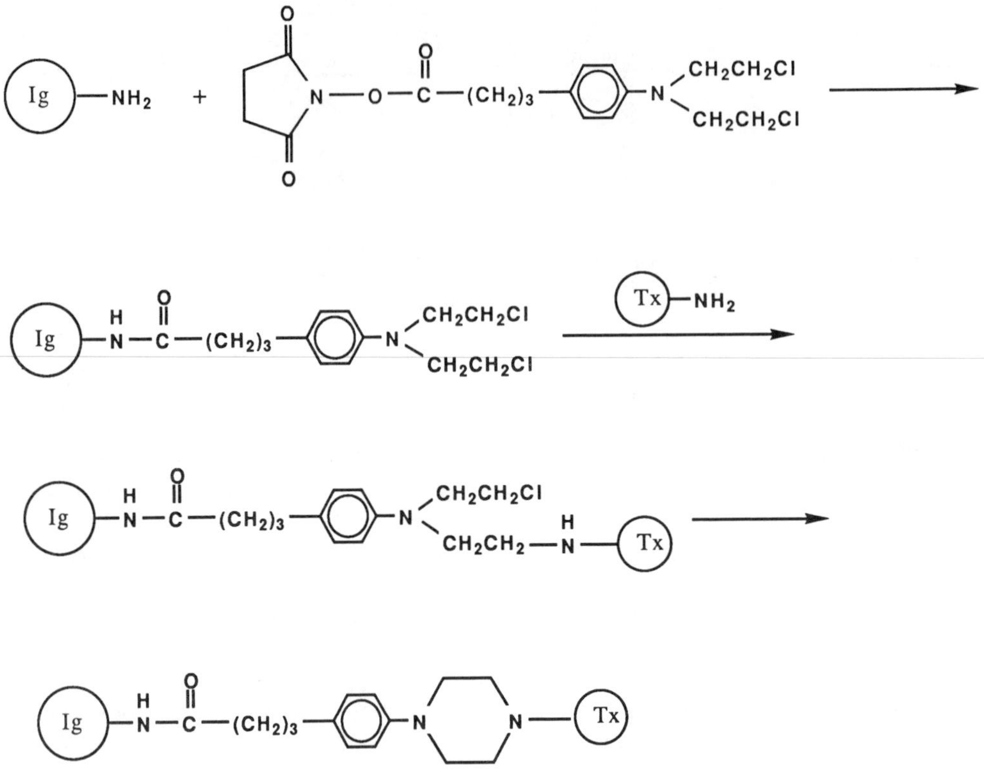

FIGURE 14. Conjugation with activated chlorambucil.

noglobulin. The overall yield of 1:1 antibody:toxin conjugate is about 5 to 10% based on the amount of antibody used. Some antibody-antibody polymers are also formed. According to Edwards et al.[97] disulfide-linked abrin was ten times more potent than linkage through chlorambucil.

VI. CONJUGATION WITH ACID-LABILE AND PHOTOCLEAVABLE CROSS-LINKERS

A. USE OF ACID-LABEL CROSS-LINKERS

A maleimide derivative of 2-methylmaleic anhydride has been used to generate an antibody-gelonin conjugate with an acid-label bond.[98] The principle of the reactions is illustrated in Figure 15A. The reaction of the amino group of gelonin with 2-methylmaleic anhydride gives a substituted maleyl derivative whose carboxamide bond is susceptible to hydrolysis under mildly acidic conditions (pH 4 to 5). The derivative is then coupled to an antibody thiolated with 2-iminothiolane.[98]

Another compound that generates an acid-cleavable amide linkage on reaction is 4-(iodoacetylamino)-3,4,5,6-tetrahydrophthalic anhydride. It has been used to conjugate interleukin 2 to gelonin.[35] The reaction is shown in Figure 15B. Although the modification reduces the activity significantly, the toxin is released in their native and fully active form by a mild acid treatment.

B. USE OF PHOTOCLEAVABLE CROSS-LINKERS

Senter et al.[99] have prepared two photolabile heterobifunctional cross-linking reagents for the conjugation of antibody to PAP. One, [4-nitro-3-(1-chlorocarbonyloxyethyl)phenyl]methyl-*S*-acetylthioic acid ester, containing a protected sulfhydryl group is first reacted with the amino group of PAP. The sulfhydryl group is then deprotected with hydroxylamine and reacted with an antibody that has been modified with a maleimide group with SMCC (Figure 16A). The second compound, [4-nitro-3-(1-chlorocarbonyloxy-ethyl)phenyl]methyl-3-(2-pyridyldithiopropionic acid) ester, is similarly reacted as shown in Figure 16B. Antibody-PAP conjugates were predominantly in the ratio of 1:1. The benzylic carbon-oxygen bond is photosensitive. On irradiation at 365 nm, fully active PAP is released after spontaneous release of CO_2.

VII. COUPLING THROUGH CARBOHYDRATE RESIDUES

A. USE OF INTRINSIC CARBOHYDRATE MOIETIES

The carbohydrate moieties of antibodies and toxins are potentially valuable sites for cross-linking. Many toxins such as ricin, CVF, and abrin as well as targeting agents including immunoglobulins and α_2-macroglobulin are glycoproteins.[100-103] Olsnes and Pihl[100] have prepared an abrin immunotoxin using the oligosaccharide side chain of the toxin. The amino groups in abrin are first blocked with formaldehyde and $NaBH_4$ before reaction with periodate. After oxidation, the oligosaccharides of the abrin B chain generate aldehyde groups which form Schiff bases with free amino groups of the antibody. Conjugates formed are stabilized with $NaBH_4$. Unfortunately, the biological activity of this conjugate is not reported. Intact ricin was also conjugated to acetylcholine receptor after periodate oxidation of the carbohydrate moiety.[104]

Similar procedure has been used to conjugate daunomycin and ouabain to melanotropin and thyrotropin.[105] The sugar residues in the drugs were first oxidized by periodate to generate the aldehydes which form Schiff bases with the peptide hormone. The Schiff bases were reduced with $NaBH_4$.

In a different approach, Susuki et al.[106] activated the glycoside moiety of antibodies by cyanogen bromide prior to the binding of drugs. In this way, mitomycin C, daunomycin and adriamycin have been linked to anti-α-fetoprotein antibodies.

B. USE OF POLYSACCHARIDE SPACERS

Binding of drugs to antibodies has been carried out via a carbohydrate spacer, dextran, by Hurwitz et al.[107-110] Dextrans of 10,000 and 40,000 Da were oxidized by sodium periodate. Drugs such as cytosine arabinoside, bleomycin, daunomycin, and adriamycin were bound through their amino groups to the generated aldehyde functions of the oxidized dextran. Antibodies were added 24 h later. The Schiff bases thus formed were reduced by sodium borohydride or cyanoborohydride, depending on the drug sensitivity to the reducing agent. The arabinoside-dextran conjugate was completely reduced, daunomycin-dextran conjugate was partially reduced and adriamycin-dextran conjugate was not reduced at all, but they were stable in phosphate-buffered saline. The antibody bound to periodate-oxidized dextran was stable without reduction.[109]

Another derivative of dextran that has been employed is carboxymethyldextran which is obtained by reaction with monochloroacetic acid. Condensation of hydrazine with carboxymethyldextran effected by carbodiimide results in the formation of dextran-hydrazide which reacts with the carbonyl group of daunomycinone moiety of either daunomycin or adriamycin.[111-113] The same hydrazide was utilized for the binding of 5-fluorouridine derivatives. Periodate oxidation of the ribose moiety of these compounds gave two aldehyde groups, which in turn reacted with dextran-hydrazide. The antibody was then cross-linked to the spacer by glutaraldehyde.[108]

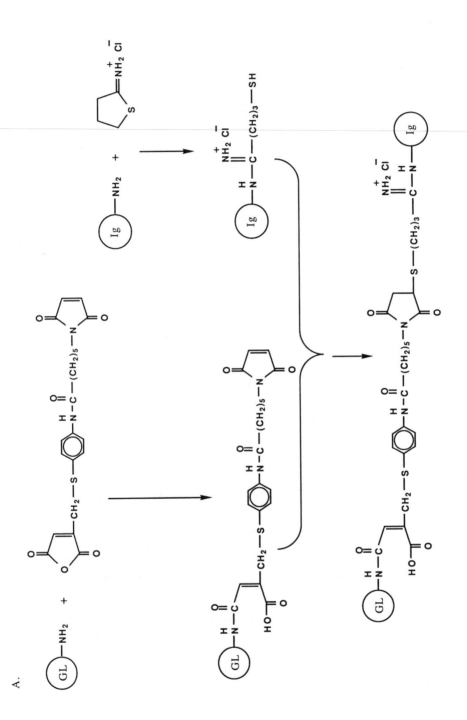

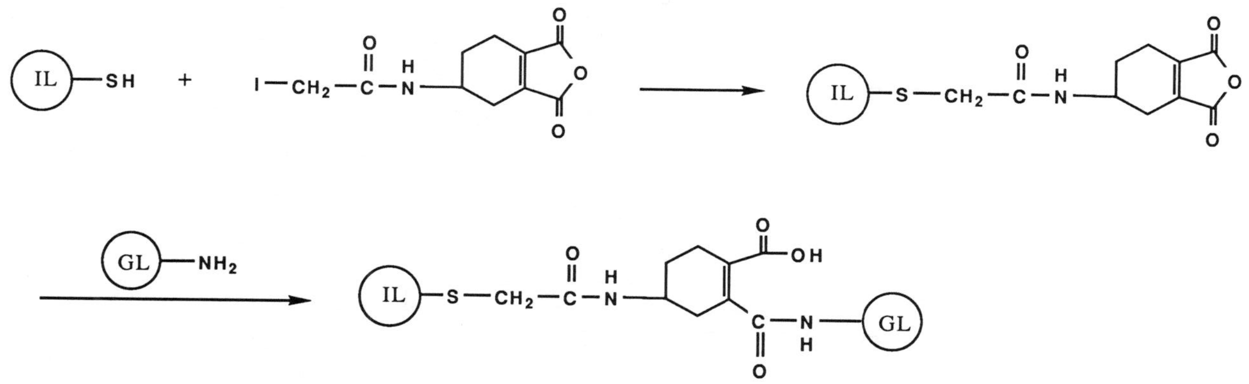

FIGURE 15. Preparation of conjugates with cross-linkers that result in acid labile bonds. (A) Cross-linking of gelonin and IgG with methylmaleic anhydride derivative. (B) Cross-linking gelonin and interleukin 2 with a tetrahydrophthalic anhydride.

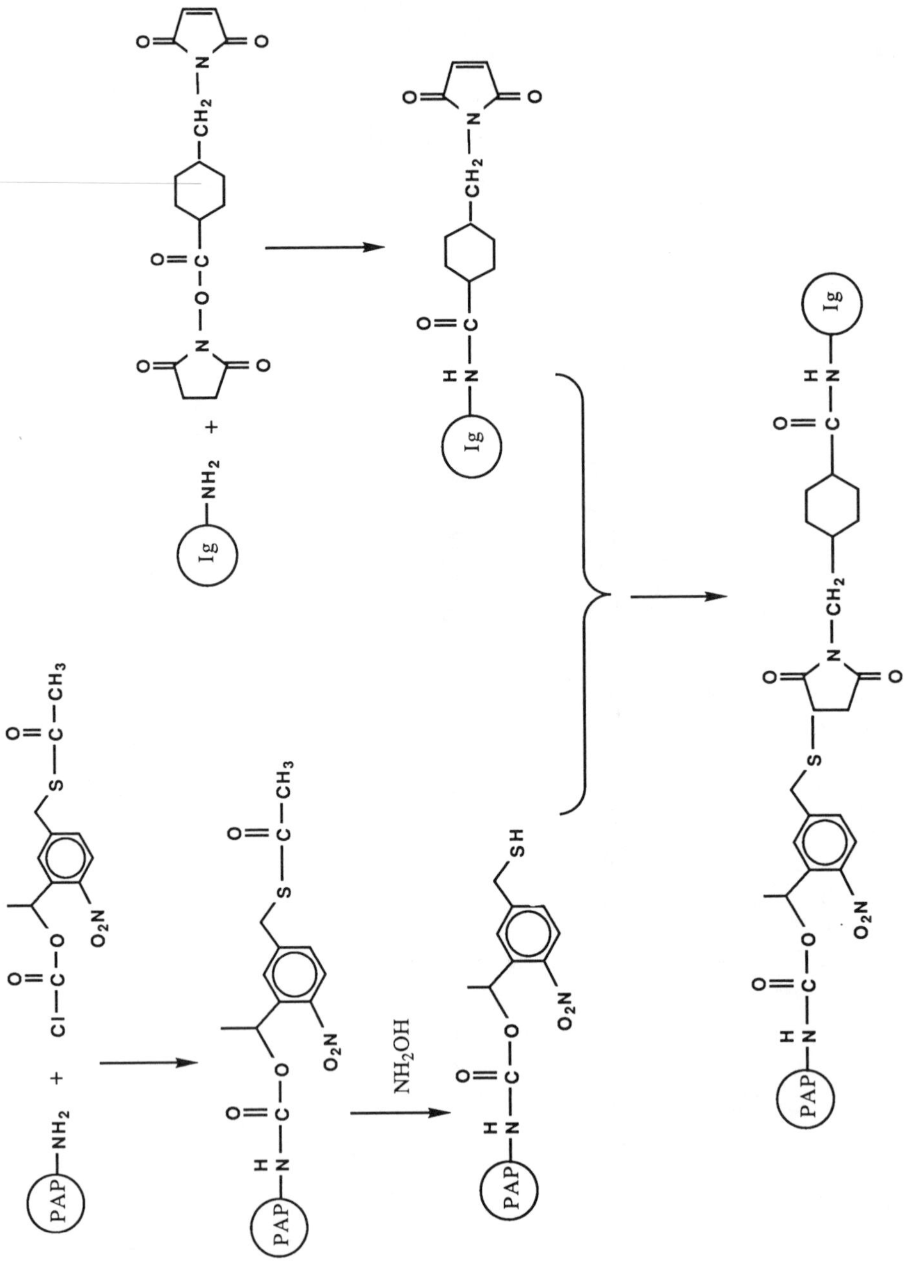

A.

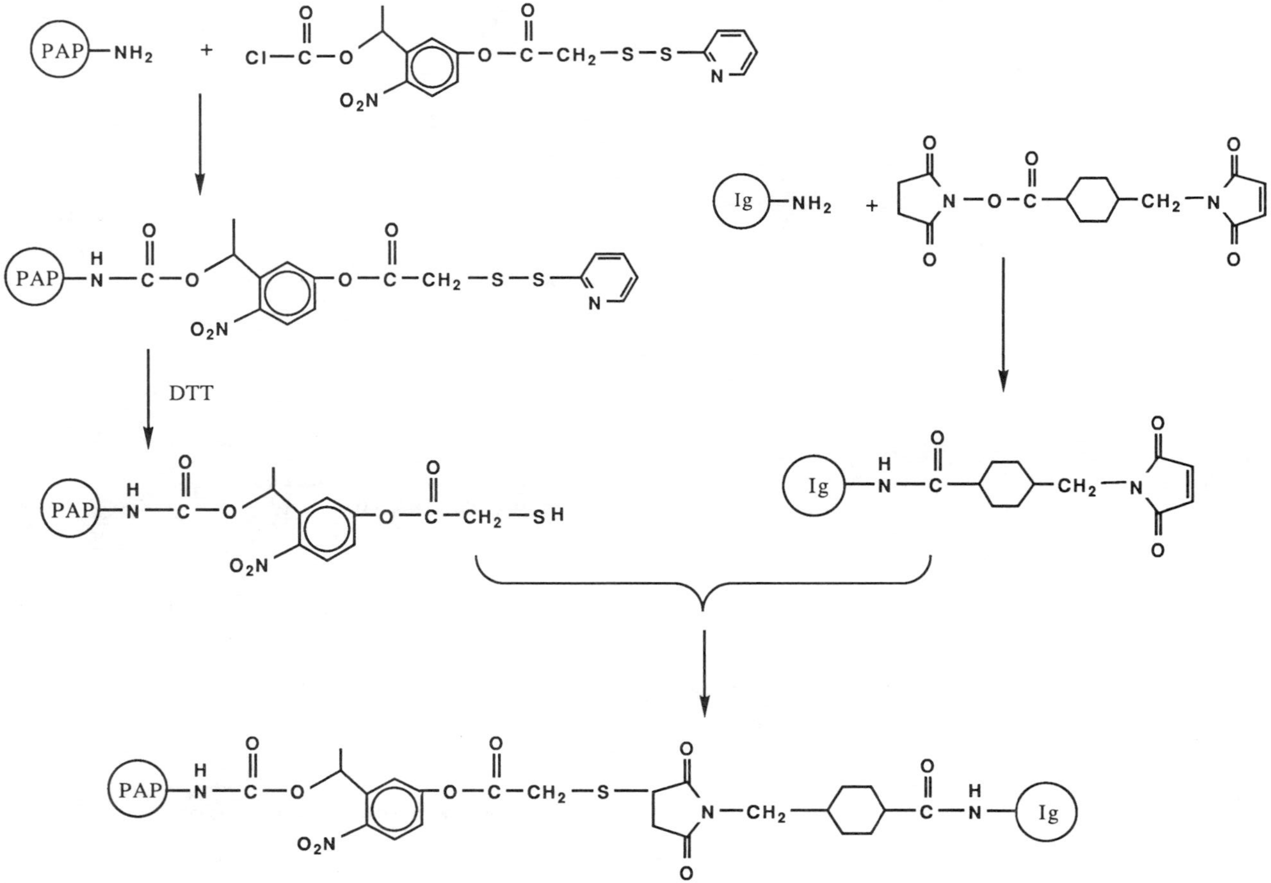

FIGURE 16. Preparation of PAP-antibody conjugates containing photocleavable bond. (A) Use of [4-nitro-3-(1-chlorocarbonylethyl)phenyl]methyl-3-acetylthioic acid ester; (B) Use of [4-nitro-3-(1-chlorocarbonyloxyethyl)phenyl]methyl-3-(2-pyridyldithiopropionic acid) ester.

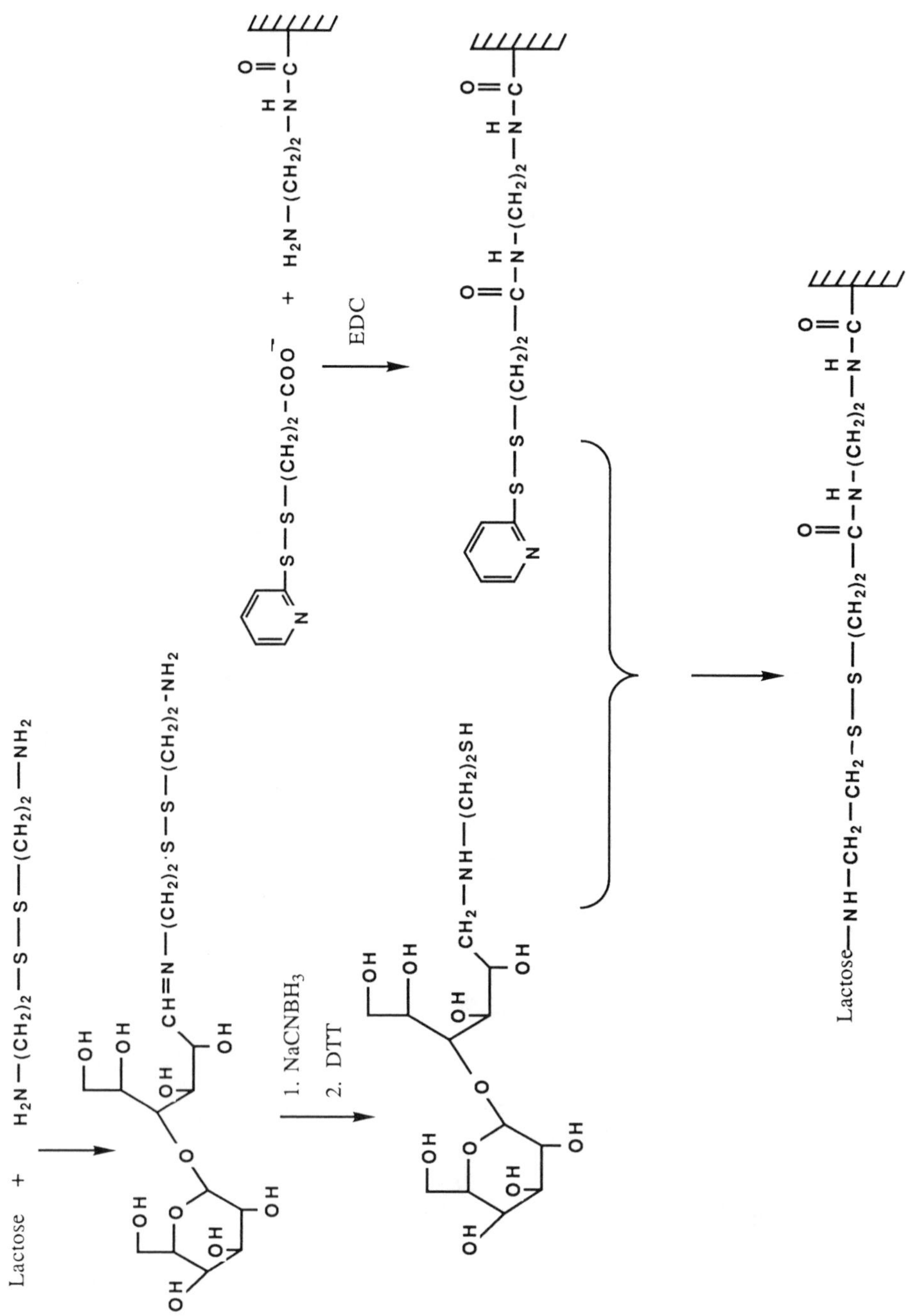
Lactose
EDC
1. NaCNBH3
2. DTT
Lactose—NH—CH2—CH2—S—S—(CH2)2—C—N—(CH2)2—N—C

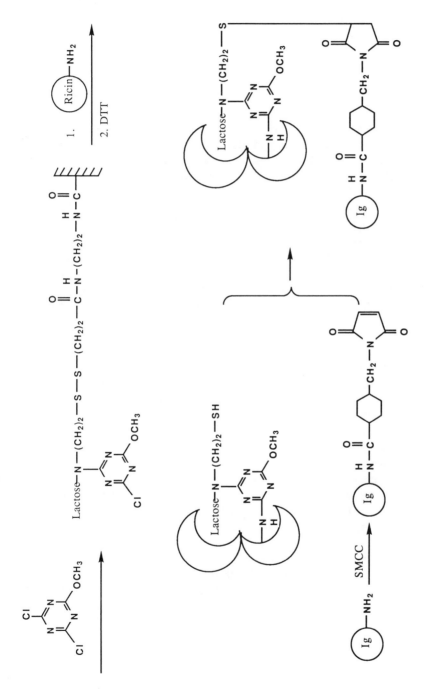

FIGURE 17. Multi-step preparation of lactose-blocked ricin-antibody conjugate.

VIII. PREPARATION OF IMMUNOTOXINS USING AVIDIN-BIOTIN LINKAGE

The strong noncovalent interaction between avidin and biotin has been exploited in the preparation of immunotoxins.[114,115] Antibodies are biotinylated with biotinyl-*N*-hydroxysuccinimide ester[116] and toxin A chain is coupled to avidin by SPDP. Association of biotin and avidin forms the immunotoxin.

IX. MISCELLANEOUS COUPLING METHODS

A. DIRECT COUPLING

One step direct coupling of toxins or drugs to antibodies and other proteins is carried out with either glutaraldehyde or water-soluble carbodiimide, particularly EDC and 1-cyclohexyl-3-[2-morpholinyl(4)-ethyl]carbodiimide (CMC). Neocarzinostatin was linked to antibodies with EDC.[117] EDC and CMC were used to conjugate amatoxins and phallotoxins to various proteins and poly(amino acids) by incubating the components at room temperature for 24 h.[118] Formyl-methionyl-leucyl-phenylalanine which is chemotactic for mononuclear and polymorphonuclear leukocytes was coupled to IgG and IgM antibodies in a CMC-mediated reaction.[119] Likewise, methotrexate was coupled to IgG with EDC.[120]

Purothionin, a low molecular weight polypeptide from barley flour that is especially toxic to dividing cells, was conjugated to an antimelanoma monoclonal antibody with EDC. Purothionin was modified with *O*-methylisourea and then conjugated to the monoclonal antibody using the water-soluble carbodiimide. The conjugate was purified by gel filtration.

Diphtheria toxin and γ-globulin were conjugated at pH 8.0 with 0.04% glutaraldehyde for 1 h at room temperature.[121] Daunomycin was bound directly to antibodies using glutaraldehyde.[122-124]

B. MULTI-STEP PROCEDURES

Moroney et al.[125] prepared a lactose-blocked ricin-monoclonal antibody conjugate through a series of reactions involving an affinity column as illustrated in Figure 17. Lactose was introduced into a free sulfhydryl group with cystamine by reductive alkylation followed by reduction with DTT. The lactose derivative was bound to pyridyldithio-activated polyacrylamide beads through the disulfide interchange reaction. The ligand was activated with the bifunctional cross-linking reagent 2,4-dichloro-6-methoxytriazine. On binding to the lactose moiety of the affinity beads, ricin was covalently bound to the cross-linker. The lactose-ricin complex was released from the solid support by DTT. The free sulfhydryl group of the complex was finally reacted with the maleimide group introduced into monoclonal antibody J5 by SMCC. The conjugate was purified on a series of affinity chromatography and was 10-fold less toxic than the native ricin.

Tsukada et al.[126] conjugated anti-α-fetoprotein antibody to daunomycin with poly-L-glutamate via a series of reactions as shown in Figure 18. The polyglutamic acid spacer was introduced 2-pyridyldithio groups with SPDP. Daunomycin was coupled to the spacer with EDC. After isolation the daunomycin labeled polyglutamate was conjugated to the antibody which contained the maleimide group after reacting with MBS.

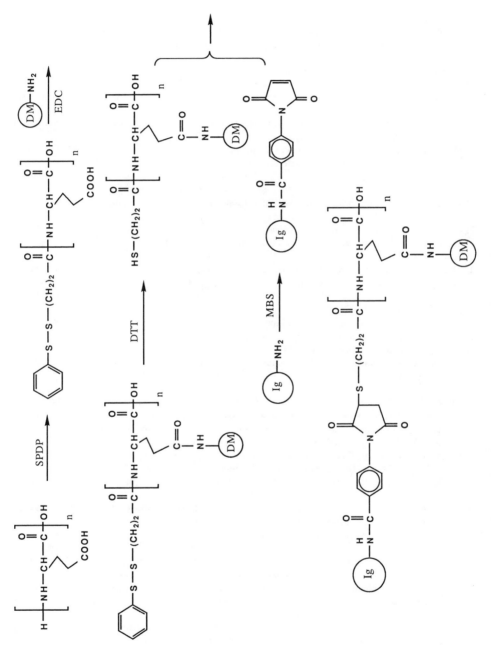

FIGURE 18. Preparation of daunomycin (DM) with IgG through a spacer polyglutamic acid.

REFERENCES

1. **Ghose, T. and Blair, A. H.,** Antibody-linked cytotoxic agents in the treatment of cancer: current status and future prospects, *J. Natl. Cancer Inst.,* 61, 657, 1978.
2. **Frankel, A. E., Ed.,** *Immunotoxins,* Klumer Academic Publishers, Boston, 1988.
3. **Vogel, C.-W., Ed.,** *Immunoconjugates: Antibody Conjugates in Radioimaging and Therapy of Cancer,* Oxford University Press, Oxford, 1987.
4. **Volkman, D. J., Ahmad, A., Fauci, A. S., and Neville, D. M., Jr.,** Selective abrogation of antigen-specific human B cell responses by antigen-ricin conjugates, *J. Exp. Med.,* 156, 634, 1982.
5. **Killen, J. A. and Lindstrom, J. M.,** Specific killing of lymphocytes that cause experimental autoimmune myathenia gravis by ricin toxin-acetylcholine receptor conjugates, *J. Immunol.,* 133, 2549, 1984.
6. **Thorpe, P. E. and Ross, W. C. J.,** The preparation and cytotoxic properties of antibody-toxin conjugates, *Immunol. Rev.,* 62, 119, 1982.
7. **Barbieri, L. and Stirpe, F.,** Ribosome-inactivating proteins from plants: Properties and possible uses, *Cancer Surveys,* 1, 489, 1982.
8. **Cumber, A. J., Forrester, J. A., Foxwell, B. M. J., Ross, W. C. J., and Thorpe, P. E.,** Preparation of antibody-toxin conjugates, *Methods Enzymol.,* 112, 207, 1985.
9. **Fulton, R. J., Blakey, D. C., Knowles, P. P., Uhr, J. W., Thorpe, P. E., and Vitetta, E. S.,** Production of ricin A1, A2 and B chains and characterization of their toxicity, *J. Biol. Chem.,* 261, 5314, 1986.
10. **Olsnes, S. and Pihl, A.,** Toxic lectins and related proteins, in *Molecular Action of Toxins and Viruses,* Cohen, P. and van Heynigen, S., Eds., New York, Elsevier Biomedical Press, 1982, 51.
11. **Houston, L. L., Ramakrishnan, S., and Hermodson, M. A.,** Seasonal variations in different forms of pokeweed antiviral protein, a potent inactivator of ribosomes, *J. Biol. Chem.,* 258, 9601, 1983.
12. **Irvin, J. D.,** Pokeweed antiviral protein, *Pharmacol. Ther.,* 21, 371, 1983.
13. **Sandvig, K., Olsnes, S., and Pihl, A.,** Chemical modifications of the toxic lectins abrin and ricin, *Eur. J. Biochem.,* 84, 323, 1978.
14. **Blakey, D. C., Wawrzynczak, E. J., Stirpe, F., and Thorpe, P. E.,** Anti-tumor activity of a panel of anti-Thy 1.1 immunotoxins made with different ribosome-inactivating proteins, in *Membrane Mediated Cytotoxicity,* Bonavida, B. and Collier, R. J., Eds., UCLA Symposia on Molecular and Cellular Biology, New Series, Vol. 45, 1986, 195.
15. **Houston, L. L. and Ramakrishnan, S.,** Immunotoxins made with toxins and hemitoxins other than ricin, in *Immunoconjugates, Antibody Conjugates in Radioimaging and Therapy of Cancer,* Vogel, C.-W., Ed., Oxford University Press, New York, 1987, 71.
16. **Masuho, Y., Kishida, K., Saito, M., Umemoto, N., and Hara, T.,** Importance of the antigen-binding valency and the nature of the cross-linking bond in ricin A-chain conjugates with antibody, *J. Biochem.,* 91, 1583, 1982.
17. **Jansen, F. K., Blythman, H. E., Carriere, D., Casellas, P., Gros, O., Gros, P., Laurent, J. C., Paolucci, F., Pau, B., Poncelet, P., Richer, G., Vidal, H., and Viosin, G. A.,** Immunotoxins: Hybrid molecules combining high specificity and potent cytotoxicity, *Immunol. Rev.,* 62, 185, 1982.
18. **Houston, L. L. and Nowinski, R. C.,** Cell-specific cytotoxicity expressed by a conjugate of ricin and murine monoclonal antibody directed against Thy 1.1 antigen, *Cancer Res.,* 41, 3913, 1981.
19. **Thorpe, P. E., Mason, D. W., Brown, A. N. F., Simmonds, S. J., Ross, W. C. J., Cumber, A. J., and Forrester, J. A.,** Selective killing of malignant cells in a leukaemic rat bone marrow using an antibody-ricin conjugate, *Nature (London),* 297, 594, 1982.
20. **Edwards, D. C., Ross, W. C. J., Cumber, A. J., McIntosh, D., Smith, A., Thorpe, P. E., Brown, A., Williams, R. H., and Davies, A. J. S.,** A comparison of the *in vitro* and *in vivo* activities of conjugates of anti-mouse lymphocyte globulin and abrin, *Biochim. Biophys. Acta,* 717, 272, 1982.
21. **Thorpe, P. E., Brown, A. N. F., Ross, W. C. J., Cumber, A. J., Detre, S. I., Edwards, D. C., Davies, A. J. S., and Stirpe, F.,** Cytotoxicity acquired by conjugation of an anti-Thy 1.1 monoclonal antibody and the ribosome-inactivating protein, gelonin, *Eur. J. Biochem.,* 116, 447, 1981.
22. **Colombatti, M., Nabholz, M., Gros, O., and Brown, C.,** Selective killing of target cells by antibody-ricin A-chain or antibody-gelonin hybrid molecules: Comparison of cytotoxic potency and use in immunoselection procedures, *J. Immunol.,* 131, 3091, 1983.
23. **Wiels, J., Junqua, S., Dujardin, P., Le Pecq, J. B., and Tursz, T.,** Properties of immunotoxins against a glycolipid antigent associated with Burkitt's lymphoma, *Cancer Res.,* 44, 129, 1984.
24. **Ramakrishnan, S. and Houston, L. L.,** Comparison of the selective cytotoxic effects of immunotoxins containing ricin A chain or pokeweed antiviral protein and anti-Thy 1.1 monoclonal antibodies, *Cancer Res.,* 44, 201, 1984.
25. **Thorpe, P. E., Brown, A. N. F., Bremner, J. A. G., Jr., Foxwell, B. M. J., and Stripe, F.,** An immunotoxin composed of monoclonal anti-Thy 1.1 antibody and a ribosome-inactivating protein from Saponaria officinalis: Potent antitumor effects *in vitro* and *in vivo, J. Natl. Cancer Inst.,* 75, 151, 1985.

26. **Pelham, J. M., Gray, J. D., Flannery, G. R., Pimm, M. V., and Baldwin, R. W.,** Interferon-α conjugation to human osteogenic sarcoma monoclonal antibody 791T/36, *Cancer Immunol. Immunother.,* 15, 210, 1983.

27. **Flannery, G. R., Pelham, J. M., Gray, J. Dl., and Baldwin, R. W.,** Immunomodulation: NK cells activated by interferon-conjugated monoclonal antibody against human osteosarcoma, *Eur. J. Cancer Clin. Oncol.,* 20, 791, 1984.

28. **Vogel, C.-W. and Müller-Eberhard, H. J.,** Induction of immune cytolysis: tumor-cell killing by complement is initiated by covalent complex of monoclonal antibody and stable C3/C5 convertase, *Proc. Natl. Acad. Sci., U.S.A.,* 78, 7707, 1981.

29. **Ross, W. C. J., Thorpe, P. E., Cumber, A. J., Edwards, D. C., Hinson, C. A., and Davies, A. J. S.,** Increased toxicity of diphtheria toxin for human lymphoblastoid cells following covalent linkage to anti-(human lymphocyte)globulin or its F(ab′)2 fragment, *Eur. J. Biochem.,* 104, 381, 1980.

30. **Thorpe, P. E., Detre, S. I., Mason, D. W., Cumber, A. J., and Ross, W. C. J.,** Monoclonal antibody therapy: model experiments with toxin-conjugated antibodies in mice and rats, *Haematol. Blood Transfus.,* 28, 107, 1983.

31. **Masuho, Y., Kishida, K., and Hara, T.,** Targeting of the antiviral protein from *Phytolacca americana* with an antibody, *Biochem. Biophys. Res. Commun.,* 105, 462, 1982.

32. **Vogel, C.-W.,** Antibody conjugate without inherent toxicity: the targeting of cobra venom factor and other biological response modifiers, in *Immunoconjugates: Antibody Conjugates in Radioimaging and Therapy of Cancer,* Vogel, C.-W., Ed., Oxford University Press, New York, 1987, 170.

33. **Reiter, Y. and Fishelson, Z.,** Targeting of complement to tumor cells by heteroconjugates composed of antibodies and of the complement component C_{3b}, *J. Immunol.,* 142, 2771, 1989.

34. **Jung, G., Kohnlein, W., and Luders, G.,** Biological activity of the antitumor protein neocarzinostatin coupled to a monoclonal antibody by *N*-succinimidyl-3-(2′-pyridyldithio)propionate, *Biochem. Biophys. Res. Commun.,* 101, 599, 1981.

35. **Lambert, J. M., Blättler, W. A., McIntyre, G. D., Goldmacher, V. S., and Scott, C. F., Jr.,** Immunotoxins containing single chain ribosome-inactivating proteins, in *Immunotoxins,* Frankel, A. E., Ed., Kluwer Academic Publishers, Boston, 1988, 175.

36. **Bjorn, M. J., Groetsema, G., and Scalapino, L.,** Antibody-Pseudomonas exotoxin A conjugates cytotoxic to human breast cancer cells *in vitro, Cancer Res.,* 46, 3262, 1986.

37. **Lambert, J. M., Senter, P. D., Yau-Young, A., Blättler, W. A., and Goldmacher, V. S.,** Purified immunotoxins that are reactive with human lymphoid cells: monoclonal antibodies conjugated to the ribosome-inactivating proteins gelonin and the pokeweed antiviral proteins, *J. Biol. Chem.,* 260l, 12035, 1985.

38. **Scott, C. J., Jr., Goldmacher, V. S., Lambert, J. M., Chari, R. V., Bolender, S., Gauthier, M. N., and Blattler, W. A.,** The antileukemic efficacy of an immunotoxin composed of a monoclonal anti-Thy-1 antibody disulfide linked to the ribosome-inactivating protein gelonin, *Cancer Immunol. Immunother.,* 25, 31, 1987.

39. **Ozawa, S., Ueda, M., Ando, N., Abe, O., Minoshima, S., and Shimizu, N.,** Selective killing of squamous carcinoma cells by an immunotoxin that recognizes the EGF receptor, *Int. J. Cancer,* 43, 152, 1989.

40. **Forrester, J. A., McIntosh, D. P., Cumber, A. J., Ross, W. C. J., and Parnell, G. D.,** Delivery of ricin and abrin A-chain to human carcinoma cells in culture following covalent linkage to monoclonal antibody LICR-LOND-Fib 75, *Cancer Drug Delivery,* 1, 283, 1984.

41. **Forrester, J. A., McIntosh, D. P., Cumber, A. J., Ross, W. C. J., and Parnell, G. D.,** The delivery of ricin and abrin A-chains to human carcinoma cells in culture following covalent linkage to a monoclonal antibody LICR/LOND/Fib 75, *Cancer Drug Delivery,* 1, 283, 1984.

42. **Hwang, K. M., Foon, K. A., Cheng, P. H., Pearson, J. W., and Oldham, R. K.,** Selective antitumor effect on L10 hepatocarcinoma cells of a potent immunoconjugate composed of the A chain of abrin and a monoclonal antibody to a hepatoma-associated antigen, *Cancer Res.,* 44, 4578, 1984.

43. **Thorpe, P. E.,** Antibody carriers of cytotoxic agents in cancer therapy: a review, in *Monoclonal Antibodies '84: Biological and Clinical Applications,* Pinchera, A., Doria, G., Dammacco, F., and Bargellesi, E., Eds., Editrice Kurtis, Milan, 1985, 475.

44. **Fulton, R. J., Uhr, J. W., and Vitetta, E. S.,** The effect of antibody valency and lysosomotropic amines on the synergy between ricin A chain and ricin B chain containing immunotoxins, *J. Immunol.,* 136, 3103, 1986.

45. **Weiner, L. M., O'Dwyer, J., Kitson, J., Comis, R. L., Frankel, A. E., Bauer, R. J., Konrad, M. S., and Groves, E. S.,** Phase I evaluation of an anti-breast carcinoma monoclonal antibody 260F9-recombinant ricin A chain immunoconjugate, *Cancer Res.,* 49, 4062, 1989.

46. **Vitetta, E. S., Cushley, W., and Uhr, J. W.,** Synergy of ricin A-chain-containing immunotoxins and ricin B-chain-containing immunotoxins in the *in vitro* killing of neoplastic human B cells, *Proc. Natl. Acad. Sci. U.S.A.,* 80, 6332, 1983.

47. **Vitetta, E. S., Fulton, R. J., and Uhr, J. W.,** Cytotoxicity of a cell-reactive immunotoxin containing ricin A-chain is potentiated by an anti-immunotoxin containing ricin B chain, *J. Exp. Med.,* 160, 341, 1984.

48. **Merriam, J. C., Lyon, H. S., and Char, D. H.,** Toxicity of monoclonal F(ab′)$_2$: ricin A conjugate for retinoblastoma in vitro, *Cancer Res.,* 44, 3178, 1984.

49. **Fizgerald, D., Morris, R. E., and Saelinger, C. B.,** Receptor-mediated internalization of Pseudomonas toxin by mouse fibroblasts, *Cell,* 21, 867, 1980.

50. **Helenius, A., Kartenbeck, J., Simons, K., and Fries, E.,** On the entry of Semliki forest virus into BHK-21 cells, *J. Cell Biol.,* 84, 404, 1980.

51. **Keen, J. H., Maxfield, F. R., Hardegree, M. C., and Habig, W. H.,** Receptor-mediated endocytosis of diphtheria toxin by cell in culture, *Proc. Natl. Acad. Sci., U.S.A.,* 79, 2912, 1982.

52. **O'keefe, D. O. and Draper, R. K.,** Characterization of a transferrin-diphtheria toxin conjugate, *J. Biol. Chem.,* 260, 932, 1985.

53. **Singh, V., Sairam, M. R., Bhargavi, G. N., and Akhras, R. G.,** Hormonotoxins: preparation and characterization of ovine luteinizing hormone-gelonin conjugate, *J. Biol. Chem.,* 264, 3089, 1989.

54. **Singh, V. and Sairam, M. R.,** Hormonotoxins. I. Strategy for synthesis of ovine luteinizing hormone-gelonin conjugate bearing the toxin in the β-subunit, *Int. J. Pept. Protein Res.,* 33, 22, 1989.

55. **Raso, V. and Basala, M.,** Monoclonal antibodies as cell-targeted carriers of covalently and non-covalently attached toxins, in *Receptor-Mediated Targeting of Drugs,* Gregoriadis, G., Ed., Vol. 2, Plenum Press, New York, 1985, 119.

56. **Raso, V., Watkins, S., Slayter, H., Fehrmann, C., and Nerbonne, S.,** Subcellular compartmentalization and the potency of ricin A chain cytotoxins, in *Membrane Mediated Cytotoxicity,* Bonavida, B. and Collier, R. J., Eds., UCLA Symposia on Molecular and Cellular Biology, New Series, Vol. 45, 1986, 131.

57. **Martin, H. B. and Houston, L. L.,** Arming α$_2$-macroglobulin with ricin A chain forms a cytotoxic conjugate that inhibits protein synthesis and kills human fibroblasts, *Biochem. Biophys. Acta,* 762, 128, 1983.

58. **Cawley, D. B., Simpson, D. L., and Herschman, H. R.,** Asialoglycoprotein receptor mediates the toxic effects of an asialofetuin-diphtheria toxin fragment A conjugate on cultured rat hepatocytes, *Proc. Natl. Acad. Sci., U.S.A.,* 78, 3383, 1981.

59. **Simpson, D. L., Cawley, D. B., and Herschmann, H. R.,** Killing of cultured hepatocytes by conjugates of asialofetuin and EGF linked to the A chains of ricin or diphtheria toxin, *Cell,* 29, 4669, 1982.

60. **Chang, T. and Kullberg, D. W.,** Studies of the mechanism of cell intoxication by diphtheria toxin fragment A-asialoorosomucoid hybrid toxins, *J. Biol. Chem.,* 257, 12563, 1982.

61. **Stirpe, F., Olsnes, S., and Pihl, A.,** Gelonin, a new inhibitor of protein synthesis, nontoxic to intact cells: isolation, characterization, and preparation of cytotoxic complexes with concanavalin A, *J. Biol. Chem.,* 255, 6947, 1980.

62. **Cawley, D. B., Herschman, H. R., Gilliland, D. F., and Collier, R. J.,** Epidermal growth factor-toxin A chain conjugates: EGF-ricin A is a potent toxin while EGF-diphtheria fragment A is nontoxic, *Cell,* 22, 563, 1980.

63. **Herschman, H. R.,** The role of binding ligand in toxin hybrid proteins: a comparison of EGF-ricin, EGF-ricin A-chain, and ricin, *Biochem. Biophys. Res. Commun.,* 124, 155, 1984.

64. **Oeltmann, T. N.,** Synthesis and *in vitro* activity of a hormone-diphtheria toxin fragment A hybrid, *Biochem. Biophys. Res. Commun.,* 133, 430, 1985.

65. **Umemoto, N., Kato, Y., and Hara, T.,** Cytotoxicities of two disulfide-bond-linked conjugates of methotrexate with monoclonal anti-MM46 antibody, *Cancer Immuno. Immunother.,* 28, 9, 1989.

66. **Jansen, F. K., Blythman, H. E., Carriere, D., Casellas, P., Diaz, J., Gros, P., Hennequin, J. R., Paolucci, F., Pau, B., Poncelet, P., Richer, G., Salhi, S. L., Vidal, H., and Voisin, G. A.,** High specific cytotoxicity of antibody-toxin hybrid molecules (immunotoxins) for target cells, *Immunol. Lett.,* 2, 97, 1980.

67. **Gros, O., Gros, P., Jansen, F. K., and Vidal, H.,** Biochemical aspects of immunotoxin preparation, *J. Immunol. Meth.,* 81, 283, 1985.

68. **Raso, V. and Basala, M.,** A highly cytotoxic human transferrin-ricin A chain conjugate used to select receptor-modified cells. *J. Biol. Chem.,* 259, 1143, 1984.

69. **Raso, V. and Basala, M.,** Study of the transferrin receptor using a cytotoxic human transferrin-ricin A chain conjugate, in *Receptor-Mediated Targeting of Drugs,* Gregoriadis, G., Ed., Vol. 2, Plenum Press, New York, 1985, 73.

70. **King, T. P., Li, Y., and Kochoumian, L.,** Preparation of protein conjugates via intermolecular disulfide bond formation, *Biochemistry,* 17, 499, 1978.

71. **Thorpe, P. E., Wallace, P. M., Knowles, P. P., Relf, M. G., Brown, A. N., Watson, G. J., Knyba, R. E., Wawrzynczak, E. J., and Blakey, D. C.,** New coupling agents for the synthesis of immunotoxins containing a hindered disulfide bond with improved stability *in vivo, Cancer Res.,* 47, 5924, 1987.

72. **Ghetie, V., Till, M. A., Ghetie, M. A., Uhr, J. W., and Vitetta, E. S.,** Large scale preparation of an immunoconjugate constructed with human recombinant CD4 and deglycosylated ricin A chain, *J. Immunol. Methods,* 126, 135, 1990.

73. **Shimisu, N., Miskimins, W. K., and Shimizu, Y.,** A cytotoxic epidermal growth factor cross-linked to diphtheria toxin A-fragment, *FEBS Lett.,* 118, 274, 1980.

74. **Miskimins, W. K. and Shimizu, N.,** Synthesis of cytotoxic insulin cross-linked to diphtheria toxin fragment A capable of recognizing insulin receptors, *Biochem. Biophys. Res. Commun.,* 91, 143, 1979.

75. **Oeltmann, T. N. and Forbes, J. T.,** Inhibition of mouse spleen cell function by diphtheria toxin fragment A coupled to anti-mouse Thy-1.2 and by ricin A chain coupled to anti-mouse IgM, *Arch Biochem. Biophys.,* 209, 362, 1981.

76. **Bacha, P., Murphy, J. R., and Reichlin, S.,** Thyrotropin-releasing hormone-diphtheria toxin-related polypeptide conjugates, *J. Biol. Chem.,* 258, 1565, 1983.

77. **Masuho, Y., Hara, T., and Noguchi, T.,** Preparation of hybrid of fragment Fab′ of antibody and fragment A of diphtheria toxin and its cytotoxicity, *Biochem. Biophys. Res. Commun.,* 90, 320, 1979.

78. **Chang, T. and Neville, D. M., Jr.,** Artificial hybrid protein containing a toxic protein fragment and a cell membrane receptor-binding moiety in a disulfide conjugate, *J. Biol. Chem.,* 252, 1505, 1977.

79. **Oeltmann, T. N. and Heath, E. C.,** A hybrid protein containing the toxin subunit of ricin and the cell-specific subunit of human chorionic gonadotropin. I. Synthesis and characterization, *J. Biol. Chem.,* 254, 1022, 1979.

80. **Ellman, G. L.,** Tissue sulfhydryl groups, *Arch. Biochem. Biophys.,* 82, 70, 1959.

81. **Pirker, R., Fitzgerald, D. J. P., Hamilton, T. Cl., Ozols R. F., Laird, W., Frankel, A. E., Willingham, M. C., and Pastan, I.,** Characterization of immunotoxins active against ovarian cancer cell lines, *J. Clin. Invest.,* 76, 1261, 1986.

82. **FitzGerald, D. J., Willingham, M. C., and Pastan, I.,** Pseudomonas Exotoxin-immunotoxin, in *Immunotoxins,* Frankel, A. E., Ed., Kluwer Academic Publishers, Boston, 1988, 161.

83. **Fitzgerald, D. J. P., Padmanabhan, R., Pastan, I., and Willingham, M. C.,** Adenovirus-induced release of epidermal growth factor and pseudomonas toxin into the cytosol of KB cells during receptor-mediated endocytosis, *Cell,* 32, 607, 1983.

84. **Fitzgerald, D. J. P., Trowbridge, I. S., Pastan, I., and Willingham, M. C.,** Enhancement of toxicity of antitransferrin receptor antibody-Pseudomonas exotoxin conjugates by adenovirus, *Proc. Natl. Acad. Sci. U.S.A.,* 80, 4134, 1983.

85. **Miyazaki, H., Beppu, M., Terao, T., and Osawa, T.,** Preparation of antibody (IgG)-ricin A chain conjugate and its biologic activity, *Gann,* 71, 766, 1980.

86. **Raso, V. and Griffin, T.,** Specific cytotoxicity of a human immunoglobulin-directed Fab′-ricin A chain conjugate, *J. Immunol.,* 125, 2610, 1980.

87. **Wawrzynczak, E. J. and Thorpe, P. E.,** Methods for preparing immunotoxins: Effect of the linkage on activity and stability, in *Immunoconjugates, Antibody Conjugates in Radioimaging and Therapy of Cancer,* Vogel, E.-W., Ed., New York, Oxford University Press, 1987, 28.

88. **Thorpe, P. E., Ross, W. C. J., Brown, A. N. F., Myers, C. D., Cumber, A. J., Foxwell, B. M. J., and Forrester, J. T.,** Blockade of the galactose-binding sites of ricin by its linkage to antibody specific cytotoxic effects of the conjugate, *Eur. J. Biochem.,* 140, 63, 1984.

89. **Youle, R. J. and Neville, D. M., Jr.,** Anti-Thy 1.2 monoclonal antibody linked to ricin is a potent cell type-specific toxin, *Proc. Natl. Acad. Sci., U.S.A.,* 77, 5483, 1980.

90. **Youle, R. J. and Neville, D. M., Jr.,** Kinetics of protein synthesis inactivation by ricin-anti-Thy 1.1 monoclonal antibody hybrids. Role of the ricin B subunit demonstrated by reconstitution, *J. Biol. Chem.,* 257, 1598, 1982.

91. **Vallera, D. A., Ash, R. C., Zanjani, E. D., Kersey, J. H., Le Bien, T. W., Beverley, P. C. L., Neville, D. M., Jr., and Youle, R. J.,** Anti-T-cell reagents for human bone marrow transplantation: Ricin linked to three monoclonal antibodies, *Science,* 222, 512, 1983.

92. **Vallera, D. A. and Myers, D. E.,** Immunotoxins containing ricin, in *Immunotoxins,* Frankel, A. E., Ed., Kluwer Academic Publishers, Boston, 1988, 141.

93. **Volkman, D. J., Ahmad, A., Fauci, A. S., and Neville, Dl. M., Jr.,** Selective abrogation of antigen-specific human B cell responses by antigen-ricin complexes, *J. Exp. Med.,* 156, 634, 1982.

94. **Bjorn, M. J., Groetsema, G., and Scalapino, L.,** Antibody-*Pseudomonas* exotoxin A conjugates cytotoxic to human breast cancer cells in vitro, *Cancer Res.,* 46, 3262, 1986.

95. **Myers, D. E., Uckun, F. M., Swaim, S. E., and Vallera, D. A.,** The effects of aromatic and aliphatic maleimide crosslinkers on anti-CD5 ricin immunotoxins, *J. Immunol. Methods,* 121, 129, 1989.

96. **Edwards, D. C., Smith, A., Ross, W. C. J., Cumber, A. J., Thorpe, P. E., and Davies, A. J. S.,** The effect of abrin, antilymphocyte globulin and their conjugates on the immune response of mice to sheep red blood cells, *Experientia,* 37, 256, 1981.

97. **Edwards, D. C., Ross, W. C. J., Cumber, A. J., McIntosh, D., Smith, A., Thorpe, P. E., Brown, A. N. F., Williams, R. H., and Davies, A. J.,** A comparison of the *in vitro* activities of conjugates of anti-mouse lymphocyte globulin and abrin, *Biochim. Biophys. Acta*, 717, 272, 1982.

98. **Blättler, W. A., Kuenzi, B. S., Lambert, J. M., and Senter, P. D.,** New heterobifunctional protein cross-linking reagent that forms an acid-labile link, *Biochemistry*, 24, 1517, 1985.

99. **Senter, P. D., Tansey, M. J., Lambert, J. M., and Blätter, W. A.,** Novel photocleavable protein crosslinking reagents and their use in the preparation of antibody-toxin conjugates, *Photochem. Photobiol.*, 42, 231, 1985.

100. **Olsnes, S. and Pihl, A.,** Chimeric toxins, *Pharmacol. Ther.*, 15, 355, 1982.

101. **Olsnes, S. and Pihl, A.,** Toxic lectins and related proteins, in *Molecular Action of Toxins and Viruses,* Cohen, P. and van Heynigen, S., Eds., New York, Elsevier Biomedical Press, 1982, 51.

102. **Pizzo, S. V. and Gonias, S. L.,** Receptor-mediated protease regulation, in *The Receptors,* Vol. 1, Conn., P. M., Ed., Academic Press, Orlando, FL, 1984, 177.

103. **Vogel, C.-W. and Müller-Eberhard, H. J.,** Cobra venom factor: improved method for purification and biochemical characterization, *J. Immunol. Meth.*, 73, 203, 1984.

104. **Killen, J. and Lindstrom, J. M.,** Specific killing lymphocytes that cause experimental autoimmune myasthenia gravis by ricin toxin-acetylcholine receptor conjugates, *J. Immunol.*, 133, 2549, 1984.

105. **Varga, J. M.,** Hormone-drug conjugates, *Methods Enzymol.*, 112, 259, 1985.

106. **Suzuki, T., Sato, E., Goto, K., Katsurada, Y., Unno, K., and Takahashi, T.,** The preparation of mitomycin C, adriamycin and daunomycin covalently bound to antibodies as improved cancer chemotherapeutic agent, *Chem. Pharm. Bull.*, 29, 844, 1981.

107. **Hurwitz, E., Maron, R., Arnon, R., Wilchek, M., and Sela, M.,** Daunomycin immunoglobulin conjugates, uptake and activity in vitro, *Eur. J. Cancer*, 14, 1213, 1978.

108. **Hurwitz, E., Kashi, R., Arnon, R., Wilchek, M., and Sela, M.,** The covalent linking of two nucleotide analogues to antibodies, *J. Med. Chem.*, 28 137, 1985.

109. **Sela, M. and Hurwitz, E.,** Conjugates of antibodies with cytotoxic drugs, in *Immunoconjugates: Antibody Conjugates in Radioimaging and Therapy of Cancer*, Vogel, C.-W., Ed., Oxford University Press, New York, 1987, 189.

110. **Manabe, Y., Tsubota, T., Haruta, Y., Okazaki, M., Haisa, S., Nakamura, K., and Kimura, I.,** Production of monoclonal antibody-bleomycin conjugate utilizing dextran T40 and the antigen-targeting cytotoxicity of the conjugate, *Biochem. Biophys. Res. Commun.*, 115, 1009, 1983.

111. **Hurwitz, E., Wilchek, M., and Phita, J.,** Soluble macromolecules as carriers for daunomycin, *J. Appl. Biochem.*, 2, 25, 1980.

112. **Hurwitz, E., Arnon, R., Sahar, E., and Danon, Y.,** A conjugate of adriamycin and monoclonal antibodies to Thy-1 antigen inhibits human neuroblastoma cells in vitro, *Ann. N.Y. Acad. Sci.*, 417, 125, 1983.

113. **Hurwitz, E., Kashi, R., Burowsky, D., Arnon, R., and Haimovich, J.,** Site-directed chemotherapy with a drug bound to antiiodiotypic antibody to a lymphoma cell-surface IgM, *Int. J. Cancer*, 31, 745, 1983.

114. **Martinez, O., Kimura, J., Gottfried, T. D., Zeicher, M., and Wofsy, L.,** Variance in cytotoxic effectiveness of antibody-toxin A hybrids, *Cancer Surveys*, 1, 373, 1982.

115. **Hashimoto, N., Takatsu, K., Masuho, Y., Kishida, K., Hara, T., and Hamaoka, T.,** Selective elimination of a B-cell subset having acceptor site(s) for T cell-replacing factor (TRF) with biotinylated antibody to the acceptor site(s) and avidin-ricin A-chain conjugate, *J. Immunol.*, 132, 129, 1984.

116. **Hofmann, K., Finn, F. M., and Kiso, Y.,** Avidin-biotin affinity columns: general methods for attaching biotin to peptides and proteins. *J. Am Chem. Soc.*, 100, 3585, 1978.

117. **Kimura, I., Ohnoshi, T., Tsubota, T., Sato, Y., Kobayashi, T., and Abe, S.,** Production of tumor antibody-neocarzinostatin (NCS) conjugate and its biological activities, *Cancer Immunol. Immunother.*, 7, 235, 1980.

118. **Faulstich, H. and Fiume, L.,** Protein conjugates of fungal toxins, *Methods Enzymol.*, 112, 225, 1985.

119. **Isturiz, M. A., Sandberg, A. L., Schiffmann, E., Wahl, S. M., and Notkins, A. L.,** Chemotactic antibody, *Science*, 200, 554, 1978.

120. **Kulkarni, P. N., Blair, A. H., and Ghose, T. I.,** Covalent binding of methotrexate to immunoglobulins and the effect of antibody-linked drug on tumor growth *in vivo, Cancer Res.*, 41, 2700, 1981.

121. **Moolten, F. L., Capparell, N. J., and Zajdel, S. H.,** Antitumor effects of antibody-diphtheria toxin conjugates. III. Cyclophosphamide-induced immune unresponsiveness to conjugate, *J. Natl. Cancer Inst.*, 55, 709, 1975.

122. **Hurwitz, E., Levy, R., Maron, R., Wilchek, M., Arnon, R., and Sela, M.,** The covalent binding of daunomycin and adriamycin to antibodies with retention of both drug and antibody, *Cancer Res.*, 35, 1175, 1975.

123. **Page, M., Belles, Isles, M., and Edmond, J. P.,** Daunomycin targeting to human colon carcinoma cells using drug anti-CEA conjugates, *Proc. Am. Assoc. Cancer Res.*, 22, 211, 1981.

124. **Page, M., Edmond, J. P., Gauthier, C., Dugour, C., and Innes, L.,** Drug targeting with monoclonal antibodies, in *Protides of the Biological Fluids,* Peeters, H., Ed., Vol. 29, Pergamon Press, Oxford, 1981, 133.
125. **Moroney, S. E., D'Alarcao, L. J., Goldmacher, V. S., Lambert, H. M., and Blättler, W. A.,** Modification of the bind site(s) of lectins by an affinity column carrying an activated galactose-terminated ligand, *Biochemistry,* 26, 8390, 1987.
126. **Tsukada, Y., Kato, Y., Umemoto, N., Takeda, Y., Hara, T., and Hirai, H.,** An anti-a-fetoprotein-daunomycin conjugate with a novel poly-L-glutamic acid derivative as intermediate drug carrier, *J. Natl. Cancer Inst.,* 73, 721, 1984.

Chapter 12

CONJUGATION OF PROTEINS TO SOLID MATRICES

I. INTRODUCTION

Proteins immobilized onto solid surfaces have found growing uses in medical[1,2] and clinical analysis,[1-3] affinity chromatography,[2-7] and synthetic chemistry applications.[2,8-16] Latex particles coated with antibodies, for example, have been increasingly employed for rapid diagnostic tests.[17] These immobilized proteins have also provided valuable information about the basic protein-protein interactions.[15,18] In addition to enzymes, many different types of proteins, including antibodies, protein antigens, enzyme inhibitors, protein toxins and peptide hormones, have been attached to insoluble carriers and have been shown to be biologically active. Quite often, immobilization of an enzyme actually results in an increased stability of the protein.[19]

Various approaches are used to insolubilize proteins onto solid supports. These methods involve either noncovalent adsorption and entrapment or covalent linkage. In this chapter, only the covalent immobilization will be considered, particularly the use of chemical cross-linkers. The reader who is interested in other methods of protein insolubilization or the applications of immobilized enzymes and proteins are encouraged to consult the many relevant books and reviews.[1-25]

Covalent attachment of a protein to a solid matrix involves the formation of a covalent bond between a functional group on the protein and a reactive group on the surface of the solid phase. The methods of such covalent bonding are as varied as organic chemistry itself. During the last 20 years, thousands of proteins have been immobilized and hundreds of matrices have become available. In the immobilization of proteins to a solid, several factors must be considered in order to retain the optimum activity of the proteins. The proteins must be attached in such an orientation that their active sites or binding domains are accessible to the surrounding milieu and not buried in or blocked by the matrix or other components on the matrix surface. During the coupling reaction, the protein must be in its active state. Royer and Uy[26] have demonstrated that an enzyme could be immobilized in an active conformation in the presence of its substrate. Another factor which must be considered in the covalent attachment of proteins is the possibility of chemically altering the protein in such a way that its reactivity is reduced. Since reactive groups on the proteins are involved in chemical bonding, it is possible that groups associated with the active site or binding site of a protein could be involved in the reaction. In addition, chemical cross-linking could take part intramolecularly causing damage to the protein, thereby reducing the attachment efficiency. In the following sections, the chemistry of protein attachment will be discussed.

II. FUNCTIONALITIES OF MATRICES

Commercial matrices used for attachment of proteins are made of different materials and contain different functional groups. In order to select the best method and reagent for coupling, the type of functional groups on the matrix must be known. Probably the most prevalent and most widely used matrices are the polysaccharides, such as agarose and cellulose and their derivatives. These materials contain vicinal hydroxyl groups. One of the derivatives of interest is chitin which consists of repeating (1-4)-linked 2-acetamido-2-deoxy-β-D-glucose moiety.[27] Another natural substance that contains hydroxyls is silica.[28] This is usually in the form of controlled pore glass. Newer hydroxylic matrices are derived from polymeric materials such as polyacrylamide.

There are hundreds of polymeric matrices that have been made of various polymers, copolymers, and terpolymers, and new ones are constantly being synthesized.[29] Many of these particles are commercially available in different sizes, forms, and colors. Fluorescent and magnetic particles are also obtainable. The surface chemistries range from plain polystyrene (benzene) to a wide variety of different surface functional groups such as sulfonate, sulfate, carboxylate, primary amine, aromatic amine, amide, hydroxyl, aldehyde, vinylbenzyl chloride and chloromethyl.[30-32] Table 1 lists the major commercially prepared solid matrices containing different functionalities that have been used for protein immobilization. This list illustrates the wide range of reactive groups that can be used for cross-linking. In addition, many activated matrices ready for protein immobilization are also available, such as CNBr-activated Sepharose, bromoacetylcellulose and N-hydroxysuccinimide ester-agarose.

Many of these groups can be further interconverted to other functionalities. For example, the surface hydroxyl groups of glass beads (silica) can be converted to amino groups by silanization with 3-aminopropyltriethoxysilane or amine-containing alkosilanes. The generated amino groups can be further converted to carboxylic acids by succinyl chloride or succinic anhydride as shown in Figure 1.[26] During the process of derivatization, a spacer is also introduced.

Amide groups can be converted to amino groups by reaction with hydrazine or other alkyldiamines. The resulting amines can be further modified to aldehydes with glutaraldehyde or to carboxylic acids with succinic anhydride (Figure 2).

Carboxylic acids can be converted to amines by coupling to diamino alkanes such as 1,6-diaminohexane or 1,7-diaminoheptane in the presence of a water soluble carbodiimide as shown in Figure 3. These reactions provide a spacer between the solid surface as well as the reactive group. In this regard, the carboxylate can be further removed from the surface by coupling to 5-aminocaproic acid or other amino acids (Figure 3).

Many other manipulations of the matrix-surface chemistry are possible. In fact the matrices can be treated in many cases as organic chemicals and organic solvents may be used for derivatization or modification of the functional groups. The reactions mentioned in Chapter 2 for protein modification are certainly applicable to the solid surfaces.

III. PROTEIN IMMOBILIZATION BY MATRIX ACTIVATION

The choice of a chemical procedure for immobilization of a protein depends on both the protein and the surface functionality of the solid matrix. As discussed in Chapter 2, many of the amino acid side chains of proteins are available for reaction. One of the most common groups used for attachment is the amino group present at the N-terminus and the side chains of lysines. Carboxyl groups from the C-terminus, aspartic and glutamic acids, are also available for bonding as well as sulfhydryl groups of cysteines, hydroxyl groups of serines and threonines, and phenyl moieties of phenylalanines and tyrosines. Because of the variety of reactive groups on the proteins, stress is placed on the functional groups on the solid surface. Thus, the method of immobilization is more or less dependent on the solid matrix chosen. Generally, the matrix is first activated by a specific chemical reaction depending on the surface chemistry. Proteins are then immobilized on mixing with the activated matrix. The following examples will demonstrate the kind of cross-linking that can be achieved through a two-step reaction.

A. ACTIVATION OF HYDROXYL GROUPS

The most commonly used method for protein immobilization involving the use of cellulose and agarose derivatives is cyanogen bromide activation. Indeed, any polymeric matrices possessing cleavable vicinal hydroxy groups can be utilized in this method. Agarose

TABLE 1
Commercially Available Solid Matrices of Different Functionalities

Solid matrices	Commercial products	Surface functional groups
Agarose	BioGel A	$-OH$
	Affi-Gel 101, 102	$-NH_2$
	Affi-Gel 201, 202	$-COOH$
	Affi-Gel 401	$-SH$
	Affi-Gel 501	$-Hg^+$
	Ultrogel A	$-OH$
	Ultrogel ACA	$-OH$
	Magnogel	$-OH$
	Act-Ultrogel ACA	$-CHO$
	Act-Magnogel ACA	$-CHO$
	AC-Ultrogel ACA	$-COOH$
	AC-Magnogel ACA	$-COOH$
	HMD-Ultrogel ACA	$-NH_2$
	HME-Magnogel	$-NH_2$
	Sea Plaque	$-OH_2$
	Sea Kem	$-OH$
	PL-agarose	$-NH_2$
	PAL-agarose	$-NH_2$
	SPL-agarose	$-COOH$
	SPAL-agarose	$-COOH$
	Sepharose 2B, 4B, 6B	$-OH$
	Sepharose AH	$-NH_2$
	Sepharose CH	$-COOH$
Cellulose	Avicel	$-OH$
	Cellex	$-OH$
	CMR cellulose	$-COOH$
	Cellex-CM	$-COOH$
	Cellex-PAB	$-ArNH_2$
	APT-paper/ABM-paper	$-ArNH_2$
	Nitrocellulose	$-NO_2$
	CM-cellulose	$-COOH$
Collagen	Collagen	$-NH_2$
Dextran	Sephadex	$-OH$
	DEAE-Sephadex	$-CH_2CH_2N(C_2H_5)_2$
	QAE-Sephadex	$-N(C_2H_5)_2C_3H_6OH$
	CM-Sephadex	$-COOH$
	SP-Sephadex	$-SO_3H$
Polyacrylamide	BioGel P	$-OH$
	BioGel hydrazide	$-CONHNH_2$
	A-E-BioGel P	$-NH_2$
	BioGel CM	$-COOH$
	Enzacryl-AA	$-ArNH_2$
	Enzacryl-AH	$-CONHNH_2$
	Enzacryl-PT	$-SH$
	Enzafix P-HZ	$-CONHNH_2$
	Enzafix P-AB	$-ArNH_2$
	Sepheron	$-OH$
Polyvinyl	Fractogel	$-OH$
Polyvinylalcohol	PVA	$-OH$

TABLE 1 (continued)
Commercially Available Solid Matrices of Different Functionalities

Solid matrices	Commercial products	Surface functional groups
Other polymeric materials (copolymers, ter-	Latex particle	$-COOH$
polymers)	Latex particle	$-CH_2NH_2$
	Latex particle	$-ArNH_2$
	Latex particle	$-CONH_2$
	Latex particle	$-CHO$
	Latex particle	$-OH$
	Latex particle	$-SH$
	Latex particle	$-CN$
Silica	CPG	$-OH$
	CPG-aminopropyl	$-NH_2$
	CPG-dextran	$-OH$
	CPG-glycerol	$-OH$
	CPG-phenylhydrazine	$-ArNHNH_2$
	Carboxy-CPG	$-COOH$
	CPG-thiol	$-SH$
	CPG-aminoaryl	$-ArNH_2$
	Unisil	$-OH$

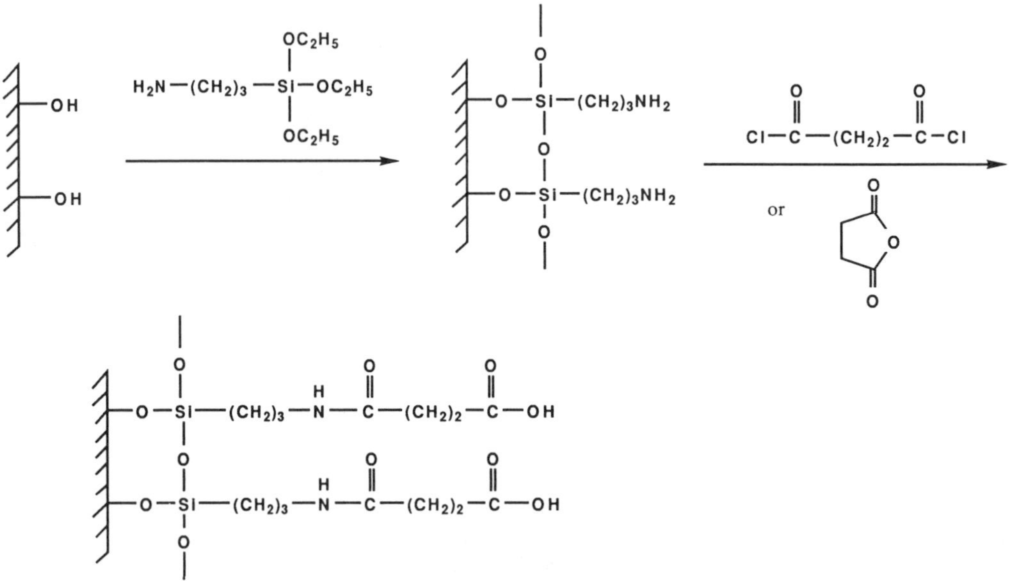

FIGURE 1.   Conversion of matrix hydroxyl groups to carboxylic acids.

is a linear polymer consisting of D-galactose and 3,6-anhydro-D-galactose. Sepharose™ is a trademark for an agarose derivative prepared by Pharmacia. Sephadex™ is a cross-linked dextran also prepared by Pharmacia. Cyanogen bromide reacts with these materials to form a cyclic reactive imidocarbonate which is susceptible to nucleophilic attack by amino groups present in the proteins to form N-substituted imidocarbonate as demonstrated in Figure 4.[33] The cyclic imidocarbonate may be converted to cyclic carbonate in aqueous solutions yielding N-substituted carbamates as the end product. A water-soluble, nonvolatile and less toxic analog, 1-cyano-4-dimethylaminopyridinium tetrafluoroborate, developed by Kohn and Wilchek[34] has also provided efficient coupling. Because of the versatility of the cyanogen bromide reaction, the activated matrix can be used for derivatization with diamines or dihydrazides.

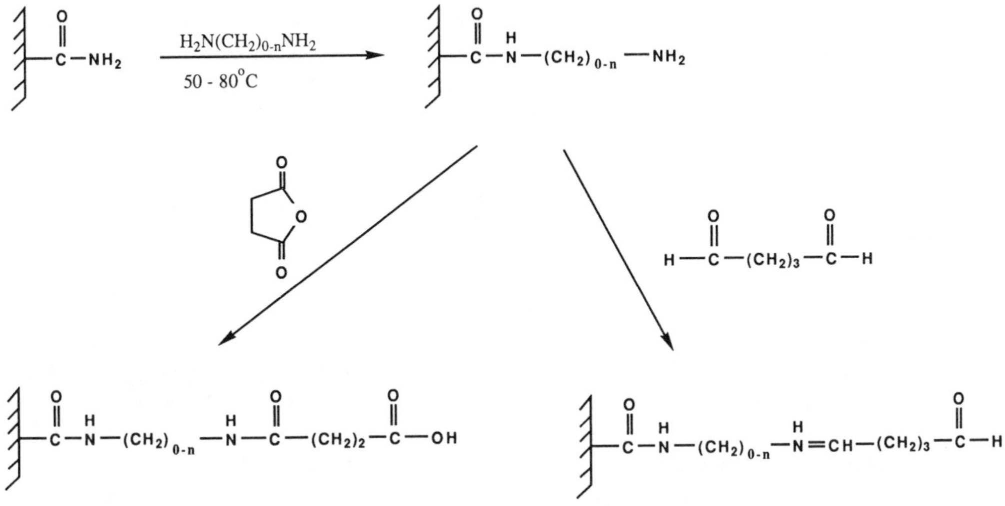

FIGURE 2.   Conversion of matrix amides to amines, carboxylic acids and aldehydes.

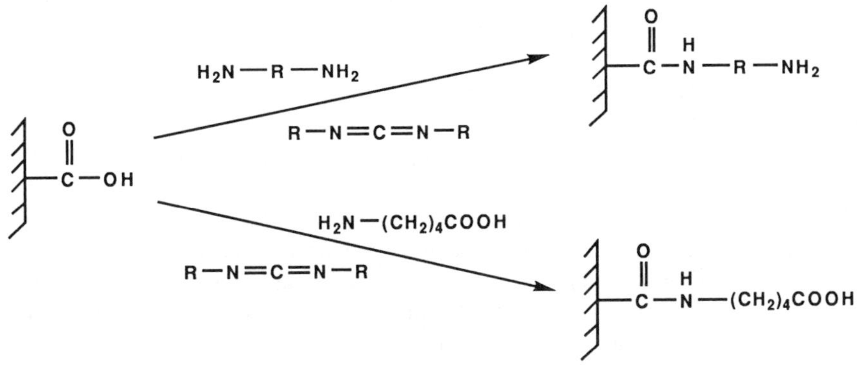

FIGURE 3.   Conversion of matrix carboxylates to amines and the introduction of a spacer.

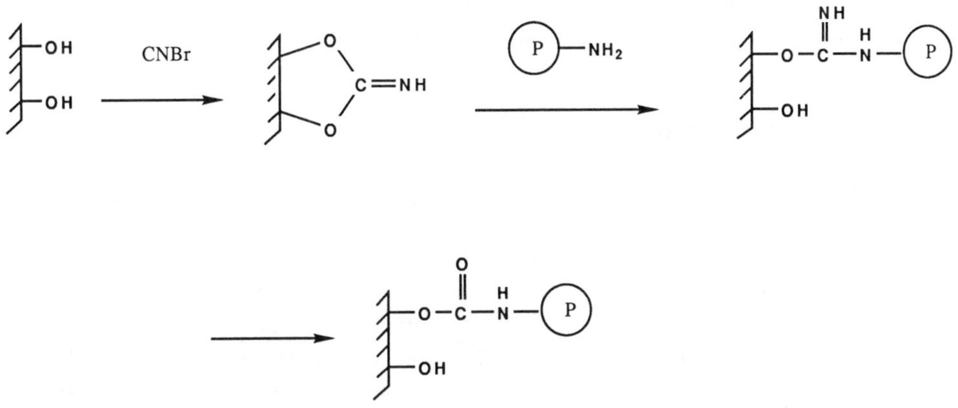

FIGURE 4.   Immobilization of proteins with cyanogen bromide activation of vicinal diols.

In practice, the reaction is quite simple. The carrier is first activated with cyanogen bromide for a brief period and then washed to remove excess reagent. It is then mixed with the protein for immobilization. The ease of the reaction and the mild conditions employed, together with the high yields of the immobilized proteins account for the popularity of this method. In fact, cyanogen bromide-activated matrices are now commercially available for immediate use. Habeeb[35] used this method to coat Sepharose℠ with gelatin, which was then linked to bovine serum albmin and lysozyme for the preparation of immoabsorbents using heterobifunctional cross-linkers as discussed below.

B. ACTIVATION OF CARBOXYL GROUPS

Carboxylate groups can be activated with *N*-hydroxybenzotriazole in the presence of a water-soluble carbodiimide. Before coupling to proteins, the activated matrix is washed to remove excess carbodiimide. The reactive ester reacts very rapidly to form stable amide bonds with amino groups of the ligand (Figure 5A). Among the other nucleophiles, only sulphydryl groups compete effectively with the amino group during the reaction.

p-Nitrophenol[36] and *N*-hydroxysuccinimide[37] are also commonly used to form active esters with carboxylic acids in the presence of a carbodiimide. These active ester derivatives, when stored in dioxane, are stable. The coupling reaction is similar to *N*-hydroxybenzo-triazole ester. As an example, Cuatrecasas and Parikh have used this method to immobilize enzymes to aminoalkyl agarose after succinylation as shown in Figure 5B.[37] The same approach was used to coat glass beads with avidin for the preparation of immunoaffinity chromatography.[38]

Carboxylate functional groups can also be directly activated to the acyl chloride by reacting with thionyl chloride. The very reactive acyl chloride reacts with proteins at low temperatures and was used to immobilize triacylglycerol lipase on Aberlite IRC-50.

C. ACTIVATION OF ACYL HYDRAZIDE

As mentioned above, reaction of a hydrazide-modified surface with glutaraldehyde generates aldehyde groups which can be used to couple to proteins by forming Schiff bases with amino groups of the proteins. Acyl hydrazides can also be converted to acyl azides with nitrous acid via the diazotization reaction. The azide groups will be replaced by amino groups on the proteins to form amide bonds (Figure 6). Thus, by a series of reactions, surface hydroxyls can be activated to acyl azides. This is illustrated in Figure 6. First, the hydroxyl is carboxymethylated with chloroacetic acid which is converted to an acyl hydrazide through an ester. The final activated acyl hydrazide then reacts with the proteins. Since matrix amide groups can also be converted to acyl hydrazide, they can be linked to proteins through this reaction sequence.

D. ACTIVATION OF AMINES
1. Use of Nitrous Acid

Aromatic amino groups from poly-*p*-aminostyrene can be diazotized with nitrous acid to form the diazonium compound. The diazonium group will react with phenolic, imidazole, and amino side chains of a protein to form a diazo bond (see Chapter 2). By this method, phenyl groups on polystyrene can be coupled to proteins through nitration and amination.[39] Reaction of the benzene ring on polystyrene with fuming red nitric acid results in nitrostyrene. The nitro benzene group is then converted to amino benzene by sodium dithionite.[40] This sequence of steps is shown in Figure 7. Hydroxyl-containing carriers, such as cellulose, can also be coupled through the diazonium group after reaction with *p*-nitrochloromethylben-zene.[41] Alternatively, the vicinal hydroxyl groups can be oxidized to dialdehyde to which 4,4′-diaminodiphenylmethane can be condensed through reduction of the Schiff bases formed.[42]

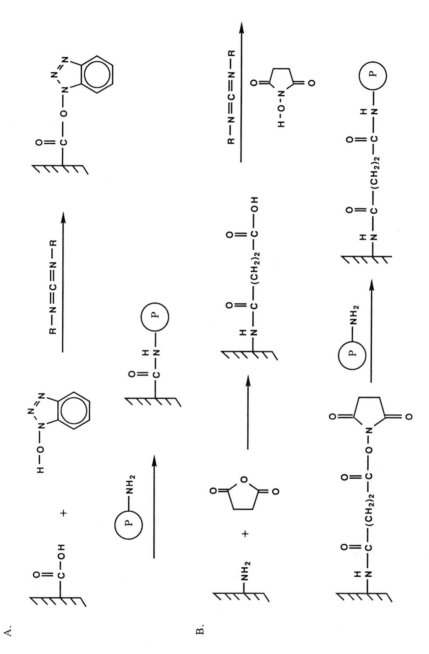

FIGURE 5. Immobilization of proteins to carboxylate matrix activated with carbodiimides and an alcohol. (A) Activation with *N*-hydroxybenzotriazole. (B) Succinylation of amino group containing matrix followed by activation with N-hydroxysuccinimide.

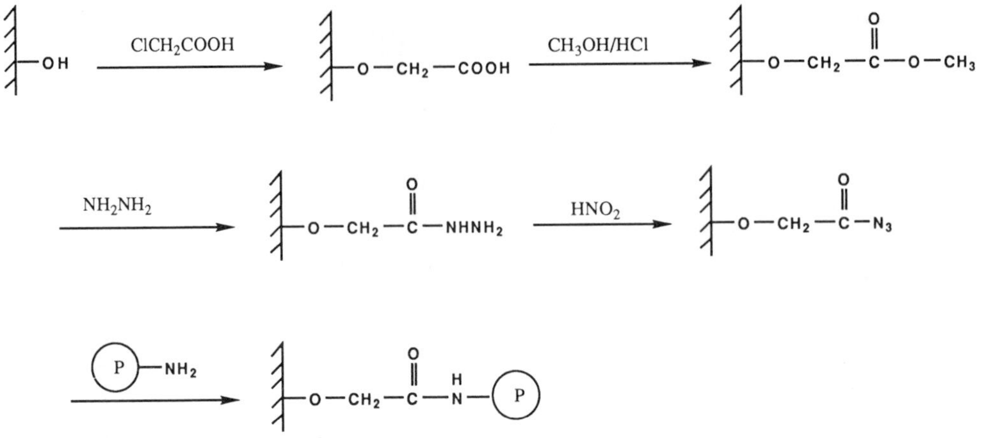

FIGURE 6.   Coupling of proteins to hydroxyl matrix via diazotization of acyl hydrazide.

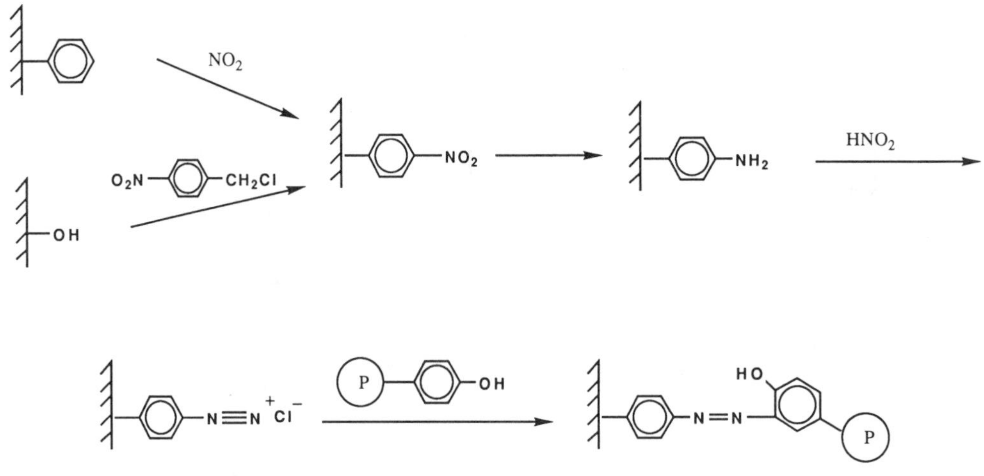

FIGURE 7.   Coupling of proteins to aromatic groups via diazotization.

2. Use of Phosgene and Thiophosgene

Aromatic amines can also be activated at alkaline pH with phosgene or thiophosgene to yield the corresponding isocyanate and isothiocyante, respectively (Figure 8).[20] These active derivatives react with free amino groups of proteins to form an amide or thioamide bond of the substituted urea or thiourea (see Chapter 2). The isothiocyanate derivative can also be obtained from aliphatic amines and acyl azides.

3. Use of Cyanogen Bromide

Like hydroxyl groups, amino group containing matrices can also be activated with cyanogen bromide to which proteins react forming guanidine linkages (Figure 9).[43] It was claimed that amine matrices yielded more stable products than the hydroxyl group based materials.

E. ACTIVATION OF ACRYLONITRILE

Polyacrylonitrile can be activated with absolute ethanol and bubbling hydrogen chloride

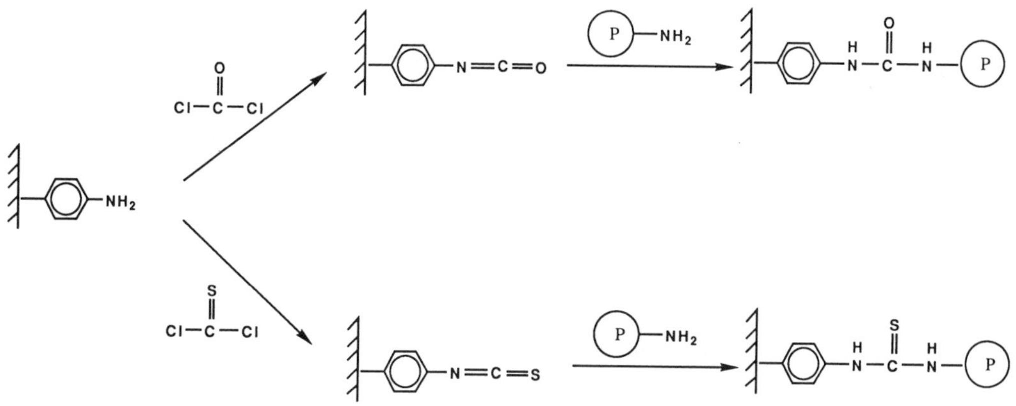

FIGURE 8. Coupling of proteins to aromatic amines through isocyanates and isothiocyanates.

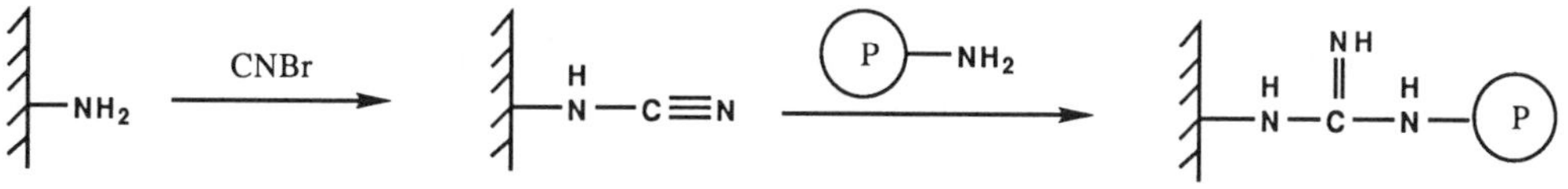

FIGURE 9. Coupling of proteins to cyanogen bromide activated amine matrices.

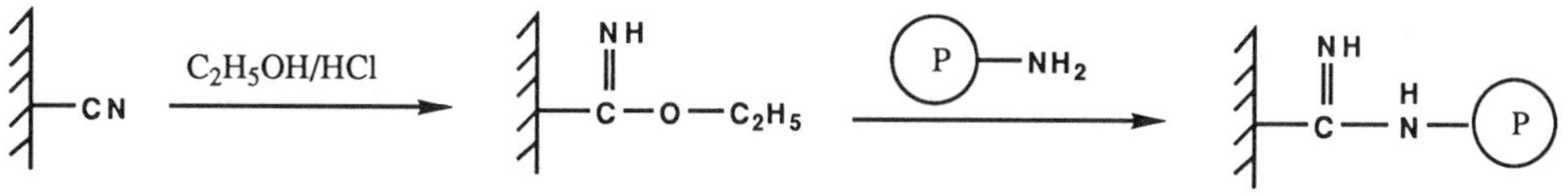

FIGURE 10. Coupling of proteins to acrylonitrile after activation to imidoester.

to an imidoester which are readily attacked by amino groups of proteins at basic pH to yield an amidine (Figure 10).[44]

IV. CROSS-LINKING REAGENTS USED IN PROTEIN IMMOBILIZATION

A. ZERO-LENGTH CROSS-LINKERS

The commonly known zero-length cross-linkers that have been used for protein immobilization include carbodiimides, Woodward's reagent K,[45] chloroformates[46] and carbonyldiimidazole.[46] While all of these compounds condense carboxyl and amino groups to form amide bonds, the latter two also activate hydroxyl groups. When hydroxyl groups are involved, a carbonyl moiety is incorporated. These compounds function as homobifunctional reagents as will be discussed below.

Different carbodiimides, including water-soluble compounds described in Chapter 6, have been used for coupling proteins to carboxyl group containing solid matrices.[48,49] The most common chloroformate used is ethylchloroformate. Others include p-nitrophenyl chlorocarbonate, 2,4,5-trichlorophenyl chlorocarbonate and N-hydrosuccinimide chlorocarbonate.[46] During the reaction, the carboxyl group is first reacted with the cross-linker to form an activated species which is attacked by the amino group nucleophile. The reaction schemes for these reagents are shown in Figure 11.

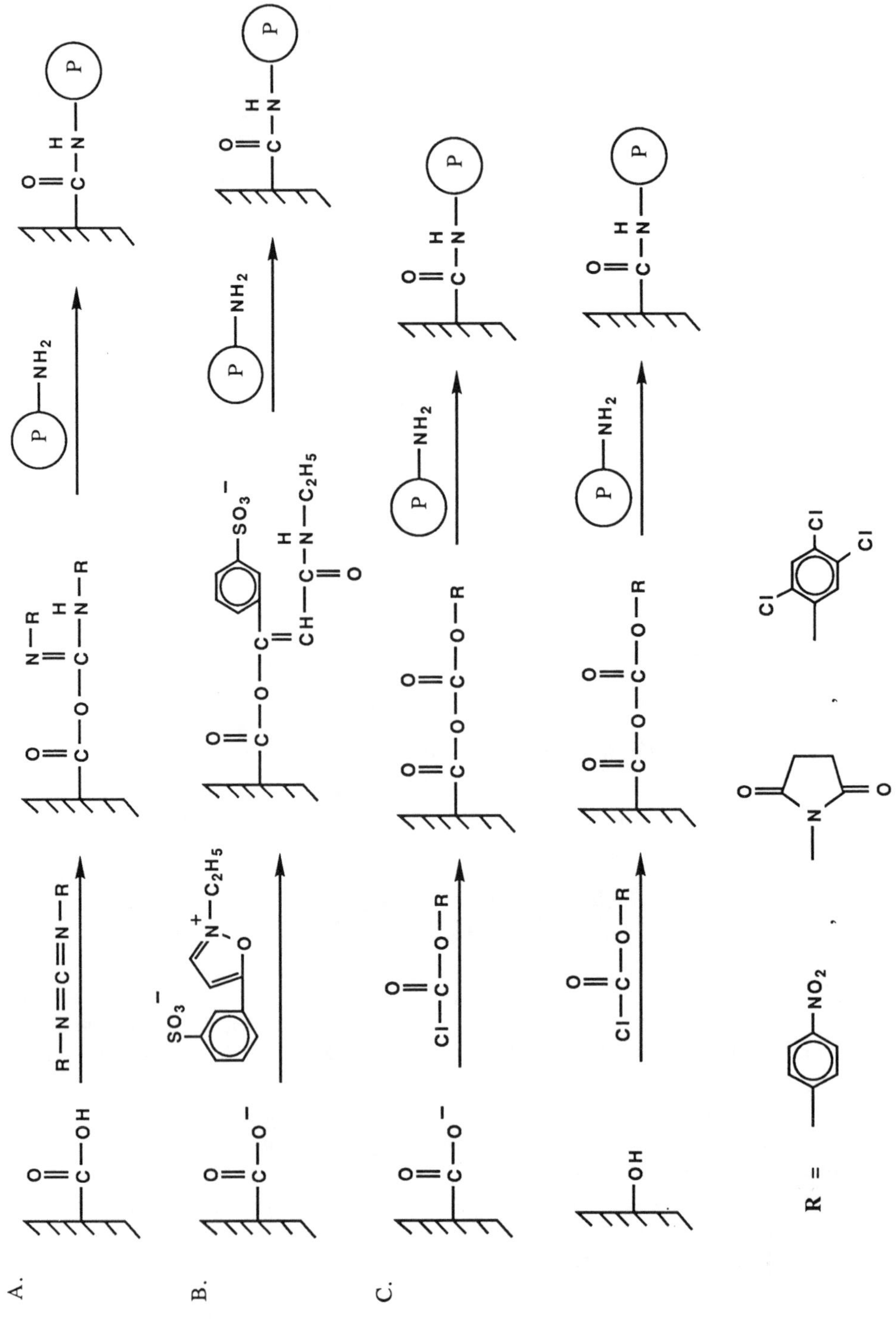
A.
B.
C.
R =

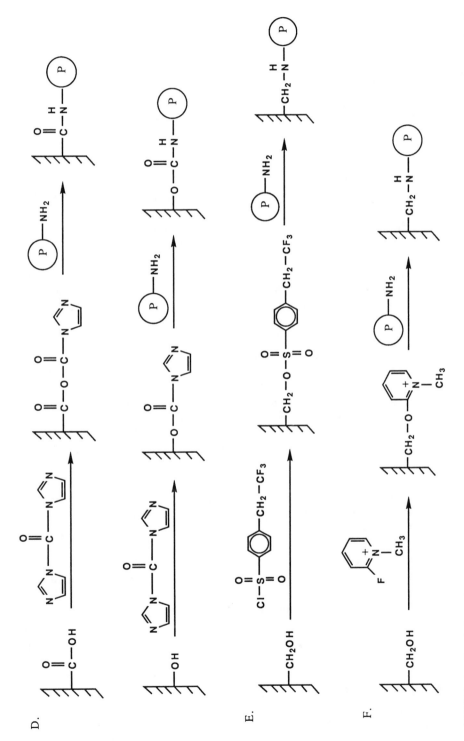

FIGURE 11. Protein immobilization by zero-length cross-linkers: (A) Carbodiimide; (B) Woodward's reagent K; (C) chlorocarbonate; (D) carbonyldiimidazole; (E) tresyl chloride and (F) 2-floro-1-methylpyridinium salt.

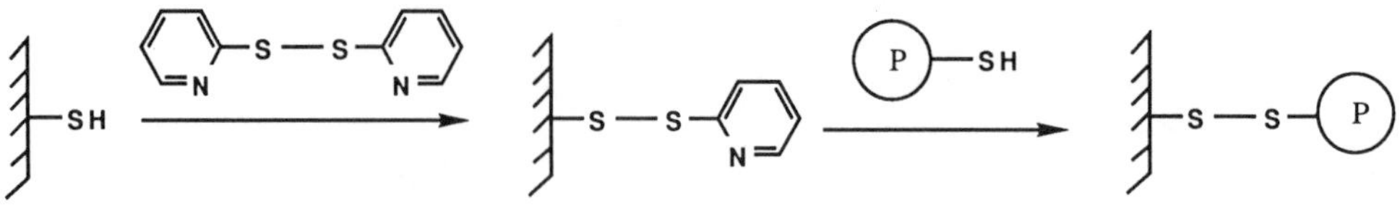

FIGURE 12.   Coupling of proteins to thiol-containing matrices by thiol-disulfide interchange.

Other reagents that can be considered as zero-length cross-linkers are sulfonyl chlorides and 2-fluoro-*N*-methylpyridinium salt. The most commonly used sulfonyl chloride is tresyl chloride,[50] although *p*-toluene sulfonyl chloride and colored sulfonyl chlorides have also been used.[51] These reagents activate the primary hydroxyl groups into good leaving groups as shown in Figures 11E and 11F. The activated function is then displaced by a protein nucleophile, chiefly the amino side chain of lysine.

It should be pointed out that not only amino groups of proteins serve as nucleophiles, free sulfhydryl, if present, can potentially be coupled. The linking between a carboxyl group and an amino group forms an amide bond, whereas a thioester bond is formed when a sulfhydryl group is involved. Similar coupling using sulfonyl chloride and 2-fluoro-1-methylpyridinium salt affords a secondary amine and a thioether bond.

Various different solid matrices have been activated by these zero-length cross-linkers and many enzymes and proteins have been immobilized this way. Carboxyl groups containing supports such as carboxymethyl cellulose, acrylamide and acrylic acid copolymer, carboxymethyl Sephadex, BioGel, carboxymethyl agarose and polyacrylic acid have been used. Polyhydroxylic matrices that have been activated by zero length cross-linkers are agarose, glycerylpropyl-silica, cellulose, and hydroxyethyl methacrylate.

Thiol-containing matrices can be activated by 2,2'-dipyridyldisulfide through the thiol-disulfide interchange reaction. The protein is bonded via its free thiol through a second thiol-disulfide interchange with the liberation of 2-thiopyridine according to Figure 12. This method generates a disulfide bond between the protein and the solid support which is stable under nonreducing conditions.[52,53] The enzyme can be released by low molecular weight thiol compounds. For proteins that do not contain a free thiol, it can be thiolated using various reagents such as *N*-acetylhomocysteinethiolactone as described in Chapter 2.

Examples of enzymes that have been immobilized by the zero-length cross-linkers are glucose-6-phosphate dehydrogenase,[54] glucose oxidase,[55] glutamate dehydrogenase,[56] peroxidase,[55,57] ribonuclease A,[55] cholinesterase,[58] alkaline phosphatase,[55] deoxyribonuclease,[55,58] apyrase,[58] trypsin,[46,55,59,60] glutamicaspartic transaminase,[61] and α-chymotrypsin.[55,56,62]

B. HOMOBIFUNCTIONAL CROSS-LINKERS

Almost any of the homobifunctional cross-linkers listed in Chapter 4 can be used one way or another to immobilize proteins. Depending on the matrix functional groups, different cross-linkers will have to be used. The following reactions demonstrate the use of these reagents.

1. Glutaraldehyde

Glutaraldehyde is the most prevalent homobifunctional reagent used for the immobilization of proteins. The reaction is dependent on pH, temperature and ionic strength.[63,64] As presented in Chapter 4, the chemistry of cross-linking by glutaraldehyde is complex. Although amino group is assumed to be the primary function to react, other functional groups, such as immidazole, thiol, and hydroxyl, have also been implicated.[64,65] Thus, glutaraldehyde has been used to couple proteins to cellulosic materials,[21] to aminoalkylsilylated glass,[66] to polyacrylhydrazide,[67] to phe-lys coated polystyrene balls,[68,69] and polyethyleneimine treated

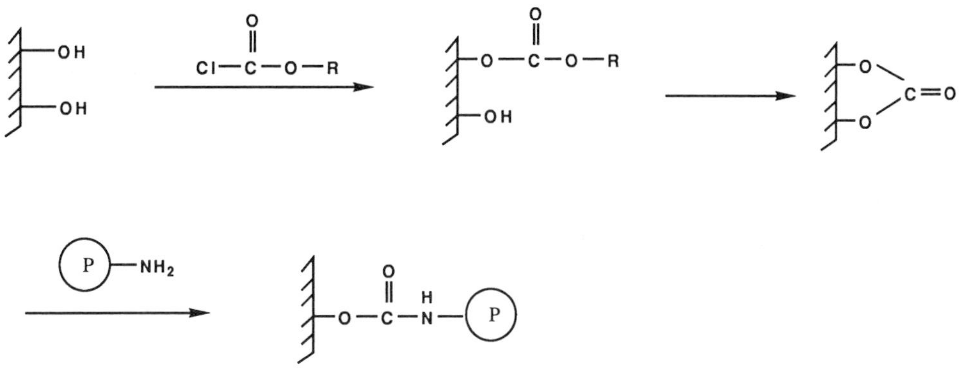

FIGURE 13.   Immobilization of proteins with alkyl or arylchloroformate.

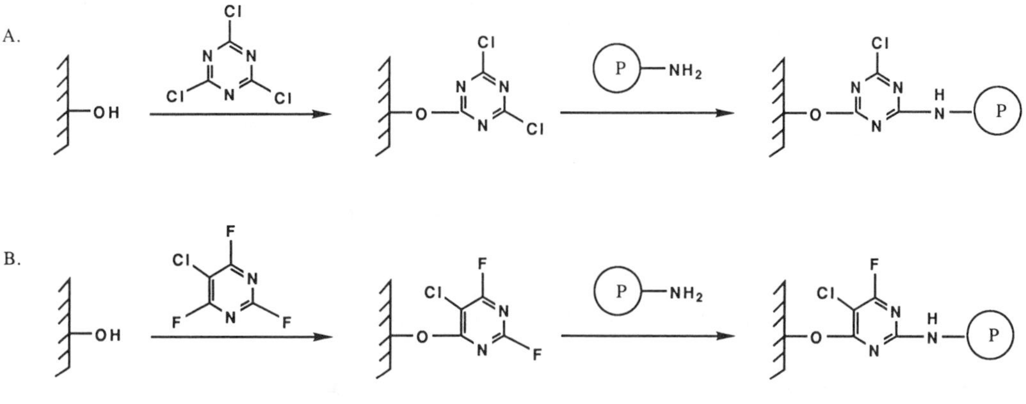

FIGURE 14.   Immobilization of proteins with heterocyclic halides: (A) Cyanuric chloride; and (B) fluoropyrimidine.

magnetite,[70] as well as in the preparation of glucose oxidase membranes for enzyme electrode use.[71,72] Both one-step and two-step procedures have been used (see Chapter 7). For the two-step process, the matrix is first activated with glutaraldehyde. The solid support is washed and then coupled with the protein.

2. Chloroformates and Carbonyldiimidazole

A reaction similar to cyanogen bromide in the activation of vicinal hydroxyl groups is the formation of cyclic carbonate derivative with 1,1'-carbonyldiimidazole, ethylchloroformate, or other alkyl or arylchloroformate in anhydrous organic solvents. The activated cyclic carbonate will react with nucleophiles on proteins at pH 7 to 8 to form substituted carbonate bonds (Figure 13).[73] When an amino group from the protein serves as the attacking nucleophile, an N-substituted carbamate bond is formed. A thiol carbonate bond results when thiol is involved. A carbonyl moiety from the cross-linker is incorporated into the bond. The carbamate and thiol carbonate bonds are not very stable and may be hydrolyzed under extreme pHs.

3. Heterocyclic Halides

s-Triazines, such as cyanuric chloride and some of its dichloro derivatives, 2-amino-4,6-dichloro-s-triazine, 2-carboxy-methylamino-4,6-dichloro-s-triazine, and 2-carboxymethyloxy-4,6-dichloro-s-triazine, have been used to couple enzymes to cellulose through the activation of the hydroxyl group as shown in Figure 14.[74-76]

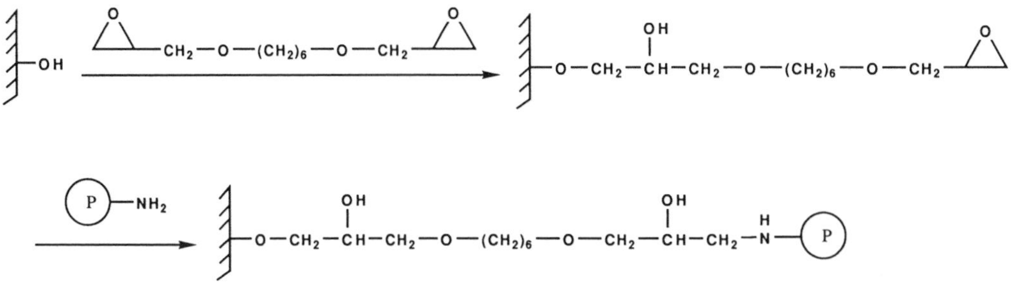

FIGURE 15.   Immobilization of proteins with bisoxirane, 1,6-bis(2,3-epoxypropoxy)hexane.

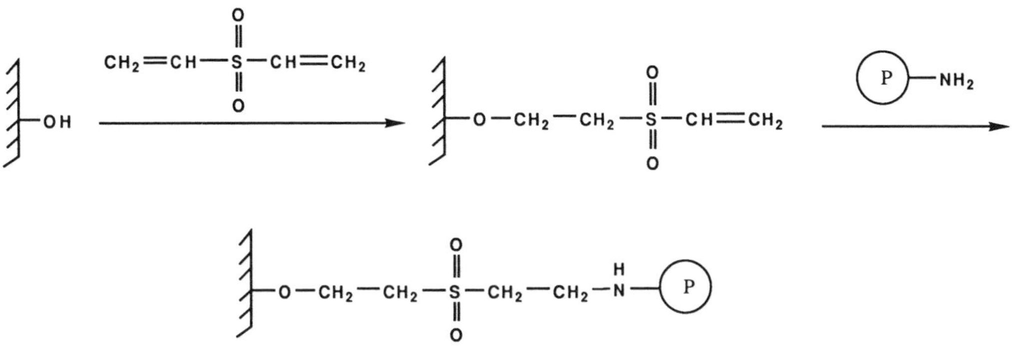

FIGURE 16.   Immobilization of proteins with vinylsulfone.

The coupling process is an alkylation reaction involving the primary amino groups of proteins with the activated carbon of the *s*-triazine molecule. In addition to cellulose, other supports containing hydroxyl or amino groups can also be used.[21,74]

Fluoropyrimidines are other heterocyclic halides that have been used for the activation of agarose. For example, 2,4,5-trifluro-5-chloropyrimidine has been employed to immobilize a wide variety of hormones, enzymes, and other proteins. Various nucleophiles on the protein, such as thiol, amino, imidazolyl, hydroxyl and guanidinyl react. The rate of the reaction is pH dependent as well as the nucleophilicity of these groups.

4. Bisoxiranes

Introduction of epoxides onto matrix surfaces can be achieved by bisoxiranes, for example, 1,4-butanediol diglycidyl ether (1,4-bis(2,3-epoxypropoxy)butane). The reaction occurs readily at alkaline pH to yield derivatives containing a long-chain hydrophilic function with a reactive epoxide. This method is suitable for hydroxyl-containing supports which form ether linkages through their hydroxyl groups. Other matrices containing the amino and thiol groups can also be modified. Immobilization occurs when nucleophilic groups on the protein react with the epoxide (Figure 15).[77] Epoxides can be directly linked to hydroxyl group-containing matrices by direct silianization with glycidoxy propyl trimethoxy organosilane.[78,79] Because of the relative stability of the epoxide group, a drawback of this technique is the difficulty to remove excess unreacted oxirane groups.[80]

5. Divinylsulfone

Divinylsulfone has been used to modify hydroxyl-containing matrices to vinylsulfone. The vinyl group is very reactive and reacts rapidly with nucleophiles of proteins.[81] However, the products are unstable. The products with a hydroxyl function is unstable above pH 9 and that with amino groups is unstable above pH 8 (Figure 16).

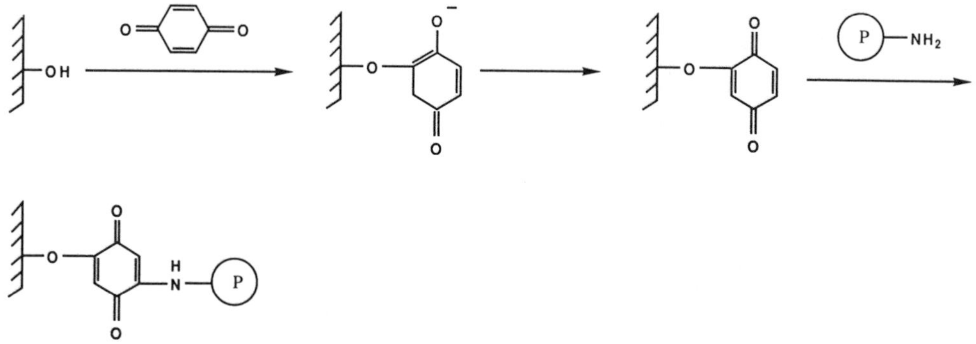

FIGURE 17. Immobilization of proteins with *p*-benzoquinone.

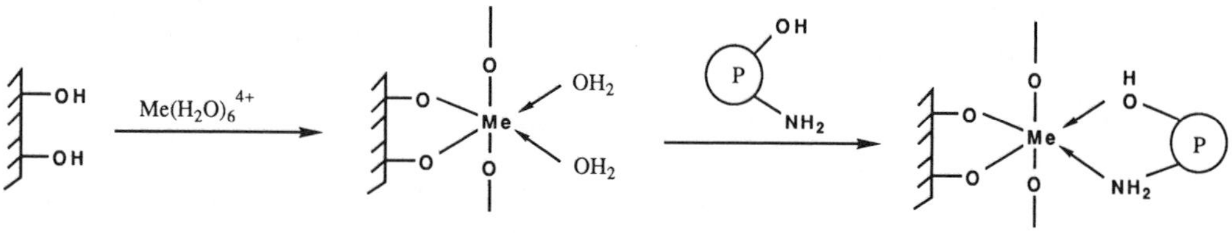

FIGURE 18. Immobilization of proteins with transition metal ions.

6. Quinones

Proteins can be immobilized onto solid supports such as agarose by quinones, for example, *p*-benzoquinone.[82] The matrix is first activated with the quinone. Proteins are then bound through their nucleophilic groups with high yields (Figure 17). The coupling reaction can occur in a broad range from pH 3 to 10. Undesirable side reactions may occur rendering a dirty color to the matrix. The reagent has also been used for the activation of polyacrylamide gels.[83]

7. Transition Metal Ions

Transition metal ions that form stable hexaqua complexes with water molecules can be used to activate hydroxyl-containing solid matrices. In this process, the support, which may consist of materials as diverse as borasilicate glass, soda glass, filter paper, cellulose derivatives and nylon 66, is first reacted with a transition metal salt, such as $TiCl_4$, $SnCl_4$, $SnCl_2$, $ZrCl_4$, VCl_3, or $FeCl_3$, for 24 h. After washing to remove unreacted metal salts, the immobilized metal ions further coordinate proteins through the carboxyl groups of C-terminus and acidic amino acids, the hydroxyl groups of tyrosyl, seryl, and threonyl residues, the free sulfhydryl groups of cysteines, or amino groups of the *N*-terminus and ε-amino groups of lysyl residues as shown in Figure 18.[84]

Several enzymes, such as α-amylase, glucoamylase, glucose oxidase, invertase, papain, nuclease, trypsin and urease have been bonded to cellulose and porous glass by this method.[84-89] Relatively high retention and operational stability are demonstrated for some of these enzymes. Recently, Moriya et al.[90,91] demonstrated that copper (II) can be used to chelate salicyladehyde and α-amino acids. Thus, a Sepharose column modified with lysine residue through the ε-amino group can be used to immobilize in the presence of copper (II) proteins modified with *N*-succinimidyl 3-formyl-4-hydroxybenzoate. The protein is specifically bound to the column but is reversible by the addition of EDTA.

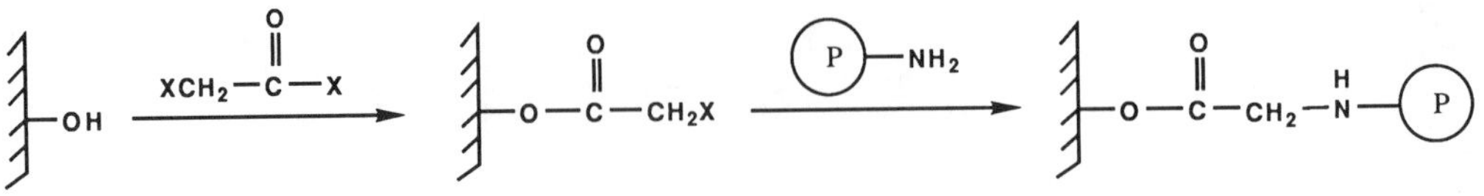

FIGURE 19.   Immobilization of proteins with monohalogenacetyl halide.

8. Other Homobifunctional Cross-Linkers

Other homobifunctional cross-linking reagents that have been used for immobilization of proteins include 4,4'-difluoro-3,3'-dinitrodiphenyl sulfone, toluenediisocyanate, 4,4'-diisothiocyanatobiphenyl-2,2'-disulfonic acid, 1,5-difluoro-2,4-dinitrobenzene, N,N'-hexamethylenebisiodoacetamide and hexamethylenediisocyanate (see Chapter 4 for structures). A two-step reaction procedure is usually followed. First the solid surface is activated by the bifunctional compound. The second step involves the reaction of the protein with the other end of the bifunctional reagent. The advantage of this two-step procedure is that it permits cleaning of the unreacted bifunctional reagent from the solid matrix before addition of the protein, thus preventing cross-linking of the protein in solution.

Bifunctional N-hydroxysuccinimide esters have been used to immobilize proteins to amino group containing solid matrices. Ethylene glycol bis(succinimidyl succinate) (EGS) and dithiobis(succinimidyl propionate) (DSP) (see Chapter 4 for structures) have been used to immobilize trypsin to hexamethylenediamine-Secharose CL-4B.[92] The immobilized protein can be released by treatment with hydroxylamine or thiol compounds, respectively, since these compounds contain cleavable bonds. A bis-sulfonated derivative of DSP, 3,3'-dithiobis(sulfosuccinimidyl propionate), which is more water soluble, has been used.

C. HETEROBIFUNCTIONAL CROSS-LINKERS

The use of heterobifunctional reagents in the immobilization of proteins to solid supports is relatively unexplored. Only a modest number of applications have been reported. These will be discussed below.

1. Monohalogenacetyl Halide

Jagendorf et al.[93] have used bromoacetylbromide to immobilize globulins to cellulose. The monohalogenacetyl halides activate the solid support with a monohalogenacetyl functional group which alkylates another nucleophile from the protein, such as an amino group (Figure 19). Of the three halides, iodide confers the best reactivity and stability to the carrier. However, bromoacetylbromide is frequently used and bromoacetyl-cellulose has also been used to immobilize enzymes.[94]

2. Epichlorohydrin

Epichlorohydrin activates matrices with nucleophiles such as amino or hydroxyl to an epoxide derivative. This epoxide derivative reacts with nucleophilic groups of proteins in the order of thiol>amino>hydroxyl, although aromatic hydroxyl, guanidino and imidazole groups also react (Figure 20). The activated matrix may be further reacted with phloroglucinol. Proteins are then coupled through the phloroglucinol bridge. This can be achieved by treating phloroglucinol intermediate with either divinylsulfone or CNBr.[80,95]

3. Amino and Thiol Group-Directed Reagents

Several heterobifunctional reagents of this type (see Chapter 5) have been used. The general reaction mechanism involves first reacting the cross-linker with the solid support. The activated matrices then react with the protein to complete the immobilization process. For example, succinimidyl 6-maleimidocaproate is reacted with an amine-derivatized Se-

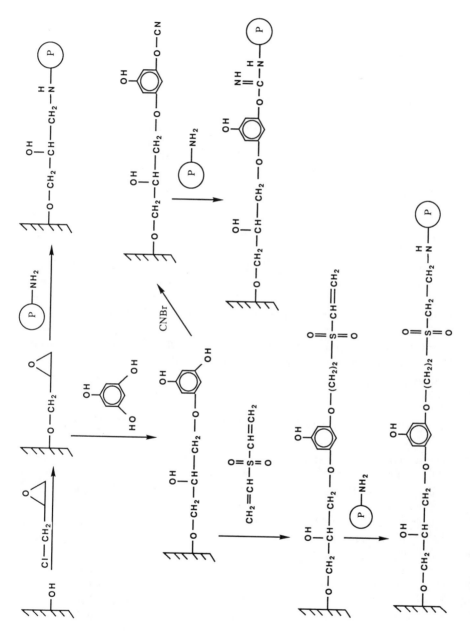

FIGURE 20. Coupling with epichlorohydrin. The protein can be directly immobilized or after reaction with phloroglucinol and divinylsulfone or cyanogen bromide.

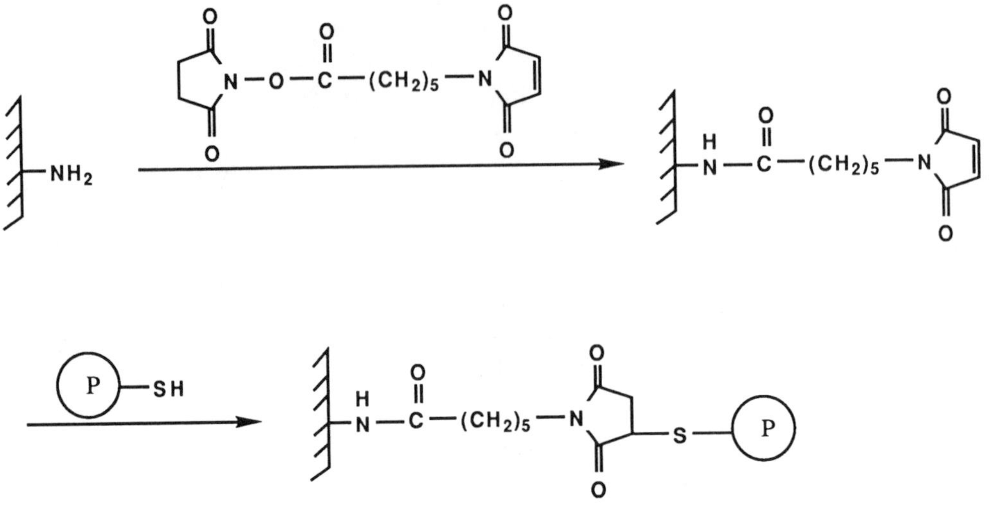

FIGURE 21. Coupling of proteins to solid supports with amino and thiol selective heterobifunctional reagents, e.g., succinimidyl 6-maleimidocaproate.

pharose to form an amide linkage. The solid-bound maleimide functions are then linked to sulfhydryl groups of polypeptides through thioether bonds as shown in Figure 21.[96,97] Another commonly used compound, *N*-succinimidyl-3-(2-pyridyldiothio)propionate (SPDP), whose *N*-hydroxysuccinimidyl ester reacts with amino groups of solid matrices and whose 2-pyridyl disulfide moiety reacts with aliphatic thiols of proteins, has been used to link calmodulin to thiol-Sepharose 4B. The immobilized protein can be cleaved with dithiothreitol.[98] Bhatia et al.[79] have used mercaptomethyldimethylethyloxysilane or mercaptopropyltrimethyloxysilane to convert the hydroxyl groups of silica to thiol groups. Heterobifunctional reagents, SPDP, *N*-γ-maleimidobutyryloxy succinimide ester, *N*-succinimidyl-(4-iodoacetyl)amino benzoate and succinimidyl 4-(*p*-maleidophenyl)butyrate were used to conjugate immunoglobulins to the support.

V. CROSS-LINKING THROUGH CARBOHYDRATE CHAINS

The use of carbohydrate moieties of proteins for their immobilization to solid matrices was demonstrated by Royer.[84] The sugar moiety is first oxidized by sodium periodate to aldehydes which form Schiff bases with either ethylenediamine or glycyltyrosine. Sodium borohydride is used to stabilize the bonds. The derivatized glycoprotein is then immobilized to activated ester of agarose or to diazotized arylamine supports as shown in Figure 22. By this procedure, glucoamylase, peroxidase, glucose oxidase and carboxypeptidase Y have been immobilized.[99]

Vicinal *cis*-hydroxyl groups of solid carriers such as agarose and many other polysaccharides are also susceptible to oxidation by periodate to yield aldehydes that can be used to insolubilize proteins by reductive amination.[100,101] Schiff bases are formed between the protein and the oxidized polysaccharide matrix. Subsequent reduction with sodium borohydride or pyridine borane stabilizes the bonds. The aldehydes can be further converted to *N*-alkyl amines by reaction with alkyl diamines or to hydrazides by reaction with dihydrazides such as adipic dihydrazide.[102,103] The use of hydrazide matrix is preferred over that of alkyl amine because the amino group is positively charged at neutral pH (pKa = 10), whereas the hydrazide moiety is uncharged at pH above 2.5 (pKa = 2.45).

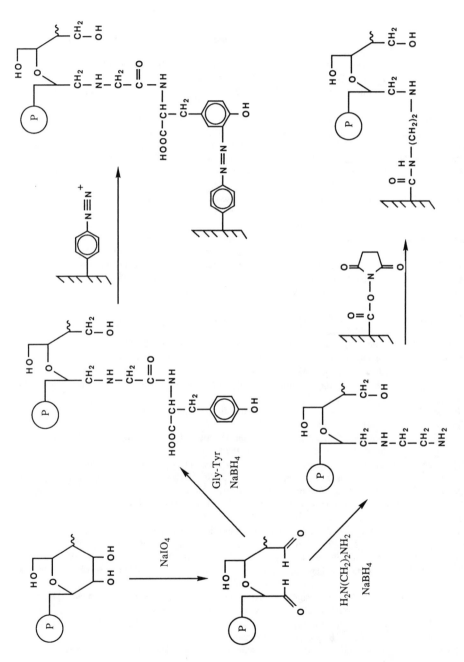

FIGURE 22.   Immobilization of glycoproteins through oxidation of their carbohydrate moiety.

REFERENCES

1. **Chang, T. M. S., Ed.,** *Medical Applications of Immobilized Enzymes and Proteins,* Vols. I and II, Plenum Press, New York, 1977.
2. **Wingard, L. B., Berezin, I. V., and Klyosov, A. A., Eds.,** *Enzyme Engineering: Future Directions,* Plenum Press, New York, 1980.
3. **Weetall, H. H., Ed.,** *Immobilized Enzymes, Antigens, Antibodies, and Peptides,* Marcel Dekker, New York, 1975.
4. **Dunlap, B. R., Ed.,** *Immobilized Biochemicals and Affinity Chromatography,* Plenum Press, New York, 1974.
5. **Mohr, P. and Pommerening, K.,** *Affinity Chromatography,* Marcel Dekker, New York, 1985.
6. **Scouten, W. H.,** *Affinity Chromatography,* John Wiley & Sons, New York, 1981.
7. **Dean, P. D. G., Johnson, W. S., and Middle, F. A., Eds.,** *Affinity Chromatography,* IRL Press, Washington, D.C., 1985.
8. **Mosbach, K., Ed.,** Immobilized enzymes and cells, Part B, *Methods Enzymol.,* 135, 1987.
9. **Mosbach, K., Ed.,** Immobilized enzymes and cells, Part C, *Methods Enzymol.,* 136, 1987.
10. **Mosbach, K., Ed.,** Immobilized enzymes and cells, Part D, *Methods Enzymol.,* 137, 1988.
11. **Pitcher, W. H., Jr., Ed.,** *Immobilized Enzymes in Food Processing,* CRC Press, Boca Raton, FL, 1979.
12. **Messing, R. A., Ed.,** *Immobilized Enzymes for Industrial Reactors,* Academic Press, New York, 1975.
13. **Weetall, H. H. and Suzuki, S., Eds.,** *Immobilized Enzyme Technology,* Plenum Press, New York, 1975.
14. **Weetall, H. H. and Royer, G. P., Eds.,** *Enzyme Engineering,* Plenum Press, New York, 1980.
15. **Kennedy, J. F. and Cabral, J. M. S.,** *Immobilized Enzymes in Solid Phase Biochemistry,* Scouten, W. H., Ed., Wiley Interscience, New York, 1983, 253.
16. **Mosbach, K.,** *Immobilized Enzymes in Organic Synthesis,* Ciba Found. Symp., 111, 57, 1985.
17. **Carney, J.,** Rapid diagnostic tests employing latex particles, *Anal. Proc.,* 27, 99, 1990.
18. **Wong, S. S., Malone, T. E., and Lee, T. K.,** Use of concanavalin A as a topographical probe for protein-protein interaction: Application to lactose synthase, *Biochim. Biophys. Acta,* 745, 90, 1983.
19. **Demers, A. G. and Wong, S. S.,** Increased stability of galactosyltransferase on immobilization, *J. Appl. Biochem.,* 7, 122, 1985.
20. **Zaborsky, O. R.,** *Immobilized Enzymes,* CRC Press, Boca Raton, FL, 1973.
21. **Mosbach, K., Ed.,** Immobilized enzymes, *Methods Enzymol.,* 44, 1976.
22. **Johnson, J. C.,** *Immobilized Enzymes,* Noyes Data Corp., Park Ridge, NJ, 1979.
23. **Gutcho, S. J.,** *Immobilized Enzymes,* Noyes Data Corp., Park Ridge, NJ, 1974.
24. **Chibata, I., Ed.,** *Immobilized Enzymes,* Kodansha, Tokyo, 1978.
25. **Salmona, M., Saronio, C., and Farattini, S., Eds.,** *Insolubilized Enzymes,* Raven Press, New York, 1974.
26. **Royer, G. and Uy, R.,** Evidence for the introduction of a conformational change of bovine trypsin by a specific substrate at pH 8, *J. Biol. Chem.,* 248, 2627, 1973.
27. **Stanley, W. H., Watters, G. G., Kelley, S. H., and Olson, A. C.,** Glucoamylase immobilized on chitin with glutaraldehyde, *Biotechnol. Bioeng.,* 20, 135, 1978.
28. **Haller, W.,** Application of controlled pore glass in solid phase biochemistry, in *Solid Phase Biochemistry: Analytical and Synthetic Aspects,* Scouten, W. H., Ed., Wiley Interscience, New York, 1983, chap. 11.
29. **Veronese, F. M., Visco, C., Massarotto, S., Benassi, C. A., and Ferruti, P.,** New acrylic polymers for surface modification of enzymes of therapeutic interest and for enzyme immobilization, *Ann. N.Y. Acad. Sci.,* 501, 444, 1987.
30. **Bangs, L. B.,** *Uniform Latex Particles,* Seragen Diagnostics, Indianapolis, 1984.
31. **Bangs, L. B.,** *Uniform Latex Particles,* American Clinical Products Review, January, 1988.
32. **Scouten, W. H.,** A survey of enzyme coupling techniques, *Methods Enzymol.,* 135, 30, 1987.
33. **Srere, P. A. and Uyeda, K.,** Functional groups on enzymes suitable for binding to matrices, *Methods Enzymol.,* 44, 11, 1976.
34. **Kohn, J. and Wilchek, M.,** 1-Cyano-4-dimethylaminopyridinium tetrafluoroborate as a cyanylating agent for the covalent attachment of ligand to polysaccharide resins, *FEBS Lett.,* 154, 209, 1984.
35. **Habeeb, A. F. S. A.,** Controlled coupling of mildly reduced proteins to Sepharose gelatin by heterobifunctional reagent, *Biochem. Biophys. Res. Commun.,* 100, 1154, 1981.
36. **Turková, J.,** Immobilization of enzymes on hydroxyalkyl methacrylate gels, *Methods Enzymol.,* 44, 66, 1977.
37. **Cuatrecasas, P. and Parikh, I.,** Adsorbents for affinity chromatography. Use of *N*-hydroxysuccinimide esters of agrose, *Biochemistry,* 11, 2291, 1972.
38. **Babashak, J. V. and Philips, T. M.,** Use of avidin-coated glass beads as a support for high-performance immunoaffinity chromatography, *J. Chromatogr.,* 444, 21, 1988.
39. **Tenoso, H. J. and Smith, D. B.,** Covalent bonding of antibodies to polystyrene latex beads: a concept, *NASA Tech. Brief,* B72-10006, 1972.

40. **Weetall, H. H.,** Covalent coupling methods for inorganic support materials, *Methods Enzymol.,* 44, 123, 1976.
41. **Campbell, D. H., Luescher, E. L., and Lerman, L. S.,** Immologic adsorbents. I. Isolation of antibody by means of a cellulose-protein antigen, *Proc. Natl. Acad. Sci. U.S.A.,* 37, 575, 1951.
42. **Goldstein, L.,** Water-insoluble derivatives of proteolytic enzymes, *Methods Enzymol.,* 19, 935, 1970.
43. **Schnapp, J. and Shalitin, Y.,** Immobilization of enzymes by covalent binding to amine supports via cyanogen bromide activation, *Biochem. Biophys. Res. Commun.,* 70, 8, 1976.
44. **Zaborsky, O.,** Immobilization of enzymes with imidoester-containing polymers, in *Immobilized Enzymes in Food and Microbial Processes,* Olson, A. E. and Cooney, C. L., Eds., Plenum Press, New York, 1974, 187.
45. **Patel, R. P., Lopiekes, D. V., Brown, S. P., and Price, S.,** Derivatives of proteins: II. Coupling of α-chymotrypsin to carboxyl containing polymers by use of N-ethyl-5-phenylisoxazolium-3′-sulfonate, *Biopolymers,* 5, 577, 1967.
46. **Miron, T. and Wilchek, M.,** Immobilization of proteins and ligands using chlorocarbonate, *Methods Enzymol.,* 135, 84, 1987.
47. **Hearn, M. T. W.,** 1,1′-Carbonyldiimidazole-mediated immobilization of enzymes and affinity ligands, *Methods Enzymol.,* 135, 102, 1987.
48. **Wu, D. L. and Walters, R. R.,** Protein immobilization on silica supports. A ligand density study, *J. Chromatogr.,* 458, 169, 1988.
49. **Sudi, P., Dala, E., and Szajani, B.,** Preparation, characterization, and application of a novel immobilized carboxypeptidase B, *Appl. Biochem. Biotechnol.,* 22, 31, 1989.
50. **Nilsson, K. and Mosbach, K.,** Tresyl chloride-activated supports for enzyme immobilization, *Methods Enzymol.,* 135, 65, 1987.
51. **Scouten, W. H., van den Tweel, W., Kranenburg, H., and Dekker, M.,** Colored sulfonyl chloride as an activating agent for hydroxylic matrices, *Methods Enzymol.,* 135, 79, 1987.
52. **Carlsson, J., Axén, R., Brocklehurst, K., and Crook, E. M.,** Immobilization of urease by thiol-disulfide interchange with concomitant purification, *Eur. J. Biochem.,* 44, 189, 1974.
53. **Axén, R., Drevin, H., and Carlsson, J.,** Preparation of modified agarose gels containing thiol groups, *Acta Chem. Scand.,* 29, 471, 1975.
54. **Mosbach, K. and Mattiasson, B.,** Matrix-bound enzymes. Part II. Studies on a matrix-bound two-enzyme-system, *Acta Chem. Scand.,* 24, 2093, 1970.
55. **Weliky, N. and Weetall, H. H.,** The chemistry and use of cellulose derivatives for the study of biological systems, *Immunochemistry,* 2, 293, 1965.
56. **Mezzasoma, I. and Turano, C.,** Binding of enzymes to insoluble supports by the carbodiimide method. Evaluation of the method, *Boll Soc. Ital. Biol. Sper.,* 47, 407, 1971.
57. **Weliky, N., Brown, F. S., and Dale, E. C.,** Carrier-bound proteins: properties of peroxidase bound to insoluble carboxy-methylcellulose particles, *Arch. Biochem. Biophys.,* 131, 1, 1969.
58. **Patel, A. N., Pennington, S. N., and Brown, H. D.,** Insoluble matrix-supported apyrase, deoxyribonuclease and cholinesterase, *Biochim. Biophys. Acta,* 178, 626, 1969.
59. **Mosbach, K.,** Matrix-bound enzymes. Part I. The use of different acrylic copolymers as matrices, *Acta Chem. Scand.,* 24, 2084, 1970.
60. **Wagner, T., Hsu, C.-J., and Kelleher, G.,** A new method for the attachment and support of active enzymes, *Biochem. J.,* 108, 892, 1968.
61. **Patramani, I., Katsiri, K., Pistevou, E., Kalogerakos, T., Pavlatos, M., and Evangelopoulos, A. E.,** Glutamic-aspartic transaminase-antitransaminase interaction: a method for antienzyme purification, *Eur. J. Biochem.,* 11, 28, 1969.
62. **Patel, R. P. and Price, S.,** Derivatives of proteins. I. Polymerization of α-chymotrypsin by use of N-ethyl-5-phenylisoxazolium-3′-sulfonate, *Biopolymers,* 5, 583, 1967.
63. **Jansen, E. F. and Olson, A. C.,** Properties and enzymatic activities of papain insolubilized with glutaraldehyde, *Arch. Biochem. Biophys.,* 129, 221, 1969.
64. **Ottesen, M. and Svensson, B.,** Modification of papain by treatment with glutaraldehyde under reducing and non-reducing conditions, *C. R. Trav. Lab. Carlsberg,* 38, 171, 1971.
65. **Habeeb, A. F. S. A. and Hiramoto, R.,** Reactions of proteins with glutaraldehyde, *Arch. Biochem. Biophys.,* 126, 16, 1968.
66. **Robinson, P. J., Dunnill, P., and Lilly, M. D.,** Porous glass as a solid support for immobilization or affinity chromatography of enzymes, *Biochim. Biophys. Acta,* 242, 659, 1971.
67. **Rapatz, E., Travnicek, A., Fellhofer, G., Pittner, F.,** Studies on the immobilization of glucuronidase (Part 1). Covalent immobilization on various carriers (a comparison), *Appl. Biochem. Biotechnol.,* 19, 223, 1988.
68. **Mattiasson, B.,** Enzyme immunoassay based on partition affinity ligand assay (PALA) system, in *Nonisotopic Immunoassay,* Ngo, T. T., Ed., Plenum Press, New York, 1988, 119.

69. **Wood, W. G. and Gadow, A.,** Immobilization of antibodies and antigens on macro solid phases. A comparison between adsorptive and covalent binding. A critical study of macro solid phases for use in immunoassay systems, Part 1, *J. Clin. Chem. Clin. Biochem.,* 21, 1983, 789.

70. **Dekker, R. F.,** Immobilization of a lactase onto a magnetic support by covalent attachment to polyethyleneimine-glutaraldehyde-activated magnetite, *Appl. Biochem. Biotechnol.,* 22, 289, 1989.

71. **Hakanson, H.,** Portable continuous blood glucose analyzer, *Methods Enzymol.,* 137, 319, 1988.

72. **Gough, D. A., Lucisano, J. Y., and Tse, P. M. S.,** Two-dimensional enzyme electrode senor for glucose, *Anal. Chem.,* 57, 2351, 1985.

73. **Baker, S. A., Doss, S. H., Gray, C. J., Kennedy, J. F., Stacey, M., and Yeo, T. H.,** β-D-Glucosidase chemically bound to microcrystalline cellulose, *Carbohydr. Res.,* 20, 1, 1971.

74. **Jakoby, W. B. and Wilchek, M., Eds.,** Affinity techniques, *Methods Enzymol.,* 34, Part B, 1974.

75. **Kay, G. and Cook, E. M.,** Coupling of enzymes to cellulose using chloro-*s*-triazine, *Nature (London),* 216, 514, 1967.

76. **Wheeler, K. D., Edwards, B. A., and Whittam, R.,** Some properties of two phosphates attached to insoluble cellulose matrices, *Biochim. Biophys. Acta,* 191, 187, 1969.

77. **Lamed, R., Oplatka, A., and Reisler, E.,** Affinity chromatography of heavy meromyosin subfragment-1 reacted with thiol reagents, *Biochim. Biophys. Acta,* 427, 688, 1976.

78. **Roederer, J. E. and Bastiaans, G. J.,** Microgranimetric immoassay with piezoelectric crystals, *Anal. Chem.,* 55, 2333, 1983.

79. **Bhatia, S. K., Shriver-Lake, L. C., Prior, K. J., Georger, J. H., Calvert, J. M., Bredehorst, R., and Ligler, F. S.,** Use of thiol-terminal silanes and heterobifunctional crosslinkers for immobilization of antibodies on silica surfaces, *Anal. Biochem.,* 178, 408, 1988.

80. **Porath, J. and Sundberg, L.,** High capacity chemisorbents for protein immobilization, *Nature (London) New Biology,* 238, 261, 1972.

81. **Porath, J.,** General methods and coupling procedures, *Methods Enzymol.,* 34, 13, 1974.

82. **Brandt, J., Andersson, L. O., and Porath, J.,** Covalent attachment of proteins to polysaccharide carriers by means of benzoquinone, J., *Biochim. Biophys. Acta,* 386, 196, 1975.

83. **Kalman, M., Szajani, B., and Boross, L.,** A novel polyacrylamide-type support prepared by *p*-benzoquinone activation, *Appl. Biochem. Biotechnol.,* 8, 515, 1983.

84. **Kennedy, J. F. and Cabral, J. M. S.,** Immobilization of enzymes on transition metal-activated supports, *Methods Enzymol.,* 135, 117, 1987.

85. **Barker, S. A., Emery, A. N., and Novais, J. M.,** Enzyme reactors for industry, *Process Biochem.,* 6, 11, 1971.

86. **Rokugawa, K., Fujishima, T., Kuminaka, A., and Yoshino, H.,** Studies on immobilized enzymes. III. Immobilization of nuclease P$_1$ on inorganic supports by titanium complex methods, *J. Ferment. Technol.,* 58, 509, 1980.

87. **Kennedy, J. F. and Pike, V. W.,** Water-insoluble papain conjugates of titanium (IV)-activated supports, *Enzyme Microb. Technol.,* 1, 31, 1979.

88. **Gray, C. J., Lee, C. M., and Barker, S. A.,** Immobilization of enzymes on Spheron: 1. Trypsin and glucoamylase by the titanium-chelation method, *Enzyme Microb. Technol.,* 4, 143, 1982.

89. **Emery, A. N., Hough, J. S., Novais, J. M., and Lyons, T. P.,** Application of solid-phase enzymes in biological engineering, *Chem. Eng. (London),* 258, 71, 1972.

90. **Moriya, K., Tanizawa, K., and Kanaoka, Y.,** Schiff base copper(II) chelate as a tool for intermolecular cross-linking and immobilization of protein, *Biochem. Biophys. Res. Commun.,* 161, 52, 1989.

91. **Moriya, K., Tanizawa, K., and Kanaoka, Y.,** Immobilized chymotrypsin by means of Schiff base copper(II) chelate, *Biochem. Biophys. Res. Commun.,* 162, 408, 1989.

92. **Royer, G. P., Ikeda, S., and Aso, K.,** Cross-linking of reversibly immobilized enzymes, *FEBS Lett.,* 80, 89, 1977.

93. **Jagendorf, A. T., Patchornik, A., and Sela, M.,** Use of antibody bound to modified cellulose as an immunospecific adsorbent of antigen, *Biochim. Biophys. Acta,* 78, 516, 1963.

94. **Sato, T., Mori, T., Tosa, T., and Chibata, I.,** Immobilized enzymes. IX. Preparation and properties of aminoacylase convalently attached to haloacetyl-cellulose, *Arch. Biochem. Biophys.,* 147, 788, 1971.

95. **Caron, M., Fabia, F., Faure, A., and Corhillot, P.,** Modified agarose derivative for affinity chromatography. Application to purification of human α-fetoprotein, *J. Chromatogr.,* 87, 239, 1973.

96. **Fauchere, J. L., and Pelican, G. M.,** Specific covalent attachment of adrenocorticotropin and angiotensin II derivatives to polymeric matrices for use in affinity chromatography, *Helv. Chim. Acta,* 58, 1984, 1975.

97. **Moeschler, H. J. and Schwyzer, R.,** Hormone-receptor interactions. Synthesis of a biologically active cysteinyl-angiotensin derivative and its use for the preparation of spin-labeled and polymer-supported molecules, *Helv. Chim. Acta,* 57, 1576, 1974.

98. **Kincaid, R. L. and Vaughan, M.,** Affinity chromatography of brain cyclic nucleotide phosphodiesterase using 3-(2-pyridyldithio)propionyl-substituted calmodulin linked to thiol-Sepharose, *Biochemistry,* 22, 826, 1983.

99. **Royer, G. P.,** Immobilization of glycoenzymes through carbohydrate chains, *Methods Enzymol.,* 135, 141, 1987.

100. **Sanderson, C. J. and Wilson, D. V.,** A simple method for coupling proteins to insoluble polysaccharides, *Immunology,* 20, 1061, 1971.

101. **Stults, N. L., Asta, L. M., and Lee, Y. C.,** Immobilization of proteins on oxidized crosslinked Sepharose preparation by reductive amination, *Anal. Biochem.,* 180, 114, 1989.

102. **Lamed, R., Leven, Y., and Wilchek, M.,** Covalent coupling of nucleotides to agarose for affinity chromatography, *Biochim. Biophys. Acta,* 304, 231, 1973.

103. **Wilchek, M. and Lamed, R.,** Immobilized nucleotides for affinity chromatography, *Methods Enzymol.,* 34, 475, 1974.

Chapter 13

OTHER APPLICATIONS AND FUTURE PROSPECTS

I. INTRODUCTION

In this book, the discussion on the application of chemical cross-linking reagents has been limited to proteins. Proteins, the most abundant and omnipresent natural product, possess the characteristic constituent suitable for chemical modification and manipulation. Other chemical entities that contain nucleophilic and electrophilic functional groups can react with these cross-linkers. Thus, nucleic acids, lipids, carbohydrates, drugs, and other chemicals can possibly be cross-linked to themselves and to each other. In fact, coupling of carbohydrates and toxin to proteins has been illustrated in various chapters of this book. The dedication of Chapter 12 to the application of cross-linking reagents in protein immobilization is an attempt to demonstrate that these reagents can be employed to cross-link various chemical species to proteins. The extension of this application is really limitless. To illustrate this visage, other potentially applicable areas are briefly described below.

II. APPLICATION TO NUCLEIC ACIDS

A. CROSS-LINKING DNA AND RNA

Cross-linking of DNA and RNA provides a means for the study of nucleic acid structure and function. Photochemical cross-linking has been achieved with derivatives of bifunctional photosensitive psoralens.[1] Psoralens are a class of furocoumarins which in the presence of long wavelength ultraviolet light (320 to 380 nm) can covalently cross-link pyrimidines in opposite strands of the DNA or RNA double helix. Various psoralen derivatives such as 4,5′,8-trimethylpsoralen (TMP), 4′-hydroxymethyl-4,5′-8-trimethylpsoralen (HMT), 4′-aminomethyl-4,5′,8-trimethylpsoralen (AMT), isopsoralen and 8-methoxypsoralen (8-MOP) (Figure 1) have been used. These compounds react with DNA and RNA by a two-step mechanism. First, the planar molecule intercalates into the double helical structure of the nucleic acids. Covalent cross-linking is effected by irradiation of the psoralen molecule which reacts primarily with thymidine in DNA and uridine in RNA, although a minor reaction with cytosine also occurs. In this way, psoralens can be used to probe both static and dynamic structural features of the nucleic acids. Psoralens have been used to probe the structure of active nucleolar chromatin,[2,3] the secondary structure of RNA in poliovirus and ribonuclease P,[4] the hairpin structure of isolated DNA,[5] and the interaction of different RNA's.[6,7] Psoralens have also been used to cross-link DNA probes to target sequences.[9] In this regard, psoralen derivatives have been developed as a nonisotopic nucleic acid probe for gene detection.[10,11] A cleavable *bis*-psoralen, dithiobis(ethylmethylamidoethylmethylaminomethyltrimethyl-psoralen), has been prepared by Welsh and Cantor[12] to investigate the DNA arrangement inside bacteriophage lambda and animal virus SV40.

In addition to the psoralens, phenyldiglyoxal has been shown to form RNA-RNA cross-links in addition to protein-RNA linkages.[13] Other compounds such as diamminedichloro-platinum (II),[14] nitrogen mustards[15] and nitroso compounds such as nitrosoureas[16] are capable of cross-linking nucleic acids. Although mostly designed as antineoplastic agents, these compounds may be useful for *in vitro* studies of the nucleic acid structures. Except phenyldiglyoxal and the nitrogen mustards, other bifunctional reagents have not been investigated for their applicability to cross-link nucleic acids. This may be an area for future development. Very interesting structural and functional properties of nucleic acid interactions may be explored by the use of bifunctional cross-linking reagents.

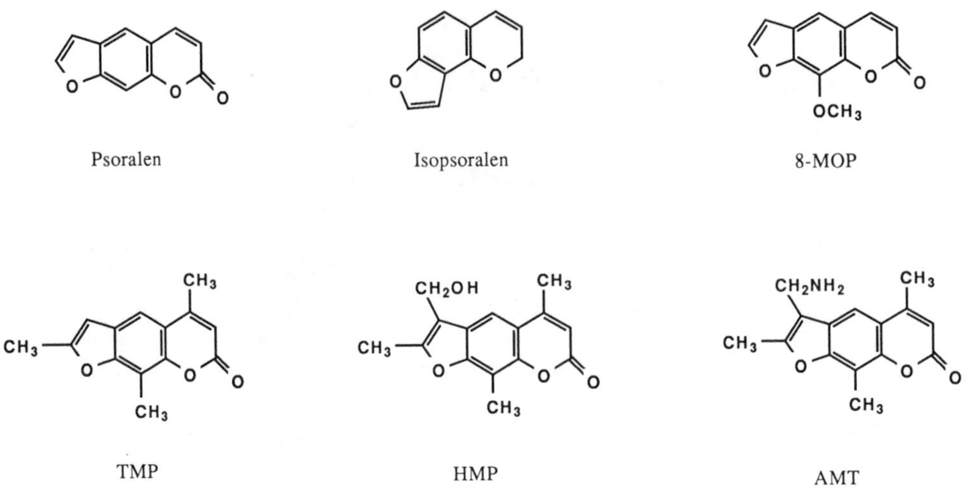

FIGURE 1.   The structure of psoralen and its derivatives.

B. DNA-PROTEIN CONJUGATES
1. DNA Probes

With the advent of nucleic acid hybridization for the detection of genes, DNA probes become an important research topic for the specific and rapid diagnosis of human genetic diseases and pathogens. Many methods have evolved for the labeling of nucleic acids. Various radioactive isotopes have been used.[17,18] Nonradioactive labels such as biotin, mercury, sulfonate, fluorescent probes as well as antigens have been developed.[10,11] The use of bifunctional cross-linking reagents to couple enzymes to nucleic acids probes has been investigated. Renz and Kurz[19] used *p*-benzoquinone to cross-link horseradish peroxidase and alkaline phosphatase to polyethyleneimine, which was then coupled to single-stranded DNA with glutaraldehyde. *Bis(N*-succinimidyl)suberate has also been used to directly couple alkaline phosphatase to 5′-deoxyuridine residue.[20,22] The nucleic acid is first derivatized on C5 with an aminoheptyl-3-acrylamido arm, the amino group of which is used to link to alkaline phosphatase through the homobifunctional cross-linker. In an alternative approach, the 3′-end of the oligonucleotide is thiolated with dithiobis(*N*-succinimidylpropionate) through dithiothreitol reduction. The sulfhydryl-labeled oligonucleotide is then coupled to alkaline phosphatase with *N*-succinimidyl 2-bromoacetate in a two-step procedure. First, the heterobifunctional cross-linker is reacted with alkaline phosphatase through the succinimidyl ester. The bromoacetyl moiety then reacts with the thiol of the nucleotide.[23] Other bifunctional cross-linkers can seemingly be used to conjugate nucleotide probes to enzymes. Indeed, a maleimido-containing cross-linker has been used to prepare horseradish peroxidase-labeled oligodeoxyribonucleotide probes.[24]

2. DNA-Protein Interactions

The application of chemical cross-linking to the investigation of DNA-protein interaction in chromatin is a challenging proposition. Use of dimethyl 3,3′-dithiobispropionimidate to cross-link nuclei and chromatin resulted the formation of H1-H1° heterodimers and H1°-H1° homodimers.[25] Since the proteins in the chromatin contains more reactive groups, DNA-directed bifunctional reagents may be necessary for this purpose. Elsner et al.[26] have synthesized a series of arylazido-cystamine psoralens for photochemical cross-linking of DNA and proteins in chromatin. The psoralen moiety of these compounds will bind to specific locations on DNA. On irradiation, they will form covalent adducts with pyrimidine residues and the arylazido end will react with proteins in the vicinity. Such DNA reagents are expected to be of great importance in the elucidation of chromatin structure. There are other reagents that have been found to cross-link proteins and DNA. These include formaldehyde, chromate and *cis*-diamminedichloroplatinum (II).[27] Solomon et al.[28] have employed formaldehyde-mediated protein-DNA cross-linking to investigate the chromatin structure of *D. melanogaster* heat shock protein 70 genes. *Cis*-diamminedichloroplatinum (II) was used to cross-link nuclear proteins to DNA in differentiating rat myoblasts.[29] While these reagents are providing valuable information on the chromatin structure, a wider selection of cross-linking reagents would be desirable.

C. RNA-PROTEIN CROSS-LINKING

Similar to DNA, the investigation on RNA-protein interactions has been limited by the availability of suitable cross-linking reagents. Use of bifunctional reagents like dimethyl suberimidate, dimethyl dithiobispropionimidate and methyl mercaptobutyrimidate in the study of bovine leukemia virus resulted in only protein-protein cross-links.[30] Only the bifunctional reagent, diepoxybutan, has been found to cross-link RNA and phosphoproteins in the virus and elongation factor G to 23S RNA in 70S ribosomes from *Escherichia coli*.[31] Of other compounds, *trans*-diamminedichloroplatinum (II) has been used to cross-link elongation factor Tu to tRNA and transcription factor TFIIIA to ribosomal RNA.[32,33] The water soluble carbodiimide, 1-ethyl-3(3-dimethylaminopropyl)carbodiimide (EDC), was found to cross-link ribosomal proteins with 16S rRNA of *Escherichia coli*,[34] and with 17S and 26S rRNAs of *Tetrahymena thermophila*.[35] RNA-protein cross-linking was accompanied by extensive protein-protein cross-links. The need for RNA specific bifunctional cross-linkers is obvious.

II. CROSS-LINKING SMALL MOLECULES TO PROTEINS

The general approach for the labeling of proteins with small organic molecules involves the synthesis of a chemically reactive analog of the desired structure followed by reaction with the proteins of interest. For example, in preparing fluorescein-tagged proteins, fluorescein isocyanate is used. Similarly, for the introduction of chelating agents, ethylenediaminetetraacetate and diethylenetriaminepentaacetate are derivatized with a *p*-isothiocyanatobenzyl moiety before their reaction with proteins.[36] The use of bifunctional reagents in these situations may avoid the complicated organic synthesis procedures. For instance, triethylenetetraminehexaacetate can be directly coupled to human serum albumin using Woodward's reagent K.[37] In Chapter 11, the conjugation of natural toxins with proteins has been described. These procedures may be extended for the preparation of drug-protein conjugates. Some new development in this area has already taken place. Hermentin et al.[38] have synthesized *p*-maleimidobenzoyl chloride and its derivatives to couple rhodosaminylanthracyclinone-type anthracyclines to the hinge region of monoclonal antibodies. The hydroxyl groups of these drugs react with the carboxyl chloride of the heterobifunctional cross-linker. The maleimido-labeled drugs are then coupled to the reduced antibody through its sulfhydryl group. The reaction scheme is shown in Figure 2. Webb and Kaneko[39] synthesized two compounds, 1-(aminooxy)-4-[(3-nitro-2-pyridyl)dithio]butane and 1-(aminooxy)-4-[(3-nitro-2-pyridyl)dithio]but-2-ene (see Chapter 5 for structures), to cross-link the carbonyl group of adriamycin to reduced monoclonal antibody. Use of these bifunctional reagents may facilitate the preparation of drug-protein conjugates.

In addition to drugs, bifunctional reagents may be used to conjugate antigens to proteins for the preparation of immunogens. Synthetic peptides have been conjugated to carrier proteins such as bovine serum albumin, bovine α-lactalbumin, and keyhole limpet hemocyanin using *N*-succinimidyl bromoacetate.[40] Other immunogens may be prepared in this manner using bifunctional reagents. Although the antigenicity of the cross-linker has to be considered, such complication is probably minimal. The *S*-carboxymethyl linkage was found not to be antigenic.

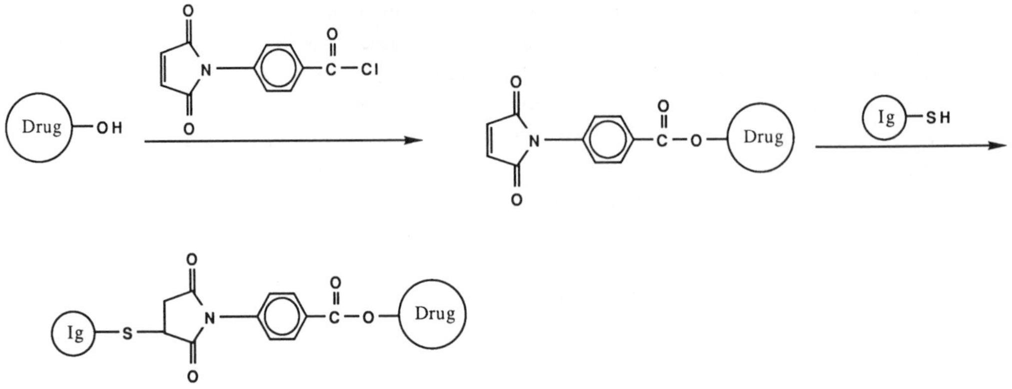

FIGURE 2.   Coupling of anthracycline drugs to reduced immunoglobulins.

III. SYNOPSIS

It should be evident from the coverage of this book that the potential usage of the bifunctional cross-linking reagents is tremendous. Indeed, the scope of application of these compounds is limitless. It ranges from the investigation of molecular interactions in basic research to the preparation of potential therapeutic drugs and genetic probes in biotechnology. The significance of these usages in protein chemistry is vividly displayed throughout this book. In conjunction with proteins, the potential use of these reagents in protein-nucleic acid interactions has been indicated. Other areas open for exploration include but not limited to protein-carbohydrate and protein-lipid interactions in membranes and lipoproteins, conjugation of colloidal gold and gold clusters to immunoglobulins or other proteins for electron microscopy and immunodiagnostics, preparation of immune complexes, formation of enzyme cofactor complexes, labeling of proteins with fluorescent, luminescent, spin and other probes, and the incorporation of various components or probes into liposomes. In some of these arenas, new cross-linking reagents will have to be developed to facilitate the research. Future amplifications of the utilization of this technique are foreseeable. The boundary is limited only by the imagination of the investigators.

REFERENCES

1. **Cimino, G. D., Gamper, H. B., Isaacs, S. T., and Hearst, J. E.,** Psoralens as photoactive probes of nucleic acid structure and function: Organic chemistry, photochemistry and biochemistry, *Annu. Rev. Biochem.,* 54, 1151, 1985.
2. **Sogo, J. M., Ness, P. J., Widmer, R. M., Parish, R. W., and Koller, T.,** Psoralen-crosslinking of DNA as a probe for the structure of active nucleolar chromatin, *J. Mol. Biol.,* 178, 987, 1984.
3. **Ross, P. M. and Yu, H. S.,** Interstrand crosslinks due to 4,5′,8-trimethylpsoralen and near ultraviolet light in specific sequences of animal DNA. Effect of constitutive chromatin structure and of induced transcription, *J. Mol. Biol.,* 201, 339, 1988.
4. **Currey, K. M., Peterlin, B. M., and Maizel, J. V., Jr.,** Secondary structure of poliovirus RNA: correlation of computerpredicted with electron microscopically observed structure, *Virology,* 148, 33, 1986.
5. **Hyrien, O.,** Large inverted duplications of amplified DNA of mammalian cells form hairpins *in vitro* upon DNA extraction but not in vivo, *Nucleic Acids Res.,* 17, 9557, 1989.
6. **Stroke, I. L. and Weiner, A. M.,** The 5′ end of U3 snRNA can be crosslinked in vivo to the external transcribed spacer of rat ribosomal RNA precursors, *J. Mol. Biol.,* 210, 497, 1989.
7. **Ruzdijic, S. and Pederson, T.,** Evidence for an association between U1 RNA and interspersed repeat single-copy RNAs in the cytoplasm of sea urchin eggs, *Development,* 101, 107, 1987.

8. **Lipson, S. E., Cimino, G. D., and Hearst, J. E.,** Structure of M1 RNA as determined by psoralen cross-linking, *Biochemistry,* 27, 570, 1988.

9. **Cebula, T. A. and Koch, W. H.,** Analysis of spontaneous and psoralen-induced *Salmonella typhimurium* hisG46 revertants by oligodeoxyribonucleotide colony hybridization: use of psoralens to cross-link probes to target sequences, *Mutat. Res.,* 229, 79, 1990.

10. **Keller, H. G. and Manak, M. M.,** Non-radioactive labeling procedures, in *DNA Probes,* Stockton Press, New York, 1989, Section IV.

11. **McInnes, J. L. and Symons, R. H.,** Preparation and detection of nonradioactive nucleic acid and oligonucleotide probes, in *Nucleic Acid Probes,* Symons, R. H., Ed., CRC Press, Boca Raton, FL, 1989, chap. 2.

12. **Welsh, J. and Cantor, C. R.,** Studies on the arrangement of DNA inside viruses using a breakable bis-psoralen crossliner, *J. Mol. Biol.,* 198, 63, 1987.

13. **Chiam, C. L. and Wagner, R.,** Composition of the Escherichia coli 70S ribosomal interface: a cross-linking study, *Biochemistry,* 22, 1193, 1983.

14. **Wilborn, Fl. and Brendel, M.,** Formation and stability of interstrand cross-links induced by *cis-* and *trans-*diamminedichloroplatinum (II) in the DNA of *Saccharomyces cerevisiae* strains differing in repair capacity, *Curr. Genet.,* 16, 331, 1989

15. **Baker, J. M., Parish, J. H., and Curtis, J. P.,** DNA-DNA and DNA-protein crosslinking and repair in *Neurospora crassa* following exposure to nitrogen mustard, *Mutat. Res.,* 132, 171, 1984.

16. **Kohn, K. W., Erickson, L. C., Laurent, G., Ducore, J., Sharkey, N., and Eurg, R. A.,** DNA crosslinking and origin of sensitivity of chloroethylnitrosoureas, in *Nitrosoureas: Current Status and New Developments,* Prestayko, A. W., Baker, L. H., Crooke, S. T., Carter, S. K., and Schein, P. S., Eds., Academic Press, New York, 1981, chap. 6.

17. **Keller, G. and Manak, M. M.,** Isotopic labeling procedures, in *DNA Probes,* Stockton Press, New York, 1989, Section III.

18. **McInnes, J. L. and Symons, R. H.,** Enzymatic and chemical techniques for labeling nucleic acids with radioisotopes, in *Nucleic Acid Probes,* Symons, R. H., Ed., CRC Press, Boca Raton, FL, 1989. chap. 2.

19. **Renz, M. and Kurz, C.,** A colorimetric method for DNA hybridization, *Nucleic Acids Res.,* 12, 3435, 1984.

20. **Jablonski, E., Moomaw, E. W., Tullis, R. H., and Ruth, J. L.,** Preparation of oligodeoxynucleotide-alkaline phosphatase conjugate and their use as hybridization probes, *Nucleic Acids Res.,* 14, 6115, 1986.

21. **McLaughlin, G. L., Ruth, J. L., Jablonski, E., Steketee, R., and Campbell, G. H.,** Use of enzyme-linked synthetic DNA in diagnosis of *Falciparum malaria, Lancet,* 1, 714, 1987.

22. **Seriwatana, J., Echeverria, P., Taylor, D. N., Sakuldaipeara, T., Changchawaht, S., and Chivoratanond, O.,** Identification of enterotoxigenic *Escherichia coli* with synthetic alkaline phosphatase-conjugated oligonucleotide DNA probes, *J. Clin. Microbiol.* 25, 1438, 1987.

23. **Li, P., Medon, P. P., Skingle, D. C., Lanser, J. A., and Symons, R. H.,** Enzyme-linked synthetic oligonucleotide probes: non-radioactive detection of enterotoxigenic *Escherichia coli* in faecal specimens, *Nucleic Acids Res.,* 15, 5275, 1987.

24. **Saiki, R. K., Chang, C.-A., Levenson, C. H., Warren, Tl. C., Boehm, C. D., Kazazian, H. H., Jr., Erlich, H. A.,** Diagnosis of sickel cell anemia and β-thalassemia with enzymatically amplified DNA and nonradioactive allele-specific oligonucleotide probes, *N. Engl. J. Med.,* 319, 537, 1988.

25. **Dashkevich, V. K., Nikolaev, L. G., Zlatanova, J. S., Glotov, B. O., and Severin, E. S.,** Chemical crosslinking of histone H1° to histone neighbors in nuclei and chromatin, *FEBS Lett.,* 158, 276, 1983.

26. **Elsner, H., Buchardt, O., Miler, J., and Nielsen, P. E.,** Photochemical crosslinking of protein and DNA in chromatin. II. Synthesis and application of psoralen-cystamine-arylazido photocrosslinking reagents, *Anal. Biochem.,* 149, 575, 1985.

27. **Miller, C. A., III. and Costa, M.,** Analysis of proteins cross-linked to DNA after treatment of cells with formaldehyde, chromate, and *cis*-diamminedichloroplatinum(II), *Mol. Toxicol.,* 2, 11, 1989.

28. **Solomon, M. J., Larsen, P. L., and Varshavsky, A.,** Mapping protein-DNA interactions in vivo with formaldehyde: evidence that histone H4 is retained on a highly transcribed gene, *Cell,* 53, 937, 1988.

29. **Wedrychowski, A., Bhorjee, J. S., and Briggs, R. C.,** In vivo crosslinking of nuclear proteins to DNA by *cis*-diamminedichloroplatinum (II) in differentiating rat myoblasts, *Exp. Cell Res.,* 183, 376, 1989.

30. **Uckert, W., Wunderlich, V., Ghysdael, J., Portetelle, D., and Burny, A.,** Bovine leukemia virus (BLV)—a structural model based on chemical crosslinking studies, *Virology,* 133, 386, 1984.

31. **Sköld, S.-E.,** Chemical crosslinking of elongation factor G to the 23S RNA in 70S ribosomes from *Escherichia coli, Nucleic Acids Res.,* 11, 4923, 1983.

32. **Metz-Boutigue, M. H., Reinbolt, J., Ebel, J. P., and Ehresmann, C.,** Crosslinking of elongation factor Tu to tRNA(Phe) by trans-diamminedichloroplatinum (II). Characterization of two crosslinking sites on EF-Tu, *FEBS Lett.,* 245, 194, 1989.

33. **Baudin, F., Romby, P., Romaniuk, P. J., Ehresmann, B., and Ehresmann, C.,** Crosslinking of transcription factor TFIIIA to ribosomal 5S RNA from *X. laevis* by trans-diamminedichloroplatinum (II), *Nucleic Acids Res.,* 17, 10035, 1989.

34. **Chiaruttini, C., Milet, M., Hayes, D. H., and Expert-Bezancon, A.,** Crosslinking of ribosomal proteins S4, S5, S7, S8, S11, S12, and S18 to domains 1 and 2 of 16S rRNA in the *Escherichia coli* 30S particle, *Biochimie,* 71, 839, 1989.

35. **Petridou, B., Guerin, M. F., and Hayes, F.,** Protein-RNA crosslinking in the subunits of the cytoplasmic ribosome of *Tetrahymena thermophila, Biochemie,* 71, 667, 1989.

36. **Westerberg, D. A., Carney, P. L., Rogers, P. E., Kline, S. J., and Johnson, D. K.,** Synthesis of novel bifunctional chelators and their use in preparing monoclonal antibody conjugates for tumor targeting, *J. Med. Chem.,* 32, 236, 1989.

37. **Subramanian, K. M. and Wolf, W.,** A new radiochemical method to determine the stability constants of metal chelates attached to a protein, *J. Nucl. Med.,* 31, 480, 1990.

38. **Hermentin, P., Hoenges, R., Gronski, P., Bosslet, K., Kraemer, H. P., Hoffmann, D., Zilag, H., Steinstraesser, A., Schwarz, A., Kuhlmann, L., Lüben, G., and Seiler, F. R.,** Attachment of rhodos-aminylanthracyclinone-type anthracyclines to the hinge region of monoclonal antibodies, *Bioconjugate Chem.,* 1, 100, 1990.

39. **Webb, R. R., II, and Kaneko, T.,** Synthesis of 1-(aminooxy)-4[3-nitro-2-pyridyl)dithio]butane and 1-(aminooxy)-4-[(3-nitro-2-pyridyl)dithio]but-2-ene, novel heterobifunctional cross-linking reagents, *Bioconjugate Chem.,* 1, 96, 1990.

40. **Bernatowicz, M. S. and Matsueda, G. R.,** Preparation of peptide-protein immunogens using *N*-succinimidyl bromoacetate as a heterobifunctional crosslinking reagent, *Anal. Biochem.* 155, 95, 1986.

INDEX

A

E